Y0-ARD-006

Springer Series in Experimental Entomology

Thomas A. Miller, Editor

Crystals of the NABA-ester bombykol (see page 25).

Techniques in Pheromone Research

Edited by
Hans E. Hummel
Thomas A. Miller

With Contributions by

H. Arn • T.C. Baker • A. Butenandt • D.G. Campion
R.T. Cardé • J.S. Elkinton • L.K. Gaston • M.A. Golub
R.R. Heath • E. Hecker • H.E. Hummel • C.E. Linn, Jr.
K. Mori • W.L. Roelofs • P.E. Sonnet • D.L. Struble
J.H. Tumlinson • L.J. Wadhams • I. Weatherston

With 125 Figures

Springer-Verlag
New York Berlin Heidelberg Tokyo

Hans E. Hummel
Harbor Branch Foundation
Fort Pierce, Florida 33450
U.S.A.

Thomas A. Miller
Department of Entomology
University of California
Riverside, California 92521
U.S.A.

Library of Congress Cataloging in Publications Data
Main entry under title:
Techniques in pheromone research.
(Springer series in experimental entomology)
Includes bibliographies and index.
1. Pheromones. 2. Insects—Physiology.
Entomology—Technique. 4. Biological chemistry—
Technique. I. Miller, Thomas A. II. Hummel, Hans. E.
III. Series.
QL495.T43 1984 595.7'059 83-20378

Typeset by MS Associates, Champaign, Illinois.
Printed and bound by Halliday Lithograph, West Hanover, Massachusetts.
Printed in the United States of America.

9 8 7 6 5 4 3 2 1

ISBN 0-387-90919-2 Springer-Verlag New York Berlin Heidelberg Tokyo
ISBN 3-540-90919-2 Springer-Verlag Berlin Heidelberg New York Tokyo

Series Preface

Insects as a group occupy a middle ground in the biosphere between bacteria and viruses at one extreme, amphibians and mammals at the other. The size and general nature of insects present special problems to the student of entomology. For example, many commercially available instruments are geared to measure in grams, while the forces commonly encountered in studying insects are in the milligram range. Therefore, techniques developed in the study of insects or in those fields concerned with the control of insect pests are often unique.

Methods for measuring things are common to all sciences. Advances sometimes depend more on how something was done than on what was measured; indeed a given field often progresses from one technique to another as new methods are discovered, developed, and modified. Just as often, some of these techniques find their way into the classroom when the problems involved have been sufficiently ironed out to permit students to master the manipulations in a few laboratory periods.

Many specialized techniques are confined to one specific research laboratory. Although methods may be considered commonplace where they are used, in another context even the simplest procedures may save considerable time. It is the purpose of this series (1) to report new developments in methodology, (2) to reveal sources of groups who have dealt with and solved particular entomological problems, and (3) to describe experiments which may be applicable for use in biology laboratory courses.

THOMAS A. MILLER
Series Editor

Preface

Throughout the history of science, technological advances—occasionally combined with a bright idea—have been setting the pace for innovation. New techniques generated new observations, questions, and answers, and ultimately improved theories about our complex natural world.

As much as Galileo's astronomical telescope opened our view to the neighboring planets and the worlds beyond, the microscope of Leeuwenhoek made biologists aware of a living microcosm previously unknown and unrecognizable by the unaided eye. As a consequence of our limited human olfactory perception, the universe of chemical interactions among organisms, or their chemical ecology remained unexplored until 1959. Although occasionally spotlighted by such eminent experimentalists as J.H. Fabre (1) in France and P. and N. Rau (2) in the United States, the field of chemical ecology continued to have the aura of the anecdotal, the exotic, and the impractical.

This situation radically changed with the discovery, isolation and chemical identification of bombykol, the silkmoth sex attractant, by Butenandt et al. in 1959 (3). Today, barely a quarter of a century after their pioneering contribution, over 800 such substances (4) are described and chemically synthesized. Aided by greatly refined microchemical techniques, by the availability of sensitive spectrometric and chromatographic methods, and by the coming of age of quantitative ecology and ethology, the rough outlines and the increasing significance of the once unknown continent of chemical ecology become apparent. On our modern maps of science this continent appears under the roughly synonymous names "pheromones," "semiochemicals," and "ecomones." For convenience, the term *pheromones* coined by Karlson and Lüscher (1959) (5) is being used throughout this volume.

Chemical communication is an information channel arthropods have perfected with mastery and exploited to a high degree. Comprising about 80% of all described animals, they use it among other things for sexual attraction, trail-following, recruitment, and defense. Outside their own species, they use it for chemically interacting with their arthropod neighbors, their plant and animal hosts, and their general environment. The quantities of chemical messengers range from a few hundred molecules (Kaissling and Priesner, 1970) to a few micrograms, an amount which is still mostly below the detection limit of our own sense of smell.

In favorable instances our most powerful instruments allow us to detect sub-nanogram (10^{-12} to 10^{-9} g) quantities of pheromone from individual insects. In other cases described only 1 to 2 decades ago, recourse was taken to collect and concentrate the pheromones of 10^3 to 10^6 insects and to arrive at milligram (10^{-3} g) quantities. As described in Chapter 1, 12 milligrams of bombykol NABA ester, in 1959 one of the most precious substances in the hands of experimental entomologists, was oxidatively cleaved to reveal the number and position of two critical C–C double bonds. In 1973, only 0.1 mg of gossyplure (Chapter 8) was needed to achieve the same goal. Today, a few micrograms will suffice for structure identifications, and nanogram quantities from individual insects will, as Gaston (Chapter 9) and Heath and Tumlinson (Chapter 11) point out, help the chemical ecologist make quantitative determinations at the lower end of the experimentally possible. Nothing will illustrate the technical progress made within the last 25 years more strikingly than this gain in sensitivity of 4–6 orders of magnitude!

Also, behavior analysis (Chapters 2 and 4) and specialized techniques like flight tunnels (Chapter 3), electroantennogram assays (Chapter 5) and tandem or coupled techniques (Chapters 6 to 8) have reached a stage of sophistication that warrants their review. Moreover, reliable techniques are now available for the synthesis of pheromones (Chapter 13) and for assignments of their absolute stereochemical configuration (Chapter 12). While some of the more refined chemical and behavioral techniques may remain in the hands of the experienced specialists, the spectrum of techniques accessible to the advanced student has vastly increased.

Over the last decade, about a dozen books (7–19) on pheromones and chemical ecology have appeared. Collectively, they relate a reasonable picture of our present knowledge of the basic biology, the chemistry, and some applications of pheromones. None of these publications, however, focuses on the equally important biological and chemical techniques that have made possible the advances of the last decade. Therefore, in this volume, 14 authors were asked to document the currently available state of the art. The product of their joint efforts may serve both as an up-to-date account to the specialist and as a guide for the advanced graduate student in chemical ecology. Because of the similarity of approach and indeed the chemical techniques involved, other fields of chemical ecology will benefit from it, as will those readers interested in chemical microtechniques, in natural products of terrestrial and aquatic origin, in the chemical ecology of microorganisms and plant-insect interactions, and in mammalian and primate pheromones.

Some schools of thought influence their students, at times with a determination bordering on religious fervor, in believing that only the molecular or the organismic approach can be valid to the exclusion of the other. Chemical ecology, as the name implies, and as the different sections of this book amply illustrate, cannot do without either of them, but can do more than either of them by itself. Chemical ecology tries to describe (and ultimately, when reaching maturity, to predict) how organisms chemically orient, behave, and interact in a chemical world accessible to our own senses. Chemical interactions between insects and their host plants, for example, have only been described in some depth for several, mostly economically important, species. Thus, an immense field for experimental investigations for covering some 700,000 insect species and an equal number of plant species waits for the student. The opportunity for major discoveries on this large and mostly unknown continent are most inviting.

Finally, a subject much on the mind of those who have access to the literature of both chemistry and biology is the current lack of cross-communication between these classical disciplines. Yet, the few laboratories in the world where cooperation and interaction between behavioral physiologists, ecologists, and organic chemists are institutionally established, have an outstanding record of achievement. Their example could provide a challenge for interdisciplinary activities elsewhere.

May this volume, in addition to its contribution to the technical knowhow, also catalyze this process and provide guides for the new generation of chemical ecologists emerging in the 1980s.

Thanks are due to the authors for their chapters and to the publisher for accepting suggestions and for providing space for ample illustrative material. Dr. T.A. Miller was both a driving force and a catalyst behind the scenes. To him the project owes momentum and timely completion.

Hans E. Hummel
Urbana, March 1984

Contents

Contributors

H. ARN
Swiss Federal Research Station for Horticulture and Viticulture, CH-8820 Wädenswil, Switzerland

T.C. BAKER
Department of Entomology, University of California, Riverside, California 92521, U.S.A.

A. BUTENANDT
Max-Planck-Institut für Biochemie, D-8031 Martinsried-Munich, West Germany

D.G. CAMPION
Centre Overseas Pest Research, Ministry of Overseas Development, College House, Wright's Lane, London W85 SJ, England

R.T. CARDÉ
Department of Entomology, University of Massachusetts, Amherst, Massachusetts 01003, U.S.A.

J.S. ELKINTON
Department of Entomology, University of Massachusetts, Amherst, Massachusetts 01003, U.S.A.

L.K. GASTON
Department of Entomology, Division of Toxicology and Physiology, University of California, Riverside, California 92521, U.S.A.

M.A. GOLUB
973 Furnace Brook Parkway, Quincy, Massachusetts 02169, U.S.A.

R.R. HEATH
Attractants, Behavior, and Basic Biology Laboratory, USDA, SEA, Gainesville, Florida 32604, U.S.A.

E. HECKER
Department of Biochemistry, Deutsches Krebsforschungszentrum, D-6900 Heidelberg 1, West Germany

H.E. HUMMEL
Harbor Branch Foundation, Fort Pierce, Florida 33450, U.S.A.

C.E. LINN, JR.
Department of Entomology, New York State Agricultural Experiment Station, Geneva, New York 14456, U.S.A.

K. MORI
Department of Agricultural Chemistry, University of Tokyo, Bunkyo-ku, Tokyo 113, Japan

W.L. ROELOFS
Department of Entomology, New York State Agricultural Experiment Station, Geneva, New York 14456, U.S.A.

P.E. SONNET
Attractants, Behavior, and Basic Biology Laboratory, USDA, SEA, Gainesville, Florida 32604, U.S.A.

D.L. STRUBLE
Agriculture Canada, Research Branch, Crop Entomology Section, Research Station Lethbridge, Alberta T1J 4B1, Canada

J.H. TUMLINSON
Attractants, Behavior, and Basic Biology Laboratory, USDA, SEA, Gainesville, Florida 32604, U.S.A.

L.J. WADHAMS
Department of Insecticides and Fungicides, Rothamsted Experimental Station, Harpenden, Herts, A15 2JQ, United Kingdon

I. WEATHERSTON
Department de Biologie, Université Laval, Quebec, Canada G1K 7P4

Chapter 1

Bombykol Revisited—Reflections on a Pioneering Period and on Some of Its Consequences

Erich Hecker[1] *and Adolf Butenandt*[2]

I. Introduction

Intelligent and careful observations made by biologists at the beginning of this century revealed that certain insects show spectacular sensitivity in scent perception: So-called "sex attractants" or, as we call them today, sex pheromones (for definition see Karlson and Lüscher, 1959) lure members of the opposite sex and release a sequence of precopulatory behavior activities. Already the early findings indicated that the active principle(s) involved may occur in extremely low amounts. Indeed research scholars exploring this exciting field of insect biochemistry must accept the challenge of developing utmost experimental skill in using extremely sensitive analytical methods of detection and special ultramicrochemical methods of handling sex pheromones and pheromones in general. Thus, especially for the newcomer to this field, it may be profitable as well as enlightening to become acquainted, from the outset, with some of the principal technical essentials required in pheromone research. Also, the following reflections on the pioneering period of pheromone research will illustrate how fortunate it was to choose, for opening this principally new area of research, just an insect species as the model (Butenandt, 1939, 1941).

In fact, *Bombyx mori* L. (Figs. 1A,B) provides another example for the remarkable list of specific insect models successfully employed in gaining basic

[1]Director, Institut für Biochemie, Deutsches Krebsforschungszentrum, Im Neuenheimer Feld 280, D-6900 Heidelberg, Federal Republic of Germany.

[2]Honorary President, Max-Planck-Gesellschaft and Emeritus Scientific Member of Max-Planck-Institut für Biochemie, 8033 Martinsried b. München, Federal Republic of Germany.

(A)

Figure 1. The silkworm moth *Bombyx mori* L. (A) Male exposing antennae.

scientific knowledge of biochemical and biological problems. Classical examples for this approach are the utilization of *Drosophila melanogaster* and *Ephestia kühniella* as model systems in early investigations of enzymological expression of genes ("one gene–one-enzyme" hypothesis) and of certain aspects of the mechanism of tumorigenesis ("somatic mutation" hypothesis). Several species of butterflies were useful in detecting the pteridines, a class of compounds later on found to be of metabolic relevance also in man. Moreover, *Calliphora erythrocephala* was the key insect used during isolation of ecdysone, one of the hormones of insect metamorphosis. In all these cases, the investigational advantages or else the uniqueness of the insect model was outstanding. They were so intriguing that all possible efforts were made to overcome the key technical problem associated with using insects in biochemical investigations: i.e., handling the tiny amounts of biochemical entities which an individual insect naturally provides.

The prospect to open up an entirely new approach into the field of physiology of olfaction together with the important practical aspect of introducing specificity into insect pest control as an alternative to insecticides provided enough motivation to search for and, finally, to find solutions to all the technical problems involved in the isolation of bombykol—the principal sex pheromone of *Bombyx mori* L.

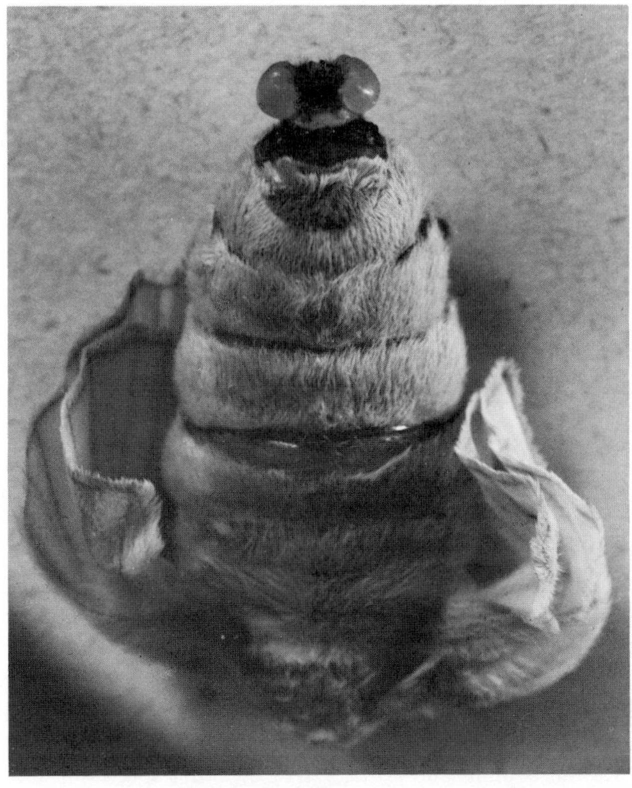

(B)

Figure 1 (cont'd). (B) Female exposing sacculi laterales (scent sacs).

The far-reaching consequences of the new area of basic research opened in 1939 are implicated in a question posed by a charming American lady after a lecture in 1951, when she had noticed that our work on sex attractants was dealing merely with insects: "Oh, Doctor B., why do you waste your time with butterflies?"

During the war times of 1939-1945 and in the subsequent years, immediate progress was severely hampered by all kinds of restrictions. But after 1950, activities in our research project were resumed gradually and developed further to isolate and fully identify bombykol (Butenandt et al., 1959; Butenandt, 1959; Hecker, 1959; Hecker, 1960) and, finally, to prove its chemical structure by total syntheses (Butenandt and Hecker, 1961). The identification of this prototype sex pheromone encouraged and stimulated greatly the worldwide endeavors to detect and identify other representatives of this new class of bioactive principle—from insect to man. As so often in science, the timely advent of new and more and more sophisticated laboratory techniques added a great deal to the success of such endeavors.

II. Spectacular Early Hypotheses on the Nature of Sex Pheromones in Moths

That species- and sex-specific "attractants" are widely distributed in the insect world was detected already toward the end of the last and the beginning of this century. Most thoroughly this phenomenon was studied in Lepidoptera, particularly in moths of spinning caterpillars and other related night-flying species. Experiments devised especially with regard to "long-distance attraction" were described and popularized by the French naturalist and eminent entomologist J. H. Fabre: His observations appeared miraculous to him and to others (e.g., Fabre, 1914).

In one of his experiments, Fabre reported that he had placed in an open window a single female of the emperor or hawk moth (*Saturnia pavonia*) kept in a screenwire cage between 8 and 10 P.M. Within the eight subsequent days this female attracted a total of 150 marked males released from kilometers away. In a similar experiment, a female oak spinner (*Lasiocampa quercus*) was exposed in a window between 3 P.M. and sunset. Up to 60 males were attracted daily. While the female within the screen cage usually was completely motionless, the males she attracted carried out a "flutterdance" on the screen. Their dances often continued until exhaustion ensued. Mell (1922) marked a few male moths of the Chinese silkworm moth (*Actias selene*) and released them from a train at different distances of 4.1 and 11 km from females retained in a gauze cage. Of the males 40 and 26%, respectively, found their way back to the female. Gypsy moth males of *Lymantria dispar* L. were claimed to find a female moth from a distance of 3.8 km (Collins and Potts, 1932). In other species of moth such as, for example, *Bombyx mori,* the attractivity of females appeared less spectacular (Kellogg, 1907).

In their experiments, Fabre and others observed that the males did not fly directly to the attracting female, but rather circled in gradually. Yet, memory for places appeared not to be important, for when the experimental cage was moved to a different location, the former site never was revisited. However, males were attracted to locations where a female had perched. The attractivity of the females could not be masked by other substances which have strong smell for humans such as, for example, naphthalene (Fabre, 1914) or—in *Bombyx*—essential oils (see Hecker, 1959). If the antennae in males were covered with laqueur, they did not take notice of females. Also, after total resection of the antennae, the capability of males to get aroused by female scent and to locate them was impaired. After partial resections, however, males still got aroused (Kellog, 1907; Fabre, 1914; see also Schwinck, 1956, 1958).

To explain the remarkable phenomenon of long-distance attraction detected by Fabre and others, various different hypotheses were developed. Fabre speculated that, for the insect, scent might have two "domains"—that of particles dissolved in the air and that of "etheric" waves; "Like light, odor has its X-rays. . . . Let science, instructed by the insect, one day give us a radiograph sensitive to odors and this artificial nose will open up a new world of marvels." The idea of holding responsible for the attractivity of females various kinds of radiation was a particularly long-lived hypothesis. For example, in 1937, based upon investigations in pyralids, especially in *Plodia interpunctella*, Barth (1937) claimed that the attracting agent might be a highly volatile and extremely labile chemical compound. Yet, the compound as such would not stimulate the males but "unmeasurably low energy" which would be released during "disintegration" of the labile chemical attractant. Later on, when, for the first time, self-recording infrared spectrometers became available as laboratory tools, Duane and Tyler (1950) tried to measure infrared emissions from females *Samia* (= *Hyalophora*) *cecropia* and *Telea polyphemus* moths. Using a female moth sitting in an infrared spectrometer, they reported: "apparently she radiated in a definite pattern in the region from 3 to 11 microns. One might say that she was broadcasting on a fixed frequency in the 100 to 273 mega-megacycle band." And they asked the question: "Is this radiation the attracting medium which guides the male moth through fog and darkness to his mate? If we could isolate the chemical that is responsible for her pattern of radiation could we attract male moths to it by warming it slightly?" While previously Fabre believed that the "attractant," whatever it might be, was situated on the underside of the female, the American scientists claimed to have recorded radiation from the region of the thorax. By photomicrographs of male moths' antennae, they found that the hairs growing from the plumage show remarkable evenness of spacing and length. They stated that all variations in the length of the hairs "appeared to be close to four microns or multiples thereof" and "it is noteworthy that four microns is one-half the wavelength of eight microns which is well within the emission band of the female." From such measurements they posed the question: "Does this mean that the male *Cecropia* moth has a tuned antenna array which is his receptor for locating the female?"

Quite apart from spectacular hypotheses of this kind [for a more complete overview see Jacobson (1972) Table 6] the fascinating field of specific sex pheromones developed gradually and calmly out of early morphological and histological investigations in moths identifying the site(s) of both perception and production of the sex pheromones, and suggesting experimental means to clarify their species specificity and their chemical nature.

III. Components of the Pheromone Communication System: Antennal Morphology, Sex Pheromone Glands, Specificity and Chemical Nature of the Message

3.1. Sex Dimorphism of Sensory Organs in Moths

A macroscopic and microscopic sex dimorphism was observed in certain but not in all species of moths and suggested sex pheromone perception to be located in the antennae of males. For example, Schenk (1903) reported that in a number of species, e.g., *Fidonia piniaria* and *Orgya antiqua* (Fig. 2), the antennae of males are considerably larger than those of females. In some species, such as, for

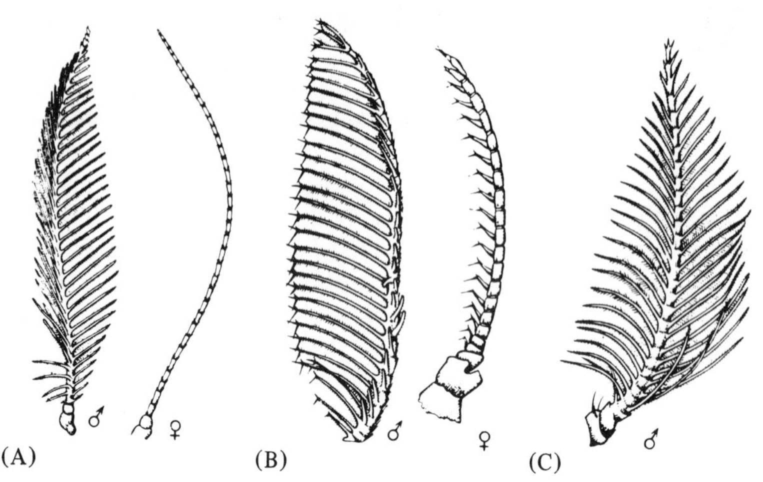

Figure 2. Morphology of antennae of male and female moths (from Schenk, 1903). (A) *Fidonia piniaria* L. (B) *Orgyia antiqua* L. (C) *Psyche unicolor* Hfn. In this species females do not have antennae.

example, *Psyche unicolor,* the females have no antennae at all! This sex dimorphism has also a microscopic counterpart (Table 1): The antennae of the males and—if present at all—those of females usually are covered with a large number of sensillae. At least five histologically different types may be distinguished. Yet, their number in males may be different from that in females. For example, males of *Fidonia piniaria* L. and of *Orgyia antiqua* exhibit much more Sensilla trichodea and Sensilla coeloconica than females. Those present in males in excess may be associated with sex pheromone perception. In contrast, for example, lack of Sensilla basiconica in males of both species may indicate that they are not associated with sex pheromone perception. Yet, the problem which of these various types of sensilla truly would be responsible for perception of sex

Table 1. Number of sensillae on the antennae of males and females in *Fidonia piniaria* L. and *Orgyia antigua* L.[1]

Morphological type of sensillae	Number of sensillae			
	Fidonia		Orgyia	
	Male	Female	Male	Female
Sensilla trichodea	Numerous	Scarce	Numerous	Scarce
Sensilla coeloconica	350	ca. 100	600	75
Sensilla styloconica	22	16	50	30
Sensilla chaetica	117	105	80	42
Sensilla basiconica	0	5	0	0

[1]From Schenk (1903).

pheromone remained unsolved until the advent of modern electrophysiological techniques. For more recent investigations of sensilla on the antennae of *Bombyx,* see Schneider and Kaissling (1957), and references in Table 5. For other insects, see references in Table 6.

3.2. Special Glandular Tissues Producing Lipophilic Material in Female Moths

Freiling (1909) found that, if in some female moths the eighth and ninth abdominal segments were removed by amputation, such an "isolated abdomen" carries the excitatory properties of the entire female moth, whereas the rest of the insect, the carcass, does not. By histologic investigations of the abdominal segments of female moths, specific "scent glands" were detected (Fig. 3). They were found to represent a variety of morphological types (Table 2). Mostly they are located in the intersegmental fold, the epidermis of the membrane forming a more or less developed glandular epithelium. The sex pheromone gland of *Bombyx* also follows these general anatomic principles, whereas deviations have been noted in some cases, e.g., *Prodenia litura* F. (Hammad and Jarczyk, 1958). In microscopic sections of many of the glandular epithelia, the plasma of its cells contains numerous vacuoles that can be visualized particularly by lipid stains. They were suspected to contain the sex pheromone secreted by subcellular structures (Freiling, 1909). Further it was pointed out that its discharge into the surrounding air may be supported mechanically: females, if undisturbed, were

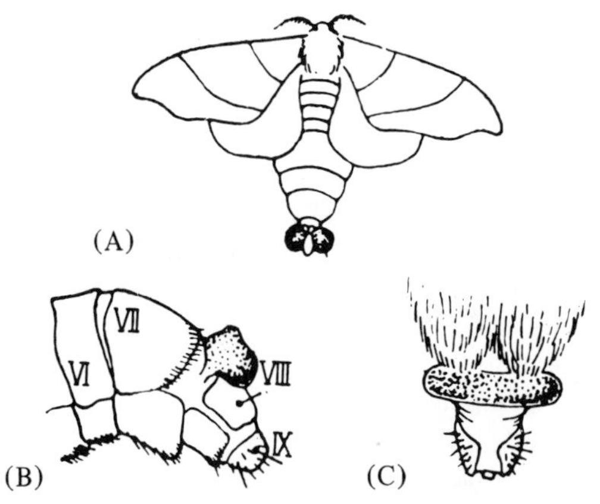

Figure 3. Morphology of scent glands located in abdominal segments of female moths (from Meisenheimer, 1921). (A) Silkworm moth (*Bombyx mori* L.): pair of intersegmental scent sacs ("sacculi laterales"). (B) *Argynnis adippe* L.: dorsal scent sac, abdominal segments with roman numerals. (C) Protrudable scent ring of *Cucullia verbasci* L.

Table 2. Morphology of sex scent organs in some female moths[1]

Morphologic types of scent organs	Species
Mobile scent areas	*Phalera bucephala* L.
Protrudable scent rings	*Cucullia verbasei* L.
	Cucullia argentea Hufn.
Dorsal and ventral scent folds	*Porthesia similis* Fuessl
	Plodia interpunctella Hb.
Intersegmental scent sacs	*Bombyx mori* L.
	Saturnia pavonia L.
	Aglia tau L.

[1] From Urbahn (1913) and Götz (1951).

observed to make "pumping movements" with their abdomen which rhythmically changes the surface of the glandular epithelium and thus may assist evaporation of the sex pheromone. For more recent investigations of the sex pheromone gland of *Bombyx* see references in Table 5; for other insects, see references in Table 6.

3.3. Species Specificity of Perception and Chemical Nature of Moth Sex Pheromones

The early morphological findings on specific perception and production of sex pheromones in moths suggested experiments for investigating their species specificity and their material nature.

By reciprocal exposure of males to females of the same and of other species, it was established that—as a rule—perception of sex pheromones usually is not only sex specific but also highly species specific. Only few exceptions of this rule were known to exist. They occur in closely related species: for example, in *Plodia interpunctella/Ephestia kühniella* (Barth, 1937; Schwinck, 1953) and in *Lymantria dispar/L. monacha* (Görnitz, 1949; Schwinck, 1955a). For more recent investigations on species specificity, see references in Tables 6 and 7.

Using lipophilic solvents, from "isolated abdomina" of females, extracts were obtained showing the entire "attractivity." Also trials were made with olfactometry and condensation traps to demonstrate the presence of sex pheromones in the air over cages containing female moths (Table 3). Experiments of this kind convincingly demonstrated the chemical nature and the volatility of the pheromones. It is interesting to see that—with the exception of *Bombyx*—the species involved in these early investigations mostly represent insect pests for which control was considered to be of prime economic importance. Indeed, serious considerations and experiments aiming at utilization of sex pheromones for selective insect pest control date back to the 1930s (e.g., Table 3 and in addition Farsky, 1938; Ambros, 1940; Hanno, 1939; Götz, 1951; Flaschenträger et al., 1957).

Table 3. Demonstrations of the chemical nature of sex pheromones in some species of moths by various means

Author(s) and year	Moth	Demonstrated by
Collins and Potts (1932)	*Lymantria dispar* L.	Extraction with xylene/benzene mixtures
Barth (1937, 1938)	*Plodia interpunctella* Hbn.	Condensation traps
Butenandt (1939, 1941)	*Bombyx mori* L.	Extraction with petroleum ether
Götz (1939a,b)	*Clysia ambiquella* Hbn.	Olfactometry
	Polychrosis botrana Schiff.	Olfactometry
Zehmen (1942)	*Lymantria monacha* L.	Extraction with various organic solvents
Flaschenträger and Amin (1949, 1950)	*Prodenia litura* Fab.	Extraction with ether, condensation trap
	Argrotis ypsilon Rott.	Extraction with ether
Inhoffen (1951)	*Euproctis chrysorrhoea* L.	Condensation trap

The biological data that accumulated gradually on the sex pheromones of moths made it apparent that these principles represent a hitherto unknown new class of highly bioactive chemical compounds which are produced in one organism, the female, and convey a specific biological message to another related organism capable of receiving it: the corresponding male. In 1939, the data on such a new biological principle appeared stimulating enough for us to try and enter the open frontier and accept the challenge of working with an appropriate species of moth for isolation and structure elucidation of a prototype sex pheromone. In fact, such a project appeared related in principle to the isolation and chemical characterization of sex hormones of higher animals and man, as accomplished previously in the late 1920s and 1930s (for a retrospective on estrone, see Butenandt, 1979). Insect sex pheromones would perhaps provide a specific and therefore a particularly useful means in trials to understand the mechanism(s) and the physiology of olfaction. Additional motivation for sincere engagement in the new field came from the expectation of using sex pheromones for highly specific insect pest control. The idea was discussed with and supported by leading scientists of the chemical industry, e.g., Schering AG, Berlin, who, at that time, were engaged in their first trials in developing synthetic chemicals for use as insecticides.

IV. Sex Pheromone Isolation: Principal Technical Essentials

In the forefront of isolation of any bioactive principle(s), two apparently trivial technical essentials are critical and must be solved: (i) availability of a sufficient amount of starting material for the isolation and (ii) availability of a quantitative bioassay to monitor the active principle(s) during the isolation. *Bombyx mori,* the only domesticated species of moths considered an adequate source for sufficient amounts of starting material, is raised commercially and in large quantities on silk farms where their pupal cocoons are processed for production of silk, the famous classical fiber. The most sensitive "detectors" for sex pheromones of female moths would be the corresponding male moths or rather, their antennae (Fig. 1A), if means can be developed to measure, in at least a semiquantitative manner, perception of the sex pheromone.

4.1. Sufficient Insect Material to Start With

At the onset of bombykol isolation (Butenandt, 1939, 1941), it was felt that at least several thousands of female moths would be required to generate an amount of sex pheromone glands sufficient to isolate the sex pheromone in measurable quantities for identification and elucidation of its chemical structure.

To raise *B. mori,* fertilized eggs, each measuring about 1 mm in diameter, are incubated under defined conditions until the small essentially monophageous larvae emerge (Cretschmar, 1947). They grow best through their four larval stages (Fig. 4A) if fed on mulberry leaves (*Morus alba* L.) (see also Watanabe, 1958). They spin their pupal cocoon (Fig. 4B) within about 4 weeks after hatching from the eggs. To obtain the female and male moths from cocoons in a controlable manner for experiments, within a few days after harvest, the pupae are cut out from their cocoons (Fig. 4C) and separated according to sex. In separate and distant rooms, on racks with shelves, the female pupae are spread on paper and covered with perforated paper sheets. This provision causes the freshly emerged moths to creep on top of the sheet from where they may be collected. After hatching of male moths in an analogous manner, they may remain sitting on the perforated paper for a day or so to allow for full development of their still soft wings, to be used as "indicators" in the bioassay for the sex pheromone (see below).

Right after hatching, the "attractivity" of the female moth appears to be optimally developed. Their paired sex pheromone glands (sacculi laterales, Figs. 1B, 3A) are collected by cutting them off the tip of the abdomen with scissors. For easy extirpation of the glands, they may be exposed by mildly squeezing the abdomen of the moth. In this way quick routine collection of the glands may be accomplished. Also this procedure is mechanically selective in that it allows collecting "pure glands," burdened only marginally by material of the carcass. The cut-off glands are dropped into methanol for preservation. Such preparations proved to be a satisfactory source to obtain a soluble active extract for

(A)

Figure 4. Raising of silkworms to obtain male and female moths. (A) Silkworms feeding on mulberry leaves.

(B)

(C)

Figure 4 (cont'd). (B) The last instar weaves the silk cocoons in which metamorphosis takes place. (C) Male and female pupae cut out from their cocoons to be separated for hatching.

fractionation (see below). As an alternative method, collection (by cooling traps of a condensate from the air above female moths) was investigated, but proved to be unsatisfactory (Butenandt, 1955a,b).

4.2. Quantitative Bioassay

The idea for use of a "behavioral" bioassay came from the observation that if a male *Bombyx* moth happens to sit in the vicinity of a female with her sex pheromone gland exposed, the male indicates arousal by a typical "flutter dance." If the female is removed or if successful copulation has taken place, the dance ends after a while, or does not take place at all. Because maintenance of *Bombyx* moths is fairly simple (see above) such behavioral response may be used for a semiquantitative bioassay. In contrast to the ever-hungry silk worms, the moth does not eat and–most convenient in a bioassay–cannot fly. If maintained singly in glass jars covered with glass plates (Fig. 5A) the males sit quietly for hours only rubbing occasionally one of their antennae with a front leg. Often this is the only indication that they are at all alive. Even if the tip of a glass rod previously dipped into a solvent, such as petroleum ether, is held in front of the antennae of one of the males (Fig. 5B) practically no reaction is observed. However, if the tip of the glass rod is dipped into the petroleum ether extract of female sex pheromone glands, the male becomes immediately aroused (Fig. 5C): It quickly and invariably presents its "flutter dance," often accompanied by searching movements of its abdomen which carries the copulatory organs. A detailed study of the behavioral response of male *Bombyx* showed that the "flutter dance" is the last of a number of rather defined behavioral stages of arousal (Schwinck, 1955b).

In a standardized setup using the "flutter dance" of the male moth as an "indicator," of a solution containing the sex pheromone a "sex pheromone unit" (SPU) may be measured semiquantitatively (Butenandt, 1939, 1941, 1955a,b). Using a glass rod as shown in Figures 5B,C a single solution of known concentration of material in petroleum ether is tested in a group of 30–60 males. In a dilution series graded by one order of magnitude, that particular solution is determined which is capable of releasing the "flutter dance" in about 50% of the moths in the group. By definition this solution contains the SPU, which is expressed in weight units per milliliter. If, for example, a solution containing 1 μg/ml of material is capable of inducing the "flutter dance" in about 30 out of 60 males, the SPU is 1 μg (see also Charts 1, 2).

To increase the sensitivity of the bioassay, the males to be used in it may be preselected for responsiveness to the sex pheromone: Only those males responding easily and quickly under otherwise strictly sex attractant-free conditions may be used in the assay. Also, as a rule, any one male may be used in the routine bioassay only once per day, in the afternoon, while its ability to respond is tested in the morning. In a late phase of bombykol isolation preselecting responsive moths introduced an operational increase in the sensitivity of the assay. Under highly standardized conditions, reproducibility and discrimination power of the assay are satisfactory for concentrations of pheromone differing by a factor of at least 10. Although much desired, especially for testing in late stages of the fractionation procedure, it was impossible within economic limits to improve the discrimination power, e.g., by increase of the group size used.

(A)

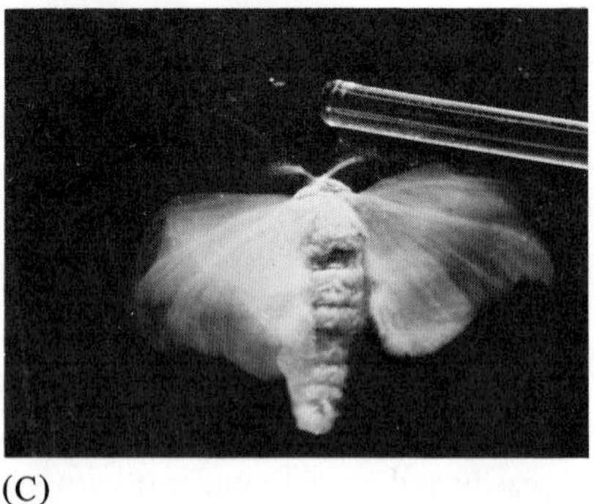

(B) (C)

Figure 5. Maintenance of males of *Bombyx mori* L. for bioassay of the sex pheromone of females. (A) Glass jars with single males arranged as used in the bioassay. (B) Negative control: the tip of a glass rod previously dipped into petroleum ether does not excite the male moth. (C) Positive reaction: the tip of the glass rod previously dipped into a solution in petroleum ether of the sex pheromone, for example, an extract of sex pheromone glands.

Chart 1. The final bioassay-monitored systematic fractionation procedure for isolation of bombykol, Part I: from 500,000 scent glands to a crude NABA-ester fraction, SPU = sex pheronome unit (from Butenandt et al., 1961a, condensed)

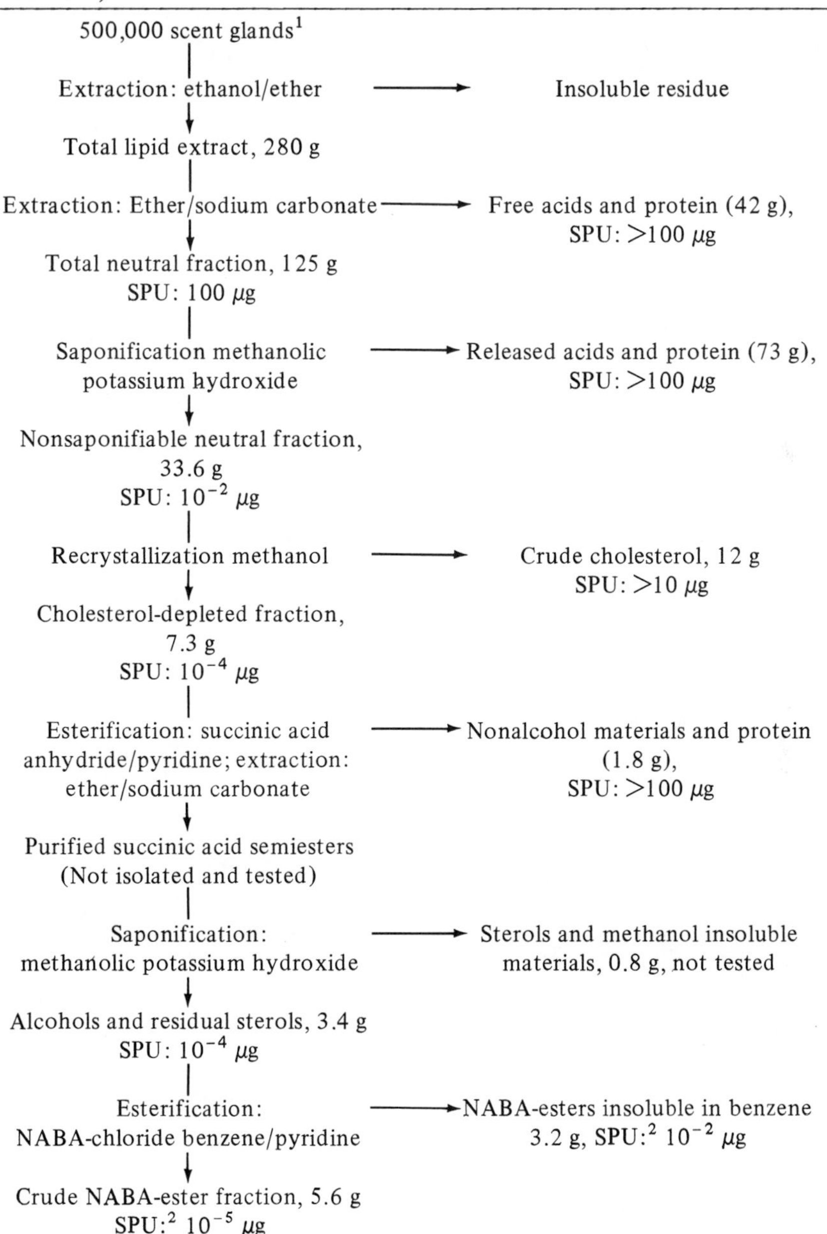

[1] Weights given in the chart refer to the sum of the individual fractions obtained in two independent fractionations, each starting with 250,000 scent glands.

[2] After standardized, mild saponification of small samples.

Chart 2. The final bioassay-monitored systematic fractionation procedure for isolation of bombykol, Part II. From the crude NABA-ester fraction to the pure NABA-ester of the sex pheromone. SPU: sex pheromone unit determined after standardized, mild saponification of small samples of NABA-ester fractions (from Butenandt et al., 1961a, condensed)

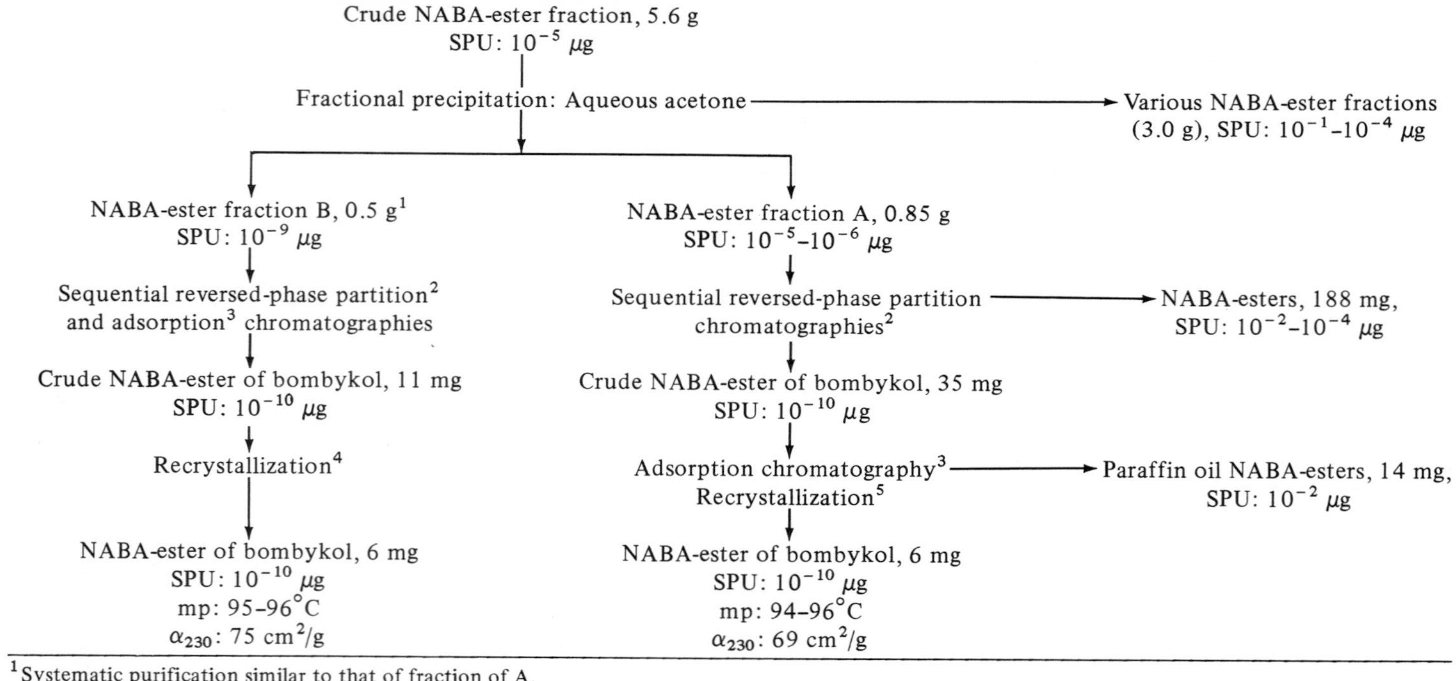

[1] Systematic purification similar to that of fraction of A.

[2] Paraffin oil/acetone, see Fig. 7.

[3] Silica gel.

[4] Acetone/water; petroleum ether.

[5] Acetone/water.

With the two principal technical essentials provided, work could be started to search for the trail through the labyrinth of numerous compounds contained in the soluble extract of sex pheromone glands. Gradually, by trial and error in biological testing, the trail to end at the pure sex pheromone of *B. mori* was endeavored.

V. Bombykol: Isolation, Derivatization, and Chemical Identification

In the first years after starting the bombykol project, it was relatively simple to raise the silk cocoons: Harvest of several thousands of glands of female moths was sufficient to produce the soluble bioactive extract and to allow for the first stages of fractionation (Butenandt, 1939, 1942). The cocoons required were purchased annually from hobby silkworm raisers in northern Germany.

In a temperate climate such as that which prevails in Germany, raising and maintenance of silkworm moths usually are restricted from the end of June to the end of October, at most. By keeping harvested mulberry twigs with leaves slightly refrigerated, the feeding period of the larvae may be extended. Also it is possible to delay hatching of moths from pupae by keeping the latter in the refrigerator for some time. By such means, it was possible to extend the assay period until around Christmas. Yet, if pupae are kept in the refrigerator for too long a time or at too low temperatures, they gradually loose their capability to successfully hatch with fully developed wings–the "indicators" of male response.

Within the years of steady improvement (1950-1959) of the systematic fractionation procedure (Charts 1, 2), it was learned gradually that–due to the immense bioactivity of the sex pheromone–tens of thousands of glands would be required if a measurable amount of the pheromone should be finally isolated. Thus, more and more silkworm growers all over Germany became involved in supplying our demand. From about 1952 onward, silk production gradually seized in Germany, due to the increasing competition by upcoming modern fabrics such as perlon and nylon together with the climatic conditions relatively unfavorable for growing silk. Hence, the resources of the country site decreased, whereas in our laboratory the demand for cocoons increased. Alternative sources were sought and found at first in Yugoslavia, subsequently in Italy, and finally in Japan. Thus, in their final stages, these investigations required logistics and financial resources, of an extent, quite unusual at that time. The male pupae obtained in mass campaigns in excess of those required for the bombykol assay were used to isolate ecdysone, one of the hormones regulating insect metamorphosis (Butenandt and Karlson, 1954).

5.1. Pure NABA-Ester of Bombykol by Bioassay-Monitored Systematic Fractionation of the Total Lip Extract of Sex Pheromone Glands

In bioassay-monitored fractionation of an active extract from biological materials, *anyone of the fractions generated* has to be tested for activity. Inactive fractions are discarded while the active fractions are retained and subjected to further fractionation to unravel, finally, the pure active entities. In such a truly "systematic" procedure the balance of the fractionation does account (i) *qualitatively*, by number, for all the active entities assayed for and of whatever structure contained in the original extract (and hence in the biological material), and (ii) *quantitatively*, by recovery, for the minimum content of any of the active entities contained in the original extract (and hence in the biological material).

In the trials to develop a systematic fractionation procedure of the total lipid extract of *Bombyx* sex pheromone glands almost every single stage of the fractionation procedure provided unexpected problems. To illustrate this, some of its key problems may be discussed briefly using, as a background, the final version of the bioassay-monitored systematic fraction procedure (see Charts 1, 2).

One of the early achievements was the finding that the Total neutral fraction may be activated by saponification to yield the Nonsaponifiable neutral fraction with an increase in activity overshooting by far the amount of inactive materials removed (see Chart 1). The Nonsaponifiable neutral fraction thus obtained—at that time without prior removal of cholesterol (see below and Chart 1)—could be esterified with succinic acid anhydride to yield a crude diethyl ether soluble fraction (not recorded in Chart 1). In the form of succinic acid semiesters, it contains all primary and secondary alcohols of the previous fraction. From ether, the semiesters may be extracted into a slightly alkaline water phase leaving essentially all nonalcohol materials and some protein in the ether phase which proved to be inactive (Chart 1). The acidic material remains in the water phase. From there, after lowering the pH, it may be reextracted into ether to yield the fraction Purified succinic acid semiesters (Chart 1). This fraction—again without isolation and testing—was subjected to mild saponification to release the fraction Alcohols and residual sterols, with a SPU of 10^{-4} μg, containing essentially all the sex pheromone activity of the Total lipid extract. Such generation of a highly active fraction of Alcohols and residual sterols in a reproducible manner indicated that the sex pheromone most likely would be a primary or secondary alcohol. Based upon the total loss of bioactivity of the Total lipid extract by catalytic hydrogenation, it was concluded that the pheromone contains C=C double bonds (Butenandt, 1939, 1941, 1955a,b, Hecker, 1956, 1958). Later on, as an advantageous further stage, selective removal of cholesterol from the Nonsaponifiable neutral fraction was introduced (Hecker, 1959) to yield in the final version of the fractionation procedure (Chart 1) the Cholesterol-depleted fraction. This fraction then was subjected to esterification with succinic acid anhydride, etc., as described above (see Chart 1).

While developing the stages of the fractionation procedure, at times, we were particularly intrigued by an apparent high volatility of the sex pheromone. For example, in early variants of the procedure, petroleum ether was used to obtain the Total lipid extract. If this solvent was removed by distillation, it was observed that the condensate was considerably active in the bioassay, exhibiting a SPU of about the equivalent of 10^{-2} μg of material. This seemed to indicate that the pheromone molecules might be rather volatile and have a low molecular weight and/or some kind of globular shape, such as camphor. Therefore a large number of known low-molecular-weight volatile compounds were prepared and tested for sex pheromone activity, e.g., mono-, sesqui-, and diterpenes, cycloheptanol and macrocyclic ketones, or simply saturated and unsaturated low-molecular-weight aliphatic alcohols, such as hexa-2*E*,4*E*-diene-1-ol (sorbinol, e.g., Butenandt et al., 1955). Yet, relative to the activity of the fraction Alcohols and residual sterols (Chart 1) obtained from the Total lipid extract of glands, the solvent condensates from active fractions as well as all of the many synthetic compounds tested may be considered practically inactive. These results, together with unsuccessful trials of combined use of condensation traps and gas chromatography, at that stage discouraged further research in this direction.

As may be seen from Chart 1, in the trial-and-error development of the stages of the fractionation procedure, the fraction Alcohols and residual sterols was reached by more or less classical group separations yielding, in each stage, essentially two fractions, one carrying the activity while the other was more or less inactive. In further stages of the purification process, separation methods had to be used generating multiple fractions. It was felt that progress in fractionation might suffer from the impossibility to discriminate by bioassay small differences in biological activities. If, for example, any subsequent stage of the fractionation procedure would provide multiple subfractions which differ in sex pheromone content between 10 and 90%, it would be impossible to discriminate these fractions by bioassay: from its outlay (see above) it is capable to distinguish reliably one order of magnitude in SPU at most. On the other hand, direct measurement by physical methods such as UV or IR spectroscopy, even of the most purified fractions obtained up to then, did not reveal characteristic physical data to be associated with biological activity.

To improve the discrimination power, also, techniques of electrophysiology were adapted. Indeed, in an arrangement of two steel microelectrodes sticking in a single antenna of a *Bombyx* male, as shown schematically in Fig. 6, an "electroantennogram" may be obtained in a reproducible manner (Schneider and Hecker, 1956; Schneider, 1957; Hecker, 1959). It may be interpreted as the summation of spontaneous discharges of many sensilla located on or within the male antenna (Fig. 6A). By exposure of the antenna to an isolated scent gland of the female moth, the frequency of discharges increases (Fig. 6B). With vapors of diethylether, the system may be narcotized reversibly (Fig. 6C). If, instead of the rather thick steel microelectrodes, glass capillaries of 1–10 μ diameter are used, it is possible to locate the tip of these "different electrodes" within the antenna at or in groups of nerve cells or even in a single nerve cell associated with sensilla. In this way the action potentials derived may be attributed to one of the histologically known structures suspected to be related to sex pheromone perception (e.g., Table 1). While this approach proved to be most promising in the long run (see Tables 5, 6), its immediate application and use in testing fractions suffered from an unexpected complication requiring more detailed investigation: through "mechanoreceptors" the antenna of the male silkworm moth also registers air currents in addition to sex pheromones. Most likely, the former

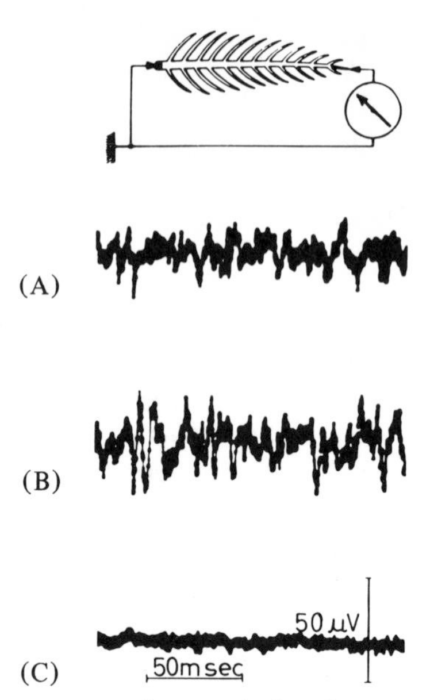

Figure 6. Electroantennogram from an isolated antenna of a Bombyx male using micromanipulator-inserted steel electrodes and recording by oscillograph (from Hecker, 1959). (A) After electrodes have been inserted a continuum of spontaneous discharges may be recorded from this system. (B) By exposure of the system to an isolated scent gland of the female, amplitudes and frequency of discharges are increased. (C) After exposure of the system to vapors of diethylether reversible "narcosis" takes place.

play an essential role in "long-distance attractivity" of insect sex pheromones as pointed out by sensible behavioral experiments (Schwinck, 1954, 1955b; Dufay, 1957). Thus, it was felt that for a solution of our testing problems, a meaningful installation of the new technique would require more time than we impatient biochemists could allow.

With this dilemma, and with the expectation that the sex pheromone would probably not crystallize, "ear marking" of the pheromone alcohol by derivatization seemed to be the most promising and quick chemical approach for sensible follow-up in purification. In addition, it was felt that it might also be helpful in this way to increase the mass of material to be handled. Yet, an appropriate derivative would have to meet a number of requirements: it should absorb UV light or be even colored for easy detection and it should crystallize easily. Moreover and most importantly, if by itself inactive, it should allow to release the pheromone quantitatively to check for activity of derivatized fractions in the bioassay.

Elaborate investigations of practically all reagents for derivatization of alcohols known up to 1953, including isocyanates (yielding urethanes of remarkably high melting points), showed that for our purposes all of them have one or more serious deficiencies. Finally it was decided to design a new reagent: 4′-nitroazobenzene-4-carboxylic acid (NABA, Hecker, 1955a). Its acid chloride proved to be an ideal reagent for derivatization of all kinds of primary and secondary alcohols: due to its specific structure NABA yields intensely red-colored and easily crystallizing NABA-esters. Their intense UV band at 330 nm allows for easy monitoring in separation procedures (Amin and Hecker, 1956a) as well as for determination of the equivalent (molecular) weight of any pure ester (and hence of the alcohol component of the ester). Last but not least, the colored, biologically inactive NABA-esters may be hydrolyzed easily in a specifically developed standardized quantitative microsaponification procedure to yield the (biologically active) alcohol.

Using the reagent, an important further stage of the fractionation procedure was made possible yielding from the fraction Alcohols and residual sterols the Crude NABA-ester fraction (Chart 1). Around that time, new and efficient, however mild multiple fractionation methods had become available, such as multiple partition in solvents (Craig distribution and other methods, Hecker, 1955b, 1963) as well as partition column and paper chromatography (see also Makino et al., 1956) and–later on–gas chromatography (e.g., Anders and Bayer, 1959). Indeed, the immediate application of some of them to our problem permitted steady progress in fractionation, the weight of active fractions becoming lower and lower and their activities higher and higher. Thus, starting with 313,000 sex pheromone glands, generated in the 1953 silkmoth campaign, in 1955 we succeeded in isolating 5.25 mg of a solid NABA-ester of the sex pheromone, for the first time. It was paper chromatographically uniform (Fig. 7) and crystallized in square platelets with a melting interval of 58–68°C. After standard saponification, the SPU was below 10^{-5} μg/ml. The UV data of the derivative together with the partitioning and chromatographic properties of NABA-esters of alcohols in general (Hecker, 1955a; Amin and Hecker, 1956a) suggested that the sex pheromone alcohol contains more than 10 carbon atoms. Moreover, the UV spectrum of the NABA-ester preparation was very similar to that of hexa-2*E*, 4*E*-diene-1-ol (sorbinol, see also Fig. 9B), thus suggesting the presence of two conjugated double bonds (Butenandt, 1955a,b; Hecker, 1956, 1958).

The breakthrough in 1955 stimulated greatly further efforts to win finally the battle of isolation of the *Bombyx* sex pheromone as a pure NABA-ester. The assay was improved regarding sensitivity by preselecting the most responsive males to be used (see above). For the final version of the fractionation procedure, almost every stage, as developed up to then, was revised and optimized with respect to efficiency of separation and to yields of activity. From more than 1 million silk cocoons purchased in Germany, Italy, and Japan, altogether 500,000 scent glands were generated. They were divided into two batches of 250,000, to be subjected individually to the final version of the fractionation procedure (Charts 1, 2; for weights of corresponding fractions, see Chart 1, footnote 1). They yielded, together, 5.6 g of the Crude NABA-ester fraction with a SPU of 10^{-5} μg (Chart 1). By careful fractional precipitation from ace-

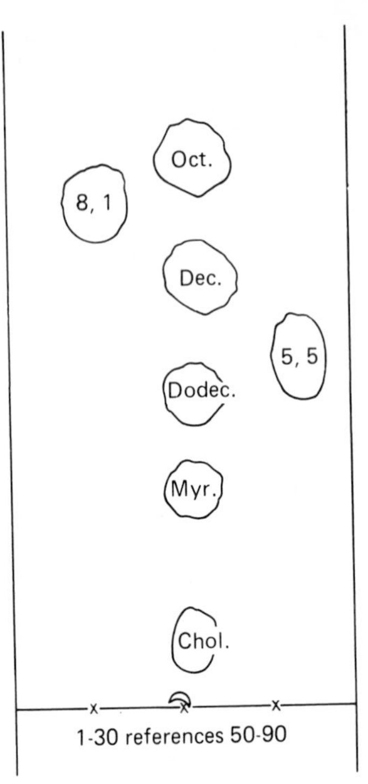

Figure 7. Comparative paper chromatography of samples of NABA-ester fractions 1–30 and 50–90 obtained by reverse-phase partition chromatography (stationary phase: nonane on hydrophobic kieselgur; mobile phase: nitromethane). Reference compounds: NABA-esters of cholesterol and *n*-tetradecanol to *n*-octanol; stationary phase: *n*-decane on hydrophobized paper; mobile phase: nitromethane; R_c migration distances relative to that of the NABA-ester of cholesterol: 5,5 NABA-ester of sex attractant, 8,1 NABA-ester of unknown companion alcohol (from Butenandt, 1955b; Hecker, 1956, 1958).

tone with water, two high-activity NABA-ester Fractions A and B were obtained (Chart 2) together with various less active fractions (Chart 2). Systematic purification of the former was accomplished in two further parallel procedures using sequential reverse-phase column partition chromatography and/or adsorption column chromatography and recrystallization, respectively (Chart 2). Of these, the stage of sequential column chromatographies is exemplified in Figs. 8A–C for 400 mg of the (less active) NABA-ester Fraction A. It yielded in a single band (Fig. 8C) 35 mg of the fraction Crude NABA-ester of bombykol (Chart 2). From these 35 mg, in the final stage (Chart 2), by recrystallization 6 mg of pure NABA-ester of bombykol was obtained, SPU 10^{-10} μg, melting point (mp) 94–96°C and an absorbance of $\alpha_{230} = 69$ cm^2/g (Chart 2). Similarly, the more active Crude NABA-ester fraction B was also purified (Chart 2) yielding another

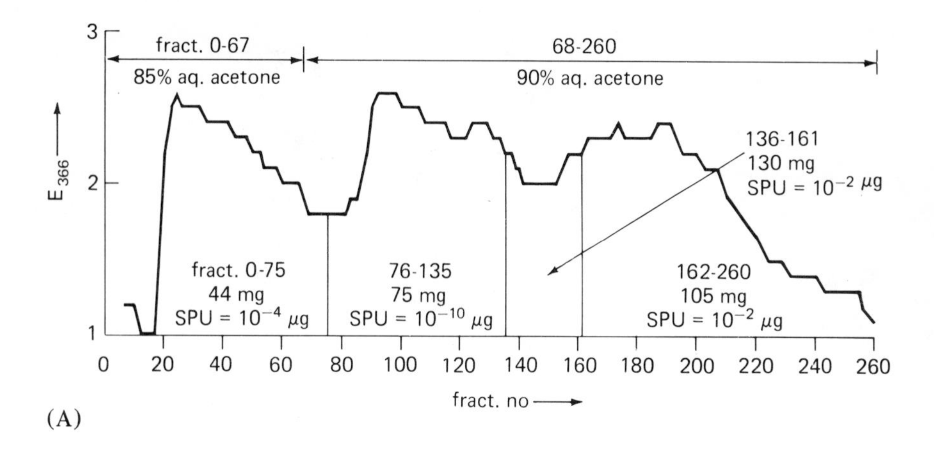

(A)

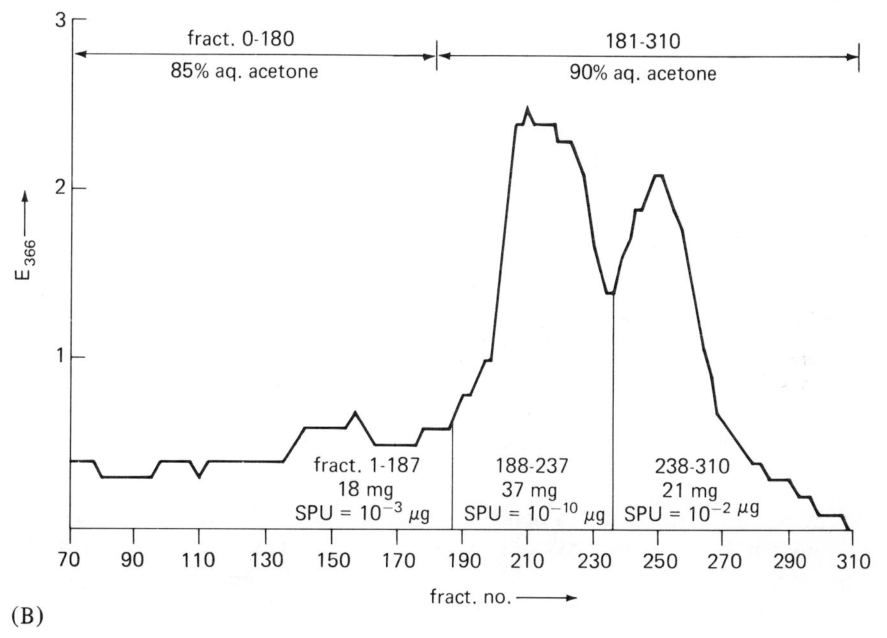

(B)

Figure 8. Reverse-phase column partition chromatography (stationary phase: paraffin oil in kieselgur; mobile phase: aqueous acetone) of 400 mg of NABA-ester fraction A from the fractionation procedure, see Chart 2; all SPU after standard saponification. (A) First run yields, besides NABA-ester fractions 76–135 with highest SPU, the lower SPU fractions 0–75, 136–161, and 162–260. (B) Second run to improve purity of fractions 76–135 (above A), yielding highest SPU fractions 188–237, together with lower SPU fractions 1–187 and 238–310.

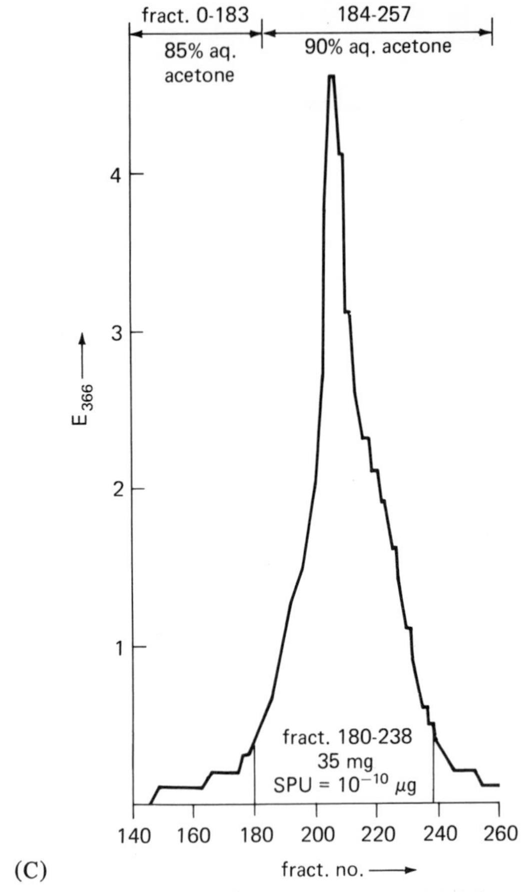

Figure 8 (cont'd). (C) Third run of fractions 188–237 (above B) to yield single band of NABA-ester of bombykol in fractions 180–238.

6 mg of Pure NABA-ester of bombykol with practically the same SPU, and similar mp and UV data (mp. 95–96°C; $\alpha_{230} = 75$ cm^2/g). Thus, altogether, *12 mg of the NABA-ester of bombykol became available.* The deep red crystalls of the pure NABA-ester of bombykol are shown in Fig. 9A. Using the molecular extinction coefficient for NABA-esters at 360 nm (Hecker, 1955a), a molecular weight of $M = 475 \pm 15$ was calculated for the NABA-ester of the sex pheromone. The elemental analysis by microcombustion of the NABA-ester was compatible with a molecular formula of $C_{29}H_{37}N_3O_4$ ($M = 491.6$). Hence the sex pheromone alcohol would meet the molecular formula $C_{16}H_{30}O$. For a straight-chained molecule, this formula would indicate two double bonds. The UV spectrum of the NABA-ester with its maximum at 230 nm (Fig. 9B) shows, as previously derived, conjugation of two double bonds in bombykol (Butenandt et al., 1961a).

(A)

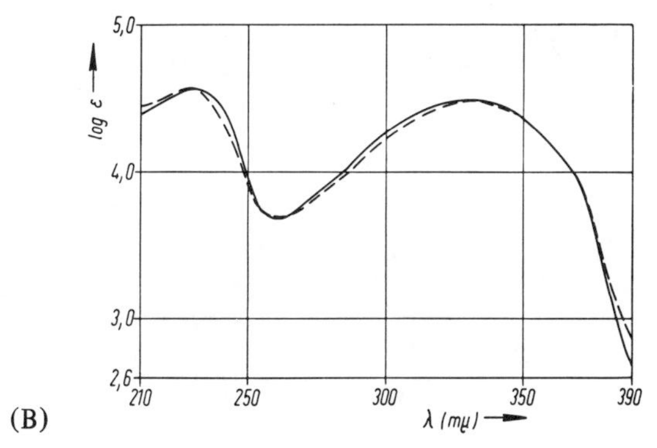

(B)

Figure 9. (A) Crystals of the NABA-ester bombykol. (B) UV spectrum of the NABA-ester of bombykol (—) as compared to that of the NABA-ester of hexa-2*E*,4*E*-diene-1-ol (sorbinol, - - -).

5.2. Structure Elucidation of the NABA-Ester of Bombykol

The infrared spectrum of free bombykol obtained by saponification of 4 mg of the NABA-ester indicated a primary alcohol which is nonallylic. *Z,E*- configuration of the double bonds was suggested by the two sharp bands in the region of 1000–900 cm^{-1}. A rocking band close to 700 cm^{-1} showed presence of a chain of at least four adjoining CH_2 groups (Butenandt et al., 1961b). Catalytic microhydrogenation of free bombykol yielded hexadecane-1-ol, establishing that a straight chain of 16 carbon atoms is present.

Such data of the pure NABA-ester provided most valuable information for further chemical micromethodology to determine its complete structure. For controlled oxidative degradation of the NABA-ester of bombykol, a micromethod was worked out using, in a closed system, potassium permanganate in water-free acetone as the oxidant: From 1 mg of the NABA-ester under carefully controlled conditions, three degradation products were obtained. They were fully identified (see Chart 3) to account for all 16 carbon atoms of the straight-chain alcohol: the red-colored NABA-ester of ω-hydroxydecanoic acid, oxalic and *n*-butanoic acid. Accordingly, bombykol was established to be a hexadeca-10,12-diene-1-ol (Butenandt et al., 1961c).

Chart 3. Controlled oxidative microdegradation of the NABA-ester of bombykol by potassium permanganate. Fragments resulting: I, NABA-ester of ω-hydroxydecanoic acid isolated as methylester; II, oxalic acid; III, butanoic acid. Reassembly: IV, NABA-ester of hexadeca-10,12-diene-1-ol.

$O_2N-C_6H_4-N{=}N-C_6H_4-CO_2-[CH_2]_9-CO_2CH_3$

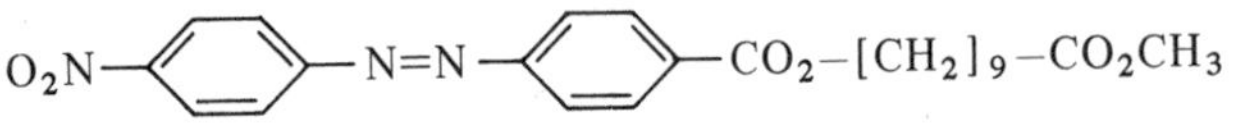

I

HO_2C-CO_2H

II

$HO_2C-[CH_2]_2-CH_3$

III

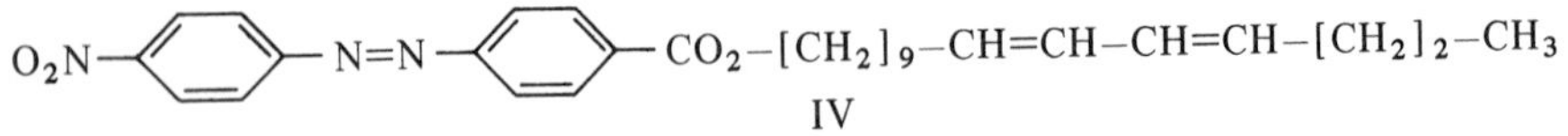

$O_2N-C_6H_4-N{=}N-C_6H_4-CO_2-[CH_2]_9-CH{=}CH-CH{=}CH-[CH_2]_2-CH_3$

IV

For a structure of this kind, theoretically four stereoisomers are possible (Figs. 10A,B) of which, according to spectroscopic data, bombykol most likely would represent one of the two possible *Z,E* configurations (Butenandt et al., 1961b,c).

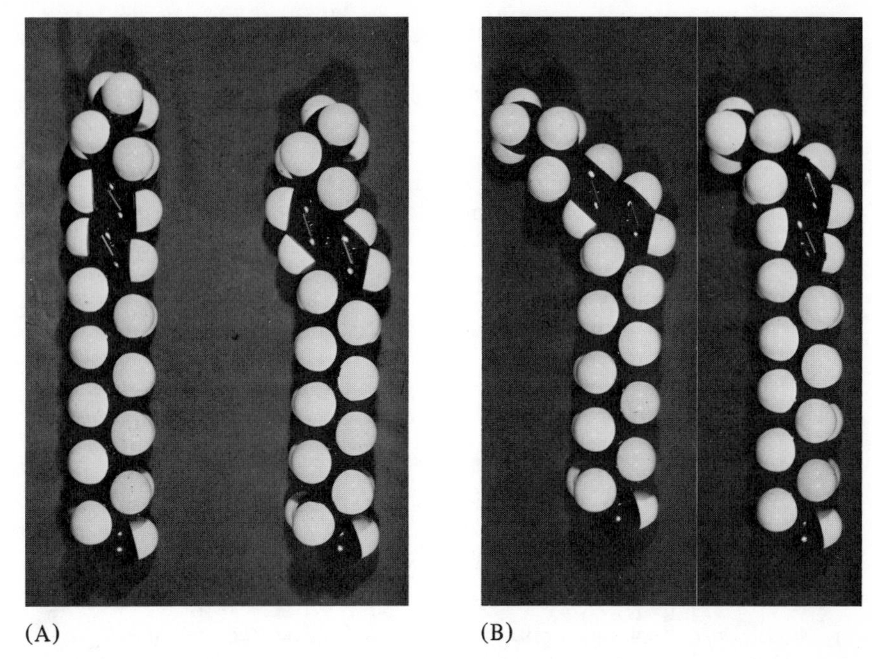

(A) (B)

Figure 10. Atomic models of the geometrical isomers of hexadeca-10,12-diene-1-ol. (A) 10-*E*,12-*E*; 10-*Z*,12-*Z*. (B) 10-*Z*,12-*E*; 10-*E*,12-*Z* (bombykol).

5.3 Final Proof of Bombykol Structure—Total Syntheses and Quantitative Bioassay of all Stereoisomers of Hexadeca-10,12-diene-1-ol

To provide final proof for the hexadeca-10,12-diene-1-ol structure of bombykol and—more specifically—to decide between the two possible *Z,E*- configurations, the total synthesis approach employing structure proofing routes was used. Some typical synthetic routes are assembled in Chart 4.

A. Route $C_7 + C_9$

Propargylbromide and *n*-butyral were condensed with zinc in ether to yield 4-hydroxy-hept-1-yne. Elimination of water yielded a mixture of the *Z,E*-isomeric heptenynes. By multiple fractional distillation using a spinning band column, the two stereoisomers were obtained in practically pure form, as verified by their infrared spectra and other criteria. With sodium amide in liquid ammonia, both heptenynes were converted simultaneously into their sodium salts which were alkylated with the tetrahydropyranyl ether of 10-bromo-nonane-1-ol. After removal of the ether groups, followed by high-vacuum distillation and recrystallizations at low temperature, the two possible *Z,E*-isomers of hexadec-10-yne-

12-en-1-ol were obtained. Partial *Z*- hydrogenation with Lindlar catalyst of each of the isomers and careful purification of the products obtained yielded the hexadeca-10*Z*,12*Z*- and -10*Z*,12*E*-diene-1-ols, respectively.

Chart 4. Typical synthetic routes for preparation of the stereoisomeric hexadeca-10,12-diene-1-ols (from Butenandt and Hecker, 1961, condensed)

Route $C_7 + C_9$ (R = tetrahydropyranyl)

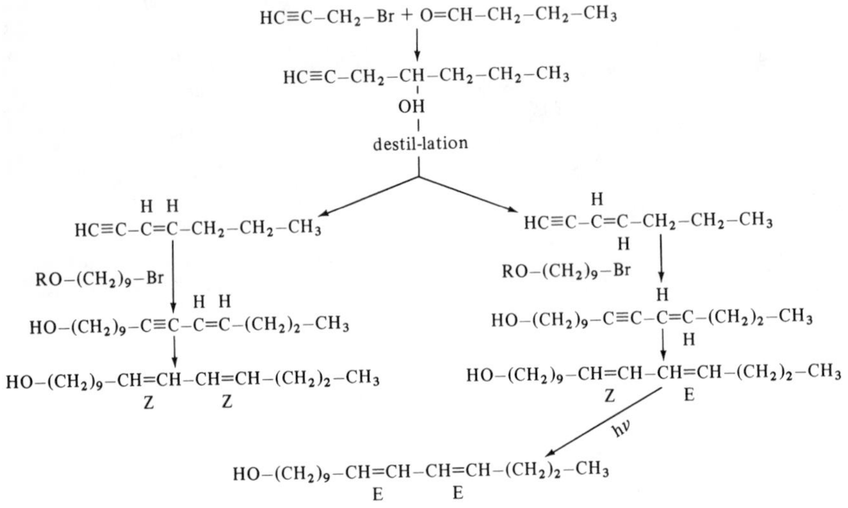

$$HC{\equiv}C{-}CH_2{-}Br + O{=}CH{-}CH_2{-}CH_2{-}CH_3$$

$$\downarrow$$

$$HC{\equiv}C{-}CH_2{-}CH(OH){-}CH_2{-}CH_2{-}CH_3$$

destil-lation

$$HC{\equiv}C{-}\overset{H}{C}{=}\overset{H}{C}{-}CH_2{-}CH_2{-}CH_3 \qquad HC{\equiv}C{-}\overset{H}{C}{=}\underset{H}{C}{-}CH_2{-}CH_2{-}CH_3$$

$RO{-}(CH_2)_9{-}Br$ ↓ $\qquad$ $RO{-}(CH_2)_9{-}Br$ ↓

$$HO{-}(CH_2)_9{-}C{\equiv}C{-}\overset{H}{C}{=}\overset{H}{C}{-}(CH_2)_2{-}CH_3 \qquad HO{-}(CH_2)_9{-}C{\equiv}C{-}\overset{H}{C}{=}\underset{H}{C}{-}(CH_2)_2{-}CH_3$$

$$\downarrow \qquad\qquad \downarrow$$

$$HO{-}(CH_2)_9{-}\underset{Z}{CH{=}CH}{-}\underset{Z}{CH{=}CH}{-}(CH_2)_2{-}CH_3 \qquad HO{-}(CH_2)_9{-}\underset{Z}{CH{=}CH}{-}\underset{E}{CH{=}CH}{-}(CH_2)_2{-}CH_3$$

$$\xrightarrow{h\nu}$$

$$HO{-}(CH_2)_9{-}\underset{E}{CH{=}CH}{-}\underset{E}{CH{=}CH}{-}(CH_2)_2{-}CH_3$$

Route $C_6 + C_{10}$ (R = phenyl)

$$CH_2O + HC{\equiv}CNa + Br{-}CH_2{-}CH_2{-}CH_3$$

$$\downarrow$$

$$HO{-}CH_2{-}C{\equiv}C{-}CH_2{-}CH_2{-}CH_3$$

$$\downarrow$$

$$Br{-}CH_2{-}C{\equiv}C{-}CH_2{-}CH_2{-}CH_3$$

$$\downarrow$$

$$Br^{\ominus\oplus}[R_3PCH_2{-}C{\equiv}C{-}CH_2{-}CH_2{-}CH_3]$$

$H_5C_2O_2C{-}(CH_2)_8{-}CH{=}O$ ↓

$$H_5C_2O_2C{-}(CH_2)_8{-}\underset{Z+E}{CH{=}CH}{-}C{\equiv}C{-}(CH_2)_2{-}CH_3$$

$$\downarrow \text{urea}$$

$$H_3C_2O_2C{-}(CH_2)_8{-}\overset{H}{C}{=}\underset{H}{C}{-}C{\equiv}C{-}(CH_2)_2{-}CH_3$$

$$\downarrow$$

$$HO{-}H_2C{-}(CH_2)_8{-}\underset{E}{CH{=}CH}{-}\underset{Z}{CH{=}CH}{-}(CH_2)_2{-}CH_3$$

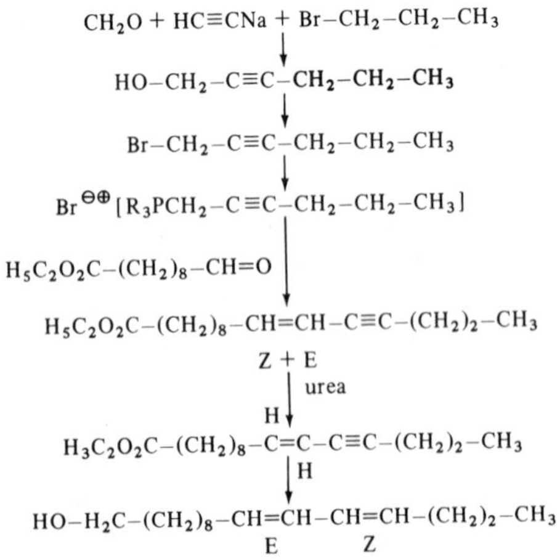

B. Route $C_6 + C_{10}$

From sodium acetylide and *n*-propylbromide pent-1-yne was prepared. Its Grignard compound was condensed with formaldehyde to yield hex-2-yne-1-ol. Bromination of the latter with phosphortribromide and reaction of the resultant 1-bromo-hex-2-yne with triphenylphosphin yielded the corresponding triphenylphosphoniumbromide. Hydrogen bromide was eliminated with phenyllithium; the resultant triphenylphosphinemethylene was reacted with ω-oxo-nonane-carboxylic acid ethyl ester. A mixture of the *Z,E*-isomers of hexadeca-10-en-12-yne-1-carboxylic acid ethylester was obtained, from which the 10*E*-isomer was separated via its inclusion compound with urea and repeated recrystallizations. After partial hydrogenation of its acetylenic group by Lindlar catalyst followed by reductive cleavage of the ester bond with lithium aluminumhydride, the alternative *Z,E*-isomer, hexadeca-10*E*,12*Z*-diene-1-ol, was obtained. By irradiation in the presence of catalytic amounts of iodine *Z*-double bonds may be converted to the corresponding *E*-configuration. In this way from both the pure *Z,Z*- and the *Z,E*-compounds obtained by routes $C_7 + C_9$ and $C_6 + C_{10}$, respectively, the same hexadeca-10*E*,12*E*-diene-1-ol was obtained, as depicted in Chart 4 for $C_7 + C_9$. Employing similar structure proofing routes, finally each of the four possible stereoisomers was prepared in pure form by two independent syntheses (Butenandt et al., 1962). As the result of cooperational efforts with Farbenfabriken Bayer, Leverkusen, simultaneously, but independently, further, but alternative pathways of syntheses of the stereoisomeric hexadeca-10,12-diene-1-ols were developed (Truscheit and Eiter, 1962). For very recent, new stereospecific syntheses of the four stereoisomers see Miyaura et al. (1983).

The four possible *Z,E*-isomers of hexadeca-10,12-diene-1-ol, their NABA-esters, and some of their properties are collected in Table 4. Already the physical properties of both, the *Z,Z*- and the *E,E*-isomers and their NABA-esters, especially their spectroscopic data, exclude them as candidates for bombykol. In addition, also their SPU is much lower than that of bombykol obtained after

Table 4: The four possible geometric isomers of hexadeca-10,12-diene-1-ol synthesized and their NABA-esters[1]

Hexadeca-diene-1-ol	Melting point (°C)	Alcohols λ_{max}[2] (nm, ε)	NABA-ester, mp[3] (°C)	SPU (μg/ml)
10*E*,12*E*	37–38	230 (31,500)	114.0–115.0	1
10*Z*,12*Z*	25.5–26.5	235 (27,500)	97.0	10
10*Z*,12*E*	Liquid	233 (26,400)	95.0–96.0	10^{-3}
10*E*,12*Z*	Liquid	232 (24,900)	95.5–96.5	10^{-12}

[1]Characterized by physiochemical data and by activity in the bioassay using *Bombyx* males (from Butenandt and Hecker, 1961).

[2]In ethanol.

[3]Koflermicroscope, corrected.

[4]Sex attractant unit.

saponification of its NABA-ester. In contrast to the *Z,Z*- and *E,E*-isomers, the 10*Z*,12*E*- and the 10*E*,12*Z*-isomers and their NABA-esters exhibit very similar physical properties (Table 4). Yet, the bioassay shows that obviously only one of them is identical with bombykol: whereas the 10*Z*,12*E*-isomer exhibits low, however, significant activity, the 10*E*,12*Z*-isomer with a SPU of 10^{-12} μg/ml is about 10^9 times as active. *Therefore bombykol is the so far unknown hexadecane-10E,12Z-diene-1-ol* (Butenandt and Hecker, 1961).

The simplicity of the structure and the order of magnitude of activity of bombykol are remarkable (Table 4) and in a sense unique. In the standard assay, a solution in petroleum ether containing, e.g., 10^{-12} μg bombykol/ml is used and the tip of a glass rod is dipped into it, removed, and, for a second or two, the solvent is allowed to evaporate from the tip: the trace of sex pheromone remaining at the tip of the "magic stick" is sufficient to arouse about 50% of the male moths used, for example, about 30 moths out of a test group of 60. From these operationally determined data (see above) in a preliminary estimation, it was calculated that a relatively small number of molecules reaches the antennae of the male moth and is capable to release its typical behavior (Butenandt, 1959, 1963; Hecker, 1959). For detailed investigations of the threshold problem of bombykol perception, see references in Table 5.

VI. New Avenues of Research Opened by Identification of the First Insect Sex Pheromone

While the results of successful key investigations certainly answer the questions which originally evoked them, usually they raise additional new—and especially more sophisticated—problems. The chemical identification of the first sex pheromone turned out to be no exception to this rule: in the last quarter of a century several hundred pheromones were detected, especially in insects. In many cases their mostly simple chemical structures were unraveled and proven by syntheses (e.g., Bestmann and Vostrowsky, 1979, 1982). The explosive extension of pheromone research, developing right after 1962 (see Tables 5, 6), was carried on by introduction of refined and highly sophisticated laboratory techniques, e.g., of high resolution and sensitivity separation methods, such as thin-layer and radio thin-layer chromatography (TLC and RTLC), gas and radio gas chromatography (GC and RGC), high-performance liquid chromatography (HPLC), or by new high-resolution spectroscopic techniques, e.g., mass and NMR spectrometry and their most sensitive versions or else by micromethods developed in electrophysiology, to mention only some key techniques employed in pheromone research today.

6.1. Bombykol and Insect Pheromone Biochemistry

Whereas a broad spectrum of insect pheromones was detected since 1962, relatively little is known on their biogenesis and metabolism in sex pheromone glands or on their metabolism in the antenna. To identify some unsolved biochemical problems, it is best to address certain qualitative and quantitative aspects of the bioassay-monitored systematic fractionation procedure which yielded the pure NABA-ester of bombykol (Charts 1, 2), the insect pheromone prototype.

With respect to the problem of *bombykol biogenesis and metabolism in the sex pheromone gland* of females it may be seen from Chart 1 of the fractionation procedure that saponification of the Total neutral fraction results in a decrease in SPU by a factor of 10^4 of the resultant Nonsaponifiable neutral fraction. This dramatic decrease of the SPU is disproportionate with the amount of inactive material sorted out in this particular stage (i.e., the fraction "released acids and protein," Chart 1). Already during our investigations this observation stimulated the idea that bombykol may be stored or held available in the sex pheromone gland of the female moth as an unknown, inactive, or only slightly active ester. Such an ester may be considered a "cryptic" source of bombykol. As part of the Total neutral fraction, in the process of preparing the Nonsaponifiable neutral fraction esters of this kind would be hydrolyzed, thus causing the dramatic decrease found in the SPU, i.e., increase in bombykol activity. Moreover, it was speculated that, according to a well-known structural principle verified in natural waxes, the acid component of such a "cryptic" source of bombykol might correspond structurally to that of its component alcohol, i.e., to bombykol: in the ideal case it might be the hexadeca-10*E*,12*Z*-diene-1-oic acid (Butenandt, 1955a,b). Indeed, such a hypothesis was supported by the finding that reduction by lithium aluminumhydride of the inactive fraction "Released acids and protein" (and also of "Free acids and protein," see Chart 1) and resulted in generation of considerable bioactivity. Preliminary trials to purify the acid postulated were promising (Amin and Hecker, 1956b; Butenandt et al., 1963).

Further, it is interesting to note that recently from sex pheromone glands of *Bombyx* by an approach designed specifically for this purpose, an aldehyde was obtained in relatively low amounts in comparison to the bombykol content; it corresponds in structure and perhaps in stereochemistry to bombykol and hence was called "bombykal" (Kasang et al., 1978a; Kaissling et al., 1978). "Bombykal" is practically inactive in the behavioral assay. Yet, obviously there are specific receptors for it on the antennae of male *Bombyx*. In the systematic fractionation procedure such an aldehyde would be expected to appear in the Total neutral fraction (see Chart 1) and may be separated from it in an "aldehyde and ketone" fraction by known methods. In fact such separation was exercised during early stages of development of our fractionation procedure but,

Table 5. Major topics in biology and physiology of bombykol, since 1962, with key references (mostly not quoted in "References" of this article)

Major topics	Author(s)	References
Sex pheromone gland, Structure and function	Steinbrecht RA (1964)	Z Zellforsch **64**:227–261
	Steinbrecht RA (1964)	Z Vergl Physiol **48**:341–356
	Kuwahara Y, Adachi S, Tsuchida N (1983)	Appl Ent Zool **18**:182–190
Antennal structures and ultrastructures	Steinbrecht RA (1970)	Z Morph Tiere **68**:93–126
	Steinbrecht RA, Müller B (1971)	Z Zellforsch **117**:570–575
	Steinbrecht RA (1973)	Z Zellforsch **139**:533–565
Threshold of perception and peripheral electrophysiology	Schneider D, Block BC, Boeckh J, Priesner E (1967)	Z Vergl Physiol **54**:192–209
	Schneider D, Kasang G, Kaissling KE (1968)	Naturwissenschaften **55**:395–396
	Schneider D (1970)	The Neurosciences: Second Study Program, Schmidt, FO (ed), Rockefeller Univ. Press, New York, pp 511–518
	Kaissling KE, Priesner E (1970)	Naturwissenschaften **57**:23–28
	Kaissling KE (1972)	In: Olfaction and Taste, IV. Schneider D (ed), Wiss Verlagsges Stuttgart, pp 207–213

	Kasang G, Kaissling KE (1972)	In: Olfaction and Taste, IV. Schneider D (ed), Wiss Verlagsges Stuttgart, pp 200–206
	Steinbrecht RA, Kasang G (1972)	In: Olfaction and Taste, IV. Schneider D (ed), Wiss Verlagsges Stuttgart, pp 193–199
	Kaissling KE (1974)	In: Biochemistry of Sensory Functions. Jaenicke L (ed) Springer-Verlag, Berlin/Heidelberg/New York, pp 243-273
	Kaissling KE (1977)	In: Olfaction and Taste, VI. le Magnen J, MacLeod P (eds), Information Retrieval, London, pp 9–16
	Kaissling KE, Thorson J (1980)	In: Receptors for Neurotransmitters, Hormones and Pheromones in Insects, Satelle DB et al. (eds), Elsevier/North-Holland Biomedical Press, Amsterdam, pp 261–282
	Steinbrecht RA, Schneider D (1980)	In: Insect Biology in the Future, Locke M, Smith DS (eds), Academic Press, New York, pp 685–703
Central nervous processing of peripheral message and behavior	Kramer E (1975)	In: Olfaction and Taste, V. Denton DA, Coghlan JP (eds), Academic Press, New York, pp 329–335
	Obara Y (1979)	Appl Ent Zool **14**:130–132
	Olberg RM (1983)	J Comp Physiol **152**:297–307
	Olberg RM (1983)	Physiol Entomol **8**:419–428

at that time, it was not pursued because that fraction did not show activity—which would be in line with the properties of bombykal detected recently. However, bioactivity might have been generated by reduction of the "aldehyde and ketone" fraction with an appropriate complex hydride in a manner similar to that exercised already in the case of the "cryptic" source of bombykol (see above). Moreover, if the "cryptic" source is structured as waxes are, bombykal might be a precursor of it: waxes may be formed by disproportionation of 2 moles of the corresponding aldehyde in a manner similar to the Cannizzaro reaction. Based upon capillary gas chromatographic analyses of the fatty acid composition of the pheromone gland, a biosynthetic route for bombykol was suggested recently (Yamaoka and Hayashiya, 1982). As pointed out by Roelofs and Brown (1982), comparative information on insect pheromone biosynthesis in related species would provide an important element in understanding phylogenetic isolation mechanisms of reproduction.

Companion compounds of bombykol in the sex pheromone gland are of additional interest if the gland is considered a physiological multicomponent system, as in some other insects (e.g., Silverstein and Young, 1976). For example, recently, by an approach designed specifically for the purpose, the *E,E*-stereoisomer of bombykol (see Table 4) was detected in the sex pheromone gland in relatively low amounts as compared to bombykol (Kasang et al., 1978b). In fact, already in the systematic fractionation procedure (see Chart 2), in the process of sorting out the NABA-ester of bombykol, it is obvious that fractions of NABA-esters were put aside (Figs. 8A,B) which, after saponification, exhibited reasonable activity in the bioassay. This may be an indication of the presence—besides bombykol and its *E,E*-stereoisomer—of additional, biologically more or less active alcohols. Such companion compounds might play an important role in inhibition and/or amplification of the interaction of the principal sex pheromone with receptors.

To clarify the problem of biogenesis and that of the role of companion compounds in bioactivity, it appears worthwhile to reinvestigate the contents of the sex pheromone gland of *Bombyx* in a comprehensive systematic manner. Such analysis might encounter a bioassay- and/or electroantennogram-monitored (see also Roelofs, 1984, this volume) *systematic* fractionation procedure. Particularly recombination testing of fractions primarily in the behavioral assay should be included to detect, per se, inactive, inhibiting, or amplifying components. Such reinvestigation could take full advantage of the most advanced present-day separation, detection, and determination techniques. It would not require derivatization, but could profit from the detailed experience available in the *Bombyx* model and revisited in this paper.

Investigation of *bombykol metabolism in the antenna* has received considerable attention after tritium-labeled bombykol was made available (Kasang, 1968). Such metabolism of bombykol (Kasang, 1971, 1973, 1974; Kasang and Weiss, 1974) is of prime importance with respect to the molecular mechanism of action of bombykol: As a chemical signal bombykol requires a quick means of metabolic deactivation.

In summary, then, continued investigation of bombykol biochemistry may increase our basic knowledge of insect pheromones for a second time, after the identification of bombykol. This time, it would contribute to more sophisticated problems of insect pheromones, such as its biogenesis, its metabolism, and its molecular mechanism of action.

6.2. Physiology of Perception of Bombykol and Insect Pheromones and Relation to General Olfaction

After recognition of biological problems and at the onset of their investigation, its chemical and biochemical aspects require clarification at first. When chemistry and biochemistry payed their tribute, usually the problem is referred back to biology. Indeed, an important avenue for investigation of the physiology of insect pheromone perception was opened by the electroantennogram technique, introduced originally with the intention to improve the bioassay used in the fractionation procedure of bombykol (Schneider and Hecker, 1956; Schneider, 1957; Hecker, 1959). While this electrophysiological approach was still in its nascent stage, bombykol was identified (see above). Meanwhile, the electroantennogram technique was developed to offer major advantages as an assay system compared to behavioral assays (see also Roelofs, 1984, this volume). On the other hand, in the last quarter century, the physiology of peripheral perception of bombykol and congeners in *Bombyx* was developed immensely and is concerned nowadays mainly with the central transformation of peripheral signals releasing behavioral responses. Thus considerable knowledge was gained in understanding how single molecules, when reaching their specialized chemoreceptors on the antenna, cause corresponding receptor cells to release those electric signals which cause, finally, the complicated behavior of the entire organism (for an overview of major topics and key references relating to bombykol, see Table 5). That in the behavior of the whole insect the brain plays an integral part was demonstrated in experiments with *Bombyx mori* already at the beginning of this century (Kellog, 1907). The brain may be responsible also for the circadian dependency (e.g., Jarczyk and Flaschenträger, 1957; Jarczyk and Hertle, 1959) of the behavioral response.

In contrast to high structural specificity of perception of insect pheromones (for an overview of key references, see Insect Olfaction, Table 6) in human olfaction, structurally quite different compounds such as nitrobenzene, benzaldehyde, benzonitrile, and hydrocyanic acid may all smell alike (bitter almonds). Thus, odors perceived by man and their molecular structure apparently show a much lower degree of correlation than in the case of insect sex pheromones. Therefore, it appears questionable if it is at all justified to compare these categories of olfaction. It is good to see investigations of pheromones and olfaction extending into the realm of aquatic invertebrates and terrestrial vertebrates, including man. Such investigations may contribute to clarify gradually mechanisms of olfaction (for an overview of key references, see General Olfaction, Table 6).

Table 6. Some monographs, monographic contributions, and books on physiology of insect and general olfaction, since 1962 (not quoted in "References" of this article)

Topics	Author(s)	References
	Insect Olfaction	
Electrophysiological Investigation of Insect Olfaction	Schneider D (1963)	In: Olfaction and Taste, I. Pergamon Press, Oxford/London/New York/Paris, pp. 85–103
Insect Olfactory Receptors	Boeckh J, Kaissling KE, Schneider D (1965)	Cold Spring Harbor Symposium of Quantitative Biology **XXX**:263–280
Insect Olfaction	Kaissling KE (1971)	In: Handbook of Sensory Physiology, Autrum FH et al. (eds), Vol IV, Chemical senses, Part 1, Beidler LM (ed), Springer-Verlag, Berlin/Heidelberg/New York, pp 351–431
Insect Sex Pheromones	Jacobson M (1972)	Academic Press, New York
Artspezifität und Funktion einiger Insektenpheromone	Priesner E (1973)	Fortschritte Zoologie **22**:49–135
Insect Pheromones	Kramer E (1978)	In: Taxis and Behaviour, Hazelbauer GL (ed), Chapman and Hall, London, pp 207–229
Evolution of antennae; their sensilla and the mechanism of scent detection in Arthropoda	Callahan PS (1979)	In: Gupta AP (ed), Arthropod Phylogeny. Van Nostrand Reinhold, New York, pp 259–298
Pheromone von Insekten, Produktion-Reception-Inaktivierung	Schneider D (1980)	Nova Acta Leopoldina N.F. **51**:249–278
Insect Pheromones	Birch MC, Hayes KF (1983)	Edward Arnold (Publishers) Ltd., London
Insect Olfaction–Our research endeavour	Schneider D (1984)	In: Foundations of the Sensory Sciences. Dawson WW, Enoch JM (eds), Springer-Verlag, Berlin/Heidelberg/New York/Tokyo, pp 381–418

Pheromone biology in Lepidoptera: Overview, some recent findings, and some generalizations	Schneider D (1984)	In: The Comparative Physiology of Sensory Systems. Bolis L, Keynes RD, Madrell SHP (eds), Cambridge University Press, Cambridge, pp 301–313
	General Olfaction	
The Science of Smell	Wright RH (1964)	Basic Books, New York
The chemical senses	Moncrieff RW (1967)	Chemical Rubber Co. Press, Cleveland, Ohio
Molecular Basis of Odor	Amoore JE (1970)	Thomas, Springfield, Illinois
The sense of smell	Engen T (1973)	Ann Rev Psychol **29**:187–206
Pheromones	Birch MC (ed) (1974)	Elsevier, North-Holland, New York
Biochemistry of sensory functions	Jaenicke L (ed) (1974)	Springer-Verlag, Berlin/Heidelberg/New York
Methods of Olfactory Research	Moulton DG et al. (eds) (1975)	Academic Press, London, G.B.
Mammalian Odors and Pheromones	Stoddart DM (1976)	Edward Arnold, London
Chemical Signals in Vertebrates	Müller-Schwarze D, Mozell JJ Jr. (eds) (1977)	Plenum, New York
Biologie des Riechens	Schneider D (1977)	Ärztliche Kosmetologie **7**:191–198
Chemical Ecology; Odor Communication in Animals	Ritter FJ (ed) (1978)	Elsevier, Amsterdam
Chemical Signals: Vertebrates and Aquatic Invertebrates	Müller-Schwarze D, Silverstein RM (eds) (1980)	Plenum, New York

Table 7. Some monographs, monographic contributions, and books relating insect pheromone research and insect pest control (not quoted in "References" of this article)

Topics	Author(s)	References
Pest Management with Insect Sex Attractants	Beroza M (ed) (1976)	American Chemical Society, Washington, D.C.
Chemical Control of Insect Behavior: Theory and Application	Shorey HH, McKelvey JJ, Jr (eds) (1977)	Wiley, New York
Die praktische Verwendung von Insektenpheromonen	Boness M (1981)	In: Chemie der Pflanzenschutz- und Schädlingsbekämpfungsmittel, Wegler R (ed), Springer-Verlag, Berlin/Heidelberg/New York, Vol 6, pp 165–184
Management of Insect Pests with Semiochemicals: Concept and Practice	Mitchell ER (ed) (1981)	Plenum, New York
Semiochemicals: Their Role in Pest Control	Nordlund DA, Jones RL, Lewis WJ (1981)	Wiley, New York
Mechanisms of Communication Disruption by Pheromone in the Control of Lepidoptera–A Review	Bartell RJ (1982)	Physiol Entomol 7:353–364
Insect Pheromone Technology: Chemistry and Application	Leonhardt BA, Beroza M (1982)	American Chemical Society, Symposium Series 190, Washington, D.C.
Introduction to Insect Pest Management	Metcalf RL, Luckmann WH (eds) (1982)	2nd ed. Wiley, New York

6.3 Perfume Instead of Poison To Be Used Against Pest Insects

One of the major driving forces of investigations in insect pheromones–those "perfumes" of insects (Hecker, 1959)–was the prospect of their utilization as a selective and inexpensive, nontoxic means of insect pest control (for an overview of key references, see Table 7). The variability in the individual complex behavior of males (and of females) triggered via the brain of individuals as well as the dynamics of entire populations may add much to some rather unexpected complications experienced in field applications of insect pheromones to control insect pests (see also Baker and Cardé, 1984, this volume). However, a quarter of a century after identification of the first sex pheromone, viable alternatives to conventional insecticides are in the process of becoming available (see Campion, this volume). Gradually the ultimate goal of protecting crops against pest insects in a highly selective manner and of simultaneously saving our environment from being spoiled with persistent poisonous chemicals will be reached. How much input of sophisticated basic research had to be raised to make possible such most-desirable present-day output!

VII. Conclusion

Nowadays, as a special class of biological messengers, the pheromones occupy their defined place in biology. The authors of this introduction to the present volume of the series observe with satisfaction and delight that the field of pheromone research, which they inaugurated and pioneered a quarter of a century ago, using *Bombyx mori* as a model organism, grew immensely both in scope and in refinement of techniques. Thus, as a consequence, the subdiscipline of yesterday, situated at the interfaces between bioorganic chemistry, biochemistry, and biology, has matured to become a genuine field of original basic research of today. Moreover it appears within the realm of present-day developments that the environmental problems generated by indiscriminate use of conventional chemical insecticides may be overcome by judicious and selective application of semiochemicals, of which sex pheromones are one subgroup. After all, Fabre's prediction became a true reality: Quantitative biology, instructed by the insect, opened up a new world of marvels!

Acknowledgment. In the beginning of our investigations, we enjoyed helpful advice and assistance by the former Institut für Kleintierzucht, Celle, in raising and collecting *Bombyx* cocoons. Without the enthusiastic assistance by numerous students from all faculties of the Universities of Tübingen and München during many annual campaigns, the masses of cocoons and the harvest of sex pheromone glands required for the isolation of bombykol could not have been handled. Moreover, engaged laboratory work of many technicians and research scholars of the Institute contributed much to develop and to maintain high standards in the behavioral bioassay as well as to work out every single stage of the bioassay-monitored systematic fractionation procedure (1939–1959) and, finally, the routes of syntheses of bombykol (1959–1962). Last, but not least, repeated substantial grants by the Deutsche Forschungsgemeinschaft, Bonn-Bad Godesberg, and generous donations by Schering AG, Berlin, by Deutsche Hoffmann-La Roche AG, Grenzach/Baden, and by Farbenfabriken Bayer AG, Leverkusen, allowed us to pursue our investigations until our goal was reached. During the preparation of this manuscript, we appreciated stimulating discussions with Dietrich Schneider, Seewiesen, and with Hans E. Hummel, Urbana. Helpful assistance in collecting some of the more recent literature references by Gerhard Kasang, München-Martinsried, is gratefully acknowledged.

VIII. References

Amin El S, Hecker E (1956a) Trennung der 4′-Nitro-azobenzolcarbonsäure-4-ester von Alkoholen-durch Verteilung und Chromatographie. Chem Ber **89**: 695–701.

Amin El S, Hecker E (1956b) Charakterisierung von Fettsäuren durch ihre p-[Nitrophenyl-azo]-phenacylester und deren chromatographische Trennung. Chem Ber **89**:1496–1502.

Ambros W (1940) Einige Beobachtungen und Untersuchungen an der Nonne im Jahre 1938. Cbl ges Forstwesen **66**:131–165, 166–176.

Anders F, Bayer E (1959) Versuche mit dem Sexuallockstoff aus den Sacculi laterales vom Seidenspinner *Bombyx mori* L. Biol Zentralblatt **78**:584–589.

Barth R (1937, 1938) Herkunft, Wirkung und Eigenschaften des weiblichen Sexualduftstoffes einiger Pyraliden. Zool Jhbuch Abtlg Allg Zool Physiol **58**:297–329.

Bestmann HJ, Vostrowsky O (1979) Synthesis of pheromones by stereoselective carbonyl olefination: An unitized construction principle. Chemistry and Physics of Lipids **24**:335–389.

Bestmann HJ, Vostrowsky O (1982) Insekten pheromone. Naturwissenschaften **69**:457–471.

Butenandt A (1939) Zur Kenntnis der Sexuallockstoffe bei Insekten. Jhbuch Preuss Akad der Wissenschaften: **97**.

Butenandt A (1941) Untersuchungen über Wirkstoffe aus dem Insektenreich. Angew Chem **54**:89–91.

Butenandt A (1955a) Über die Wirkstoffe des Insektenreiches, II. Zur Kenntnis der Sexuallockstoffe. Nat Wiss Rdschau **8**:457–464.

Butenandt A (1955b) Wirkstoffe des Insektenreiches. Nova Acta Leopodina (N.F.) **17**:445–471.

Butenandt A (1959) Geschlechtsspezifische Lockstoffe der Schmetterlinge. Max-Planck-Gesellschaft Jhbuch 23–32.

Butenandt A (1963) Bombykol, the sex-attractive substance of the silk worm, *Bombyx mori.* Annual Lecture 1962 before the Society of Endocrinology, Proceedings of the Society of Endocrinology. **27**:IX–XVI.

Butenandt A (1979) 50 years ago—The discovery of oestrone. Trends Biochem Sci **4**:215–216.

Butenandt A, Beckmann R, Hecker E (1961a) Über den Sexuallockstoff des Seidenspinners, I. Der biologische Test und die Isolierung des reinen Sexuallockstoffs Bombykol. Hoppe-Seyler's Z Physiol Chem **324**:71–83.

Butenandt A, Beckmann R, Stamm D (1961b) Über den Sexuallockstoff des Seidenspinners, II. Konstitution und Konfiguration des Bombykols. Hoppe-Seyler's Z Physiol Chem **324**:84–87.

Butenandt A, Beckmann R, Stamm D, Hecker E (1959) Über den Sexuallockstoff des Seidenspinners Bombyx mori. Reindarstellung und Konstitutionsermittlung. Z Naturforsch **14b**:283–884.

Butenandt A, Hecker E (1961) Synthese des Bombykols, des Sexual-Lockstoffes des Seidenspinners, und seiner geometrischen Isomeren. Angew Chem **73**: 349–353.

Butenandt A, Hecker E, Hopp M, Koch W (1962) Über den Sexuallockstoff des Seidenspinners, IV. Die Synthese des Bombykols und der cis-trans-Isomeren Hexadecadien-(10,12)-ole(1). Liebigs Ann Chem **658**:39–64.

Butenandt A, Hecker E, Zachau HG (1955) Über die geometrischen Isomeren des 2.4-Hexadienols-(1). Chem Ber **88**:1185–1196.

Butenandt A, Hecker E, Zayed MAD (1963) Über den Sexuallockstoff des Seidenspinners, III. Ungesättigte Fettsäuren aus den Hinterleibsdürsen (Sacculi laterales) des Seidenspinnerweibchens. Hoppe-Seyler's Z Physiol Chem **333**: 114–126.

Butenandt A, Karlson P (1954) Über die Isolierung eines Metamorphosehormons der Insekten in kristallisierter Form. Z Naturforsch **9b**:389–391.

Butenandt A, Stamm D, Hecker E (1961c) Mikromethode zur Konstitutionsermittlung ungesättigter Alkohole und Säuren. Chem Ber **94**:1931–1942.

Collins CW, Potts SF (1932) Attractants for the flying gypsy moth as an aid in locating new infestations. US Dep Agric Techn Bull **336**:1–43.

Cretschmar M (ed) (1947) Leitfaden für den deutschen Seidenbauer. M and H Schaper Verlag, Hannover.

Duane JP, Tyler JE (1950) Operation Saturnid. Interchem Rev **9**:25–28.

Dufay C (1957) Sur l'attraction sexuelle chez *Lasiocampa queraus* L. Bull Soc Entomol (France) **62**:61–64.

Fabre JH (1914) "Hochzeitsflüge der Nachtpfauenaugen", S 80-86; "Aus dem Liebesleben des Eichenspinners", S 86-92; Duft- und Geruchsinn der Insekten, S 92-98; in Bilder aus der Insektenwelt, Verlag Kosmos, Stuttgart (authorized translations in German by Fabre's "Souvenirs entomologiques" Paris: Delagrave 1879, "Moers des Insectes" and "La vie des Insectes").

Farsky O (1938) Nonnenkontroll- und Vorbeugungsmethode nach Prof. Forst. Ing. Ant. Dyk. Anz Schädlingskunde **14**:52–56, 65–67.

Flaschenträger B, Amin El S (1949) Über Anlockungsstoffe von Baumwollschädlingen. Angew Chem **61**:252.

Flaschenträger B, Amin El S (1950) Chemical attractants for insects: Sex- and food-odors of the *cotton leaf worm* and the *cut worm*. Nature (London) **165**:394.

Flaschenträger B, Amin El S, Jarczyk HJ (1957) Ein Lockstoffanalysator (odor analyser) für Insekten. Mikrochim Acta 385–389.

Freiling HH (1909) Duftorgane der weiblichen Schmetterlinge nebst Beiträgen zur Kenntnis der Sinnesorgane auf dem Schmetterlingsflügel und der

Duftinsel der Männchen von *Danais* and *Euploea*. Z Wiss Zool **92**:210–290.

Görnitz K (1949) Anlockversuche mit dem weiblichen Sexuallockstoff des Schwammspinners (*L. dispar*) und der Nonne (*L. monacha*). Anz Schädlingskunde **22**:145–148.

Götz B (1939a) Untersuchungen über die Wirkung des Sexualduftstoffes bei den Traubenwicklern *Clysia ambiguella* und *Polychrosis botrana*. Z Angew Entomol **26**:143–164.

Götz B (1939b) Über weitere Versuche zur Bekämpfung des Traubenwicklers mit Hilfe des Sexuallockstoffes. Anz Schädlingskunde **15**:109–114.

Götz B (1951) Die Sexualduftstoffe an Lepidopteren. Experienta (Basel) **7**: 406–418.

Hammad SM, Jarczyk HJ (1958) Contributions to the biology and biochemistry of the cotton leaf worm, *Prodenia litura* F. III. The morphology and histology of the sexual scent glands in the female moth of Prodenia litura F. Bull Soc Entomol Egypte **XLII**:253–261.

Hanno K (1939) Anlockversuche bei *Lymantria monacha* L. Z Agnew Entomol **25**:628–641.

Hecker E (1955a) 4′-Nitro-azobenzol-carbonsäure-4-chlorid als Reagenz auf Alkohole. Chem Ber **88**:1666–1675.

Hecker E (1955b) Verteilungsverfahren im Laboratorium. Monographie Nr. 67 zu Angewandte Chemie und Chemie-Ingenieur-Technik, Verlag Chemie, Weinheim/Bergstrasse.

Hecker E (1956, 1958) Isolation and characterization of the sex attractant of the silk worm moth (*Bombyx mori* L.). Proc X. Int Congr Entomol **2**:293–294.

Hecker E (1959) Sexuallockstoffe–hochwirksame Parfüms der Schmetterlinge. Umschau 465–467, 499–502.

Hecker E (1960) Chemie und Biochemie des Sexuallockstoffes des Seidenspinners (*Bombyx mori* L.). XI. Int Kongr f Entom Wien, Symp 3, Verh B III Wien 69–72.

Hecker E (1963) Der Mechanismus multiplikativer Verteilungsverfahren. Naturwissenschaften **50**:165–171, 290–299.

Inhoffen HH (1951) Versuch zur Isolierung eines Sexuallockstoffes. Arch Pharm **284/56**:337–341.

Jarczyk HJ, Flaschenträger B (1957) Contribution to the biology and biochemistry of the cotton leafworm, *Prodenia litura* F. (Noctuidae). Z ang Entomol **45**:84–102.

Jarczyk HJ, Hertle P (1959) Contributions to the biology and biochemistry of the cotton leafworm, *Prodenia litura* F. (Noctuidae). Z. Angew Entomol **45**:94–102.

Kaissling KE, Kasang G, Bestmann HJ, Stransky W, Vostrowsky O (1978) A new pheromone of the silk worm moth *Bombyx mori*, sensory pathway and behavioural effect. Naturwissenschaften **65**:382–384.

Karlson P, Lüscher M (1959) "Pheromones," a new term for a class of biologically active substances. Nature (London) **183**:55–56.

Kasang G (1968) Tritium-Markierung des Sexuallockstoffes Bombykol. Z Natforsch **23b**:1331–1335.

Kasang G (1971) Bombykol reception and metabolism on the antennae of the silk worm moth *Bombyx mori* L. In: Gustation and Olfaction. Ohloff G, Thomas AF (eds) Academic Press, New York/London, pp 245-250.

Kasang G (1973) Physikochemische Vorgänge beim Riechen des Seidenspinners. Naturwissenschaften **60**:95-101.

Kasang G (1974) Uptake of the sex pheromone ^{3}H-bombykol and related compounds by male and female *Bombyx* antennae. J Insect Physiol **20**:2407-2422.

Kasang G, Kaissling KE, Vostrowsky O, Bestmann HJ (1978a) Bombykal, eine zweite Pheromonkomponente des Seidenspinners *Bombyx mori* L. Angew Chem **90**:74-75; Int ed. **17**:60.

Kasang G, Schneider D, Schäfer W (1978b) The silk worm moth *Bombyx mori.* Presence of the (E,E)stereoisomer of bombykol in the female pheromone gland. Naturwissenschaften **65**:337-338.

Kasang G, Weiss N (1974) Dünnschichtchromatographische Analyse radioaktiv markierter Insektenpheromone. Metaboliten des [^{3}H]-Bombykols. J. Chromatrog **92**:401-417.

Kellogg VL (1907) Some silkworm moth reflexes. Biol Bull (of the Marine Biol Laboratory Stanford Univ, Calif) **XII**:152-154.

Makino K, Satoh K, Inagami K (1956) Bombixin, a sex attractant discharged by female moth, *Bombyx mori.* Biochim Biophys Acta **19**:394-395.

Meisenheimer J (1921) Geschlecht und Geschlechter im Tierreich. G. Fischer, Jena.

Mell R (1922) Biologie und Systematik der chinesichen Sphingiden. Friedländer, Berlin.

Miyaura N, Suginome H, Suzuki A (1983) New stereospecific syntheses of pheromone bombykol and its three geometrical isomers. Tetrahedron **39**: 3271-3277

Roelofs WL, Brown RL (1982) Pheromones and evolutionary relationships of Tortricidae. Ann Rev Ecol Syst **13**:395-422.

Schenk O (1903) Die antennalen Hautsinnesorgane einiger Lipidopteren und Hymenopteren mit besonderer Berücksichtigung der sexuellen Unterschiede. Zool Jb (Anat) B 17, 573-616.

Schneider D (1957) Elektrophysiologische Untersuchungen von Chemo- und Mechanorezeptoren der Antenne des Seidenspinners *Bombyx mori* L. Z Vergl Physiol **40**:8-41.

Schneider D, Hecker E (1956) Zur Elektrophysiologie der Antenne des Seidenspinners *Bombyx mori* bei Reizung mit angereicherten Extrakten des Sexuallockstoffes. Z Naturforsch **11b**:121-124.

Schneider D, Kaissling KE (1957) Der Bau der Antenne des Seidenspinners *Bombyx mori* L. II. Sensillen, cuticulare Bildungen und innerer Bau. Zool Jhb Abt. Anat Ontog Tiere **76**:223-250.

Schneider D, Steinbrecht RA (1968) Checklist of insect olfactory sensilla. Symp Zool Soc (London) **23**:279-297.

Schwinck I (1953) Über den Sexualduftstoff der Pyraliden. Z Vergl Physiol **35**: 167-174.

Schwinck I (1954) Experimentelle Untersuchungen über Geruchssinn und Strömungswahrnehmung in der Orientierung bei Nachtschmetterlingen. Z Vergl Physiol **37**:19-56.

Schwinck I (1955a) Freilandversuche zur Frage der Artspezifität des weiblichen

Sexualduftstoffes der Nonne (*Lymantria monacha* L.) und des Schwammspinners (*Lymantria dispar* L.). Z Angew Entomol **37**:349–357.

Schwinck I (1955b) Weitere Untersuchungen zur Frage der Geruchsorientierung der Nachtschmetterlinge: Partielle Fühleramputation bei Spinnermännchen, insbesondere am Seidenspinner, *Bombyx mori* L. Z Vergl Physiol **37**:439–458.

Schwinck I (1956, 1958) A study on olfactory stimuli in the orientation of moths. Proc X Int Congr Entomol **2**:577–582.

Silverstein RM, Young JC (1976) Insects generally use multicomponent pheromones. In: Pest Management with Insect Sex Attractants and Other Behaviour-Controlling Chemicals. Beroza M (ed) Am Chem Soc Symposium, Series 23:1–29, Washington D.C.

Truscheit E, Eiter K (1962) Synthese der vier Isomeren Hexadeca-dien-(10,12)-ole(1). Liebigs Ann Chem **658**:65–90.

Urbahn E (1913) Abdominale Duftorgane bei weiblichen Schmetterlingen. Jenaische Z Naturw. 50(NF 43):277–358.

Yamaoka R, Hayashiya K (1982) Daily changes in the characteristic fatty acid (Z)-11-hexadecenoic acid of the pheromone gland of the silk worm pupa and moth, *Bombyx mori* L. (Lepidoptera: Bombycidae). Jap J Appl Ent Zool **26**:125–130.

Watanabe T (1958) Substances in mulberry leaves which attract silk worm larvae (*Bombyx mori*). Nature (London) **182**:325–326.

Zehmen H von (1942) Ein Beitrag zur Frage der Anlockstoffe weiblicher Falterschädlinge. Cbl Ges Forstwesen **68**:57–64.

Chapter 2

Techniques for Behavioral Bioassays

T. C. Baker[1] *and R. T. Cardé*[2]

I. Introduction

Insect pheromone research involves tremendous effort and exacting techniques, not only for isolating and identifying chemicals, covered in earlier and later chapters in this volume, but also in recognizing which chemicals are behaviorally active. Tests to demonstrate the behavioral activity of a compound are essential to proving that a compound is a pheromone component, i.e., that it is used in intraspecific communication. In addition, behavioral tests, or bioassays, of chemicals' effects on receiving individuals can identify the type of response they elicit. Therefore, well-designed bioassays can be invaluable for deducing the communicative function (alarm, aggregation, sexual communication, etc.) of a chemical identified from an insect. They can also give information as to the mechanisms that are used by responding insects to move toward or away from the chemical source.

The researcher must determine the objectives of the behavioral tests before choosing among the possible bioassay setups. Perhaps none of the types of assays described in this chapter will be useful in meeting the objectives, in which case a new type will be devised. There is no single "correct" assay, and elaborate ones may be just as "incorrect" for an objective as types that are too simple. The assay must answer the key questions as quickly and efficiently as possible. Hence

[1]Division of Toxicology and Physiology, Department of Entomology, University of California, Riverside, California.

[2]Department of Entomology, University of Massachusetts, Amherst, Massachusetts.

it is important to design assays so that they discriminate unequivocally among a variety of chemical fractions, synthetic analogs, or types of possible response.

Behaviorally discriminating assays (Kennedy, 1977) need not be expensive or technologically complex. However, their design should make use of the mechanisms of response used by the individuals receiving the chemical message. These responses may involve direct or indirect reactions to concentration gradients (Bell and Tobin, 1982), taxes as opposed to kineses (Fraenkel and Gunn, 1940), or the interplay of response with a cue from another modality (Kennedy, 1978).

Direct reactions are caused by the chemical concentration gradient itself (Bell and Tobin, 1982), eliciting taxes (responses steered with respect to the gradient) (Fraenkel and Gunn, 1940). Indirect reactions (Bell and Tobin, 1982) may involve a heightened reactivity to another stimulus, for instance, when sex pheromone elicits steering with respect to wind direction (Kennedy, 1978). Or, they may simply involve an internal program of movements that are self-steered, either using proprioceptive feedback (idiothetically controlled), or feedback from the environment (allothetically controlled, Kennedy, 1978) or both.

Reactions previously labeled as kineses (ortho- and klino-) appear to fall into the category of indirect responses but either together or integrated with other indirect responses the movements can result in displacement in the "correct" direction regardless of the lack of a "direct" response. In general, bioassays that make use of the entire pattern of natural responses and do not restrict the behavior to one or two simple activities will be most discriminating. The principles and strategies involved in designing discriminating assays involving orientation reactions have been reviewed by Kennedy (1977).

II. General Considerations for Conducting Bioassays

In the course of designing a bioassay, the experimenter must make several major decisions that influence the type of information generated. These include the type of apparatus in which to house and observe the insects, the way to obtain reproducible responses, whether to observe groups of insects or individuals, the type of substrate used to release the chemicals, the strength of the stimulus, the way in which two or more chemicals should be mixed, the types of behavioral responses to be scored, and how best to record these responses.

2.1. Type of Dispenser for Chemicals

Test chemicals or fractions can be dispensed from a variety of materials such as glass rods, metal discs, filter paper discs, rubber septa, etc. A major consideration is to ensure that the surface emits the compound at a fairly constant rate over the course of the assay. This may be more difficult for very volatile chemicals of low molecular weight, and so the dispenser as well as the length of time it is used before reloading and using a fresh dispenser must be chosen properly.

Unfortunately, without actually measuring the emission rate of chemical from the surface, one suspects that either of these factors should be altered only when there is a change in the level of response over the course of using the same treatment. Such variation in response due to the release of chemical should be avoided.

2.2. Quantity of Stimulus Used

Another decision is the quantity of material to use. For a natural extract, this is usually expressed in terms of insect-equivalents, but for synthetics there may be a wide range of possibilities if the "natural" dosage of the synthetics is unknown. Ideally a dosage first should be found that is the minimal amount needed to produce a level of response useful for the purpose at hand. Usually, lower dosages are best for discriminating among the responses to various treatments.

For instance, in assaying natural fractions of moth sex pheromone, it is best first to find the minimum dosage from the complete set of fractions or crude extract that will elicit the maximum level of response from males. Then when separate fractions are tested alone or in various combinations at this concentration (in terms of female equivalents), the observer can be confident that a response as great as to the crude extract means that the full complement of necessary compounds is present. Likewise, if the optimal level of response is not generated by recombining the fractions, it can be assumed either that the fractionation process has left out a necessary compound, or that one or more compounds was altered by fractionation and rendered less active.

Another decision concerning the quantity of stimulus is whether to keep the total quantity of compounds, or the quantity of a single compound, constant across treatments. For instance, for a two-component sex pheromone blend, an increased response to the two components together might be explained as purely a quantitative effect—more total molecules are present—but if the total dosage had been held constant, the quality of the new blend could be responsible for the increased response. Of course there are other ways to clarify the interpretation. If adding increments of a second compound to a constant amount of a first compound causes first an increase of response, then a decrease as the increments increase, then it is the quality of the specific blend, not the total quantity of chemical, which is the cause of higher levels of response. Each researcher is faced with these interpretation problems, and must decide what is happening on the basis of the particular bioassay and method of blending the chemicals, plus the blends chosen as controls.

2.3. Standardization of "Responsiveness" (Internal State)

In trying to obtain reproducible responses from insects, it should be realized that behavior is a result of the interaction of external (stimulus) and internal (physiological state) factors. The insect's internal state may fluctuate throughout a 24-

hr period due to an underlying circadian rhythm or a periodicity triggered by regularly occurring cues in the environment, such as dawn and dusk (Corbet, 1966). Other environmental factors such as temperature may also have immediate modulating effects on the internal state, by modifying either periodicity (e.g., Cardé et al., 1975) or the threshold of responsiveness (Cardé and Hagaman, 1983). Fluctuations in responsiveness are the rule and thus the time of assay should be first optimized to the period when the insect demonstrate the most intense response to the pheromone.

Successive presentations of a pheromone stimulus have been shown in many species to raise the threshold for subsequent responses or even entirely eliminate responsiveness. Both effects are presumably the result of CNS habituation, rather than peripheral sensory adaptation, a rather transient phenomenon (Bartell and Lawrence, 1977; Kuenen and Baker, 1981). Additionally, responsiveness may vary with age, particularly in species that are not reproductively mature upon eclosion. Standardization of the level of responsiveness for behavioral assays thus necessitates a single stimulus presentation per 24-hr interval, unless it is first demonstrated that more than one presentation does not alter the threshold. Ideally insects could be used for an assay only once. Similarly, the age of the tested insects should be held constant, unless it is shown that the ages to be used are comparable.

As noted earlier, the time of the assay should be optimized to the period when the most intense response (or lowest threshold) is exhibited. Among nocturnal species, particular attention should be given to the ambient light levels. When the light levels exceed those of full moonlight (ca. 0.3 lux), then the threshold for responses may be raised (e.g., Shorey and Gaston, 1964). "Red" light has been used with success in many bioassays, but its use should be based upon experiments showing that the behavioral reactions parallel those under noctural conditions. The search for the quintessential environment for assay of a particular species may seem entirely superfluous, but development of a useful assay usually requires some description of the optimal set of environmental conditions.

A series of assay treatments should be standardized so that each will have an equal opportunity to be presented to the insects during the interval of peak responsiveness. This is especially important when a large number of treatments is to be tested; there is a danger that the internal state of the insects will have changed drastically by the time the last treatments are tested. This type of variation can be accounted for and factored out of the bioassay by using a randomized, complete-block design, but in addition, it can be avoided by designing the assay so that a complete series of treatments is tested during a period of relatively shallow fluctuations of the internal state. These procedures will enhance the observer's ability to discriminate among responses to treatments.

2.4. Recording the Responses

In recording bioassay responses, it is important to optimize the time spent recording data. Recording equipment can range from a pencil and paper to high-resolution video cameras and recorders. The recording method must be matched with the objectives of the assay. Do not record more data than you need to answer the question for your assay. In general, the more complex the recording device, the more work it takes to analyze the results.

Audio- and video-taped observations take at least twice as long to analyze as the original assay takes to conduct. If a simple presence or absence of a behavior needs to be scored, it would be much quicker to record the results immediately with a pencil and paper. Video recording is not always necessary; for instance, the positions of a group of moths in a sex pheromone olfactometer tube can be incremented by marking the tube in five sections or so, and recording at set intervals the numbers of moths in each section. If these position numbers are tape recorded, transcribing the tape would more than double the time involved.

In many assays, particularly those (cf. Section III below) in which displacement is not monitored, only a single behavior, often termed the "key" response, is scored. In the past this was judged to be suitable in the Lepidoptera because the pheromone was thought to consist of a single chemical and the *sequence* of behavioral response to pheromone was thought to be mediated in part by increases in pheromone concentration (Shorey, 1970). Now, we know that pheromones more typically are blends of components and that in some species (see Chapter 3 by Baker and Linn, this volume) early and late behaviors in the normal sequence of response can be elicited by different combinations of pheromone components. The possibility that a critical behavior in the normal sequence of response to pheromone will be missed by monitoring only a single reaction, such as preflight wingfanning, is thus a legitimate concern.

2.5. Number of Insects in the Assay

Another consideration in recording data is whether to use groups of insects or individuals. This decision must be made case by case, and again rapidity and discrimination among treatments can be optimized by the correct choice. Group effects may alter the results and so the experimenter must decide whether such a danger exists, and if so, whether it will increase, decrease, or not affect discrimination. Again, to optimize the speed with which bioassay information is gathered, groups of insects may be best, but this decision always must be balanced with the type of recording method to be used and how much detailed information the experimenter is willing to loose.

III. Examples of Bioassays

Bioassays may be divided into two categories, those with and those without moving air. In addition, in both of these classes displacement of the insects in space may or may not be monitored, or even allowed. In still air, there is usually an insufficiently steep gradient to allow preferential movement toward or away from the stimulus unless the insect is very close to the source. In moving air, movement toward or away from the source along the windline may be restricted, as in some apparati with vertical airflow. In the remainder of the chapter, we will examine examples of bioassay devices that have been used in pheromone research.

3.1. Bioassays without Airflow, Displacement Not Monitored

Bioassays without airflow are very useful and simple, but they generally rely on diffusion to transport the test chemical toward the receiving insects. In one category of windless bioassay, the insects are stationed such that they cannot make use of a spatial concentration gradient to move toward or away from the source, and of course they cannot steer with respect to wind because there is none. Hence, the insect cannot use a direct response to chemical in such cases and often these bioassays do not allow much movement of the insect in space anyway. A second class of bioassays without wind is the type in which a sufficient concentration gradient is present, and direct responses to the test chemicals' gradient can occur. These assays are often larger and more elaborate because the researcher is interested in the insect's displacement in space.

An example of the first type of windless bioassay system is the one used by Baker et al. (1976) in studies of the sex pheromone of the red-banded leafroller moth, *Argyrotaenia velutinana.* One of the objectives of the study was to see whether (*E*)-11 tetradecenyl acetate showed activity as a sex pheromone component along with the (*Z*) isomer, and if so, whether there was an optimal *Z:E* ratio. Small plastic boxes (12.4 × 9.0 × 7.0 cm) were used and each had a small hole through which 10 males were introduced about 1 hr before assay. The hole was then plugged with a cork, and a filter paper tab containing the chemical blend was introduced into the box through a narrow slit at the time of testing.

The optimal time of bioassay had been determined by placing a standard amount (10 female equivalents) of extract on the filter paper and presenting it to a different box of 10 males every 2 hr and observing them for 1 min to count the maximum number of males simultaneously exhibiting wing fanning while walking. This behavior is performed by *A. velutinana* males just prior to copulation. Only this behavior was monitored because (a) it was easy to observe and therefore the percentage of males performing it could be deduced quickly; (b) the number of males assayed simultaneously did not allow the observer to monitor the percentages of males performing other behaviors; (c) the chambers themselves were so small that they prohibited prolonged flight and other reponses that usually occur during response to sex pheromone; (d) it became clear that moths did not accumulate regularly on the filter paper tab at the end of the 1-min period as anticipated, and so this extra measurement was unnecessary.

Baker and Cardé (1979) later showed for another moth species, *Grapholita molesta,* that preflight wing fanning while walking was the behavior most highly correlated with ability to locate the pheromone source in a laboratory wind tunnel (Fig. 1). Although these behaviors may not be correlated in other species (Cardé and Hagaman, 1979), wing fanning has been a useful response for bioassays of tortricid moths.

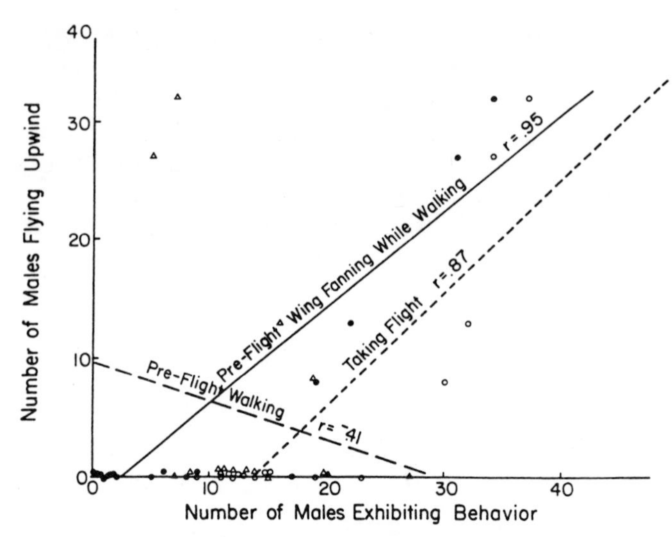

Figure 1. Correlations between pre-upwind flight behaviors and the number of *G. molesta* males flying upwind in the pheromone plume in a wind tunnel. Note the high correlation coefficient between wingfanning while walking and upwind flight (from Baker and Cardé, 1979).

Through other experiments it became apparent that the optimal response period for *A. velutinana,* normally occurring during scotophase (darkness) of the photoperiod at 24°C, could be advanced into the photophase several hours before lights-off by lowering the temperature to 16°C. Now full room illumination could be used, simplifying observation of behavior. A series of ratios of (*Z*) and (*E*)-11-tetradecenyl acetate were prepared in solution so that the same amount (2 ng) of (*Z*) was present in each. The filter paper tabs were each stored in a separate sealed vial and impregnated with the test blend only minutes before the bioassay was to begin. The vials were chosen randomly for assay, and the code number on each vial corresponding to the ratio of isomers was hidden until after all ratios had been tested. Pheromone blends were tested in a randomized, complete-block design to try to factor out any unexpected time-dependent variation in responsiveness. Before each box of males was tested, it was observed for 60 sec, and the maximum number of males simultaneously fanning their wings was recorded as the "spontaneous response." Then males were observed for 60 sec after introducing the paper tabs. The maximum number of moths simultaneously fanning their wings at any time during the 60 sec was recorded. This number was corrected for spontaneous levels of wing fanning according to the formula:

$$\frac{\text{response to stimulus} - \text{spontaneous response}}{10 - \text{spontaneous response}}$$

For optimum discrimination, the importance of using a minimal dosage that elicits a maximum response became apparent here. High quantities (1–100 μg) of (*E*) or (*Z*) alone could elicit high levels of wing fanning; however, at a dosage of 2 ng of (*Z*), only the optimal (8% *E*) ratio could cause more than 80% of the males to fan their wings simultaneously (Baker et al., 1976). This assay, despite its simplicity, could discriminate among some blends that differed by only a few percent (*E*), and the results from the laboratory agreed well with field trapping experiments showing 8% (*E*) to be the optimal blend.

Another simple bioassay without wind in which displacement was not monitored was that used by Vick et al. (1970) for sex pheromones of the dermestid beetle species of the genus *Trogoderma*. Each beetle was housed alone for 1 hr in a 3.7-ml glass vial. Then the vial's top was removed and 0.01 female-equivalents of pheromone on a 12.7-mm-diameter antibacterial assay disc was placed in the vial. The assay disc was held at the end of a glass rod, which itself was affixed to a rubber stopper. When the stopper was seated in the mouth of the vial, the assay disk was suspended 1 cm above the male. Each beetle was observed for 60 sec for evidence of running in circles beneath the disc and stretching toward it. This assay again, despite its simplicity and lack of displacement, was able to discriminate among several *Trogoderma* pheromones and demonstrate that some were species specific.

3.2. Bioassays Without Airflow, Displacement Is Monitored

Some bioassays without wind have been used successfully to test the displacemet of insects in response to various chemicals. In such cases, the apparatus itself has been designed to permit movement along a chemical gradient.

A good example is the method of assaying for alarm pheromone activity in aphids. Montgomery and Nault (1977a,b) allowed groups of 20-30 aphids, all 7-9 days old, to develop on plants by removing adult females immediately after they had deposited about 30 young. Standard conditions of 21°C, 55-75% relative humidity, and light intensity of 2900 lux were used for all assays. A dilution series of the synthetic, purported alarm pheromone, (*E*)-β-farnesene was made in methanol. Then a filter paper triangle, 8 × 8 × 2 mm, was touched to a solution and allowed to saturate by capillarity. The paper was held 0.5 cm from the center of a cluster and the proportion responding by falling from the plant or walking away from the paper was recorded. The gradient was steep enough for the aphids to move away from the source, and also the intensity of the response (those falling rather than walking) decreased according to the distance from the source. This assay revealed sharp differences in sensitivity to this alarm pheromone among species and tribes of aphids (Fig. 2), and also pointed to innate differences in type of response according to whether species are usually tended by ants or not.

An example of another assay in two dimensions without wind is the type used by Hawkins (1978) and Bell and Tobin (1981) to test the activity of the natural sex pheromone extract of the American cockroach, *Periplaneta americana*. The males' movements were monitored in a 2.5-m-diameter circular arena

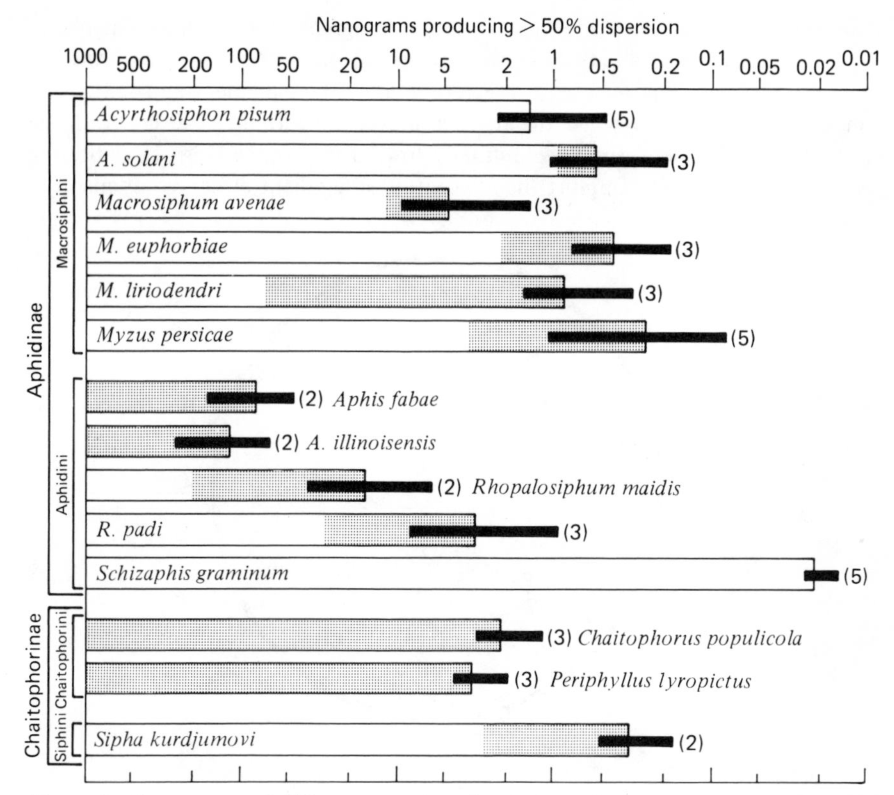

Figure 2. Responses of different species of aphids to the alarm pheromone component (E)-β-farnesene using a simple bioassay without wind (from Montgomery and Nault, 1977a).

with the test stimulus located in the center. In Bell and Tobin's procedure, males were kept isolated from females for at least 2 weeks in groups of 10 on a 12:12 light:dark photoperiod regime. For the photography used in the assay, a flat triangular 9 × 9 × 9-mm tab that reflected UV light was glued to each male's pronotum. The arena was painted flat black and had 20-cm-high walls which were coated with petroleum jelly to prevent the roaches from leaving the recording area. Illumination was provided by a UV light (365 nm). The males' movements were photographed from above the arena in time-lapse fashion by means of a 35-mm camera with its shutter held in the open position and a "stroboscopic" slit rotating at 60 rev/sec providing on a single frame successive shots of the roach's position 0.08 sec apart.

A single roach was placed in the arena and allowed to adjust for 15 min before its movements were photographed with only a solvent control present. Then a 5.5-cm-diameter filter paper disc loaded with extract equivalent in activity to 10^{-5} μg of synthetic periplanone B was introduced into the center of the arena and the males' movements were photographed for 10 min. The pheromone increased the velocity of movement even at the farthest distances from the source (indirect response), and direct response to the gradient apparently was more often employed when the roaches moved to within ca. 40 cm from the source (Fig. 3). Here, turns were usually toward the source (Fig. 4), indicating that the males were sampling the concentration gradient either simultaneously

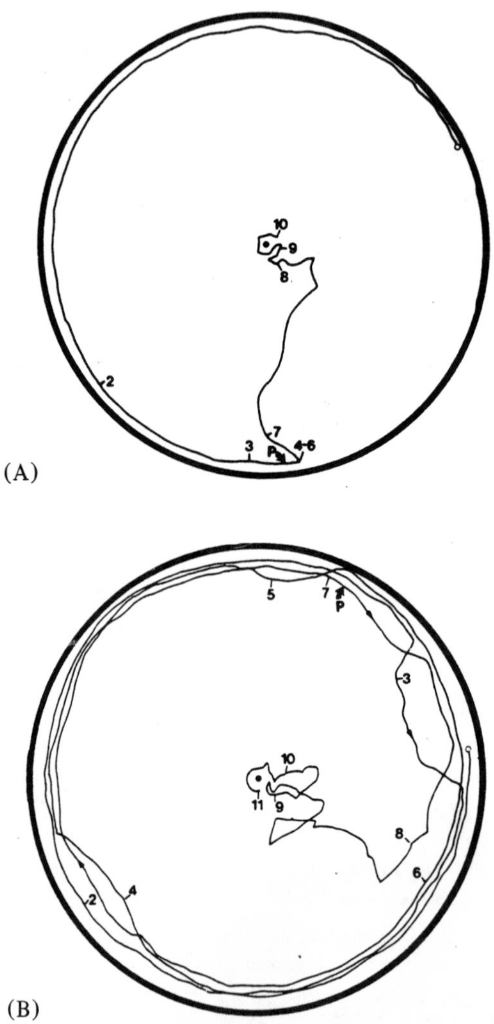

Figure 3. Typical tracks of male *P. americana* in a 2.5-m-diameter arena in response to sex pheromone located in the center (black dot). (A) Male with two antennae; (B) male with one antenna. Pheromone was introduced into the center when the males were at point P, whereupon they no longer walked around the periphery, but rather headed toward the center, of the arena.

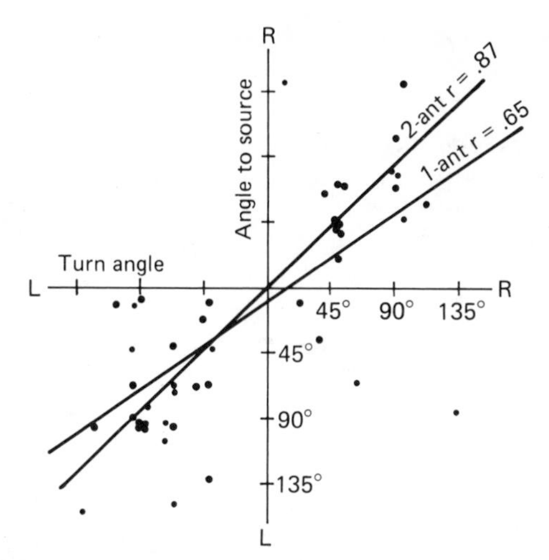

Figure 4. Correlation between the angle the roaches turned and the actual "correct" angle toward the source when males were <40 cm from the source in the arena shown in Fig. 3. The high coefficients of correlation especially for two antennae-males indicates a direct response to the concentration gradient (from Bell and Tobin, 1981).

(using their long antennae) or sequentially, and the gradient was sufficiently steep to allow them to move accurately toward the higher concentration using the direct orientation mechanisms, tropotaxis or klinotaxis.

A similar type of assay was used by Von Keyserlingk (1982) to test the activity of the aggregation pheromones of *Scolytus scolytus* as well as their responses to host odor and water. Although the "arena" was much smaller, a 5-cm-diameter petri dish, a similar type of direct response to pheromone was recorded when the beetles walked away from the walls toward the arena's center where the source was located (Fig. 5). Here the responses were recorded simultaneously with two high resolution video cameras, one stationary for track recording and a second, mobile camera with a macro lense for magnified viewing of behavioral events. Both video signals were blended onto the same videotape and later played back and analyzed with a microcomputer. The slow motion and single frame facilities of the video recorder and the use of the computer as both track analyzer and 30-channel event recorder made uninterrupted analyses of behavioral details over extended periods of time comparatively fast and easy. The computer plotted the insects' tracks and printed out a statistical analysis after each run. Again, care had to be taken to standardize the time of assay and physiological state of the beetles.

Assaying putative trail pheromone components, such as those used by ants for recruiting nestmates, also involves monitoring spatial displacement, and so these assays must allow for movement in two dimensions as in the roach assay described above. Again, with no moving air, the pheromone must be presented

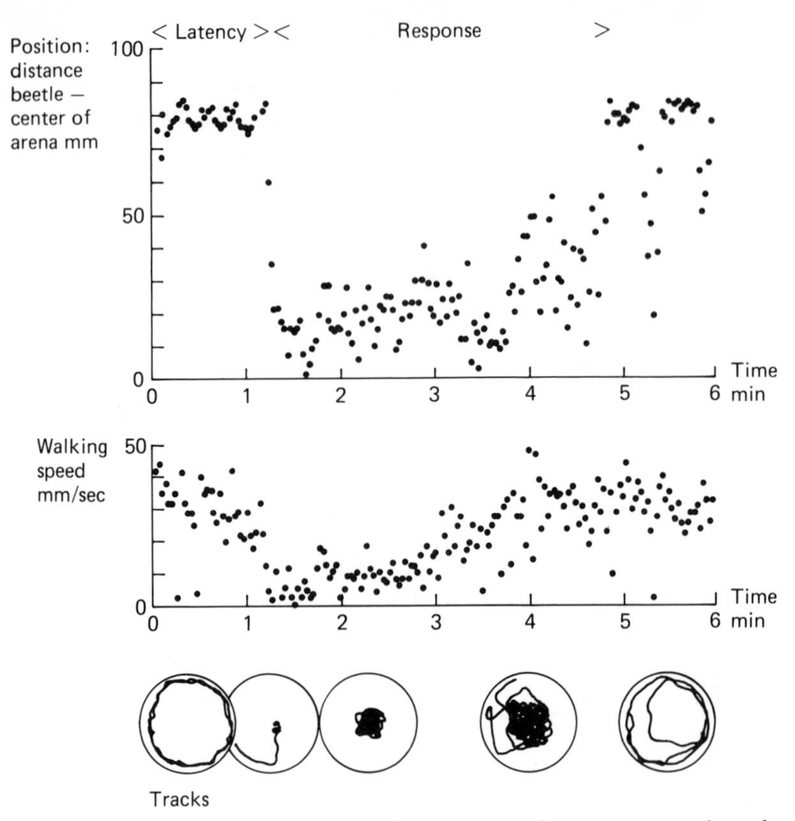

Figure 5. Tracks of *Scolytus scolytus* beetle responding to aggregation pheromone in the center of the petri dish arena (bottom). Computer-assisted analysis includes simultaneous readout of the beetle's velocity and distance from the center, and these values can be correlated with event recordings of the behavior video-recorded on a second camera equipped with a macro-lens (from Von Keyserlingk, 1982).

to the insect with a sufficiently steep gradient so that displacement aided by the gradient can occur. The trail pheromone bioassay used by Van Vorhis Key and Baker (1981) for the Argentine ant, *Iridomyrmex humilis,* utilized a circular deposition of test chemical on a large piece of filter paper. The test compounds were pipetted onto the paper as it revolved on a phonograph turntable. This allowed the chemicals to be deposited in a narrow line and with an even concentration along the circle; both of these deposition characteristics were important to achieving reproducible responses.

Another factor crucial to reproducibility of responses was standardizing the internal state of the ants. Workers that had been fed only water for a few days were given sugar water. They returned to the colony to recruit other workers, and it was these recruits that were diverted individually through a movable

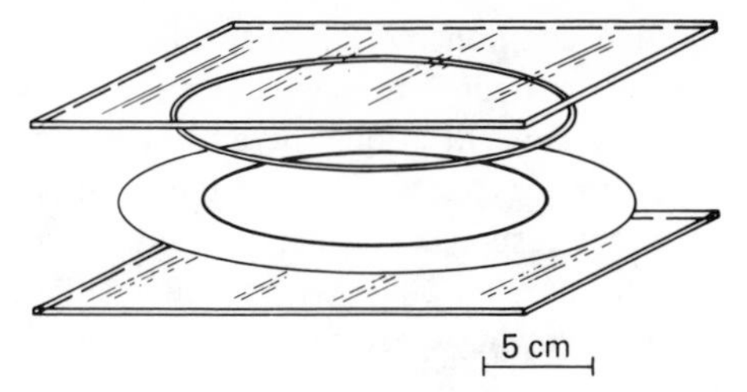

Figure 6. The assay disc (wide, lower circle) containing the circular test trail (dark circle) was housed between two glass plates separated by a Teflon spacer ring (upper circle). Thus there was no wind, and ants were scored for the durations of continuous contact with the trail (from Van Vorhis Key and Baker, 1981).

plastic (Teflon) tube onto the treated paper. Recruited workers were most likely to respond to the pheromone with continuous locomotion around the circular trail. Wind was negligible because a glass plate was placed over the trail and held above the paper by a circular Teflon spacer ring (Fig. 6). Ants were scored as trail-following when they were moving inside of two lines drawn in pencil 0.5 cm to either side of the deposited trail.

The average duration of each bout of continuous trail-following was calculated by dividing the total time spent in the trail vicinity (within the lines) by the number of times the ant crossed out of the trail's scoring area. This assay discriminated among several positional and geometrical isomers of a trail pheromone component of the Argentine ant (Van Vorhis Key and Baker, 1981, 1982).

A choice bioassay was also created to test preference of the ant for its trail pheromone extract compared to the single synthetic component. Two circular trails were deposited by using the phonograph apparatus so that the trails intersected in two places (Fig. 7). An ant following one trail encountered several choice points where it had a chance to begin following the alternate trail. The

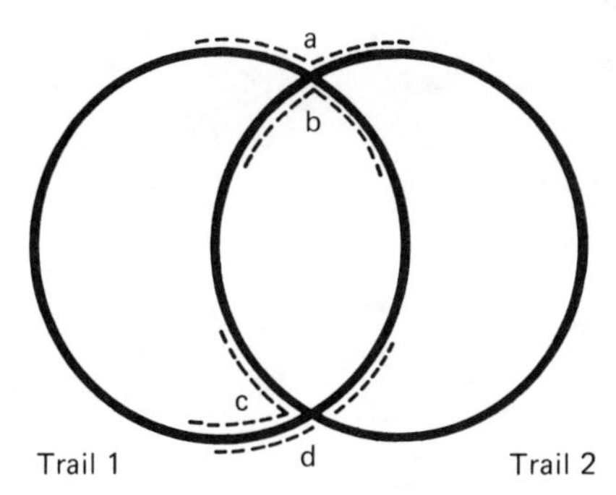

Figure 7. A choice test arrangement of circular trails deposited so that they overlapped and ants periodically had to choose to either remain on the same trail or switch to a new one (from Van Vorhis Key and Baker, 1982).

synthetic trail pheromone component (*Z*)-9-hexadecenal was found to be preferred by ants over gaster extract if the concentration of synthetic compound was high enough. At lower concentrations the ants switched equally as well between synthetic and natural extract trails. In these assays, the gradient was sufficient to cause a direct response because the ants were never more than a few millimeters from the applied trail, where the gradient was steepest. Of course indirect responses involving speed of locomotion (orthokinesis) were also integrated into the response to result in displacement along the trail.

The tent caterpillars' (*Malacosoma* sp.) responses to trail pheromones deposited between the nest and foraging sites has been studied in detail by means of several innovative assays (Fitzgerald and Gallagher, 1976; Fitzgerald and Edgerly, 1979). With one setup (Fig. 8), the responses to new trails were shown to be higher than those to old trails, and pheromone extracted from silk deposited by walking larvae was shown to be active in eliciting trail-following apart from the effect of silk alone.

Second-instar larvae were allowed to construct a silk nest on an inverted tripod, and they foraged on a small cherry tree connected to the nest by a bridge (Fig. 8) (Fitzgerald and Gallagher, 1976). As often as four times a day, the larvae moved from the nest to the cherry tree to feed, and if observations needed to be

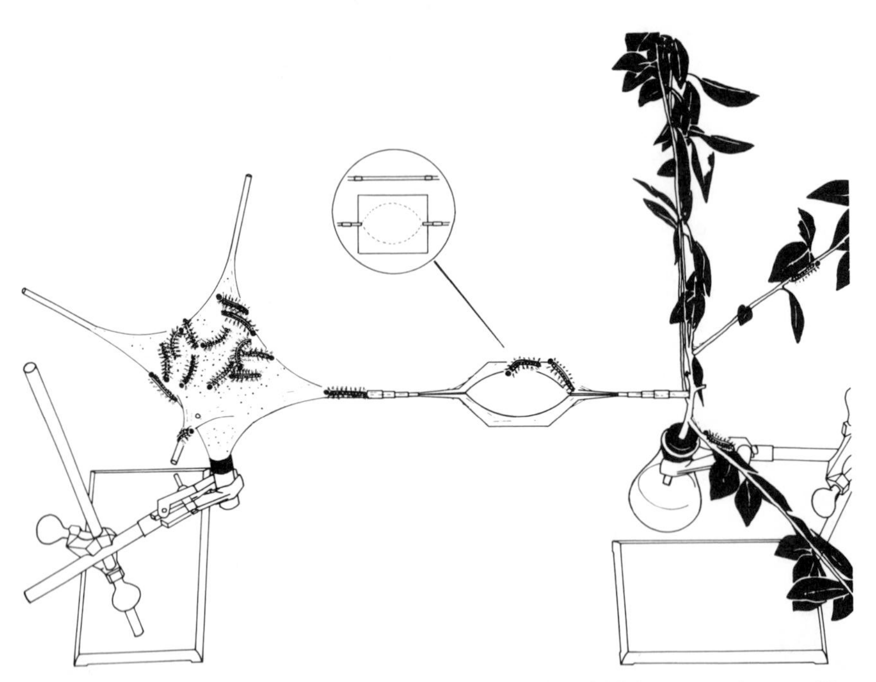

Figure 8. An assay system for the trail pheromone of *Malacosoma* larvae. The setup allows larvae to forage in cherry foliage, but to do so they must walk across a removable section of the bridge that can be replaced by an experimentally altered section (from Fitzgerald and Gallagher, 1976).

performed during the dark phase of the 15:9 light:dark photoperiod, a subdued red light was used. The key to the assay was a removable section of the bridge that could be treated experimentally. For example, when an established trail on a glass rod was extracted with methylene chloride and replaced the larvae would not cross it. When the extract was added back to the silk on the glass rod, trail-following resumed.

In another experiment, a plate, rather than a rod was used to test in choice fashion the response of larvae to extract-alone trails compared to washed silk alone. The larvae usually chose the extract trail and deposited new silk trails on it. They even could be induced to follow elaborate, curved trails that had been drawn with methylene chloride extract on larger plates that were inserted into the bridge. A Y-maze, choice section was also inserted into the bridge to test old versus new trails.

In another type of choice test whole silk trails were deposited on filter paper and cut up to form a "Y" (Fig. 9) (Fitzgerald and Edgerly, 1979). Larvae crawl-

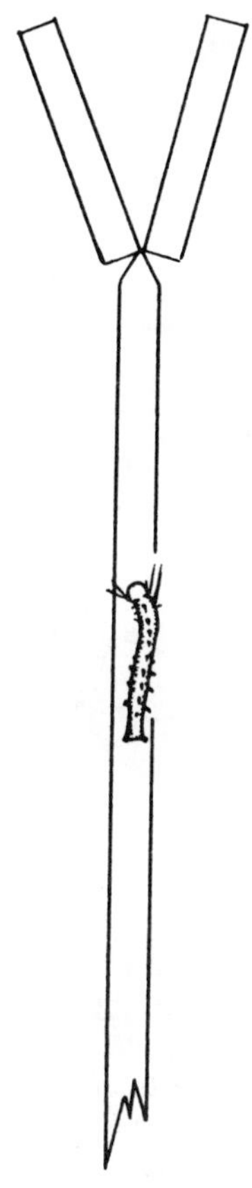

Figure 9. A choice test for the trail pheromones of larvae *Malacosoma sp.* and *Archips cerasivoranus* (from Fitzgerald and Edgerly, 1979).

ing up the main trail had to choose a trail at the branch point. This assay was able to show that *Malocasoma distria* and *americana* readily followed each other's trails, but trails of a tortricid species, *Archips cerasivoranus,* were not followed by these two tent caterpillars.

3.3. Bioassays with Airflow, Displacement Not Monitored

By the use of moving air in bioassays one can examine anemotaxis, or steering with respect to wind direction. This indirect response to pheromone (Bell and Tobin, 1982) can be triggered by odor and used by the insect to move toward or away from the source (Kennedy, 1977, 1978). The researcher can use this wind-

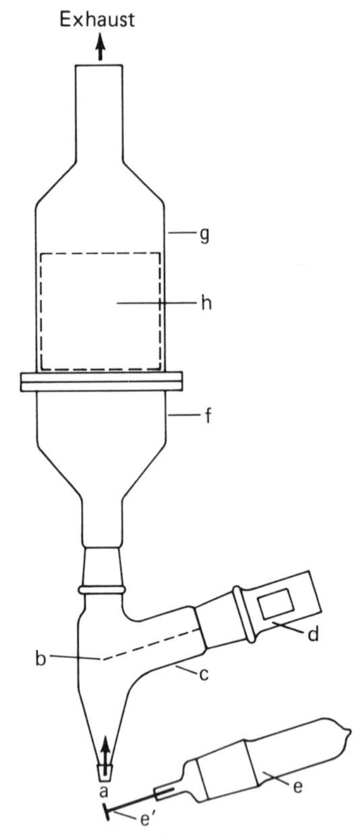

Figure 10. A stimulation-type bioassay chamber with airflow but not allowing the insects to displace. Air enters the apparatus at (a) and flows over the pheromone-impregnated brass disc (e′) attached to a ground-glass stopper (e) which is normally inserted at (c) (the disc rests in the airstream at point (b)). The pheromone-laden air enters the glass cylinder (f, g) in which a cage of males (h) is housed. The observer scores the males for the "key" response of activation (from Bartell and Shorey, 1969).

steered displacement as yet another way to discriminate among treatments. There is no need for a steep gradient emanating outward from the source, and the wind delivers the odor quickly to the test insects.

However, airflow can be utilized without displacement, either by restricting the insects' movements in all directions, or more importantly, by placing the wind direction perpendicular to the only plane in which the insect is allowed to move. This technique was used by Bartell and Shorey (1969) in bioassaying the sex pheromone of the light-brown apple moth, *Epiphyas postvittana.*

Groups of 10 unmated male moths were placed in copper wire-mesh cylindrical cages, 7.6 cm diameter × 7.0 cm high. The cages then were placed individually into vertically standing glass cylinders (Figs. 10, 11), into which compressor-generated air was blown from the bottom at a rate of 5 liters/min. All assays were conducted at ca. 2 hr after lights-off on a 14:10 photoperiod regime with transitional dawn and dusk light intensities. The sex pheromone extract was impregnated onto a brass disc which was introduced into the air-stream through a port in the inlet tube at the bottom of the cylinder (Fig. 10). Pheromone and air mixed in a small chamber before entering the chamber con-

Figure 11. A group of the assay chambers depicted in Fig. 10, containing males ready for assaying (from Bartell and Shorey, 1969).

taining the moths, where movements were limited to a plane perpendicular to the airflow or only minimally up and down in the chamber. The observer focused only on one "key" response and scored the group of moths for the percentage of individuals moving. The overall measure, therefore, was a general "stimulation" due to pheromone without displacement with respect to the wind direction.

This type of assay has been used for identifying sex pheromone blends of other moth species such as *Heliothis virescens* (Roelofs et al., 1974). Other assays using wind but not permitting displacement along it have been successful in identifying sex pheromones from a large number of species. The use of "activation" bioassay chambers connected directly to a split gas chromatographic outlet is described in Chapter 8.

3.4. Bioassays with Airflow, Displacement Is Monitored

When wind is parallel to the plane of displacement, it can aid both the speed of delivery of odor to the insects and the insects' movements along the windline toward or away from the source. The simplest arrangement allows movement along the windline, but restricts it perpendicular to this line.

A very useful assay using such a one-dimensional displacement was developed by Sower et al. (1973) to document the reaction of *Sitotroga cerealella,* the Angoumois grain moth to sex pheromone. Groups of 8–12 males were placed in 1.9-cm-inside diameter × 44-cm-long Plexiglas tubes. Charcoal-filtered air entering a manifold was distributed to 15 such tubes (Fig. 12), and a system of stoppers and screens at each end prevented the moths from escaping while allowing air to move through. The tubes were lighted from below by diffuse light of 0.3 lux. Pheromone-laden air was exhausted through a fume hood. The pheromone treatment, either female extract or a synthetic analog, was deposited from solution onto a 0.5-cm glass applicator which was inserted through a stopper

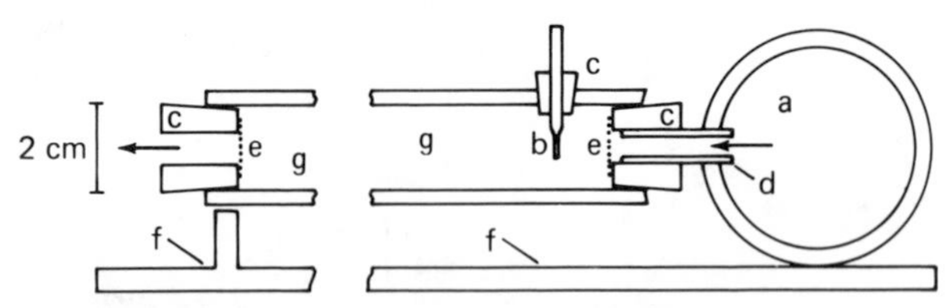

Figure 12. Cutaway view of one tube (g) of a 15-tube array for assaying displacement of male moths in one dimension in response to wind plus pheromone. A manifold (a) feeds air into the 15 tubes, and it enters each tube through a hole (d) covered by a screen (e). Pheromone evaporates into the airstream from a glass rod (b). A screen (e) at the downwind end prevents males from leaving that end of the tube. Response is measured as the net displacement of males upwind in the tube (from Sower and Vick, 1973). Structure (f) is a supporting base.

into the airstream through a port at the upwind end of the tube (Fig. 12). The number of males that had moved to within 4 cm of the source was counted after 15 and 30 sec, and from this number was subtracted the number within 4 cm of the source before introduction of the treated rod. This method of scoring allowed quick counts of the responses to be taken by hand, and the comparison of moth positions before and after pheromone introduction measured the activity of the treatment.

An advantage to measuring net movement up and down a tube is that effects of concentration can be observed. Males may accumulate farther down the tube when concentrations are too high. Daterman (1972), working with the European pine shoot moth, *Rhyacionia buoliana,* further increased discrimination among treatments by pitting positive phototaxis against the response to pheromone. He placed the downwind end of the tube in a box illuminated with dim light. The moths tended to accumulate at the downwind end of the tube before testing, and only the most attractive treatments induced them to move away from the light to the upwind end of the tube. However, the possible suppression of responsiveness by light levels above moonlight, noted earlier in the chapter, must be considered in this system. The orientation tube bioassay was used in identifying the sex pheromone components of the oak leafroller moth, *Archips semiferanus* (Miller et al., 1976).

Tobin et al. (1981) used a tube-type bioassay, modeled after one designed by Persoons (1977), to assess the activity of synthetic *Periplaneta americana* sex pheromone. A kind of choice test was performed in this assay against a blank control. Two parallel plastic tubes, 3.8 cm diameter × 30 cm, were connected to a 5.0-liter container holding 20 males. Air was drawn through the tubes at 50 cm/sec by a vacuum pump attached to the container of males. Pheromone was introduced into one of the tubes by means of a disposable pipet inserted in a stopper, and the other tube contained only a clean pipet and stopper. After 6 min, the numbers of cockroaches in the test and control tubes were counted, and for analysis the latter were subtracted from the former.

Tobin et al. (1981) were able to discriminate among several concentrations of pheromone using this assay. They also showed the importance of the release surface in influencing "threshold" values. Greater quantities of pheromone needed to be loaded onto filter paper compared to glass in order to evoke equivalent levels of response (Fig. 13).

The next increase in discrimination in moving air is gained from assays in two dimensions, usually for walking insects on a flat surface. Here, the lateral movements of the insect can take it out of contact with the pheromone, adding to the power of the assay. Under natural conditions, for instance, a male insect not only has to advance toward a sex pheromone source but also has to maintain lateral contact with it. Such a two-dimensional bioassay in wind has been utilized in studies of bark beetle aggregation pheromone by Payne et al. (1976), who modified Wood and Bushing's design (1963).

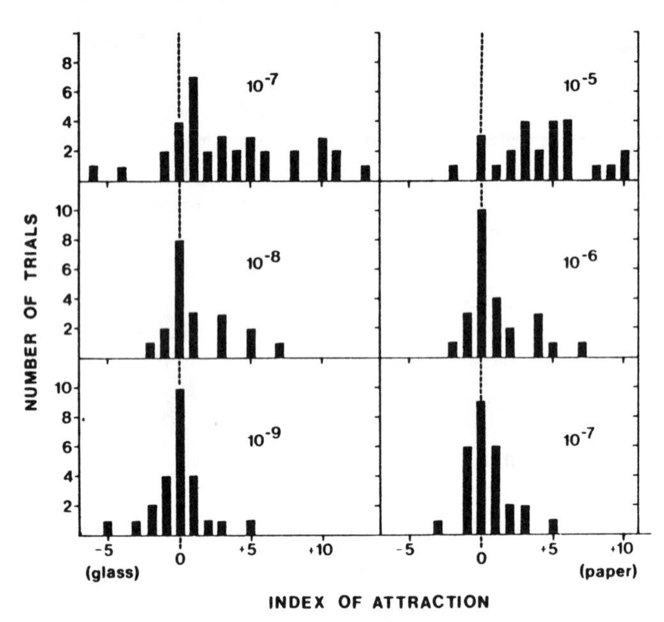

Figure 13. A demonstration of the importance of the pheromone release substrate. Ca. 100 times more pheromone needs to be loaded onto filter paper compared to glass to elicit similar levels of *P. americana* male attraction (from Tobin et al., 1981).

The apparatus of Payne et al. consisted of a flat arena (Fig. 14), 22.5 X 28 cm covered with a sheet of paper that could be discarded after each assay. The arena was housed in a specially constructed 183 X 122 X 214 cm controlled-environment room to optimize assay conditions for the beetles. The room was maintained at 20°C to keep the beetles from flying from the arena and relative humidity was kept at ca. 80%. The room also had adequate exhaust to remove pheromone from the room, plus an air diffuser to minimize turbulence on the arena surface. Activated charcoal-filtered airflow was provided at 1.5–2 liters/min from a compressor, and pheromone was introduced at a constant rate from a motor-driven syringe loaded with pentane solutions of the pheromone treatments to be tested. Air velocity was recorded by an anemometer at the far end of the arena. Groups of 10 beetles were placed on the arena at a starting point 9 cm downwind of the pheromone source and positive responses were recorded for those walking to within 1 cm of the source. Beetles leaving the pheromone stream were collected and rereleased for a second try. If they did not respond on this try, they were recorded as negative responses.

The internal state of the beetles was kept as constant as possible through a variety of procedures. They were collected daily from an emergence chamber, examined for presence of all appendages (antennae, etc.), and held individually in No. 10 (ca. 1-cm-long) gelatin capsules to prevent injury. Also, daily fluctuations in internal state were monitored first before performing assays on test

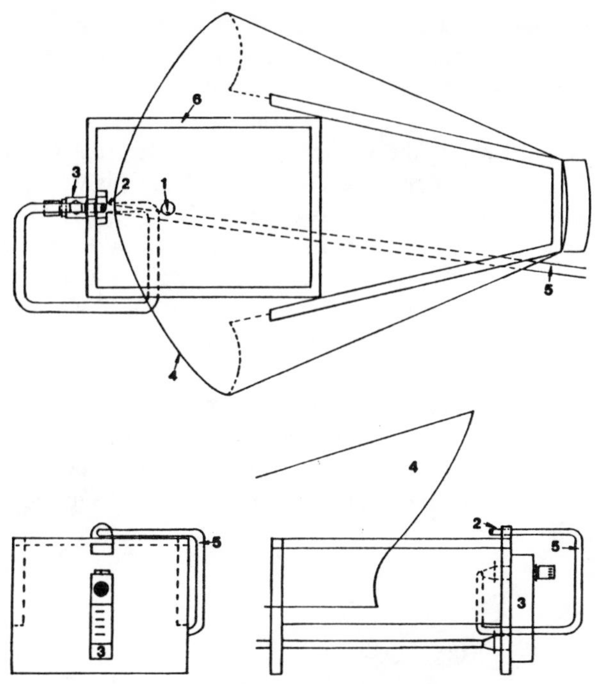

Figure 14. Top view of a bioassay arena (6) for bark beetles walking in a pheromone-laden stream. The beetles are placed on the rectangular arena at (1), from downwind of the source of the airstream (2). Beetles walking to within 1 cm of the source are scored as positive responders (from Payne et al., 1976). Bottom right is sideview. 4 is exhaust hood; 5 is a connecting tube; and 3 is an adjustable mixing chamber for pheromone and air.

materials. Groups of 10 males and 10 females were tested for their response to solvent alone and to a standard pheromone stimulus. If the positive response level to solvent was greater than 20% or if the response to the standard treatment was less than 50%, assays on test chemicals were not performed that day.

Tobin et al. (1981) used a low-airspeed, 2.4 × 1.2 × 0.6 m wind tunnel to test the activity of synthetic *Periplaneta americana* sex pheromone against natural extract. Airspeed was 22 cm/sec, and males were held individually in a wire cage for 20 min prior to assay at the downwind end of the tunnel to allow them to acclimate to the tunnel conditions. The cage was then opened and the pheromone sample introduced on filter paper suspended 2 cm from the floor. The percentage of males locating the filter paper was scored. Using this assay plus a variety of other assays and trapping experiments, they concluded that the synthetic pheromone, periplanone B, elicited the complete range of sexual behaviors in males, from long-distance orientation to close-range courtship behaviors, such as wing-raising. These were the same behaviors evoked by natural extract at comparable concentrations.

With such two-dimensional assays, measuring spatial displacement becomes desirable and often necessary to gain maximal discrimination between treatments. But the types and patterns of movements used by the insect to gain this displacement can also be monitored. Usually for this, only photographic or video records will capture the movements in enough detail to allow analysis. Of course, the amount of time needed to record and analyze these recordings (usually of individual insects) increases dramatically.

A very simple and effective type of assay was used by Rust et al. (1976) to monitor turns in response to pheromone by a walking insect, the American cockroach. Individual male roaches were tethered to a 10-cm wooden applicator stick and labeled with a mark to the back of the pronotum. The sticks were then held in a clamp on a ring stand and the cockroaches given a styrofoam Y-maze globe to "hold" (Fig. 15). Responses to sex pheromone were tested in the dark. The rate of locomotion was measured by the amount of time taken for 20 revolutions of the globe to occur, and the turning tendency was measured by recording the number of left or right 60° turns taken by the males.

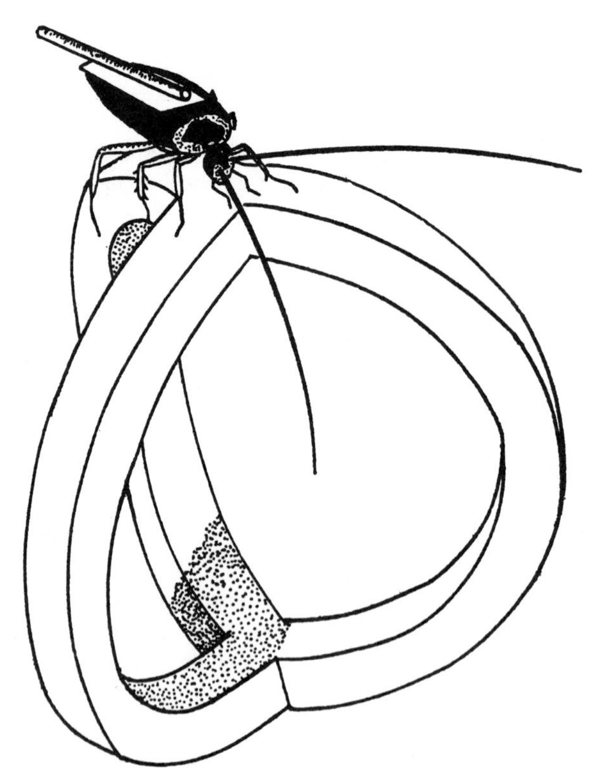

Figure 15. Y-Maze globe used to measure the turns and locomotory rate of American cockroach males in response to pheromone in still air and in air currents (from Rust et al., 1976).

Observations were recorded on audio tape and later transcribed for analysis. Head movements of the cockroaches were filmed against a background grid (10 mm^2) using a 16-mm movie camera. From these assays, Rust et al. (1976) found that cockroaches tend to turn away from air currents without pheromone and upwind in air containing pheromone. Furthermore, by presenting the pheromone in still air to one side or the other, they found that the males could orient accurately toward the source by a simultaneous sampling of the gradient by the antennae (chemotropotaxis). If one antenna was removed, the males adapted within a few days and started sampling the gradient sequentially (chemoklinotaxis) by waving their lone antenna from side to side before executing a turn toward the pheromone.

A more elaborate record of walking insects' movements was obtained using the so-called servosphere apparatus. This was used by Kramer (1975) to monitor the pheromone-mediated movements of *Bombyx mori,* and by Bell and Kramer (1980) for *Periplaneta americana.* The apparatus consists of a Plexiglas sphere, 50 cm diameter, that is mounted so that it can be rotated in two different planes by two low-inertia servomotors (Fig. 16) (Kramer, 1975). The insect, placed on top of the sphere with a disc of reflective material attached to its dorsum, is kept in the field of an infrared light beam by the corrections of the motors on the sphere. The sphere's counter-movements required to keep the running insect in the beam are recorded and can be plotted as a record of the insect's movements.

One nice feature of the servosphere assay, and the previous one by Rust et al.

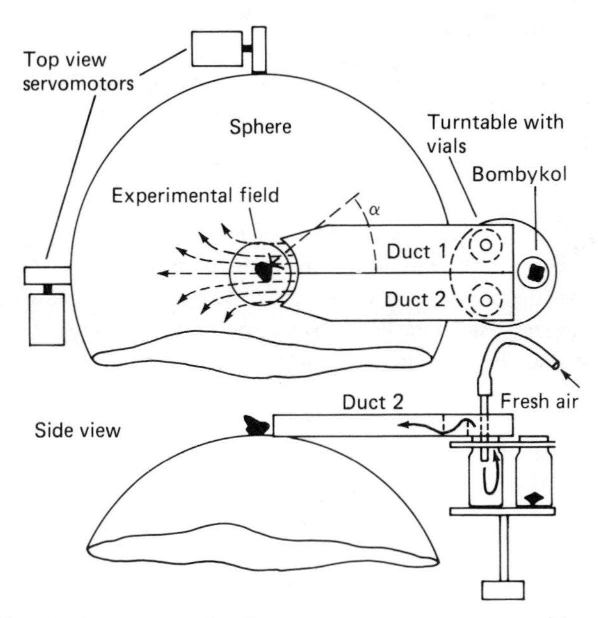

Figure 16. Servosphere apparatus for measuring movements of insects walking in response to sex pheromone plus moving air (from Kramer, 1975).

(1976) is that the insect remains in the same spot relative to the odor source, and cannot enter a new stimulus situation as in the previous type of two-dimensional test for bark beetles. In the latter, test concentration will increase to some degree as the insect approaches the source, and this may not always be desirable. In addition, the servosphere technique may allow for better testing between a choice of treatments than in the previous type of assay. This is because the insect, held at the same location between two odor streams, must continually choose between them and cannot take itself out of contact with one or the other. Thus the assay will measure a more *continuous* discrimination by the insect, not just an initial brief judgment after which it only moves in response to one of the stimuli. Just such a choice test was performed by Kramer, who partitioned the airstream into two halves and presented pheromone of differing concentrations in each side. He found that *B. mori* males could discriminate between concentrations of pheromone that differed by a ratio of only 5:3.

Monitoring insect movements in three dimensions really pertains only to cases in which insects are allowed to fly or swim. In this case, wind is created in a large chamber, usually 1 or more m long, called wind tunnels, or sustained-flight tunnels. Kennedy and Marsh (1974) used such a tunnel for demonstrating the optomotor anemotactic response of flying *Anaghasta kuhniella* males. Miller and Roelofs (1978a) demonstrated the usefulness of the wind tunnel for discriminating among several pheromone treatments in the red-banded leafroller moth. This technique has been used since then to discern differences between pheromone blends for several other moths, including *G. molesta,* the oriental fruit moth (Baker and Cardé, 1979; Baker et al., 1981), and the noctuid moth, *Euxoa ochrogaster* (Palaniswamy et al., 1983), the gypsy moth, *Lymantria dispar* (Miller and Roelofs, 1978b; Cardé and Hagaman, 1979), *Heliothis virescens* (Vetter and Baker, 1982), and the cabbage looper moth, *Trichoplusia ni* (Linn and Gaston, 1981).

There are two major advantages of wind tunnels for assaying pheromone blends. First, the insect must perform a series of movements similar to those used in the field to locate a pheromone source. Second, apart from the degree of spatial displacement to be monitored by the experimenter, the *duration* of the response can be measured by making use of the moths' optomotor response to the ground pattern. The duration of sustained in-plume flight at a fixed point over a moving floor can discriminate among treatments that, using only flight to the source as a measure, would otherwise not have appeared to be different.

Miller and Roelofs' tunnel was a clear, somewhat flattened cylinder formed from the inverted "U" of two bowed sheets of Plexiglas joined at the middle, and included a treadmill of canvas, painted with a striped pattern (Fig. 17). Wind was supplied by a small window fan connected to the Plexiglas by a flexible plastic tube. Several layers of muslin were stretched across the tunnel's entrance to smooth the airflow and make it essentially laminar. Lights were mounted overhead and could be varied in intensity from bright levels for day-flying insects (fluorescent lights) to moonlight intensities using a bank of

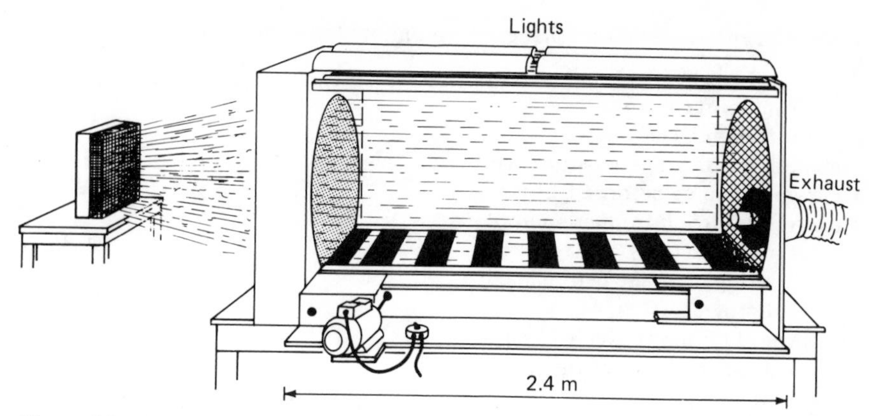

Figure 17. A typical wind tunnel for assaying pheromone compounds (from Miller and Roelofs, 1978a).

voltage-regulated incandescent bulbs covered with red cellophane. The floor pattern on the canvas treadmill was alternating 15-cm-wide red and black stripes, and the treadmill could be rotated in the direction of the wind at speeds ranging from 2 to 60 cm/sec by means of a joy stick connected to a sliding-gear variable transmission and electric motor.

Moths were introduced into the downwind end of the tunnel through a hole in the metal screening that provided a barrier to escape for moths but allowed the passage of air. Groups or individual moths were caged in screen cylinders that fit into the hole; the screen lid to the cylinder was removed just before placing it through the hole. Four different types of measurements were used to demonstrate the location and diameter of the pheromone plume at the moths' release cage in the hole. There was remarkable correspondence among titanium tetrachloride-generated smoke, a plume of hydrochloric acid on indicator paper, apple odor sensed by the human nose, and an electroantennogram as it scanned through a plume of (±) disparlure; all showed the time-averaged plume to have horizontal and vertical diameters of ca. 15 and 18 cm, respectively. In addition to the discrimination provided by sustained flight caused by the rotating floor (which showed differences between racemic and (+) disparlure, Miller and Roelofs, 1978b) many measurements could be performed with the floor stationary that may be useful in discriminating among treatments (Table 1). These parameters are easily recorded with a pencil, paper and stopwatch, but tape-recording or use of multichannel event recorders are also very useful for recording wind tunnel behaviors (Baker et al., 1981). An exhaust tube (Fig. 17) scavenged the pheromone plume from the room while allowing the rest of the air to recirculate into the tunnel.

Choice tests can also be performed on insects flying in a wind tunnel. Linn and Gaston (1981) were able to show the importance of a two-component *Trichoplusia ni* pheromone blend at close range compared to a single com-

Table 1. Types of responses that can be monitored in a flight tunnel[1]

Type of behavior	$\bar{x} \pm SD$	Range
Time from source introduction until first visible response (sec)	6.9 ± 2.1	3–9
Time walking or running before flight (sec)	8.3 ± 4.4	3–15
Time wing-fanning before flight (sec)	4.2 ± 3.3	0–9
Time from source introduction until flight initiation (sec)	27.3 ± 26.8	12–85
Time hovering in plume before making upwind progress (sec)	5.1 ± 3.3	1–11
Flight speed (ground) over a 70-cm sector in the downwind half of the tunnel (cm/sec)	17.1 ± 7.1	10.8–31.8
Flight speed (ground) over a 70-cm sector in the upwind half of the tunnel (cm/sec)	14.4 ± 3.8	9.0–21.3
Number of approaches to within 2–3 cm of the source before landing	1.9 ± 0.9	1–3
Number of times dropping back more than 30 cm and reorienting before landing	0.3 ± 0.8	0–2
Spot where insect lands:		
Directly on chemical source	71%	
On structures supporting source	29%	
Time landing after first coming within 45 cm of source (sec)	10.9 ± 5.3	4–19
Elapsed time from flight initiation until landing on source (sec)	24.3 ± 5.7	15–32
Time walking or running on source (sec)	8.0 ± 9.3	3–29
Time wing-fanning on source (sec)	8.2 ± 9.7	3–30
Time genital claspers extended (sec)	8.2 ± 9.7	3–30
Number of copulatory attempts	0.1 ± 0.4	0–1
Time quiescent on source (sec)	0.1 ± 0.2	0–0.5
Total time on source	9.9 ± 14.2	3–42

[1] Data are taken from experiments with redbanded leafroller males (from Miller and Roelofs, 1978a).

ponent, *Z*7-12:OAc, by creating two plumes that intersected at ca. 70 cm downwind of the sources. *T. ni* males were released at their optimal period of responsiveness into the blended, single plume. As they flew up-tunnel, they eventually had to choose to fly in one plume or the other. If one plume was the two-component blend containing both 12:OAc and *Z*7-12:OAc, they continued to fly along it all the way to the source. If the two plumes were each component alone, the males would fly only as far as the limits of the blended plume, or perhaps a little farther along the *Z*7-12:OAc plume alone, but not all the way to the source. *Z*7-12:OAc alone did not elicit continued advancement toward the source.

The use of wind tunnels in studies of orientation to sex pheromone has blossomed in the past few years. Their many uses will be explored in more detail in the following chapter by Baker and Linn.

IV. References

Baker TC, Cardé RT (1979) Analysis of pheromone-mediated behavior in male *Grapholitha molesta*, the oriental fruit moth (Lepidoptera:Tortricidae). Environ Entomol **8**:956–968.

Baker TC, Meyer W, Roelofs WL (1981) Sex pheromone dosage and blend specificity of response by oriental fruit moth males. Ent Exp Appl **30**:269–279.

Baker TC, Cardé RT, Roelofs WL (1976) Behavioral responses of male *Argyrotaenia velutinana* (Lepidoptera:Tortricidae) to components of its sex pheromone. J Chem Ecol **2**:333–352.

Bartell RJ, Lawrence LA (1977) Reduction of responsiveness of male apple moths, *Epiphyas postvittana* to sex pheromone following pulsed pheromonal exposure. Physiol Entomol **2**:1–6.

Bartell RJ, Shorey HH (1969) A quantitative bioassay for the sex pheromone of *Epiphyas postvittana* (Lepidoptera) and factors limiting male responsiveness. J Insect Physiol **15**:33–40.

Bell WJ, Kramer E (1980) Sex pheromone stimulated orientation responses by the American cockroach on a servosphere apparatus. J Chem Ecol **6**:287–295.

Bell WJ, Tobin T (1981) Orientation to sex pheromone in the American cockroach: Analysis of chemo-orientation mechanisms. J Insect Physiol **27**: 501–508.

Bell WJ, Tobin TR (1982) Chemo-orientation. Biol Rev **57**:219–260.

Cardé RT, Comeau A, Baker TC, Roelofs WL (1975) Moth mating periodicity: Temperature regulates the circadian gate. Experientia **31**:46–48.

Cardé RT, Hagaman TE (1979) Behavioral responses of the gypsy moth in a wind tunnel to air-borne enantiomers of disparlure. Environ Entomol **8**: 475–484.

Cardé RT, Hagaman TE (1983) Influence of ambient and thoracic temperatures upon sexual behaviour of the gypsy moth, *Lymantria dispar*. Physiol Entomol **8**:7–14.

Corbet PS (1966) The role of rhythms in insect behaviour. In: Insect Behaviour. Haskell PT (ed), Symp R Ent Soc Lond, Vol **3**, pp 13–28.

Daterman GE (1972) Laboratory bioassay for sex pheromone of the European pine shoot moth, *Rhyacionia buoliana*. Ann Entomol Soc Am **65**:119–123.

Fitzgerald TD, Gallagher EM (1976) A chemical trail factor from the silk of the Eastern tent caterpillar *Malacosoma americanum* (Lepidoptera:Lasiocampidae). J Chem Ecol **2**:187–193.

Fitzgerald TD, Edgerly JS (1979) Specificity of trail markers of forest and Eastern tent caterpillars. J Chem Ecol **5**:565–574.

Fraenkel GS, Gunn DL (1940) The Orientation of Animals. Revised edition (1961) Dover, New York.

Hawkins WA (1978) Effects of sex pheromone on locomotion in the male American cockroach. J Chem Ecol **4**:149–160.

Kennedy JS (1977) Behaviorally discriminating assays of attractants and repellents. In: Chemical Control of Insect Behavior: Theory and Application. Shorey HH, McKelvey JJ (eds), Wiley, New York, pp 215–229.

Kennedy JS (1978) The concepts of olfactory arrestment and attraction. Physiol Entomol **3**:91-98.

Kennedy JS, Marsh D (1974) Pheromone-regulated anemotaxis in flying moths. Science **184**:999-1001.

Kramer E (1975) Orientation of the male silkmoth to the sex attractant bombykol. Olfaction and Taste V. Dento D, Coghlan JP (eds), Academic Press, New York, pp 329-335.

Kuenen LPS, Baker TC (1981) Habituation versus sensory adaptation as the cause of reduced attraction following pulsed and constant sex pheromone pre-exposure in *Trichoplusia ni*. J Insect Physiol **27**:721-726.

Linn CE Jr, Gaston LK (1981) Behavioral function of the components and the blend of the sex pheromone of the cabbage looper, *Trichoplusia ni*. Environ Entomol **10**:751-755.

Miller JR, Baker TC, Cardé RT, Roelofs WL (1976) Reinvestigation of oak leaf roller sex pheromone components and the hypothesis that they vary with diet. Science **192**:140-143.

Miller JR, Roelofs WL (1978a) Sustained-flight tunnel for measuring insect responses to wind-borne sex pheromones. J Chem Ecol **4**:187-198.

Miller JR, Roelofs WL (1978b) Gypsy moth responses to pheromone enantiomers as evaluated in a sustained-flight tunnel. Environ Entomol **7**:42-44.

Montgomery MES, Nault LR (1977a) Comparative response of aphids to the alarm pheromone (E)-β-farnesene. Ent Exp Appl **22**:236-242.

Montgomery MES, Nault LR (1977b) Aphid alarm pheromones: Dispersion of *Hyadaphis erysimio* and *Myzus persicae*. Ann Entomol Soc Am **70**:669-672.

Palaniswamy P, Underhill EW, Steck WF, Chisholm MD (1983) Responses of male redbacked cutworm, *Euxoa ochrogaster* (Lepidoptera:Noctuidae) to sex pheromone components in a flight tunnel. Environ Entomol **12**:748-752.

Payne TL, Hart ER, Edson LJ, McCarty FA, Billings PM, Coster JE (1976) Olfactometer for assay of behavioral chemicals for the Southern pine beetle, *Dendroctonus frontalis* (Coleoptera:Scolytidae). J Chem Ecol **2**:411-419.

Persoons CJ (1977) Structure elucidation of some insect pheromones: A contribution to the development of selective pest control agents. PhD thesis, Agricultural University, Wageningen, The Netherlands.

Roelofs WL, Hill AS, Cardé RT, Baker TC (1974) Two sex pheromone components of the tobacco budworm moth, *Heliothis virescens*. Life Sci **14**: 1555-1561.

Rust MK, Burk T, Bell WJ (1976) Pheromone stimulated locomotory and orientation responses in the American cockroach. Anim Behav **24**:52-67.

Shorey HH (1970) Sex pheromones of Lepidoptera. In: Control of Insect Behavior by Natural Products. Wood DL, Silverstein RM, Nakajima M (eds), Academic Press, New York, pp 249-284.

Shorey HH, Gaston LK (1964) Sex pheromones of noctuid moths III. Inhibition of male responses to the sex pheromone in *Trichoplusia ni* (Lepidoptera: Noctuidae). Ann Entomol Soc Am **57**:775-779.

Sower LL, Vick KW, Long JS (1973) Isolation and preliminary biological studies of the female-produced sex pheromone of *Sitotroga cerealella* (Lepidoptera: Gelechiidae). Ann Ent Soc Am **66**:184–187.

Tobin TR, Seelinger G, Bell WJ (1981) Behavioral responses of male *Periplaneta americana* to periplanone B, a synthetic component of the female sex pheromone. J Chem Ecol **7**:969–979.

Vetter RS, Baker TC (1982) Behavioral responses of male *Heliothis virescens* in a sustained-flight tunnel to combinations of the 7 compounds identified from the female sex pheromone gland. J Chem Ecol **9**:747–759.

Vick KW, Burkholder WE, Gorman JE (1970) Interspecific response to sex pheromones of *Trogoderma* species (Coleoptera:Dermestidae). Ann Entomol Soc Am **63**:379–381.

Van Vorhis Key SE, Baker TC (1981) Effects of gaster extract trail concentration on the trail following behaviour of the Argentine ant, *Iridomyrmex humilis* (Mayr). J Insect Physiol **27**:363–370.

Van Vorhis Key SE, Baker TC (1982) Specificity of laboratory trail following by the Argentine ant, *Iridomyrmex humilis* (Mayr), to (Z)-9-hexadecenal, analogs, and gaster extracts. J Chem Ecol **8**:1057–1063.

Von Keyserlingk H (1982) The measurement of attractant or repellent effects of chemicals on walking insects. Med Fac Landbouww Rijksuniv Gent **47**: 547–555.

Wood DL, Bushing RW (1963) The olfactory response of *Ips confusus* (LeC.) (Coleoptera:Scolytidae) to the secondary attraction in the laboratory. Can Entomol **95**:1066–1078.

Chapter 3

Wind Tunnels in Pheromone Research

T. C. Baker[1] *and C. E. Linn, Jr.*[2]

I. Introduction

During the dramatic growth in research on insect sex pheromones over the past 10 years, it has been demonstrated that both the chemical signal and the precopulatory behaviors exhibited by males and females are much more complex than originally thought. Chemical studies of sex pheromones have shown that, with few exceptions, females release a blend of several chemical components in a specific ratio and release rate. This specificity of the chemical signal is in part a result of the behavioral reactions of males under "natural" conditions: flight through space from a distance of several meters or more in shifting wind fields. Upwind progress in the plume of chemicals continues if the blend is that of a conspecific female, and the maintenance of contact with the plume is one of the most complex behavioral responses, involving visual feedback, a chemically modulated self-steered program of zigzags, and changes in linear velocity of flight, all performed as the insect samples the odor environment sequentially (Kennedy, 1983; Kuenen and Baker, 1983).

Because the responses made in-flight are so highly integrated, bioassays utilizing flight in wind are probably the most discriminating in pheromone research. Field bioassays involving the capture of males, discussed in Chapter 4 by Cardé and Elkinton, are the ultimate tests of a pheromone blend's "activity," due in a

[1] Division of Toxicology and Physiology, Department of Entomology, University of California, Riverside, California.

[2] Department of Entomology, New York State Agricultural Experiment Station, Geneva, New York.

large part to the complexity of the environment and of the shifting wind fields. Sexual communication systems evolved in the field, and one would expect that insects brought indoors and placed in more simplified environments would encounter less demanding conditions. Performance indoors, then, could result in less "specificity" of response compared to field behavior.

In-flight bioassays in wind are performed indoors using wind tunnels, or "sustained-flight" tunnels, and they have some definite advantages over outdoor flight bioassays involving field capture of insects. First and most important, the flight tunnel is a physical model of the environment, allowing the experimental manipulation of one variable at a time. Temperature, humidity, wind velocity, and chemical plume conditions can be reproduced day after day, and the experimenter does not encounter the daily variation in results common to field tests that must be factored out by replication and experimental design (see Cardé and Elkinton, Chapter 4). Cause-and-effect relationships from these variables can be gained more easily in a wind tunnel than in the field.

Although wind tunnels cannot duplicate the combination of wide plume dimensions and low concentration that occur in the field at great distances from a source, the use of low, as well as high, emission rates in a wind tunnel can shed light on the responses that occur there, at both ends of the active space (Baker et al., 1981; Linn and Roelofs, 1981). The prolonged (sustained) flight over a distance of many tens or hundreds of meters can be mimicked in the wind tunnel by rotating a visual floor pattern beneath the insects as they fly. The insects compensate for this higher velocity of optomotor stimulation by reducing their airspeed and hence can be made to fly for many minutes, even hours, at zero net up-tunnel ground speed while in the pheromone plume. This can increase discrimination among treatments (Miller and Roelofs, 1978a,b). Another key advantage of wind tunnels over field tests is that experiments can be performed throughout the year, and especially for pheromone identifications, much progress can be made under inclement conditions in preparation for the time when field tests can ultimately be performed.

Because flying while maintaining contact with pheromone while progressing upwind is such a highly integrated behavior and is sensitive to even tiny changes in thrust and attack angle, wind tunnels have come to be used for a wide variety of studies requiring detection of subtle behavioral changes. They are currently being used for pheromone identifications, blend quality testing, in the design of pheromone traps, in tests for habituation, for orientation studies, and most recently to detect sublethal, behavioral effects from insecticide intoxication (Linn and Roelofs, 1984; Haynes and Baker, unpublished).

Wind tunnels are not strictly an invention of researchers in the field of sex pheromones. Rather, their use grew out of studies on orientation responses of insects to visual feedback from the environment (Kennedy, 1940), and to host odors (Kellogg and Wright, 1962; Kellogg et al., 1962; Daykin, 1967). These early studies are well documented by Kennedy (1977a). With the increased interest in, and identification of, sex pheromones, they became used for investigating orientation mechanisms used by male moths flying to sex pheromone (Traynier, 1968; Farkas and Shorey, 1972; Kennedy and Marsh, 1974).

The scientific exchange of ideas between Farkas and Shorey (1972) and Kennedy and Marsh (1974) (also Shorey, 1973; Kennedy, 1977a; Farkas and Shorey, 1976) concerning the relative importance of anemotaxis and chemotaxis in the pheromone-mediated upwind flight of male moths was a direct result of experiments performed in wind tunnels, and these brought the technique to the attention of researchers in the field of pheromone identification and chemistry. Realizing its potential value in pheromone research, Miller and Roelofs (1978a) designed a sustained-flight tunnel and demonstrated its versatility and utility in discriminating among blends of incipient pheromone components.

The need for more discriminating bioassay techniques and for a more complete knowledge of the behavioral reactions to pheromones was also stressed at the same time in reviews by Kennedy (1977a,b, 1978). These papers set a standard for future studies of pheromone-mediated behaviors and occurred slightly after a dramatic increase had begun in the number of pheromone identifications involving multiple components (Roelofs and Cardé, 1977).

This chapter will discuss a number of pheromone-related studies involving wind tunnels and emphasizing design features and techniques that are necessary for achieving optimal sensitivity, discrimination, and experimental flexibility. We will illustrate the diversity of uses for wind tunnels in trying to dissect the behavioral responses that result in the displacement outcomes "attraction," "arrestment," and "repellency," and that contribute to the evolutionary outcome, reproductive isolation.

II. Wind Tunnel Design and Operation

We find it helpful to categorize wind tunnels by mode of air movement. As described in Chapter 2 on bioassays by Baker and Cardé, the most common form of wind tunnel in pheromone research involves horizontally moving air, and horizontal displacement using anemotaxis (steering with respect to wind direction) is possible. But just as it is possible for bioassay tubes and chambers to employ moving air that does not permit anemotaxis by moving the air perpendicular to the plane in which displacement is permitted, so too can wind tunnels measure movements of flying insects that are not steered anemotactically. Such a situation is found in vertical wind tunnels, in which air moves either straight up or down and the insect's horizontal displacement in or out of pheromone has no wind-induced drift component. Also, air movement can be stopped in a horizontal tunnel to observe pheromone-mediated movements lacking an anemotactic component. Finally in at least one tunnel, horizontal air flow has been superimposed on vertical air flow to create an elaborate assay for a pheromone's behavioral effects (Phelan and Miller, 1982).

2.1. Moving the Air

The two fundamental decisions to make regarding wind tunnels are these: how to get the air to move and how to keep the air in the room and in the tunnel clean and free of contamination. Contamination is bad because the insects waiting to be tested can slowly become habituated, or they may respond poorly because the air in the tunnel can become tainted with an antagonistic compound. The second decision is somewhat related to the first in that if you choose to pull the air through the tunnel rather than push it, you can accomplish air cleansing by having the exhaust fan (centrifugal blower) pull the entire volume of the tunnel's air out of the room, and usually, out of the building. This may be the easiest way to ensure that all the pheromone, especially from a tunnel permeated with a cloud of pheromone (see below), is exhausted from the room. If a chemical exhaust hood is present in the laboratory, usually it is easy to adapt it for air pulling by building boxes to fit it to the end of the tunnel or by running a tube from its duct to hook up to the end of the tunnel.

Other than for pheromone permeation tunnels, however, we do not recommend using a fan to pull the air. It is better to push it, for the simple reason that access to the inside of the tunnel can be gained through doors, holes, slits, etc., without worrying about disturbing the pheromone plume's position or structure. With titanium tetrachloride-generated smoke, one quickly sees that even with access doors to push-fan-type tunnels left open, actions such as sticking hands, head, arms, etc., into the tunnel do not disturb the plume, unless of course they are placed directly upwind of the smoke source. If the tunnel is housed in a small room, though, walking rapidly past the open doors can push the plume off course momentarily, and so it is always wise to operate the tunnel with the doors closed if the operator is expected to do much moving around while the insects are in flight.

With pulling-type tunnels, however, even without observer movement any crack, hole, pinhole leak, or especially any open tunnel door creates visible, severe perturbations of the (smoke) plume. Therefore these tunnels must always be operated completely sealed along the side, and this makes them less adaptable for photography and for in-flight experimental manipulations of all types. This is true even during the act of releasing an insect into a plume, because the turbulence caused by opening a door at the downwind end creates uncertainty as to where the plume has moved.

Our experience with designing and constructing wind tunnels has led us to formulate the "First Law of Wind Tunnels:" *Do not follow the advice of engineers.* In our view, engineers have been taught to be concerned about wind tunnel features that have no bearing on their use as an instrument to study insect flight. For one thing, they have traditionally used wind tunnels to study the behavior of objects in winds of very high velocities, usually 100–640 km/h. Under these velocities, they have learned to be concerned about events occurring near the walls of the tunnel and at the intake end, such as friction, shearing, and turbulence. Entomologists are interested in the center of the tunnel, near which the insects fly, and the ultralow velocities needed for such flights are not only quite foreign to engineers, but also they make many of the wall–wind interactions insignificant.

Engineers nearly always will recommend building a pull-type tunnel, because this is what they are used to using, and for their studies at high velocities this type has been most reliable. We are aware of three out of three cases in which entomologists have first consulted engineers about constructing a tunnel and have been advised to pull, rather than push, air. The two who followed the advice have had many problems and eventually changed over to push-type tunnels. In the third case, R. T. Cardé and T. C. Baker at Michigan State University went ahead and constructed a push-type from the start, and were extremely happy with it.

A smoke source of some type is essential for making sure of the placement of pheromone sources and insect release cages, demarcating time-averaged plume boundaries, checking for efficacy of wind-stopping procedures, and for visualizing the movement of air throughout all areas of the tunnel. Smoke plume movement is also the easiest and most accurate way to measure the wind velocity, as pointed out by Kellogg and Wright (1962). The observer only needs to time with a stopwatch the passage of a smoke filament down a known length of the tunnel. J. S. Kennedy and his colleagues (personal communication) light up a few cigarettes and place them in a stoppered flask, waiting a bit for the smoke to build up, and then turn on a gentle flow of air through the flask. The smoke flows out of the flask through flexible tubing whose tip can be positioned easily anywhere in the tunnel.

Another way of generating smoke is by using an airstream laden with the combined vapors from concentrated HCl and ammonium hydroxide. Kellogg and Wright (1962) found that ammonium acetate smoke, generated by mixing the vapors from NH_4OH and acetic acid, was the smoke least irritating to flying insects that they tested, and so they could use it combined with a host attractant to visualize the passage of flying insects through and up the plume.

Titanium tetrachloride-generated smoke, made by placing a small quantity of this material straight out of the bottle onto a cotton wick or rubber septum, is mildly irritating to flying moths. We have been able to combine the smoke with a pheromone plume, and oriental fruit moths will fly in apparently normal fashion upwind in the plume until they reach a critical point near the source and drop to the ground, apparently dead. At least they do not move any more. We find this white smoke to be the thickest and easiest to produce. Very thick black smoke can be produced by burning acetylene from a tank, but this produces much heat and seems best to be used outdoors (Von Keyserlingk, 1983).

There are many ways to generate wind to push air down a tunnel, but perhaps the simplest way is with a simple, rotary-blade fan. Whether the tunnel is round or rectangular, air can be easily conducted from a fan to the tunnel by means of a duct constructed from thick, flexible plastic sheeting, such as plastic bag material, taped together and sealed to form a tube. Usually the duct must expand from the smaller fan's size to the larger tunnel's opening.

The duct of R. T. Cardé's tunnel at Michigan State (Fig. 1) was made of Plexiglas and expanded from a smaller, round fan opening to the larger rectangular opening of the tunnel. At the tunnel's opening, some sort of mixing chamber is needed to dampen the turbulence created by the fan's blades and to balance the wind velocities throughout the tunnel's middle, sides, top and bottom. The chamber, usually a box shaped like the tunnel, creates resistance by means of several layers of narrow-mesh cloth or screening, or both.

Miller and Roelofs' tunnel used several layers of cheesecloth stretched tightly across the opening and separated from each other by several centimeters or more. A final layer of silk was the last layer the air passed through as it entered the tunnel. The mixing chamber of one of the author's (TCB) wind tunnels is a box 125 X 110 X 31 cm made of plywood, and inside there are three particle board panels that slide into the box from the top along grooved tracks. Although the chamber and the panels are rectangular and the tunnel is U shaped, made from bowed Plexiglas (Miller and Roelofs, 1978a), the panels have a U-shaped section carved out of their center so air can pass through them in that configuration. Both a single layer of cheesecloth and an aluminum window screening (to support the cloth) are stretched and stapled across each panel's U-shaped opening. The panels are easily slid out and removed when the cloth and screens need cleaning to remove moth scales and dirt (see Section 2.3 below). A final layer of brushed nylon is stretched across the box where air exits the box and enters the tunnel.

The mixing chambers have usually been described as essential so that laminar flow can be created. We do not think that laminar flow is necessary to get successful insect flights to a pheromone source. Indeed air that is too smooth has been cited at least once as a hindrance to successful orientation (Marsh et al., 1978), and special "spoilers" had to be installed to create a little turbulence and

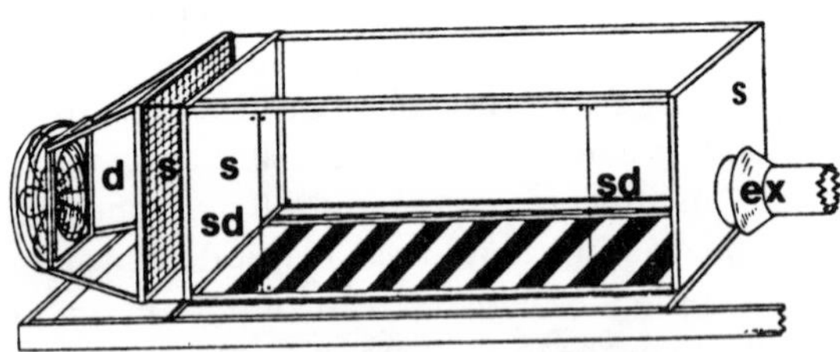

Figure 1. Rectangular, Plexiglas wind tunnel with push-type fan. Plexiglas duct (d) expands from circular fans' size to larger, rectangular tunnel's size, and conducts air to screens (s) which smooth the airflow. Access to tunnel is gained by two sliding doors (sd), and pheromone plume is scavenged from tunnel by exhaust tube (ex). Insects fly from right to left up a pheromone plume, and moving striped cloth floor pattern located beneath the Plexiglas floor can be used to sustain flight (from Cardé and Hagaman, 1979).

make the plume wider and choppier. Notwithstanding the fact that laminar flow is not essential, we think that it is desirable for most pheromone studies because it creates a simplified, moving air environment.

As experimental biologists, we prefer fewer variables to worry about during analysis and interpretation of results. Of course, during studies of in-flight orientation mechanisms, air blowing predictably in one direction at a known velocity and laminar structure is essential in interpreting steering in wind, but for most studies where tunnels are used, as when comparing different blends and dosages of pheromone, getting the flow to be laminar is of less concern. It is more important to be able to pinpoint where the plume is so that every insect can be placed in it without a doubt that it is in fact receiving optimal exposure to the pheromone plume. We have both had the experience of getting suboptimal responses from male moths and then upon checking the plume's location with smoke (which we should have done beforehand), found that the release-cage platform had been moved slightly so that the moths were not really in the plume's center. We cannot stress this enough: consistency of the plume's position and that of the release station for insects are of the utmost importance regardless of the tunnel design that is being used.

2.2. Cleaning the Air

Given that a push-type fan is the best way to move air and that it allows the most freedom in experimentation, how best to clean the air? If it is not likely that you will ever be able to install an exhaust blower to remove air from the room in which your wind tunnel is housed, the air can be scrubbed by means of a bed of activated charcoal. This technique was demonstrated by Kellogg and Wright (1962) in their tunnel (Fig. 2), and later used with much success by J. S. Kennedy and his colleagues at Imperial College, Silwood Park, England, for many years. The advantages are that the tunnel can be aligned anywhere in the room without having the worry about fighting exhaust tubes and hoods. Probably more importantly, the entire room's air is recirculated and therefore it is relatively easy to keep the temperature and humidity constant.

In wind tunnel rooms having exhaust tubes and hoods, large volumes of air are continually being lost, and efforts to heat or moisturize the air often are made futile. If the climate-controlling capabilities are great, the room is large, and the volume of exhausted air is relatively small, then the problem of climate control versus exhausting of pheromone is minimized, as in the case of Wendell Roelofs' tunnel at Geneva, New York.

Disadvantages to the bed of activated charcoal are that it can have high resistance to wind flow, and it may be difficult to force enough air through the bed to create wind velocities other than the very lowest. Kellogg and Wright's charcoal bed was 5 cm thick but of unspecified width, height, and weight. Because it was housed in the inlet tube to the tunnel, it was likely no more than 30 cm in diameter, which would have reduced the weight considerably. The

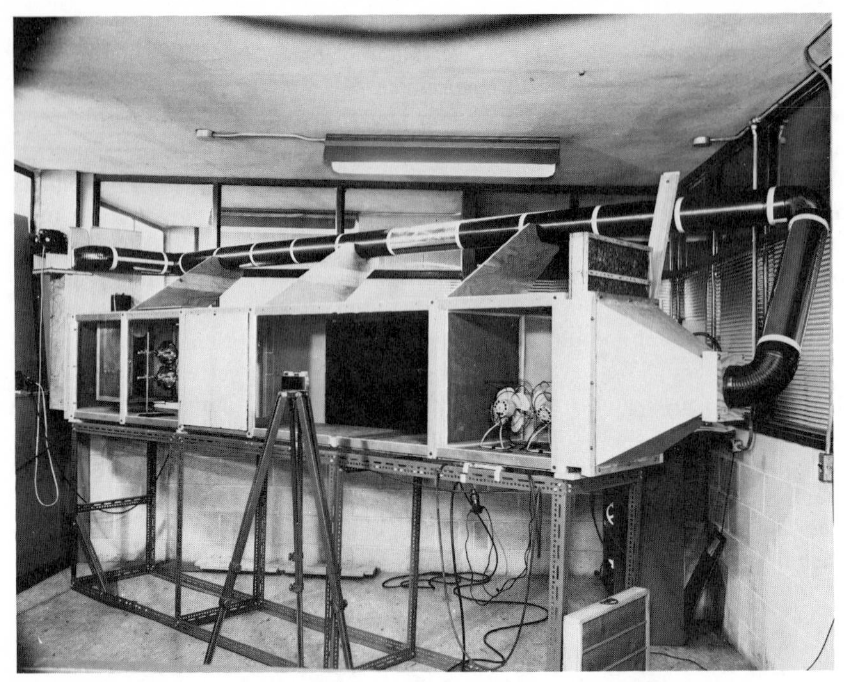

Figure 2. Kellogg and Wright's wind tunnel, which used a 5 cm-thick bed of activated charcoal at the inlet end (right) to cleanse the air. Air was drawn through the tunnel by a centrifugal fan at downwind end (left), and air could be recirculated into the tunnel's inlet end, if necessary, by the return pipe shown on top of the tunnel. Tunnel is made of plywood, with front and rear walls made of glass. Access to tunnel's interior was through either the top or the bottom of the center, working section. Stereo cameras are positioned horizontally for photographing tracks of flying insects against a black background (from Kellogg and Wright, 1962).

charcoal bed described by Kennedy and Marsh (1974) and Marsh et al. (1978) was contained between two 2.4 × 1.2 × 0.45 m vertical screens and was 0.5 cm thick. One problem with a vertical bed of charcoal such as this is that the material tends to settle toward the bottom and create a higher density (and more wind resistance) at the bottom of the bed compared to the top (J. S. Kennedy and C. T. David, personal communication).

A later modification reduced the charcoal bed's thickness to facilitate higher wind velocities, and positioned the bed horizontally, not vertically, nearer the fan. The horizontal placement solved the charcoal settling problem, and even this relatively thin layer of charcoal has been sufficient to adsorb all the gypsy moth pheromone that would have recirculated into the tunnel over several years. There has been no apparent sign of "breakthrough" of pheromone due to saturation of the active sites in the bed (J. S. Kennedy and C. T. David, personal communication).

When plumes are the usual mode of using pheromone, the best way to remove the pheromone using an exhaust system is the technique described by Miller and Roelofs (1978a). A wide exhaust tube connected to the outside of the building, ca. 20-40-cm diameter with an exhaust flow much faster than the tunnel's wind speed, is placed just outside the end of the tunnel at the plume's height (Fig. 1). The pheromone is effectively scavenged and the rest of the tunnel's (pheromone-free) air is allowed to recirculate without filtering. One advantage to removing only the plume-air from the room is that wind-induced pressure changes at the exhaust outlet outside the building will affect only the rate at which pheromone is scavenged, not the tunnel's wind velocity. We know of at least two cases where exhaust fans pulling wind through the tunnel have been affected significantly by outdoor wind, so much so that in one case the wind flow in the tunnel would reverse direction every so often!

2.3. Contamination of Surfaces

Another type of contamination concerns surfaces within the tunnel. All surfaces that may possibly come in contact with the pheromone source or the plume should be of a material able to withstand rinsing with a solvent. Glass and metal are best, but a tunnel made of glass is extremely heavy and fragile. Most types of Plexiglas (polycarbonate, acetate, etc.) can withstand rinsing with ethanol; acetone is usually not worth the risk, although it is the solvent of choice for rinsing all metal surfaces such as release cages for insects, the platform on which the source rests, etc.

Most of the working wind tunnels in pheromone research are constructed of Plexiglas and metal. The table on which our wind tunnel rests and which is not likely to ever be exposed to much pheromone, is made of wood. For studies using point source plumes, pheromonal adsorption onto the sides of the tunnel can be minimized by positioning the plume so that even with lateral and vertical spreading, it does not touch the sides. Special care should be taken if an end-of-tunnel screen is used, as described by Miller and Roelofs (1978a), to keep moths from exiting the back of the tunnel. This arrangement, which employs a hole in the end screen at plume height into which the release cages for moths are inserted, has a special problem in that the area of end screen in the plume can become contaminated from previous plumes (J. R. Miller, personal communication). Thus, even if the release cages are cleaned with solvent after each use, anomalous results can occur if the screening immediately surrounding the release cage is not also rinsed whenever a new treatment is tested.

The surface most likely to be contaminated is the one on which the pheromone source rests. Certainly the rubber septum, filter paper, glass rod, etc., can be suspended by a thread (see Baker and Cardé, Chapter 2, for types of pheromone sources). However, most often it is easiest to place it on some type of platform which will need rinsing after every change of treatment.

We have found that a simple stand with a sheet-metal base and four sheet-metal legs oriented with their edges parallel to the wind line is a very workable setup for quickly changing pheromone sources. The stand creates minimal turbulence because of the edge-on orientation to the wind, and the plume is thus minimally disturbed by mechanical turbulence. A clean sheet-metal plate is placed on this stand whenever a new source is to be tested and the base and legs of the stand remain in the tunnel throughout the experiment. This plate becomes the table top of the platform when in place, there being just four legs sticking into the air when the plates are absent. The entire stand is rinsed with acetone after each experiment, and the plates are rinsed before each use. If a permanent table top to the stand is left in place, even if clean plates are placed on top of it the bottom plate can become contaminated, presumably due to slight turbulence at the airspace between the plates, and can affect responses to subsequent treatments.

One final source of contamination should be mentioned, but it is not clear whether it is pheromonal or not. Over time, enough escaped males and their scales enter the mixing box's intake, and scales and dirt can build up on the layers of cloth stretched across the mixing chamber. In several laboratories, we have experienced and heard reports of poor flights correlated with scale-laden screens and cloths. When the screens are washed with acetone and the cloths are cleaned or replaced, the responses improve significantly, usually back up to their normal levels.

In these instances airflow as checked by smoke plumes appeared to be normally laminar but possibly the velocity was slightly reduced. The scales and dirt may possibly interfere chemically with optimal behavioral responses, although this is pure conjecture. When poor responses occur day after day, we suggest checking the screens for scales and cleaning them for good luck.

2.4. Movable Visual Patterns

To have stands of any type resting on the tunnel's floor would seem to prohibit moving the floor. It may be important for some studies to have the ability to rotate a ground pattern beneath a flying insect to make use of the optomotor compensatory responses they perform. Long-duration flights of zero net up-tunnel ground speed can then be executed for increased discrimination among treatments. We use a movable floor pattern stationed just below a clear Plexiglas floor, and the insects can see the pattern because they respond nicely to it by changing their airspeeds when it moves beneath them. Depending on the species, lighting, and the height above the floor that the plume is stationed and at which moths are forced to fly, the moving floor pattern will have varying degrees of effect on flying insects.

For instance, Sanders et al. (1981) found that their spruce budworm males responded better to the striped pattern's motion after it was mounted on the ceiling of the tunnel and the glass floor was given a black backing, making it into a mirror to reflect the pattern upward. Earlier they had suspected that the stripes, when mounted below the glass floor, were being obscured by the fluorescent lights' reflection from the floor, because it was often difficult to slow males' progress by moving the pattern. Kennedy and Marsh (1974) stationed their plume ca. 15 cm above the floor pattern to obtain good optomotor responses to its movement. Kuenen and Baker (1982a) found that for *Grapholita molesta* and *Heliothis virescens* males, increasing the height of the plume to 40 cm or more above the floor pattern eliminated detectable optomotor response to the floor's movement. Above 30 cm the males seemed to be increasingly watching other cues in the wind tunnel room.

Interestingly, Farkas and Shorey (1972) might not have discarded optomotor anemotaxis and embraced chemotaxis alone as an orientation mechanism in wind had they stationed their pink bollworm pheromone plume closer to the floor. They reported in a footnote that they had tried rotating the floor beneath the males, and because there was no observable effect on the males' flight speeds they concluded optomotor anemotaxis was not being used. After inspecting their 0.6 × 0.6 × 1.5 m tunnel, we conclude that it is likely that they had stationed their plume 30 cm or so above the rotating pattern. The males were using visual cues in their optomotor reactions to wind-induced drift, but unknown to Farkas and Shorey, at that height, the cues must have been predominantly from the stationary objects in the room.

Because Kennedy and Marsh (1974) used a treadmill-type floor pattern with transverse (cross-tunnel) stripes, most others that have been used since then have also been striped treadmills, e.g., Miller and Roelofs (1978a) who used alternating black and orange 10-cm-wide transverse stripes. R. T. Cardé used alternating green and white stripes (Fig. 1) (Michigan State U.'s colors) and all species of insects tested, even a fungus gnat *Bradysia impatiens* (Alberts et al., 1981) and Oriental fruit moths originating from Cornell, responded nicely to its movement.

There are some problems with moving treadmill-type cloth patterns (see below), and the first patterns used successfully were projected onto the floor (Fig. 3) (Kennedy, 1940; Kellogg and Wright, 1962). These clearly evoked optomotor compensation in both free-flying insects and those flying in odor, and in the future perhaps more use will be made of this method of moving a pattern beneath, above, or beside insects.

Kennedy (1940) used a slatted cylinder rotating around a light (Fig. 3) and Kellogg and Wright (1962) used a continuous loop of film through which light was transmitted. One advantage to projected patterns is that very small changes in velocity or direction of the pattern's flow result in large changes on the tunnel's floor. It is therefore much easier to play with these variables to observe movement changes in flying insects than for cloth treadmills. This is especially true if the moving pattern needs to be rotated to flow at some oblique angle with respect to the wind.

One disadvantage of such patterns would seem to be that for diurnal species it would be difficult to get the pattern bright enough to compete with the room lights that are at daylight intensities. For nocturnal insects, too, the problem of

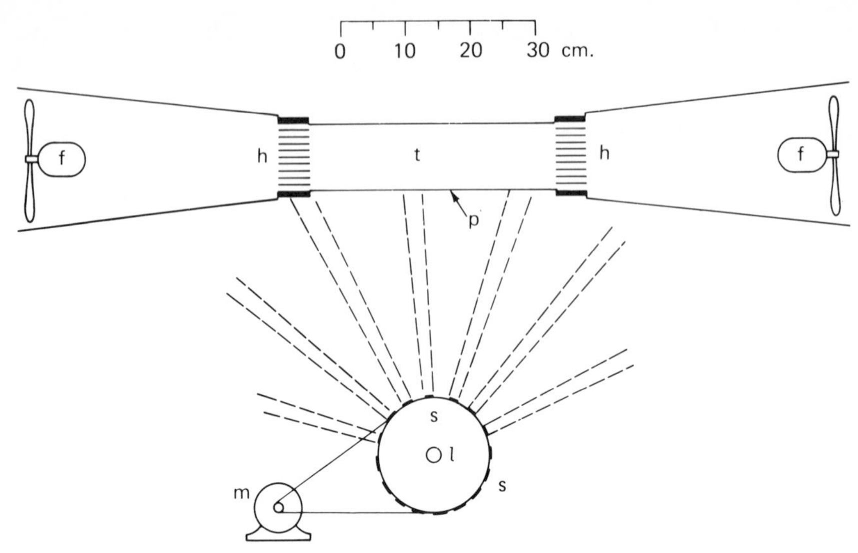

Figure 3. Kennedy's (1940) method for projecting moving, visual striped pattern onto the floor of his wind tunnel, 38 cm long, 7.5 cm wide, and 10 cm high. Strips of metal (ss) driven by motor (m) rotate around light bulb (l) and project a pattern onto waxed paper sheet (p) on the floor of the tunnel. Fan (f) can draw air in either direction through honeycombs (h) which smooth the air.

balancing the intensities, especially for floor-projected patterns, would seem to be critical. If the light below is too bright, the insects' dorsal light response can cause them to end up diving toward the floor rather than flying up-tunnel (K. F. Haynes, personal communication).

Regardless of whether the pattern is projected or not, recent findings indicate it might be safer not to use transverse stripes, but instead use spots or some other pattern lacking long, straight lines. This caution would seem to be applicable mainly for orientation studies in which track and course angles are being measured. J. S. Kennedy, whose tunnel originally had transverse stripes on the floor, recently switched to a pattern of solid pink circles on a yellow background (Kennedy et al., 1980, 1981).

Charles David, also at Silwood Park, recently demonstrated that stripes can have some unexpected effects on the orientation of flying insects, in this case free-flying *Drosophila* (David, 1982a). He recommended that for orientation studies, spotted instead of striped patterns be used because of these findings and because long straight lines are seldom found in nature.

The colors of the patterns should be chosen carefully, especially if photography of the insects' flight tracks is to be performed. In our tunnel (TCB's), we started with black and white 10-cm-wide transverse stripes, but quickly found that in video recordings, the oriental fruit moths disappeared whenever they were over black. Cardé and Hagaman (personal communication) had the same problem, and they analyzed only the one-half of the data where their gypsy moth males were over the white stripes, not the green (Cardé and Hagaman, 1979). To tackle our problem, we then softened the contrast by placing a layer of cheesecloth on the Plexiglas over the floor, so that now the moths could be seen over the grayish stripes. They continued to respond to the stripes' movement (Baker and Kuenen, 1982).

To make it even easier to track our moths in frame-by-frame playback and to remove possible bias caused by stripes, we have recently converted the pattern to 10-cm-diameter red circles on a white background. The red pattern influences the males' movements because the red should appear fairly dark to them against the white background (insects in general are not as sensitive to long wavelengths as to short ones).

On our black and white video recordings we obtain a nearly uniform white background because we place a red filter over the lens. A similar effect can be gained by illuminating the tunnel with red lights. Making spots of any color appear white on black and white video tape is possible by using a filter or room lighting of the spots' color.

One of the problems with cloth-treadmill patterns is that they frequently come off-center, sag, or wrinkle. It is best to make at least one of the two roller-axels capable of being adjusted backward or forward in tiny increments so that the belt can be realigned when necessary. It is extremely frustrating to have to stop repeatedly in the middle of an experiment and yank the tightly seated cloth or canvas back into position to keep it from slipping off the side of the roller or binding and wrinkling at the sides. If you are lucky, a good axel alignment can last many months and many revolutions of the pattern. For tautness, J. S. Kennedy and colleagues use a third spring-loaded roller below one of the end rollers that presses on the lower fabric and takes up the slack that will always occur there.

The belts can be driven in a variety of ways, but it is always best to have a motor or drive system capable of a continuum of speeds without the motor burning up. We (TCB) use a sewing machine motor which was built for variable speeds, whereas Miller and Roelofs scavenged a motor from a counter-current distillation apparatus that had rarely been used. Their variable speeds resulted from the use of a worm-gear-clutch arrangement controlled by a joy stick. The motor runs at a constant speed, and another nice feature of their pattern (and J. S. Kennedy's and R. T. Cardé's) is that its direction can be reversed at the flip of a switch.

2.5. Creating Pheromone "Clouds"

For some studies, such as those in which the researcher wants to make sure that the insect is in continual contact with pheromone, point source-generated plumes are not adequate. Rather, the pheromone needs to be diffused into a nearly homogeneous cloud by creating turbulence around the pheromone source before the pheromone-air mixture is made laminar by the layers of fine-mesh screening. Traynier (1968) was the first to attempt this with a sex pheromone as the odor, although in vertical wind tunnels Daykin and Kellogg (1965), Daykin (1967), and Daykin et al. (1965) generated air uniformly permeated with water vapor, CO_2, or repellents. Two separate columns of permeated air could be formed, one in each half of the tunnel, and the tracks of insects between the halves of the "choice-chamber" were recorded and analyzed for the nonanemotactically guided horizontal movements modulated by the odors.

Traynier (1968) first used pheromone-permeated air in the horizontal tunnel of Kellogg and Wright (1962) (Fig. 2). He placed the abdominal tips of 30 *Anaghasta kuhniella* females into the air-inlet tube of the tunnel and then recirculated the air. The upwind tracks of males flying in air permeated with pheromone certainly appeared different from those flying to a point source, but they were not quantified, smoke was not used to visualize the cloud, and it is not clear how uniform the air really was.

This simple experiment, though, stimulated much thought on the anemotactic–anemomenotactic model of orientation to a pheromone source (Kennedy et al., 1980, 1981; Kuenen and Baker, 1983; Kennedy, 1983). Kennedy et al. (1980, 1981) created a cloud of pheromone uniformly permeating the air by means of a grid of turbulence-producing strips of masking tape immediately downwind of a grid of pheromone sources, all affixed to a brushed nylon screen. The strips of tape were 2.5 cm wide and applied to the upwind side of the screen at 6.0-cm intervals vertically and horizontally, leaving seven rows of 20 squares of uncovered nylon screen, each 3.5 X 3.5 cm. Every alternate square was then covered with tape, leaving a total of 70 unoccluded squares. The grid of polyvinyl chloride rod pheromone sources was formed by attaching the 0.4 X 1.0 cm rods to the center of each of the 70 occluded tape squares on the upwind side. A second, "smoothing" screen was located 50 cm further downwind, effectively creating a mixing section for the cloud before the flow was smoothed before entering the tunnel's working section (Fig. 4).

The ability of this system to permeate the air uniformly with pheromone was checked with smoke. Side "corridors" of air uniformly permeated with pheromone were neatly created by placing only a partial grid of pheromone sources at one side of the tunnel. Smoke source visualization confirmed that there would have been a sharp edge between the clean air and the pheromone cloud. Although Kennedy et al. used their clouds to investigate orientation movements, Sanders (1982) used a nearly identical system of creating clouds to examine habituation and background odor "noise" effects on males' ability to locate successfully a point source.

Further orientation studies in uniformly permeated air were performed by Willis and Baker (1984) using a system of turbulence strips again similar to Kennedy et al.'s (1980, 1981). Their system differed slightly, however, in that only vertically oriented 4.5-cm-wide sheet-metal strips separated from each other by 0.5 cm were used, and the pheromone sources were located downwind of the strips. Experimentation with cardboard prototypes revealed that the distance between the grid of rubber septum pheromone sources and the strips

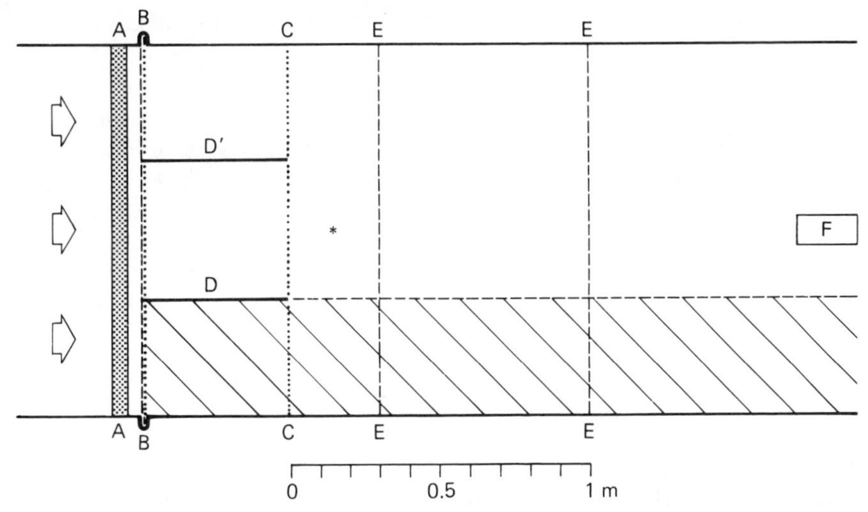

Figure 4. Wind tunnel used by Kennedy et al. (1980, 1981) to create uniform, permeated clouds of pheromone to study flight tracks of *Adoxophyes orana* males (this tunnel was same one used by Marsh et al., 1978). In this depiction, a side corridor of pheromone (hatched lines) was created by lowering a "spoiler" grid of tape affixed to a nylon screen into the airstream with pheromone pellets attached to its upwind side only along the one section in which the corridor was to appear. A, charcoal filter; B, brushed nylon fabric screen carrying the "spoiler" grid of tape and pheromone-containing PVC pellets; C, plain brushed nylon smoothing screen; D, partition preventing cross-tunnel mixing of the pheromone-bearing air (D′ serving the same purpose when the pheromone was released on the other side of the tunnel); E, video-recording zone; F, release cage for moths.

was critical, with the strips and grid of sources needing to be securely fastened in place with a separation of 1.5 cm. Another feature which aided in homogenizing the pheromone and air was two layers of fine-mesh brass screening located farther downwind in the sheet-metal mixing box just before the pheromone entered the Plexiglas working area of the tunnel. A side corridor of pheromone was created by only partially filling the grid with pheromone sources, plus a sheet-metal partition in the mixing box kept the clean air and pheromone from mixing before reaching the tunnel's working section.

Another way of premeating the air with pheromone is to coat a fabric screen uniformly with pheromone solution. There is at least one drawback to this method, though. Over the course of an experiment, the concentration of the cloud is likely to drop very rapidly due to the rapid loss of pheromone from the screen. Therefore the movements of insects even an hour after the start of observations are likely to be caused by an unknown, lower concentration than that used at the start. Also, it will be very difficult to be sure of the initial concentration actually adsorbed onto an area of screening. If the screen is dunked into the solution, the loading rate may depend on the duration of the dunking.

This way of applying the pheromone will also require a large volume of solution, and is likely to be quite messy with regard to contamination of the person doing the dunking and of the room. If the solution is sprayed through an atomizer or other such device, the pattern of application must be quite carefully performed to prevent overlapping layers of higher concentration strips, and again the initial amount actually absorbed by the fabric's fibers compared to the quantity passing through the holes will vary from application to application. On the positive side, this method probably is the one most likely to create the most uniform cloud of pheromone.

2.6. Stopping the Wind

For some experiments, usually for studies of orientation, the researcher may need to stop the wind to create various chemical stimulus situations in which the insect must move without the benefit of wind drift. Over the years, several methods have arisen for stopping wind. Kennedy (1940) used a system of louvers at the downwind and upwind ends of the tunnel which were snapped shut quickly. Farkas and Shorey (1972) used two solid panels which were abruptly slid into place along grooved tracks simultaneously at the upwind and downwind ends of the tunnel. Smoke visualization confirmed that the wind stopped at the instant the panels closed, and only lateral spreading and gradual dissipation of the plume's filaments were observed without significant up- or downwind displacement.

Baker and Kuenen (1982) used a leather cinch attached to the axel of their rotary-blade fan to stop the rotation of the blades after the power had been switched off. However, it took over 2 sec for the blades to stop rotating and the wind (as visualized by smoke) to stop. A cardboard cover was placed over the exhaust tube simultaneously with turning off the fan in order to keep a vortex from rolling up the tunnel and disturbing the plume.

Baker et al. (1983) later improved on this technique by using two window shades, one at each end of the tunnel, which on cue were lowered simultaneously. The one at the upwind end was located inside the mixing chamber, and its cord for lowering it exited the chamber's bottom through a small hole. It was essential that it be pulled down while the fan was still on, because the air pushed the shade against the downwind end of the chamber, sealing it against the sides and completely blocking air flow. A piece of weather stripping was also essential to seal the bottom of the shade against the bottom of the box. The air was now diverted out through the chamber's open top and into the room housing the wind tunnel. The shade at the downwind end was necessary to keep slight air movements of all types from entering the tunnel and disturbing the plume.

Smoke was extensively used outside the tunnel to find out where air movements that might cause within-tunnel disturbances were originating, and this was how it was discovered that the down-tunnel shade was needed. Smoke was also used to check for the complete cessation of up- or down-tunnel displacement of the plume within the tunnel, which in actuality could be limited to less than 0.5 cm/sec. Smoke also revealed another important feature of wind stoppage. If the tunnel's doors were not completely sealed, even just a tiny crack would result in a puff of turbulence at the moment the shades were pulled down, causing the plume to snake sideways and dissipate. The plume remained completely intact when the doors were closed during stoppage (Baker et al., 1983).

Kennedy, David, and Ludlow (personal communication) found another simple way to stop the wind suddenly. They created a small door on the duct leading from the fan to the bed of activated charcoal before the entrance to the tunnel. When the door was abruptly opened by the operator, who pulled on a long cord from his position near the observation area several meters away, the air now dumped out into the room. The wind in the tunnel stopped immediately, and the fan was then switched off. One key feature which allowed this technique to work was the relatively high resistance to airflow by the charcoal, and the very low resistance through the open door.

2.7. Recording and Analyzing the Data

As discussed in the previous chapter on bioassays by Baker and Cardé, data can be recorded in wind tunnels by a variety of methods, depending on the objectives of the experiment and the speed with which the data need to be analyzed, conclusions drawn, and new experiments designed. If only the presence or absence of a behavior needs to be scored, a pencil and paper often are the only required tools. For testing responses to different blends of components, we often have a data sheet composed of a checklist in our wind tunnel, where, after each moth is tested, the observer checks off the behaviors performed by the moth and notes the closest approach to the source.

If temporal features are of interest, such as latencies and duration of response, then audio recording of the events as described by the observer will likely be most useful. A multichannel strip chart event recorder can also be useful here for gaining an immediate "hard copy" of the timing and sequence of behaviors (Baker et al., 1981). The recent availability and low cost of microcomputers have made these more appealing as event-recording tools. Different keys can code for different behaviors, and the advantage is that the data are immediately stored on disc for further analysis by programs written for the needs of the experiment. These may be increasingly used for wind tunnel studies and other behavioral experiments.

Studies of the orientation of flying insects, including a fine-grained dissection of the movements, requires more complex instruments, such as multi-image, still-frame photographic techniques, movie cameras, or video-recording devices. Video recording is probably the most popular device at present and for good reasons. A video recording gives immediate feedback to the researcher as to whether the images that were recorded are of good enough quality to be analyzed.

Films require much more time to develop. Video tapes can also be reused, and much expense is saved over a long period of time. Finally, there are now many low-light-level-sensitive video cameras available that make recording under moonlight levels or lower very easy, without requiring any special alterations. In contrast, film photography under low-light intensities requires expensive, fast lenses and special films.

One of the disadvantages to video has been, until recently, that motion resolution has been low compared to that of films. One video scan occurs in 1/60 sec (1/50 sec in Europe) and thus a fast-moving flying insect in still-frame will often appear as a faint streak, or worse, become invisible, over that span of time. For slower-moving species, the ordinary video camera has proven to be adequate for single-frame tracking of the insect's motion.

Baker and Kuenen (1982) found, however, that a special rotary-shutter camera combined with a special video disc playback device improves motion resolution significantly. They used a Sony RSC 1050 camera that comes equipped with a high-sensitivity Newvicon tube to record moths' tracks onto ordinary cassette tape. The internal, rotating shutter provided 1/500 sec "snap shots" of the moth on the tape, and then when rerecorded onto the Sony SVM 1010 motion analyzer's video disc and replayed, discrete point images of the moths, not streaks, were visible in still-frame. This system is now used routinely in TCB's lab.

The Newvicon tube is advisable because much light is lost when the shutter is rotating and the more sensitive tube allows low levels of light to be used even with the shutter rotating (another model, RSC 1010, equipped with a less sensitive tube is also available). If slower insects are used, one advantage to the Newvicon-equipped camera is that the shutter can be turned off and ultralow levels of light used successfully. This type of tube is very sensitive to the near-infrared, and recordings can be made under (visible) light levels that are too low for human eyes if even an incandescent bulb covered with an I.R. filter is used. By viewing the monitor, the system can double as a night-vision device for watching live behavior. We see no need to use color cameras for motion recording and analysis; the wide array of black and white high-sensitivity cameras currently available gives them the best flexibility for all sorts of conditions under which insects may be responding to sex pheromone.

Of course, the positioning of the camera(s) in the wind tunnel depends on the experiment being performed. We (TCB) place ours right on top of the tunnel on a tripod, at the back of the tunnel looking up the plume toward the source (Von Keyserlingk, 1983), or from the side, depending on the experiment. J. S. Kennedy and his colleagues and R. T. Cardé's group have successfully used a large mirror mounted on top of the tunnel at an angle such that the video camera can obtain a straight-down image of the moth while shooting from the side while mounted on the wind tunnel room's floor, or aimed horizontally at the mirror while fixed to the tunnel. The reversed images obtained by this technique need correcting only if the absolute, not relative, directions of movement are of importance in the study.

Analysis of the recorded tracks is another matter entirely. The minimal requirement for orientation studies is a video deck that is capable of slow-motion and still-frame playback of the recordings. The tracks may be traced by hand from the monitor onto clear acetate sheets for a hard copy, and then analyzed by hand for velocity and angular features of the movement. But clearly this is an extremely slow and tedious method, prohibitively so for large numbers of tracks and experiments.

The advent of microcomputers and X, Y digitizing and plotting devices has made the job of analyzing such tracks not quite as hopeless as it once was. We (TCB) increased our analysis speed and accuracy by taking our acetate-sheet tracings and digitizing them onto a microcomputer by means of a HiPad digitizer pad (Houston Instruments). The coordinates were stored on disc on a Radio Shack TRS-80 microcomputer and analyzed for turning and velocity features by means of a program written especially for our purposes by a number of graduate students and postdoctorates.

We next eliminated the time-consuming step of tracing onto acetate by using a T-bar type of digitizer from Radio Shack whose cursor can be maneuvered on the video screen's surface. As the operator advances the video image frame by frame, he also moves the cursor over the images on the screen and enters the moths' coordinates onto the computer disc with each advance. A hard copy of the track is immediately obtained and examined by having the track plotted by a Radio Shack flatbed plotter. We believe that it is important to have an intimate knowledge of each recording, as gained by the experimenter physically digitizing the moths' positions and entering the coordinates.

Instruments will soon be available that can electronically track and digitize a moving image, and this may be fine for future studies where we already know much about the orientation of flying insects. But for right now, we feel that the time spent over at least one frame-by-frame playback of a track is extremely valuable in producing new ideas and questions about orientation. On the other hand, too much time can be spent, as in hand measuring and hand tracing of tracks. Some degree of computer assistance is definitely needed for this type of research.

2.8. Optimizing Stimulus and Response Variables

We will not spend time in this chapter outlining the techniques for optimizing the responsiveness (internal state) of the insects to be tested, or with matters concerning the reliable, consistent emission of pheromone from the source. In these regards, wind tunnels are just like any other type of bioassay, and these techniques are covered in the previous chapter by Baker and Cardé. Suffice it to say, that because wind tunnels are more discriminating, sensitive types of assays, any inconsistencies in the performance of such techniques would be more likely to result in inconsistencies in the flight responses in a wind tunnel than in other types of assays.

A number of studies have shown, though, that apart from the usual isolation of males from females and optimization of diel responsiveness, best results were achieved when males were acclimated to flight tunnel conditions for at least 15-30 min before testing (Linn and Gaston, 1981; Turgeon and Linn, unpublished). In the case of the nocturnally active cabbage looper moth, acclimation to the tunnel's light intensity was most important. In the absence of acclimation moths seldom flew more than 30 cm from the release point. In addition, "spontaneous" flight activity inside the cages was very high; many of these males could not be tested (Linn and Gaston, 1981). Thus for some insects proper acclimation and careful handling to avoid mechanical and auditory disturbance may be critical to obtaining optimum, consistent results in the wind tunnel.

III. Uses of Wind Tunnels in Pheromone Research

3.1. Pheromone Component Isolation and Identification

Horizontal wind tunnels have proven to be very useful for sex pheromone identifications, either at the stage where gas–liquid chromatography fractions are recombined, or when synthetic compounds purported to be components need to be validated for their ability to attract flying insects all the way to the source. Usually the latter is needed either because field trapping is impossible due to the time of year, or evaluation of the mixture's activity is needed before involving a lot of people in a large, expensive field test.

Hill and Roelofs (1981) identified three chemical components from the salt marsh caterpillar moth, *Estigmene acrea,* using Miller and Roelofs' horizontal flight tunnel to make sure complete activity was present in the recombined fractions and synthetic components ((*Z,Z*)-9,12-octadecadienal, (*Z,Z,Z*)-9,12,15-octadecatrienal, and (*Z,Z*)-3,6-*cis*-9,10-epoxyheneicosadiene. Males fanned their wings and began walking to the epoxide alone, but successful upwind flight to the source 1.5 m away occurred only to binary or tertiary combinations, not to any component presented alone. At the filter paper source, males hovered for long periods, and extended their claspers while wing fanning on the paper. More examples of pheromone identifications which utilized horizontal wind tunnels are Hill et al. (1979, 1982) and Roelofs et al. (1982).

3.2. Behavioral Roles of Pheromone Components and Blends

The sex pheromone of the gypsy moth, *Lymantria dispar,* is (+)(*Z*)-7,8-epoxy-2-methyl octadecane (Bierl et al., 1970). Although the (+) enantiomer is produced by females and is very active, addition of the (−) enantiomer significantly reduced captures of males in traps in the field (Cardé et al., 1977; Miller et al.,

1977). The reduction in trap capture could not be explained by wing-fanning levels of males exposed to (+) and (±) enantiomers (Yamada et al., 1976). Using their horizontal wind tunnel for choice tests between (+) and (±) disparlure, Miller and Roelofs (1978b) found no significant differences in the number of in-flight orientations to either source. It was only in paired, flight duration tests in which sustained flights in one location were induced by moving the floor pattern beneath the males that significant differences in the behavior began to unfold. Males flew for considerably shorter periods of time to the (+) enantiomer when exposed first to the (−) enantiomer. Under optimal conditions, males exhibited continuous flight for 30 min to the (+) enantiomer alone, and significantly shorter durations were recorded for flights to the (±) mixture.

The cabbage looper moth, *Trichoplusia ni,* was also the subject of a study in a horizontal wind tunnel (Linn and Gaston, 1981) to determine the behavioral effect of a recently identified secondary component, dodecyl acetate. Bjostad et al. (1980) determined that females release a 93:7 ratio of *Z*7-12:OAc and 12:OAc. The males were found to exhibit a sequence of behaviors similar to that described by Baker and Cardé (1979) in wind tunnel studies of *G. molesta,* and this was utilized as a framework for comparing different treatments. The results supported two important points concerning the behavioral function of pheromone components. First, the minor component, 12:OAc, did not elicit any behavior when presented alone. Second, the optimal attraction was obtained to the synthetic blend and dosage (10^{-3} μg of 93:7 ratio *Z*7-12:OAc/12:OAc) that most closely mimicked the natural blend and emission rate. With the optimum blend and dosage, significantly more males flew upwind to the source and spent significantly longer periods within 25 cm of the source.

Choice tests similar to those of Miller and Roelofs for the gypsy moth (1978b) were performed to determine the interaction of the two pheromone components. Two copper disc pheromone emitters were placed upwind and separated by either 12 or 8 cm. Smoke plume visualization showed that pheromone component plumes should have merged 85 or 35 cm downwind, respectively. Males flew upwind in the merged, two-component plume, and never flew in the 12:OAc plume when they reached the "choice" point. They did fly in the *Z*7-12:OAc plume after that point, but not nearly as close to the source as when both components were present on one of the discs as the choice (Fig. 5). Therefore, 12:OAc could not be called a "close-range" component, because it did not elicit upwind flight by itself at close range. Rather, it was clear that the two-components together was a good close-range blend compared to *Z*7-12:OAc, which alone evoked optimal levels of upwind flight from long-range, with or without the addition of 12:OAc.

3.3. Response Profiles to Blends and Concentrations

In studies by Baker and Cardé (1979), Baker et al. (1981), and Linn and Roelofs (1981a, 1983a), the flight behavior of individual male *G. molesta* in a horizontal wind tunnel was observed. In some of the studies, the ratio of components and

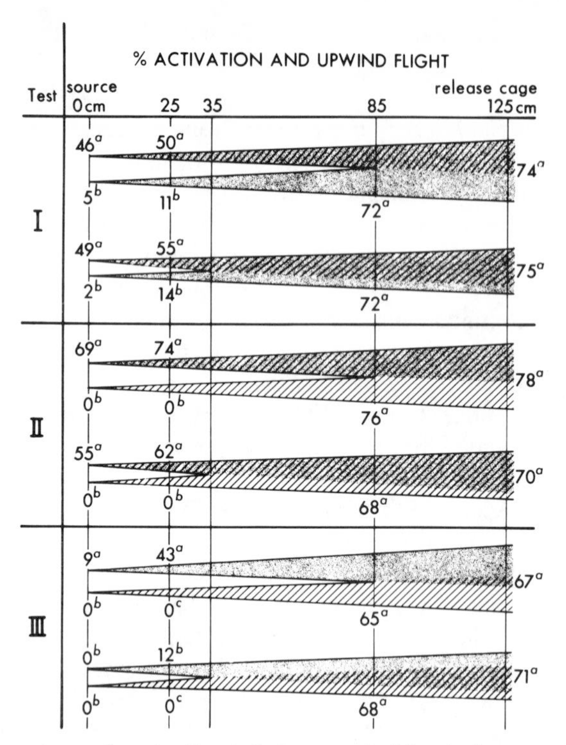

Figure 5. Percentage of male *T. ni* flying upwind in choice tests between two sources positioned in a wind tunnel so that a common plume formed 35 or 85 cm downwind of the sources. Time-averaged plumes are drawn to show the presumed boundaries (as determined by NH_4Cl smoke) and the contents of the two plumes (12:OAc = slashed lines; *Z*7-12:OAc = shading). For each test, $n = 65$ for each pair of sources (from Linn and Gaston, 1981).

the dosage were varied around the natural levels emitted by females (Baker et al., 1980). Males were found to be very sensitive to changes in blend and concentration, with optimal attraction to the source occurring to a narrow range of concentrations of the natural 6% *E*8-12:OAc blend containing 3 to 10% of *Z*8-12:OH. The ability of the experimenter to discriminate among the treatments was clearly increased when moths were required to fly close to the source (Fig. 6) (Linn and Roelofs, 1983; Baker et al., 1981). Males became activated and took flight in nearly equal numbers to a wide range of blends and dosages, but only a narrow range of treatments elicited high levels of completed flights to the source. Increasing the proportion of (*E*) isomer from 6 to 10% (*E*), caused significant arrestment of upwind flight within the plume apparently due to increased turning and decreased linear velocity. This suggested that the (*E*) isomer may function as a "turning" component, which may explain what happens when the "arrestment threshold" is reached (Baker and Roelofs, 1981) either with too

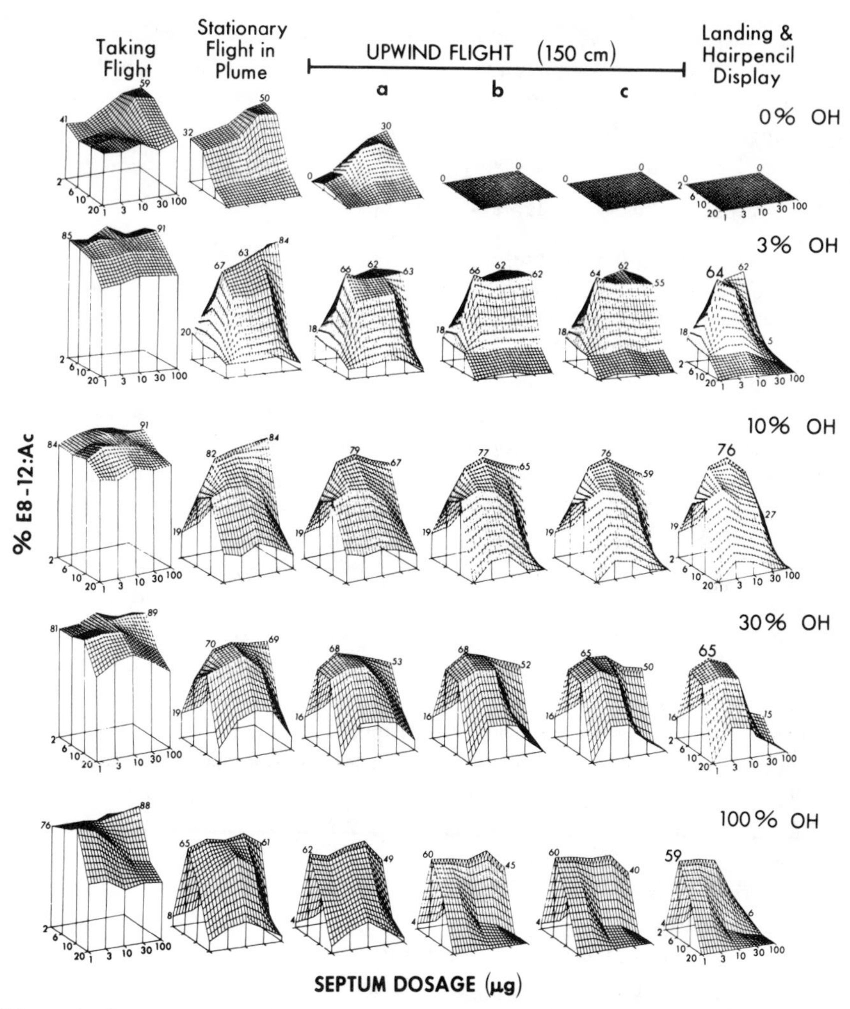

Figure 6. Response profiles (percentage response) of male *G. molesta* to 20 blend–dosage combinations of *Z*8- and *E*8-12:OAc and five proportions of *Z*8-12:OH. Response surfaces for each behavior in the sequence are based on the number of responding males to each treatment ($N = 100$). Response values for selected treatments are shown along with peak values for the number of hairpencil displays at the source within each percentage of *Z*8-12:OH (from Linn and Roelofs, 1983).

much (*E*) in the blend or concentrations of the optimal blend that are too high. In the latter case we now know that arrestment is not only a function of lower flight velocity and higher turning frequency, but also the single most important change involved in within-plume arrestment is that the males steer more obliquely across the wind, allowing more lateral drift (Kuenen and Baker, 1982b).

In a subsequent study, Linn and Roelofs (1983) varied all three components. Using hierarchical clustering techniques, they were able to define an area of optimal response around the natural blend ratio and emission rate, partitioned by "threshold" regions affecting specific behaviors in the sequence.

3.4. Habituation, Sensory Adaptation, Disruption

Habituation and sensory adaptation to sex pheromone are phenomena that have long interested researchers, but the traditional ways of measuring them have been by means of observing "key" responses (see Baker and Cardé, Chapter 2; Bartell and Shorey, 1969). In two instances, habituation has now been measured in horizontal wind tunnels, and new unexpected results have come from the greater ability of the wind tunnels to monitor subtle behavioral changes.

Kuenen and Baker (1981) found that a pulsed pheromonal preexposure of cabbage looper males did not reduce their wing-fanning reaction in a small chamber over the course of an hour. However, this same regime reduced the percentage of males that flew all the way to the pheromone source in horizontal wind tunnels. Continual preexposure, however, of males in the chamber did not result in as severe a reduction of flight to the source in the tunnel, but nevertheless the decrease was significant compared to those receiving no preexposure whatsoever. Because electroantennogram responses of males preexposed to both pulsed and continual pheromone recovered within minutes after the exposure, habituation rather than sensory adaptation was proposed as the mechanism for reducing flights in the tunnel. This corresponded well to bioassay studies and predictions by Bartell and Lawrence (1977), but contradicted the results of Farkas et al. (1975), who thought pulsed exposure did not affect subsequent responses. Farkas et al.'s responses were measured in small bioassay chambers, and in this respect, their lack of reduced wing-fanning responses was similar to that found by Kuenen and Baker.

Linn and Roelofs (1981) preexposed male *G. molesta* to *E*8-12:OAc and were able to demonstrate a prolonged effect (likely habituation) on the males' pheromone quality perception. After such preexposure, males readily flew all the way to sources containing high percentages of the (*E*) isomer, something they would not do without being habituated to (*E*). A combination of (*E*)-isomer preexposure duration and dosage directly determined the subsequent tendency to fly all the way to sources emitting as high as 20% (*E*).

Rather than preexposing moths to pheromone, simultaneously presenting pheromone point sources in the midst of other pheromone sources has resulted in a greater understanding of the mechanisms or disruption of communication in air permeated with pheromone. Sanders (1982) created a grid-array of nine synthetic spruce budworm (*Choristoneura fumiferana*) pheromone point sources at the upwind end of his horizontal wind tunnel and placed a cage of calling females at various points within that grid. He also created a cloud of uniformly permeated air by the tape-grid-turbulence method of Kennedy et al. (1980,

1981), against which he placed the cage of calling females allowed to create a discrete plume. He found that the uniform cloud was less effective in preventing location of the females than the grid of discrete point sources of comparable overall emission concentration. He hypothesized that two different mechanisms were responsible for reduced location of females under the two regimes. The cloud produced habituation or adaptation which could be overcome by the higher concentration bursts of pheromone in the filamentous plume from the females, whereas the discrete point sources often elicited upwind flight of males, who switched over from the female's plume. The fact that the discrete plumes caused a greater reduction of orientation to females could also be due to a higher overall effective concentration of preexposure caused by the peak concentrations within filaments and the condensed nature of the narrow plumes compared to diffuse clouds.

Disruption of upwind flight to sex pheromone was also performed in a horizontal wind tunnel by Baker and Cardé (1978), only here the disruption was created by ultrasound. Gypsy moth males (*L. dispar*) were allowed to begin flying upwind in a plume of their synthetic sex pheromone and then an ultrasonic burst was generated that caused males to fly out of the plume (Fig. 7). Males experimentally deafened on one side nearly always turned away from the plume toward their deaf side. The wind tunnel provided a nice experimental environment to study the interaction between sex pheromone-stimulated flight and

Figure 7. Stroboscopically photographed track of male *L. dispar* flying upwind to synthetic sex pheromone emitted from filter paper disc (left). When a burst of ultrasound was generated outside the wind tunnel, the male abruptly changed course and flew out of the plume (from Baker and Cardé, 1978).

evasive flight from ultrasound, which apparently evolved as a defense against predation by bats and overrides mate location. For studies of in-flight responses to sex pheromone by noctuoid moths, sources of ultrasound in wind tunnel rooms such as machinery that "clicks" on and off, or any metal-on-metal impact or friction, should be silenced for greatest experimental consistency.

An elaborate wind tunnel utilizing horizontal and vertical wind was used by Phelan and Miller (1982) to try to disrupt landing on host plants or host plant models by aphids by introducing clouds of the aphids' alarm pheromone over the plants (Fig. 8). The part of the tunnel with vertically moving air was

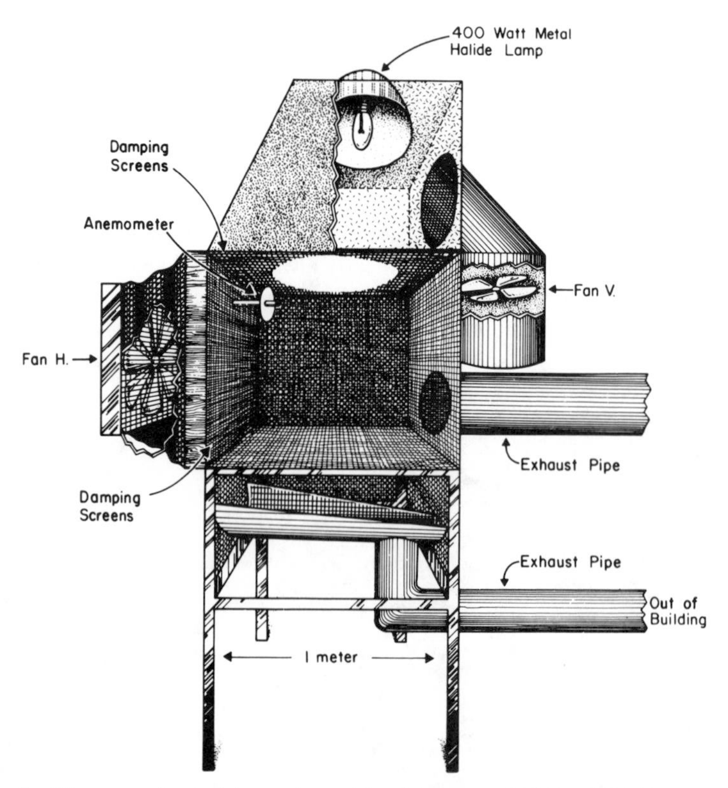

Figure 8. Diagram of wind tunnel used by Phelan and Miller (1982) to generate both vertical and horizontal wind to study the responses of flying *Myzus persicae* aphids to their alarm pheromone and various fatty acids. Aphids fly toward light in the ceiling, but the operator counteracts their upward movement with downward-moving air generated by fan at upper right. After aphids have flown for awhile, their tendency to land on a green plate introduced at the bottom is heightened, and potentially repellent or deterrent odors preventing landing and feeding are introduced as a cloud moving horizontally from left to right in bottom third of tunnel by means of second fan at left. Odor is removed by exhaust tube at right.

modeled after that of Kennedy (1966), in which the velocity of downward-moving air could be increased quickly by the experimenter to counteract the aphid's upward flight toward a light in the tunnel's center, placed there to elicit such flight. Phelan and Miller's tunnel was used precisely like Kennedy's, to elicit prolonged dispersal flight toward the light, and then, when the aphids were most responsive to vegetative stimuli (plants), they would tend to fly instead toward a green plate placed below them in the tunnel. Phelan and Miller's tunnel enabled them to introduce over the plate a horizontally moving layer of air permeated with (*E*)-β-farnesene, the aphid's alarm pheromone, and to observe the possible repellent or deterrent effects of this pheromone on landing and feeding. Although they found no reduction in landing to the alarm pheromone, behavioral reductions were observed when the certain fatty acids were presented, and this unusual tunnel demonstrates that two columns of air moving in different directions can be superimposed for pheromone studies of flying insects.

3.5. Orientation Studies

Entomologists constructed the first wind tunnels in order to determine the mechanisms of orientation used by insects flying in wind with or without odor. However, it took a long time, along with the growth of the pheromone field, for their use to become widespread. Kennedy (1940) developed several models for steering with respect to wind after constructing a horizontal wind tunnel and observing the odor-free flight of yellow fever mosquitos in wind and in still air, plus visually imposed "drift" through ground pattern movement. Kellogg and Wright (1962) developed a horizontal wind tunnel to study the in-flight maneuvers of insects flying to various odors, and these studies really marked the beginning of quantitative, detailed studies of olfactory-mediated flight. Wright, Kellogg, and colleagues not only developed sophisticated, three-dimensional photographic imaging techniques involving stereoscopic cameras or simultaneous side and plan photography but also demonstrated an awareness of the need to simplify the odor environment. They were the first to present odor as a homogeneous cloud and, using smoke sources, demonstrated repeatedly the highly complex nature of an odor plume. They were able to find a less irritating smoke, ammonium acetate, that could be presented along with an attractant with minimal effect on flying insects. They also demonstrated an awareness of the role of visual wind-drift information on the orientation of insects flying to odors, by experimenting with projected, moving floor patterns and through experiments in which wind was absent. Finally, they were the first to use vertically moving air to study the nonanemotactic, odor-mediated horizontal movements of insects flying in choice chambers of odor-permeated columns of air. A pheromone orientation study by Traynier (1968) in a horizontal wind tunnel developed earlier for the host-odor work now set the stage for the growth of wind tunnels in pheromone research.

Farkas and Shorey (1972) used a horizontal tunnel to create a pheromone plume in wind, induce males to fly within a pheromone plume, and then stop the wind in order to prove, so they believed, that only chemotaxis was used for pheromone source location in wind. Kennedy and Marsh (1974) used their tunnel to negate neatly part of this hypothesis, that anemotaxis was not used, by demonstrating the optomotor-anemotactic response to a moving floor pattern. But they did not address the part of Farkas and Shorey's hypothesis concerning a chemotactic mechanism for maintaining contact laterally with the plume. Support for a chemotactic mechanism superimposed on anemotaxis came from wind tunnel studies by Baker and Kuenen (1982) and Kuenen and Baker (1983) in which they repeated and extended Farkas and Shorey's experiment of stopping the wind and observing whether males could locate the source. However,

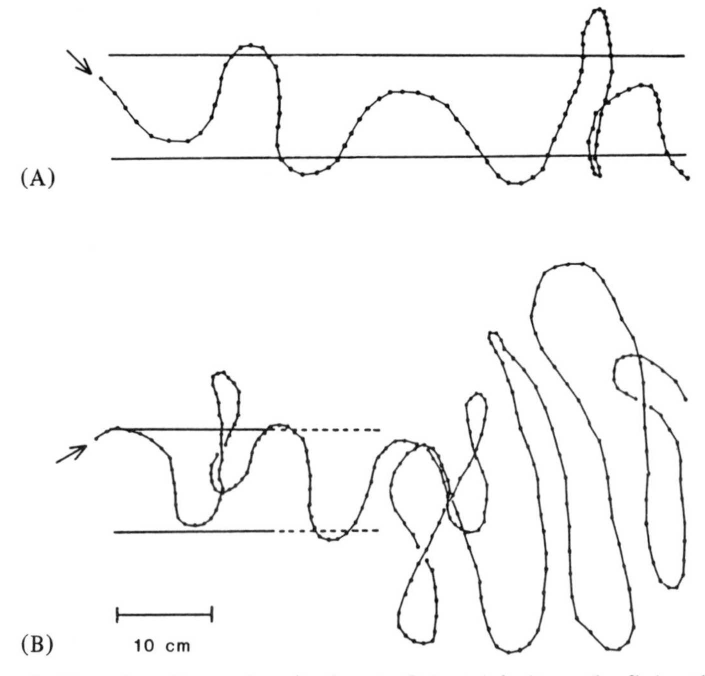

Figure 9. Plan view (from above) of male Oriental fruit moths flying through a 65-cm-long section wind tunnel. Dots represent each 1/60 sec of elapsed time. Solid straight lines denote the boundaries of the time-averaged pheromone plume as visualized by smoke. The pheromone source was 105 cm from the right edge in each illustration, and the wind, before stoppage, was from the right. Arrows denote the direction of flight in each track. (A) The wind stopped 0.25 sec before the male entered the field of view, and the male continued to the septum. (B) The pheromone plume was removed, and the wind stopped just as the male entered the field of view; the approximate up-tunnel end of the plume (by smoke visualization) is indicated by the dashed, straight lines; on entering clean air, the male's track changed significantly, with a closest approach to the septum of only 90 cm (from Baker and Kuenen, 1982).

they made more detailed measurements of the flight tracks before and after wind stoppage, discovering that the tracks up the plume with no wind remained quite similar to those in wind. The definitive experiment demonstrating a chemotactic program of zigzagging in zero wind was one in which the pheromone source was removed and the wind then was stopped to create a truncated plume in zero wind. Males zigzagging along the plume in still air changed the amplitude, frequency, and angle of their zigzags upon entering clean air as they flew out the end of the plume, showing that their movements were in fact pheromone mediated, independent of anemotaxis (Fig. 9).

Later it was found that in-flight experience with wind-induced drift tended to polarize the zigzags of males in the toward-source direction, even when wind was later stopped when males were part way to the source. Males released into a stationary plume in zero wind performed their zigzag flight movements, but the zigzags meandered with no consistent directional component as they had in the cases where the wind was stopped after they had launched themselves in the plume (Fig. 10) (Baker et al., 1984). This was more evidence for a self-steered program of zigzagging mediated by pheromone concentration.

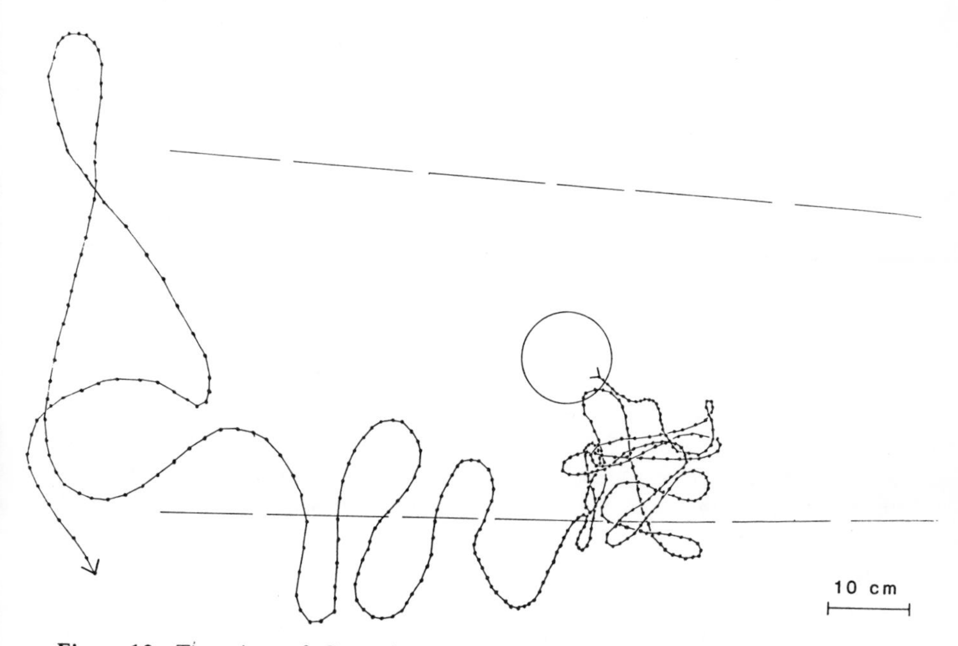

Figure 10. Top view of *G. molesta* male, introduced into plume in zero wind, taking flight from release cage (circle) 5 sec after wind had stopped. Zigzagging flight began immediately in the plume, but meandered in the down-tunnel direction (source is to right, 1.5 m away from release cage). With no pheromone in zero wind, males seldom take off, but when they do, no such zigzagging occurs. Dashed, straight lines indicate plume boundaries 5 sec after wind stoppage, and dots on moth's track indicate 1/60 sec elapsed time (from Baker et al., 1984).

Horizontal wind tunnels provided even more support for a chemotactic program of zigzagging. Using a tunnel uniformly permeated with a cloud of pheromone or with only a side corridor-cloud, Kennedy et al. (1980, 1981) showed that for male *Adoxophyes orana* casting in clean air, upon encountering the cloud males decreased the width and increased the frequency of their zigzagging. However, after several seconds of this tonic stimulation, the reversals became wide again and the surge of upwind progress that had accompanied the narrow zigzags now ceased and the moths became arrested once more in wide casting flight.

Their hypothesis was that adaptation to the constant stimulus caused the narrow reversals to wane, because the superpositioning of a point source plume of pheromone in the cloud readily elicited rapid upwind zigzagging to the source. This demonstrated that the cloud's concentration had not been too high, but rather adaptation may have developed, which the phasic arrival of higher peak concentrations then overcame.

These results were supported with another species, *G. molesta,* using a similar set of pheromone clouds in a horizontal tunnel (Willis and Baker, 1984). Males casting in clean air briefly narrowed their zigzags upon entering a homogeneous cloud of pheromone, but rapid upwind zigzagging was elicited at the "edge" of a side corridor of pheromone. Evidence came from another experiment that it was the phasic stimulation from the non-uniform mixing of air and pheromone at the edge that initiated the self-steered program, as well as the males' excursions and incursions from the corridor that perpetuated turning responses along the corridor's edge. The tunnel was rotated 90° so that the side corridor now became a corridor filling the bottom half of the tunnel, with clean air above. The males zigzagged up the tunnel at the horizontal edge of the bottom corridor in a manner similar to males zigzagging up the vertical edge of the side corridor (Fig. 11) (Willis and Baker, 1984).

The mechanisms of counterturning in walking insects that may be analogous to zigzagging in flying insects has also been studied in horizontal wind tunnels (Tobin, 1981; Tobin et al., 1981). Using a plume of periplanone-B positioned 2 cm off the ground at the upwind end of a 2.4 × 1.2 × 0.6 m tunnel, the tracks of *Periplaneta americana* were recorded from above. The males walked upwind in the plume by means of a combination of anemotaxis, endogenously triggered counterturns, and counterturns back into the plume triggered by a decrease in concentration at the boundary of the time-averaged plume. Thus again an integrated system of chemically mediated movements and anemotaxis was implicated, this time for a walking insect.

New advances in the understanding of the orientation of insects flying in response to pheromone will be gained through the use of wind tunnels in which flight tracks are recorded and analyzed in three dimensions. R. H. Wright and his colleagues in the later 1950s and early 1960s realized the importance of three-dimensional analysis, although until recently technical and data-handling procedures have been major deterrents to quantitative rather than descriptive

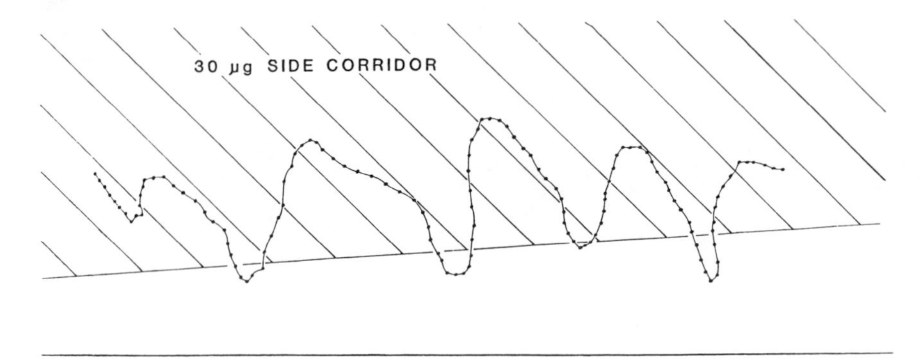

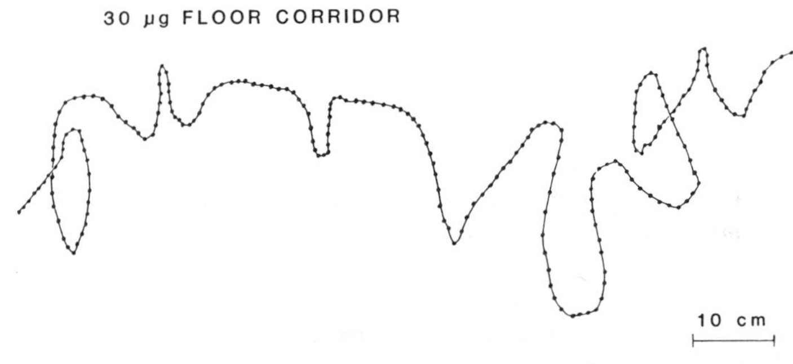

Figure 11. (Top): Top view of male *G. molesta* flying along the vertical edge of a side corridor (hatched lines) of air moving left to right uniformly permeated with pheromone emitted from a grid of rubber septa each loaded with 30 µg. (Bottom): Same cloud, except now it has been rotated so it occupies bottom half of tunnel along the floor. Male zigzags upwind left to right, maintaining an altitude consistent with the horizontal edge's location. Distance between dots denotes 1/60 sec elapsed time (from Willis and Baker, 1984).

studies. Advances will be made on the vertical components of flight movements and how they are integrated with the horizontal zigzagging components. Height control is obviously integrated into the zigzagging component, but how this is accomplished remains to be discovered. Certainly more ingenious studies involving wind tunnels will need to be performed such as C. T. David's (1982b) (Fig. 12) in which a visual surround of "barber pole" stripes was made to appear to move past flying insects. In this case the *Drosophila* adult was allowed to fly in a plume of banana odor, and the angular velocity of apparent image movement, the frequency, and the pitch of the stripes in the barber pole cylinder were varied systematically to determine the effect upon upwind flight speed. The advantage of the moving visual surround is that contradictory visual information from stationary objects is absent, unlike when moving floor or ceiling patterns are used. In order to relate wind tunnel effects to those in the field, further adaptations will have to be made on existing tunnels once the simpler questions are answered. Wind in the field can shift direction and velocity very

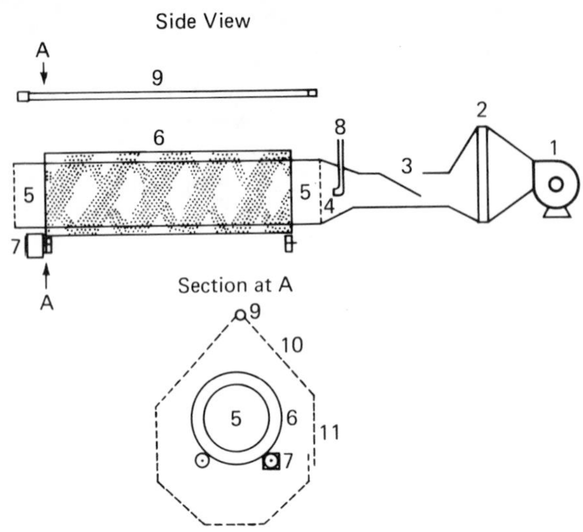

Figure 12. "Barber's pole" wind tunnel of David (1982b), with cylindrical cellulose acetate working section 1.4 m long X 0.25 m diameter. 1, fan; 2, charcoal filter; 3, wind speed control valve; 4, brushed nylon air-smoothing screen; 5, working section; 6, outer cylinder with helical pattern; 7, stepper motor to rotate cylinder; 8, tube for introducing attractant odors; 9, fluorescent light; 10, brushed nylon screen; 11, muslin screen.

rapidly (David et al., 1982, 1983), and tunnels will need to have some of these features incorporated into their design.

Certainly the use of wind tunnels in pheromone research will continue to expand and, as this is done, further understanding of odor perception and orientation will be gained. In the meantime, wind tunnels will retain their utility to entomologists as useful tools in which to test potential field formulations of pheromones.

IV. References

Alberts SA, Kennedy MK, Cardé RT (1981) Pheromone-mediated anemotactic flight and mating behavior of the Sciarid fly *Bradysia impatiens.* Environ Entomol **10**:10–15.

Baker TC, Cardé RT (1978) Disruption of gypsy moth male sex pheromone behavior by high frequency sound. Environ Entomol **7**:45–52.

Baker TC, Cardé RT (1979) Analysis of pheromone-mediated behavior in male *Grapholitha molesta,* the oriental fruit moth (Lepidoptera:Tortricidae). Environ Entomol **8**:956–968.

Baker TC, Kuenen LPS (1982) Pheromone source location by flying moths: A supplementary non-anemotactic mechanism. Science **216**:424–427.

Baker TC, Roelofs WL (1981) Initiation and termination of Oriental fruit moth male response to pheromone concentrations in the field. Environ Entomol **10**:211–218.

Baker TC, Cardé RT, Miller JR (1980) Oriental fruit moth pheromone component emission rates measured after collection by glass-surface adsorption. J Chem Ecol **6**:749–758.

Baker TC, Meyer W, Roelofs WL (1981) Sex pheromone dosage and blend specificity of response by Oriental fruit moth males. Entomol Exp Appl **30**: 269–279.

Baker TC, Willis MA, Phelan PL (1984) Pre-wind-lull optomotor anemotaxis contributes to successful pheromone source location by flying moths. Physiol Entomol **9** (in press).

Bartell RJ, Lawrence LA (1977) Reduction of responsiveness of male apple moths, *Epiphyas postvittana,* to sex pheromone following pulsed pheromonal exposure. Physiol Entomol **2**:1–6.

Bartell RJ, Shorey HH (1969) A quantitative bioassay for the sex pheromone of *Epiphyas postvittana* (Lepidoptera) and factors limiting male responsiveness. J Insect Physiol **15**:33–40.

Bierl BA, Beroza M, Collier CW (1970) Potent sex attractant of the gypsy moth: Its isolation, identification, and synthesis. Science **170**:87–89.

Bjostad LB, Gaston LK, Noble LL, Moyer JH, Shorey HH (1980) Dodecyl acetate, a second pheromone component of the cabbage looper moth *Trichoplusia ni.* J Chem Ecol **6**:727–734.

Cardé RT, Hagaman TE (1979) Behavioral responses of the gypsy moth in a wind tunnel to air-borne enantiomers of disparlure. Environ Entomol **8**: 475–484.

Cardé RT, Doane CC, Baker TC, Iwaki S, Marumo S (1977) Attractancy of optically active pheromone for male gypsy moths. Environ Entomol **6**: 768–772.

David CT (1982a) Competition between fixed and moving stripes in the control of orientation by flying *Drosophila.* Physiol Entomol **7**:151–156.

David CT (1982b) Compensation for height in the control of groundspeed by *Drosophila* in a new, "Barber's Pole" wind tunnel. J Comp Physiol **147**: 485–493.

David CT, Kennedy JS, Ludlow AR, Perry JN, Wall C (1982) A re-appraisal of insect flight towards a point source of wind-borne odor. J Chem Ecol **8**: 1207–1215.

David CT, Kennedy JS, Ludlow AR (1983) Finding of a sex pheromone source by gypsy moths released in the field. Nature (London) **303**:804–806.

Daykin PN (1967) Orientation of *Aedes aegypti* in vertical air currents. Can Entomol **99**:303–308.

Daykin PN, Kellogg FE (1965) A two-air-stream observation chamber for studying responses of flying insects. Can Entomol **97**:264–268.

Daykin PN, Kellogg FE, Wright RH (1965) Host-finding and repulsion of *Aedes aegypti.* Can Entomol **97**:239–263.

Farkas SR, Shorey HH (1972) Chemical trail-following by flying insects: A mechanism for orientation to a distant odor source. Science **178**:67–68.

Farkas SR, Shorey HH (1976) Anemotaxis and odour-trail following by the terrestrial snail *Helix aspersa.* Anim Behav **24**:686–689.

Farkas SR, Shorey HH, Gaston LK (1975) Sex pheromones of Lepidoptera. The influence of prolonged exposure to pheromone on the behavior of males of *Trichoplusia ni.* Environ Entomol **4**:737–741.

Hill AS, Roelofs WL (1981) Sex pheromone of the saltmarsh caterpillar moth, *Estigmene acrea.* J Chem Ecol **7**:655–668.

Hill AS, Rings RW, Swier SR, Roelofs WL (1979) Sex pheromone of the black cutworm moth, *Agrotis ipsilon.* J Chem Ecol **5**:439–457.

Hill AS, Kovalev BG, Nikolaeva LN, Roelofs WL (1982) Sex pheromone of the fall webworm moth, *Hyphantria cunea.* J Chem Ecol **8**:383–396.

Kellogg FE, Frizel DE, Wright RH (1962) The olfactory guidance of flying insects. IV. *Drosophila.* Can Entomol **94**:884–888.

Kellogg FE, Wright RH (1962) The olfactory guidance of flying insects. III. A technique for observing and recording flight paths. Can Entomol **94**:486–493.

Kennedy JS (1940) The visual responses of flying mosquitoes. Proc Zool Soc Lond A **109**:221–242.

Kennedy JS (1966) The balance between antagonistic induction and depression of flight activity in *Aphis fabae* Scopoli. J Exp Biol **45**:215–228.

Kennedy JS (1977a) Olfactory responses to distant plants and other odor sources. In: Chemical Control of Insect Behavior: Theory and Application. Shorey HH, McKelvey JJ (eds), Wiley, New York, pp 67–91.

Kennedy JS (1977b) Behaviorally discriminating assays of attractants and repellents In: Chemical Control of Insect Behavior: Theory and Application. Shorey HH, McKelvey JJ (eds), Wiley, New York, pp 215–229.

Kennedy JS (1978) The concepts of olfactory arrestment and attraction. Physiol Entomol **3**:91–98.

Kennedy JS (1983) Zigzagging and casting as a programmed response to wind-borne odour–A review. Physiol Entomol **8**:109–120.

Kennedy JS, Marsh D (1974) Pheromone-regulated anemotaxis in flying moths. Science **184**:999–1001.

Kennedy JS, Ludlow AR, Sanders CJ (1980). Guidance system used in moth sex attraction. Nature (London) **288**:475–477.

Kennedy JS, Ludlow AR, Sanders CJ (1981) Guidance of flying male moths by wind-borne sex pheromone. Physiol Entomol **6**:395–412.

Kuenen LPS, Baker TC (1981) Habituation versus sensory adaptation as the cause of reduced attraction following pulsed and constant sex pheromone pre-exposure in *Trichoplusia ni.* J Insect Physiol **27**:721–726.

Kuenen LPS, Baker TC (1982a) Optomotor regulation of ground velocity in moths during flight to sex pheromone at different heights. Physiol Entomol **7**:193–202.

Kuenen LPS, Baker TC (1982b) The effects of pheromone concentration on the flight behavior of the Oriental fruit moth, *Grapholitha molesta.* Physiol Entomol **7**:423–434.

Kuenen LPS, Baker TC (1983) A non-anemotactic mechanism used in pheromone source location by flying moths. Physiol Entomol **8**:277–289.

Linn CE Jr, Gaston LK (1981) Behavioral function of the components and the blend of the sex pheromone of the cabbage looper, *Trichoplusia ni*. Environ Entomol **10**:751–755.

Linn CE Jr, Roelofs WL (1981) Modification of sex pheromone blend discrimination in male oriental fruit moths by pre-exposure to (*E*)-8-dodecenyl acetate. Physiol Entomol **6**:421–429.

Linn CE Jr, Roelofs WL (1983) The effect of varying proportions of the alcohol component on sex pheromone blend discrimination in male oriental fruit moths. Physiol Entomol **8**:291–306.

Linn CE Jr, Roelofs WL (1984) Sublethal effects of neuroactive compounds on pheromone response thresholds in male Oriental fruit moths. (In preparation).

Marsh D, Kennedy JS, Ludlow AR (1978) An analysis of anemotactic zigzagging flight in male moths stimulated by pheromone. Physiol Entomol **3**:221–240.

Miller JR, Roelofs WL (1978a) Sustained-flight tunnel for measuring insect responses to wind-borne sex pheromones. J Chem Ecol **4**:187–198.

Miller JR, Roelofs WL (1978b) Gypsy moth responses to pheromone enantiomers as evaluated in a sustained-flight tunnel. Environ Entomol **7**:42–44.

Miller JR, Mori K, Roelofs WL (1977) Gypsy moth field trapping and electroantennogram studies with pheromone enantiomers. J Insect Physiol **23**: 1447–1453.

Phelan PL, Miller JR (1982) Post-landing behavior of alate *Myzus persicae* as altered by (*E*)-β-farnesene and three carboxylic acids. Ent Exp Appl **32**: 46–53.

Roelofs WL, Cardé RT (1977) Responses of Lepidoptera to synthetic sex pheromone chemicals and their analogues. Ann Rev Entomol **22**:377–405.

Roelofs WL, Hill AS, Linn CE (1982) Sex pheromone of the winter moth, a geometrid with unusually low-temperature precopulatory responses. Science **217**:657–658.

Sanders CJ (1982) Disruption of male spruce budworm orientation to calling females in a wind tunnel by synthetic pheromone. J Chem Ecol **8**:493–506.

Sanders CJ, Lucuik GS, Fletcher RM (1981) Responses of male spruce budworm (Lepidoptera:Tortricidae) to different concentrations of sex pheromone as measured in a sustained-flight wind tunnel. Can Entomol **113**: 943–948.

Shorey HH (1973) Behavioral responses to insect pheromones. Annu Rev Entomol **18**:349–380.

Tobin (1981) Pheromone orientation: Role of internal control mechanisms. Science **214**:1147–1149.

Tobin TR, Seelinger G, Bell WJ (1981) Behavioral responses of male *Periplaneta americana* to periplanone B, a synthetic component of the female sex pheromone. J Chem Ecol **7**:969–979.

Traynier RMM (1968) Sex attraction in the Mediterranean flour moth *Anagasta kuhniella*: Location of the female by the male. Can Entomol **100**:5–10.

Von Keyserlingk H (1983) Vertical and horizontal counterturning of male *G.*

molesta flying in interrupted pheromone plumes in constant wind in a flight tunnel and in shifting wind in the field. (In preparation).

Willis MA, Baker TC (1984) The effects of phasic and tonic pheromone stimulation on the flight behavior of the oriental fruit moth, *Grapholitha molesta* (Busck). Physiol Entomol (In preparation).

Yamada M, Saito T, Katagiri K, Iwaki S, Marumo S (1976) Electroantennogram and behavioral responses of the gypsy moth to enantiomers of disparlure and its *trans* analogues. J Insect Physiol **22**:755–761.

Chapter 4

Field Trapping with Attractants: Methods and Interpretation

Ring T. Cardé [1] *and J. S. Elkinton* [1]

I. Introduction

Numerous insects communicate with pheromones that induce attraction or directed movement toward the pheromone source. Documentation of orientation to such chemicals and their use to monitor insect pests typically require development of traps and a trapping protocol. Field trapping also has served as the bioassay in the characterization and identification of many pheromones. Among the attributes sought in a trapping system are low cost, sensitivity to and specificity for the target species and user convenience.

Our review will not attempt to summarize all the approaches tested, nor will it suggest that there are 'generic' solutions to the development of a trapping system. Indeed, the behavioral reactions, nonchemical cues, and the trapping system all potentially differ from one species to another. Even though each species must be examined individually, it will be useful to describe some of the successful approaches and define areas where principles remain elusive.

II. Verification of Pheromone Identity

Trapping experiments are commonly used to compare the behavioral activity of synthetic compounds to natural pheromone, typically emitted by caged insects. If the magnitude of trap catch in the synthetic pheromone-baited traps matches or exceeds the catch evoked by the natural source, then the identification of the pheromone is assumed to be correct and complete with certain reservations.

[1] Department of Entomology, University of Massachusetts, Amherst, Massachusetts 01003.

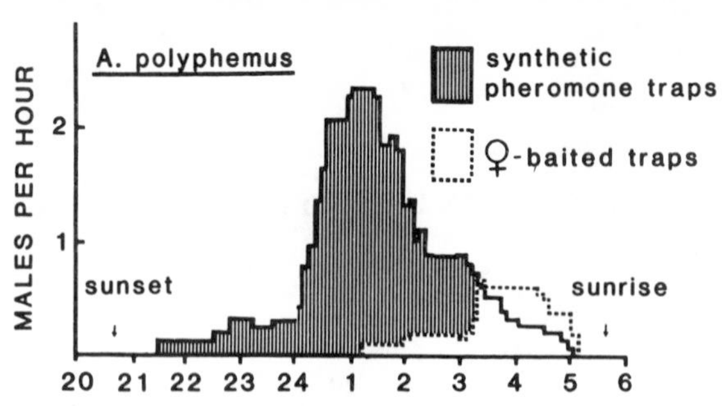

Figure 1. Diel rhythms of male *Antheraea polyphemus* attraction to synthetic and female-baited traps (after Kochansky et al., 1977).

It is assumed that both the synthetic and the natural pheromone sources emit at nearly identical rates and that the natural insects emit pheromone over the same time interval as the response to pheromone occurs. Obviously, releasing synthetic pheromone at a higher rate than natural emission potentially increases trap catch [although in some species, there may be a decrease in the performance of "late" orientation behaviors close to a stimulus at concentrations above that released by a natural source, potentially diminishing or even cancelling trap catch as in *Grapholitha molesta,* the oriental fruit moth (Cardé et al., 1975a; Baker and Roelofs, 1981).

In many insects, the responder has a broader diel activity rhythm than the emitter. Without a somewhat detailed knowledge of the timing of these events, one might conclude that synthetic pheromone was considerably more effective than a caged emitting insect, when indeed the increased catch in a trap baited with pheromone was caused by the continuous emission of pheromone and the promiscuous sexual activity rhythm of the responders, as shown in Fig. 1.

III. Trap Design

Trap design is vital in developing trapping systems. It would be difficult to review comprehensively the numerous types that have been tested. Instead, we will cover some of the major types and consider some of their attributes and deficiencies.

The most common types of traps in use employ a sticky surface to retain the attracted insect. Three of the most common sticky traps are the delta trap, the tent trap, and the wing trap (Figs. 2A-2C). If the sticky surface on these traps is fresh, then such traps ensnare most insects that contact their sticky surface. As the sticky surface ages, becomes covered with debris, or, most commonly, is paved with the target insect, its ability to retain new arrivals is reduced or even eliminated. For example, with the codling moth, *Cydia pomonella,* trapping efficiency begins to decline once the catch exceeds 0.2 moths/cm^2 of the sticky sur-

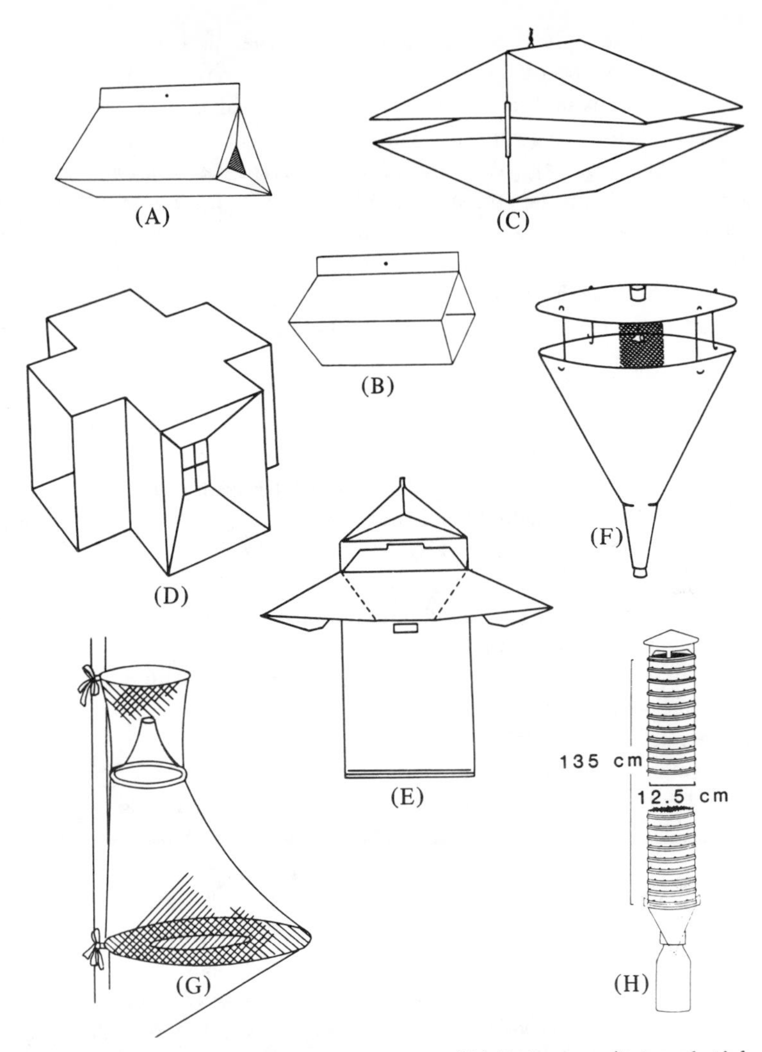

Figure 2. Various types of pheromone traps. (A) Delta trap (internal sticky surface). (B) Tent trap (internal sticky surface). (C) Wing trap (sticky surface on inside trap bottom). (D) Granett trap (vaporous insecticide inside). (E) Milk carton trap (vaporous insecticide inside). (F) Covered funnel trap (vaporous insecticide inside). (G) Double cone trap (insects trapped in top compartment). (H) Cylinder trap.

face (Riedl, 1980). Thus, the suitability of sticky traps is dependent upon the population levels of the target species and the necessity of trapping a high proportion of the attracted insects.

In survey and detection programs, the foremost concern is sensitivity of the system, i.e., detecting the presence of the target species. Similarly, when trap catch is used to time phenological events such as the onset of adult emergence, trap saturation may be of no concern. But in other trapping programs, saturation can pose serious problems. When trap catch is used as an indicator of population density, the limited capacity will make it difficult to develop predictable relationships between trap catch and density, unless sticky traps are baited with very low dosages of attractant, or changed frequently. In mass trapping applications, superior trap efficiency necessary for management would appear to eliminate sticky traps, except at very low pest densities.

A variety of nonsticky traps with large capacity and unvarying efficiency (at least until capacity is reached) offer alternative approaches. For example, Granett (1973, 1974) developed a no-exit trap with small entrance ports (Fig. 2D) for the gypsy moth, *Lymantria dispar.* Its capacity exceeds several thousand gypsy moths, in contrast to the delta trap which becomes saturated with about 25 males. A standard trap for gypsy moth in generally infested areas is the "milk carton" trap (Fig. 2E) (Schwalbe and Paszek, 1978). Both of these no-exit traps kill the trapped moths with a vaporous insecticide (dichlorvos); however, to date, little attention has been given to the possibility that the insecticide repels the moth before it can enter the trap.

A similar approach has been used in developing a trapping system to monitor population density of the spruce budworm, *Choristoneura fumiferana.* Sticky traps, even in subdefoliation population levels, can fill up with moths in several minutes. One approach to circumvent this difficulty is to use a variety of lure dosages (Sanders, 1978, 1981), but this tactic would require deploying many more traps than if a single type of trap were used. A no-exit trap (Fig. 2F) developed by Ramaswamy and Cardé (1983) in preliminary field trials showed an excellent correlation between the larval density and the ensuing male trap catch ($R^2 = 0.93$) (Ramaswamy et al., 1983). The no-exit trap used was *not* as efficient as a wing sticky trap, but in this use pattern, high sensitivity (the ability to detect low population levels) is relatively unimportant compared to capturing a constant proportion of the males (Cardé, 1979; Daterman, 1980).

Electrocution traps have been employed in some studies (e.g., Mitchell et al., 1972; Stanley et al., 1976). They offer high efficiency in "retaining" those animals contacting the surface of the electrocution grid, but the electrocution itself results in an audible (to us) sound and, as well, a burst of ultrasound, which can cause moths with ultrasound sensitivity to veer away from the trap, as in evident response to bat cries (Webb et al., 1978). Sensitivity to ultrasound only occurs in a few insect groups. Among the Lepidoptera it is found in the Noctuoidea and the Sphingidae. Electrocution traps require a power source and are comparatively expensive.

The effect of trap color has not been systematically investigated in many species. In the day-flying gypsy moth it has not been shown to influence catch (Granett, 1973), but in the day-flying sesiid moth *Podosesia syringae,* the lilac borer, black, brown, or red sticky traps were more effective than white traps and other colors (Timmons and Potter, 1981). Trap color also influences trap catch in the noctuids, *Trichoplusia ni* (cabbage looper) and *Pseudoplusia includens* (soybean looper) (McLaughlin et al., 1975). In pheromone-baited double-cone traps which rely upon the upward flight of the moth (possibly an escape reaction) after the male flies below the cone, traps painted with colors exhibiting low spectral reflectance in the 360- and 550-nm regions captured the most moths. These regions correspond well to the peak spectral sensitivities of many noctural Lepidoptera and as well the peak energy levels of night sky radiation under moonless conditions (McLaughlin et al., 1975).

In species in which attraction involves aggregation on the host, as in the boll weevil, *Anthonomus grandis,* a visual mimic of the host might well be expected to improve attraction to pheromone. The optimal color of pheromone-baited boll weevil traps is yellow (500–525 nm) (Cross et al., 1976), the color most typically used in traps to mimic the peak wavelength reflectance of green foliage.

Trap color, then, may be important to maximize trap catch. This may come about because of either high or low reflectance in particular regions. Systematic investigation of trap color (and shape) is a potentially important and somewhat neglected aspect of trap design.

IV. Effects of the Structure of the Pheromone Plume

A consideration of the various orientation mechanisms supposedly used by flying insects to discover the location of a chemical source is important because some of the various strategies proposed (Cardé, 1984; see also Baker and Cardé, this volume) rely, to varying extents, upon the spatial structure of the stimulus. In wind (defined here as an airspeed above the *anemotactic* threshold) pheromone induces flight upwind for several moth species (Kennedy and Marsh, 1974). An alternative mechanism, verified in two moth species to date, evidently involves sampling the pheromone plume's spatial structure, and steering a course along the plume's path (Farkas and Shorey, 1972; Baker and Kuenen, 1982). This guidance system has been shown to operate under windless conditions and relies upon perception of the spatial structure of the chemical stimulus (Cardé, 1984). Hence we could expect that the overall shape of the generated plume, and possibly the distribution of filaments of pheromone within the plume, could affect the orientation process.

Indirect evidence that plume structure modifies orientation comes from field trapping experiments of Lewis and Macaulay (1976) in which the magnitude of trap catch seemed to decrease as the plume became more diffuse (Fig. 3). As yet this process has not been systematically studied, but the results with the pea moth, *Cydia nigricana,* suggest that attention to how the trap design modifies the dimensions of pheromone plume could yield substantial improvements in trap efficiency. Such improvements would be most relevant to mass trapping programs in which maximal trap efficiency could be a critical factor in treatment efficacy and economic feasibility and to the sensitivity and performance of traps

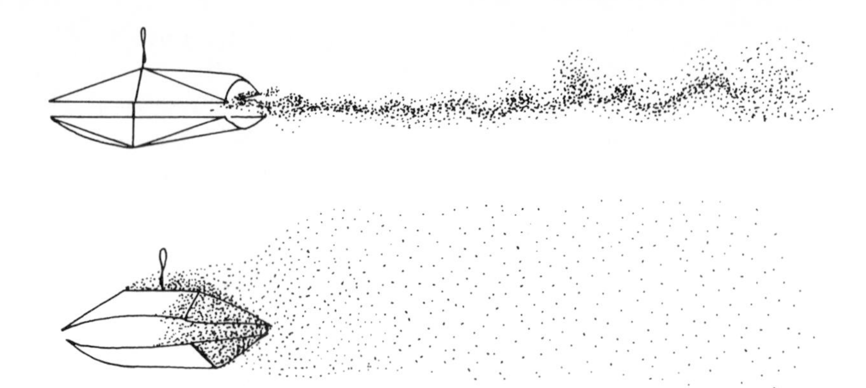

Figure 3. Effect of wind direction on pheromone dispersion from a wing-type trap (after Lewis and Macaulay, 1976).

in survey and detection. This direction of approach leads to a relatively uncharted but potentially fruitful area.

V. Trap Placement

The precise position of a trap within a habitat can have a substantial effect on the levels of trap catch. For example, Riedl et al. (1979) have demonstrated that male codling moth trap catch is highest when traps are positioned near the top of the apple tree canopy. One may test to determine optimal position for maximal sensitivity, although it is clear that user convenience is also important in the placement scheme.

In some cases, insects will only be attracted to traps when they are placed in the natural host tree, as appears to be the case in the yellow-headed spruce sawfly, *Pikonema alaskensis* (Bartlett et al., 1982). Sex pheromone extracts attracted from 25 to 188 males/trap when traps were positioned in spruce trees but only 0 to 2 males/trap in traps 2 m away. Visual cues from the trees would seem to be ruled out as an explanation because traps in nonhost trees also caught few males. The close relationship between the host tree and the sex attraction in *P. alaskensis* is not noted in the pine sawfly, *Diprion similus,* in which males can be "lured" to pheromone in an open field at least 60 m from the host trees (Coppel et al., 1960).

The turbulence generated by foliage, tree trunks, or other features of the habitat near the trap modify the dispersion of the pheromone plume, creating a more diffuse plume, and possibly lowering the proportion of insects attracted, as discussed by Lewis and Macaulay (1976) and in Section VIII on Factors Inducing Landing later in this chapter. Forest dwelling moths such as the gypsy moth are usually trapped in highest numbers when the trap is on trees (Cardé et al., 1977a,b; Elkinton and Childs, 1983).

In a comprehensive field study of codling moth attraction to traps positioned at various sites in pear and apple trees, Riedl et al. (1979) found that trap placement on the north side of a tree produced about 30% less trap catch than placement in other cardinal directions, and trap catch in the upper portion of the tree canopy was highest. These findings bear on placement of traps in monitoring adult activity, mass trapping, and monitoring of mating disruption, techniques to be discussed in Section XII below.

VI. Avoiding Contamination

Contamination of lures and traps with an incorrect treatment is a major concern in field evaluations of different traps or lures. Some practical comments on the techniques used in deploying a field test to avoid this problem may be useful.

Traps should be baited in the field. Prior baiting and transport to a field site certainly can result in cross-contamination of traps. We typically prepare lures in the laboratory and transport them to the field in sealed glass vials. Pheromone dispensers should not be handled directly. We have used forceps, rinsed with acetone between different treatments, or disposable plastic gloves to transfer lures in the field from the vials to the traps.

These suggestions may appear superfluous to experienced investigators, but a surprising number of workers neglect these simple precautions. In cases where contamination of lures or use of an incorrect chemical is suspected, it would be wise to retrieve and repackage the suspect lures for possible chemical analysis.

VII. Rate of Pheromone Release

Generally a high rate of pheromone emission will produce an increase in trap catch, although at some high rates performance of late behaviors may be diminished. This effect is pronounced in *G. molesta,* a species that exhibits "normal" attraction and mating responses to a comparatively narrow range of pheromone concentrations (Cardé et al., 1975a; Baker and Cardé, 1979; Baker and Roelofs, 1981). Thus, "supernormal" concentrations, although eliciting attraction well downwind, cause arrestment of upwind progress near the source, resulting in virtually no trap catch. Concentrations of pheromone from dispensers baited with 30-fold less pheromone than a dose that evokes maximal catch produce male catch at the same levels as unbaited traps. Although in *G. molesta* the effects of pheromone concentration upon trap catch are pronounced, similar trends presumably exist in other species.

A complicating factor is that pheromone dispensers do not release pheromone at a constant rate over time, at different temperatures or wind speeds. Methods to determine these rates are given in Chapter 10 by Golub and Weatherston (this volume). In cases where traps are used in survey and monitoring, a constant relationship between catch and population levels facilitates interpretation of the results; an even rate of pheromone emission (zero order decay) removes a major source of "noise." Development of data bases for interpretation of trap catch will be greatly hindered, and in some cases be impossible, without dispensers that emit pheromone over time as evenly as practicable. The lack of relationship between trap catch and conventional populations measures, observed in many field studies, could be attributable in some cases to variations in the release of pheromone during the course of the field trial.

VIII. Factors Inducing Landing

Most traps require the insect to land before capture can occur. The specific cues inducing this behavior have been poorly documented and remain speculative. For some moths, it is known that the presence of specific pheromone components elevates the proportion of males landing on the trap surface (e.g., Baker et al., 1976; Baker and Cardé, 1979). For other species, we may speculate that the plume's spatial dimensions are important to landing. For some moths that tend to call and mate on tree trunks, landing and attraction within decimeters are enhanced markedly by the presence of the tree trunk, presumably due to its silhouette (Cardé, 1981).

The "cylinder trap" (Fig. 2H) used for *Ips typographus*, the pine bark beetle, presumably is partially reliant upon its visual resemblance to a tree for its effectiveness (Bakke and Riege, 1982). Tilden et al. (1983) have shown that the presence of a silhouette enhances the pheromone trap catch of the bark beetle *Dendroctonus brevicomis*. Only direct observation of landing in the vicinity of the trap will reveal its importance.

IX. Trap Efficiency

Given that the responding insect is lured to the vicinity (e.g., within a meter) of the trap, the likelihood of capture in the trap or *trap efficiency* will depend upon several factors: retention of the insect in the trap, visual characteristics of the trap, etc. If the effects of these trap characteristics upon capture are to be determined (and the trapping system improved by increasing trap efficiency), then counting insects captured at the experiment's end will not prove illuminating. Instead, direct observation of the incoming responders and tallying whether they are ultimately captured or not is necessary. Even release of tagged insects and a subsequent high recapture in baited traps may not be indicative of high trap efficiency, because individual captures may reflect many unsuccessful and one successful visit to a trap.

Such assessments in day-flying species can be comparatively straightforward. Among moths, trap efficiencies with synthetic pheromone baits have been reported as 88% for *Argyrotaenia velutinana* (Baker et al., 1976), 93% for *G. molesta* (Cardé et al., 1975b), both for fresh sticky traps, and 47% for *L. dispar* in a no-exit trap (Cardé et al., 1977a). We might define trap efficiency as the percentage captured of those individuals that approached to within some arbitrary distance of the trap such as 2 m or the percentage captured of those individuals that exhibited oriented behavior toward the trap within that distance as in Cardé et al., 1977a). A less subjective definition would be the percentage captured of those individuals that contacted the trap as in Mastro et al. (1977), but this might be a small percentage of those that approached the trap. Furthermore, when observations are conducted in areas of high population density, which is usually the case, many individuals may approach the trap but never orient to or contact the pheromone plume.

Elkinton and Childs (1983) have shown that the USDA Milk Carton trap (Fig. 2E) captured 9.6% of the male gypsy moths that approached within 2 m, 21% of the males that oriented to the trap or the tree it was attached to, and 76% of the males that contacted the traps. The technique for collecting data in these studies involved observers positioned to the side of the trap, generally 1-3 m crosswind, describing the behaviors into an audio tape recorder for subsequent analysis. There was no evident effect of the observation process upon moth behavior. Males of *L. dispar* respond to ultrasound with "escape" maneuvers as a bat evasion strategy (Baker and Cardé, 1978). For those insects that respond similarly, care must be taken in the field to avoid generating ultrasound (by breaking twigs, etc.) because ultrasound will readily terminate response to pheromone in *L. dispar* and potentially other insects sensitive to bat cries.

There are formidable difficulties in observing activity at night. Image intensification (Lingren et al., 1978) and IR vision equipment (Conner et al., 1980) offer sophisticated solutions to this difficulty and can allow observation of trap efficiency for nocturnal Lepidoptera.

X. Design and Interpretation

First, we will consider field experiments intended to determine relative activities of different treatments of trap or lure type. In these comparisons the objective is to distinguish statistically among those treatments differing in ability to lure and retain the target species. The ideal experimental design would minimize variation in trap catch caused by differences in population density in different areas of the test site (or account for these differences in replicate blocks) and as well minimize or eliminate interactions between different treatments.

Wall and Perry (1978, 1980, 1981) have investigated trap interactions in some detail in *C. nigricana* and they have concluded that "poaching" of one trap by another occurs with intertrap distances up to 100 m. Poaching is suppression of trap catch relative to the level achieved by a single isolated trap; it is a composite of the effects of the overlapping of the active spaces generated by the traps and general activity levels of the responders which result in movement between the active spaces.

In testing different lures or traps when interactions exist at typical intertrap distances, the object of the experimental design should be to ensure that interactions affect all treatments as equally as practicable. A Latin-square array of traps achieves this quite readily (Perry et al., 1980), but it has practical limitations in the number of treatments that can be tested simultaneously. In species possessing a relatively short seasonal interval for trapping (so that few sequential tests can be conducted) or where "position effects" are strong as when highly clumped dispersion patterns exist, a fixed array may not offer an efficient scheme.

A randomized complete block design is an alternative that offers the opportunity to test simultaneously a large number of treatments while keeping the number of replicates (e.g., 5) in bounds. However, the arrangement of interactions will be uneven, so that some treatments will receive a greater degree of poaching than others. This undesirable effect can be mitigated to a large extent by rerandomization of the treatments within blocks as often as the traps are sampled. Upon each rerandomization, the catches can be considered as separate replicates.

It should be noted that Wall et al. (1981) have documented in *C. nigricana* "contamination" of the vegetation near a trap with pheromone. Presumably, such emission of pheromone from vegetation could render a trap site unsuitable for a new treatment in a rerandomization program. This phenomenon ought to be explored in other species and pheromone systems.

An additional advantage of the complete randomized block design is that it allows blocks to be positioned according to population levels, when such information is available. The intertrap distance used in many experiments with Lepidoptera is about 10 m, even though this is typically much less than the interaction distance.

The effects of wind direction on trap catch is another factor to consider in experimental design. Wall et al. (1981) showed that in an array of traps the trap farthest upwind captures the most moths. Any design where treatments remain in the same location throughout are subject to differences in trap catch caused by their position relative to the wind direction. If tests are conducted in areas where wind direction is relatively constant, this problem can be partially alleviated by deploying a line of traps crosswind. However, wind direction in most locations changes markedly. Regardless of wind direction, it is likely that trap catch is greater in the outermost traps in any line or array. This problem can be alleviated by placing at the end of each line an extra baited trap whose catch is not tabulated with the rest of the data.

Establishing positive catch requires a treatment of unbaited (and uncontaminated) traps for comparison to catches of putative attractive treatments. Otherwise, even simple "attraction" cannot be verified, because some insects can blunder into unbaited traps, especially in epidemic population levels. Wall and Perry (1982) also advocate the use of a weakly attractive treatment against which other treatments can be compared, but the unbaited trap is still a useful benchmark.

Statistical tests to separate treatments of course fall into two general categories: parametric and nonparametric. Parametric tests make assumptions about the underlying distribution of the data. For instance, analysis of variance, which is undoubtedly the most widely used procedure, assumes that the trap catch data are normally distributed about the true mean value for each treatment and that the variances of these distributions are equal. These assumptions are rarely addressed or verified, although statistical validation tests such as Bartlett's test for equal variance are available. Usually the variance associated with high trap catch is much greater than that for traps that catch few individuals.

If analysis of variance is comtemplated, transformations, such as the square-root transformation $(x_i + 0.5)^{1/2}$ where x_i is the number caught in trap i, can be used to stabilize the variance. A discussion of the appropriate transformation to select can be found in Snedecor and Cochran (1980). If the effects of additional factors on trap catch such as overall differences between blocks or between days are to be investigated then a two-way or n way analysis of variance is required. This entails the additional assumption of no interaction between factors, an assumption which can be tested statistically.

Associated with analysis of variance for three or more treatments are tests for multiple pairwise comparisons such as Duncan's new multiple range test or the Student-Neuman-Keuls (SNK) test (see Sokal and Rohlf, 1969). These tests enable the investigator to determine which of the treatments fall into significantly distinct categories. These tests are designed to hold the overall error rate or the probability of finding differences when they do not exist at some specified level (α) such as 0.05 for all comparisons conducted. As the number of treatments tested increases the likelihood of detecting small but real differences between treatments declines. For this reason the number of treatments tested in any one experiment should be minimized. These multiple comparison tests also assume underlying normality of the trap catch data.

Nonparametric tests have the advantage of making no assumptions about the underlying distribution of data and therefore obviate the need for transformations. For comparisons of two treatments the Wilcoxon rank sum test or the signed ranks test for matched pairs is easily calculated and detects differences between treatments with almost as much power as the alternative parametric tests. Unfortunately, nonparametric multiple comparison tests for three or more treatments are not widely available, although they do exist. For instance, Bedard et al. (1980) compare catch at a series of pheromone trap treatments using a nonparametric multiple comparisons test devised by Sen (1968).

An additional limitation on nonparametric tests appears when "ties" occur in the data such as multiple occurrences of 0 or 1 individual captured. Although corrections for tie values exist, nonparametric comparisons are inadvisable if there are a large number of them.

Failure to use unbaited traps, to replicate treatments, and to subject the data to statistical analysis, all of which were common in the early studies of attraction, are now infrequent, but nonetheless still exist in the recent literature.

XI. Competition of Traps with Natural Sources

Besides trap interactions or poaching, traps of course compete with natural pheromone emitters. Male moths, for example, obviously can be lured to females instead of pheromone-baited traps. Depending upon the ratio of males to calling females, the ratio of traps to females, the time spent in copulation, and, after mating, the refractory period until male competence for attraction, the proportion of males captured in traps will be altered. If the emergence patterns of the sexes are not entirely synchronous, as would be the case in protandrous species, then in the early part of the flight, males may be attracted to traps; when the females emerge in numbers and many of the available males mate, then trap catch should decline. Competition, as outlined here, would be most important to document and understand in trapping applications where trap catch is to be related to either population levels or flight phenology.

The phenomenon of competition has been invoked in two species, the summer fruit tortrix, *Adoxophyes orana* (Minks and DeJong, 1975), and the codling moth, *C. pomonella* (Howell, 1974; Riedl et al., 1976), to explain the drop in catch that occurs during the peak of female emergence. Clearly, males that are engaged in mating cannot be lured to traps, but it is not simply explicable (Minks, 1977; Cardé, 1979) why trap catch should drop so precipitously unless the proportion of the calling sex is relatively high.

Another possible explanation is that traps are relatively less attractive than the calling insect. In the cases of *A. orana* (Den Otter and Klijnstra, 1980) and *C. pomonella* (Bartell and Bellas, 1981) there is evidence that the entire pheromone bouquets are not yet known. It is tempting to speculate that complete pheromone blends would restore the attractiveness of the synthetic-baited traps during the peak female emergence.

In the Oriental fruit moth, *G. molesta,* Baker et al. (1980) found no suppression of trap catch during the peak of female eclosion during the first adult flight. Experimental manipulation of the numbers of virgin female and male gypsy moths in a grid of traps demonstrated a minor competition effect (Elkinton and Cardé, 1984).

The competition effect warrants investigation in those species in which the interpretation of trap catch necessitates a constant relationship with population density.

XII. Use Patterns in Pest Management

12.1. Survey and Detection

The most vexing problems in survey and detection are knowing how the observed captures of the target insect relate to the actual numbers and dispersion of the animals in the survey area. Because the target organism presumably is present in very low numbers and other life stages are exceedingly difficult to detect, much less sample with precision, it is usually impossible to calibrate the trap catch by means of other detection techniques.

One method that can be used is to release marked individuals and document their capture pattern in a survey grid (Elkinton and Cardé, 1980, 1981; Schwalbe, 1981). A key assumption, of course, is that the dispersal and attraction behaviors of the released animals are essentially identical to those of the native insects. Verifying such comparability in behavior necessitates separate and often elaborate studies, including simultaneous release of both types of insects in a trap grid.

A second consideration is the dispersion pattern of release. Releasing all animals from a "point" scource (a relatively small area) will provide excellent information on dispersal, but if the intertrap distance is greater than the typical dispersal distance, then the percentage of recapture will be lower than that from native insects, presumably because the latter will be distributed initially in no systematic fashion with respect to the traps. Values of recapture (but not dispersal) applicable to the average that will occur across a grid of survey traps then require that released insects be distributed "uniformly" (at a number of release sites) across an area bounded by four traps in the grid system. A full discussion of the assumptions and interpretation of such a simulation in the gypsy moth, *L. dispar,* is given in Elkinton and Cardé (1980, 1981). This system can be used as well to calculate daily survival of adults in the field.

12.2. Population Estimation

Trap catch in conjunction with weather data can be used to initiate phenology models (trap catch serves to initiate the model) (e.g., Riedl et al., 1976; Welsh et al., 1981), to establish thresholds for predicting economic damage and when pesticides are warranted (e.g., Minks and DeJong, 1975), or to estimate population levels (e.g., Daterman, 1978) and population trends. The methods used to generate models that allow these predictions and interpretations vary with the intended use pattern, but in all cases they require relatively detailed information on the biology of the insect. Of particular importance are the dispersal behaviors of the pheromone-responsive and nonresponsive insects (Cardé, 1979; Daterman, 1980) and the foraging behavior of the female.

Because males in many species may disperse much farther than females, trap catch of males in a given site may not accurately reflect density of foraging (i.e., ovipositing) females. Further, as noted earlier in the section on rate of pheromone release, a comparatively high rate may lure males from a considerable distance, probably on the order of 10 to 100 m, thereby sampling populations potentially *outside* the crop. This effect can be mitigated by employing low rates of pheromone emission, and thus tailoring the sampling to the immediate vicinity of the trap as has been demonstrated in *G. molesta* (Baker and Roelofs, 1981); such a tactic must be weighed against a possible loss in sensitivity of the system.

12.3. Mass Trapping

If traps can be deployed at a density sufficient to eliminate a very large proportion of the population [e.g., both sexes as in the Japanese beetle, *Popillia japonica* (Ladd and Klein, 1982) or the boll weevil (Hardee, 1982)], then damage may be reduced. In species in which females lure males, generally the trap must remove a sufficient proportion of the males to greatly lower the percentage of fertile females. The theory of mass trapping is covered comprehensively by Knipling (1979). The development and promise of this technique have been described for the redbanded leafroller by Roelofs et al. (1970), the gypsy moth by Webb (1982), and the pine bark beetle by Bakke and Riege (1982) and Bakke (1982). Key factors in efficacy are the efficiency of the trap, the differences in the response rhythms of the responding and the emitting insects, seasonal differences in the emergence rhythms of the sexes, and the population levels to be suppressed.

12.4. Evaluation of Mating Disruption

An important use of pheromone-baited traps is to monitor the efficacy of mating disruptant applications. For example, catch in baited traps is employed as an indicator of the efficacy of aerial application of pink bollworm pheromone, *Pectinophora gossypiella* (Doane and Brooks, 1980). In most cases it would seem most efficient to determine the rate of pheromone emission from the calling insects and duplicate this rate in the trap. A "supernormal" rate may lure pink bollworm males in disruptant plots, whereas females and traps baited with dispensers yielding a natural emission rate may not (Doane and Brooks, 1980). The latter traps then give an accurate reading of the suppression of attraction and mating. This suggests that the most useful approach to monitoring the efficacy of disruption with traps is to duplicate the natural emission rate.

XIII. Conclusion

Field trapping with pheromones serves manifold ends but often initially it is employed in the characterization of the behavioral reaction and in aiding identification of the semiochemicals involved. Once lures baited with synthetic pheromone are available, traps can be important management tools in population survey and monitoring, direct population suppression by mass trapping, and development and monitoring of the efficacy of mating disruption achieved via atmospheric permeation with pheromones (see Campion, Chapter 14, this volume).

Success in these endeavors may be quite simply achieved by adapting an existing trapping system developed for another species, typically one closely allied to the target insect. But as considerable value often rests on the efficacy of the trapping system, development efforts in lure characteristics, trap configuration, deployment protocols, and, most important, models for interpretation of the trapping results ought not to be neglected. Unfortunately, the potential of pheromone traps has not yet been realized in many cases because traps have been deemed "operational" for management merely because they could trap the target species.

Pheromone traps are in general many times more sensitive and specific than other types of sampling systems, but it is doubtful that these advantages will be fully utilized unless we can couple the trapping system with a detailed knowledge of the target insect's behavior and population dynamics. Up to now, such information has not been necessary in pest management programs, largely because control has been based upon broad spectrum insecticides. The rapidity with which we implement these new technologies will be in large measure dependent upon the integration of the trapping system with models of the insect's behavior and population ecology.

XIV. References

Baker TC, Cardé RT (1978) Disruption of gypsy moth male sex pheromone behavior by high frequency sound. Environ Entomol **7**:45–52.

Baker TC, Cardé RT (1979) Analysis of pheromone-mediated behavior in male *Grapholitha molesta,* the Oriental fruit moth (Lepidoptera:Tortricidae). Environ Entomol **8**:956–968.

Baker TC, Cardé RT, Croft BA (1980) Relationship between pheromone trap capture and emergence of adult Oriental fruit moths, *Grapholitha molesta* (Lepidoptera:Tortricidae). Can Entomol **112**:11–16.

Baker TC, Kuenen LPS (1982) Pheromone source location in flying moths: A supplementary nonanemotactic mechanism. Science **216**:424–427.

Baker TC, Cardé RT, Roelofs WL (1976) Behavioral responses of male *Argyrotaenia velutinana* (Lepidoptera:Tortricidae) to components of its sex pheromone. J Chem Ecol **2**:333–352.

Baker TC, Roelofs WL (1981) Initiation and termination of Oriental fruit moth male response to pheromone concentrations in the field. Environ Entomol **10**:211–218.

Bakke A (1982) Mass trapping of the spruce bark beetle *Ips typographus* in Norway as part of an integrated control program. In: Insect Suppression with Controlled Release Pheromone Systems. Kydonieus AF, Beroza M (eds), CRC Press, Boca Raton, Florida, Vol 2, pp 17–25.

Bakke A, Riege L (1982) The pheromone of the spruce bark beetle *Ips typographus* and its potential use in the suppression of beetle populations. In:

Insect Suppression with Controlled Release Pheromone Systems. Kydonieus AF, Beroza M (eds), CRC Press, Boca Raton, Florida, Vol 2, pp 3–15.

Bartell RJ, Bellas TE (1981) Evidence for naturally occurring, secondary compounds of the codling moth female sex pheromone. J Aust Entomol Soc **20**: 197–199.

Bartlett RJ, Jones RL, Kulman HM (1982) Evidence for a multicomponent sex pheromone in the yellowheaded spruce sawfly. J Chem Ecol **8**:83–94.

Bedard WD, Wood DL, Tilden PE, Lindahl KQ, Silverstein RM, Rodin JO (1980) Field response of the western pine beetle and one of its predators to host- and beetle-produced compounds. J Chem Ecol **6**:625–641.

Cardé RT (1979) Behavioral responses of moths to female-produced pheromone and the utilization of attractant-baited traps for population monitoring. In: Movement of Highly Mobile Insects: Concepts and Methodology in Research. Rabb RL, Kennedy GG (eds), North Carolina State University, pp 286–315.

Cardé RT (1981) Precopulatory sexual behavior of the adult gypsy moth. In: USDA Tech Bull 1584. Doane CC, McManus ML (eds), pp 572–587.

Cardé RT (1984) Chemo-orientation: Flying insects. In: Chemical Ecology of Insects. Bell WJ, Cardé RT (eds), Chapman and Hall, London, pp 111–124.

Cardé RT, Baker TC, Roelofs WL (1975a) Ethological function of components of a sex attractant system for Oriental fruit moth males, *Grapholitha molesta* (Lepidoptera:Tortricidae). J Chem Ecol **1**:475–491.

Cardé RT, Baker TC, Roelofs WL (1975b) Behavioral role of individual components of a multichemical attractant system in the Oriental fruit moth. Nature (London) **253**:348–349.

Cardé RT, Doane CC, Baker TC, Iwaki S, Murumo S (1977a) Attractancy of optically active pheromone for male gypsy moths. Environ Entomol **6**:768–772.

Cardé RT, Doane CC, Granett J, Hill AS, Kochansky J, Roelofs WL (1977b) Attractancy of racemic disparlure and certain analogues to male gypsy moths and the effect of trap placement. Environ Entomol **6**:765–767.

Conner WE, Eisner T, Vander Meer RK, Guerrero A, Ghiringelli D, Meinwald J (1980) Sex attractant of an arctiid moth (*Utethesia ornatrix*): A pulsed chemical signal. Behav Ecol Sociobiol **7**:55–63.

Coppel HC, Casida JE, Dauterman WC (1960) Evidence for a potent sex attractant from the introduced pine sawfly *Diprion similus.* Ann Entomol Soc Amer **53**:510–512.

Cross WH, Mitchell HC, Hardee DD (1976) Boll weevils: Response to light sources and colors on traps. Environ Entomol **5**:565–571.

Daterman GE (1978) Monitoring and early detection. In: The Douglas-fir Tussock Moth: A Synthesis. USDA Tech Bull 1505. Brookes MH, Stark RW, Campbell RW (eds), pp 99–102.

Daterman GE (1980) Pheromone responses of forest Lepidoptera: Implications for dispersal and pest management. In: Dispersal of Forest Insects: Evaluation, Theory and Management Implications. Berryman AA, Safranyik L (eds), Washington State University, pp 251–265.

Den Otter CJ, Klinjnstra JW (1980) Behaviour of male summerfruit tortrix moth, *Adoxophyes orana* (Lepidoptera:Tortricidae), to synthetic and natural sex pheromone. Entomol Exp Appl **28**:15–21.

Doane CC, Brooks TW (1980) Research and development of pheromones for insect control with emphasis on the pink bollworm, *Pectinophora gossypiella*. In: Management of Insect Pests with Semiochemicals: Concepts and Practice. Mitchell ER (ed), Plenum, New York, pp 285–303.

Elkinton JS, Cardé RT (1980) Distribution, dispersal, and apparent survival of male gypsy moths as determined by capture in pheromone-baited traps. Environ Entomol **9**:729–737.

Elkinton JS, Cardé RT (1981) The use of pheromone traps to monitor distribution and population trends of the gypsy moth. In: Management of Insect Pests with Semiochemicals: Concepts and Practice. Mitchell ER (ed), Plenum, New York, pp 41–55.

Elkinton JS, Cardé RT (1984) The effect of wild and laboratory-reared female gypsy moths (Lepidoptera:Lymantriidae) on the capture of males in pheromone-baited traps. Environ Entomol **13** (in press).

Elkinton JS, Childs RW (1983) Efficiency of two gypsy moth (Lepidoptera: Lymantriidae) pheromone-baited traps. Environ Entomol **12**:1519–1525.

Farkas SR, Shorey HH (1972) Chemical trail-following by flying insects: A mechanism for orientation to a distant odor source. Science **176**:67–68.

Granett J (1973) A disparlure-baited trap for capturing large numbers of gypsy moths. J Econ Entomol **66**:359–362.

Granett J (1974) Estimation of male mating potential of gypsy moths with disparlure baited traps. Environ Entomol **3**:383–385.

Hardee DD (1982) Mass trapping and trap cropping of the boll weevil, *Anthonomus grandis* Boheman. In: Insect Suppression with Controlled Release Pheromone Systems. Kydonieus AF, Beroza M (eds), CRC Press, Boca Raton, Florida, Vol 2, pp 65–71.

Howell JF (1974) The competitive effect of field populations of codling moth on sex attractant trap efficiency. Environ Entomol **3**:803–807.

Kennedy JS, Marsh D (1974) Pheromone-regulated anemotaxis in flying moths. Science **184**:999–1001.

Kochansky JP, Cardé RT, Taschenberg EF, Roelofs WL (1977) Rhythms of male *Antheraea polyphemus* attraction and female attractiveness, and an improved pheromone systhesis. J Chem Ecol **3**:419–427.

Knipling EF (1979) The basic principles of insect population suppression and management. Agriculture Handbook 512 USDA, Washington.

Ladd TL Jr, Klein MG (1982) Trapping Japanese beetles with synthetic female sex pheromone and food-type lures. In: Insect Suppression with Controlled Release Pheromone Systems. Kydonieus AF, Beroza M (eds), CRC Press, Boca Raton, Florida, Vol 2, pp 57–64.

Lewis T, Macaulay EDM (1976) Design and elevation of sex-attractant traps for pea moth, *Cydia nigricana* (Steph.) and the effect of plume shape on catches. Ecol Entomol **1**:175–187.

Lingren PD, Sparks AN, Raulston JR, Wolf WW (1978) Applications for nocturnal studies of insects. Bull Entomol Soc Am **24**:206–212.

McLaughlin JR, Brogdon JE, Agee HR, Mitchell ER (1975) Effect of trap color on captures of male cabbage loopers and soybean loopers in double-cone pheromone traps. J Georgia Entomol Soc **10**:174–179.

Mastro VC, Richerson JR, Cameron EA (1977) An evaluation of gypsy moth pheromone-baited traps using behavioral observations as a measure of trap efficiency. Environ Entomol **6**:128-132.

Minks AK (1977) Trapping with behavior-modifying chemicals: Feasibility and limitations. In: Chemical Control of Insect Behavior. Shorey HH, McKelvey JJ (eds), Wiley Interscience, New York, pp 385-394.

Minks AK, DeJong DJ (1975) Determination of spraying dates for *Adoxophyes orana* by sex pheromone traps and temperature recordings. J Econ Entomol **68**:729-732.

Mitchell ER, Webb JC, Baumhover AH, Hines RW, Stanley JM, Endris RG, Lindquist DA, Masuda S (1972) Evaluation of cylindrical electric grids as pheromone traps for loopers and tobacco budworms. Environ Entomol **1**:365-368.

Perry JN, Wall C, Greenway AR (1980) Latin-square designs in field experiments involving insect sex-attractants. Ecol Entomol **5**:385-396.

Ramaswamy SB, Cardé RT (1982) Nonsaturating traps and long-life lures for monitoring spruce budworm males. J Econ Entomol **75**:126-129.

Ramaswamy SB, Cardé RT, Witter JA (1983) Relationship between catch in pheromone-baited traps and larval density of the spruce budworm (Lepidoptera:Tortricidae). Can Entomol **115**:1437-1443.

Riedl H (1980) The importance of pheromone trap density and trap maintenance for the development of standardized monitoring procedures for the codling moth (Lepidoptera:Tortricidae). Can Entomol **112**:655-663.

Riedl H, Croft BA, Howitt AJ (1976) Forecasting codling moth phenology based on pheromone trap catches and physiological time models. Can Entomol **108**:449-460.

Riedl H, Hoying SA, Barnett WW, Detar JE (1979) Relationship of within-tree placement of the pheromone trap to codling moth catches. Environ Entomol **8**:765-769.

Roelofs WL, Class EH, Tette J, Comeau A (1970) Sex pheromone trapping for red-banded leaf roller control: Theoretical and actual. J Econ Entomol **63**: 1162.

Sanders CJ (1978) Evaluation of sex attractant traps for monitoring spruce budworm populations (Lepidoptera:Tortricidae). Can Entomol **110**:43-50.

Sanders CJ (1981) Sex attractant traps: Their role in management of spruce budworm. In: Management of Insect Pests with Semiochemicals. Mitchell ER (ed), Plenum, New York, pp 75-91.

Schwalbe CP (1981) Disparlure-baited traps for survey and detection: In: The Gypsy Moth: Research Toward Integrated Pest Management. Doane CC, McManus ML (eds), USDA Tech Bull 1584, pp 542, 549.

Schwalbe CP, Paszek EC (1978) Preliminary evaluation of large capacity traps. Progress Report, APHIS, Gypsy Moth Methods Development Laboratory, April 1, to September 30, 1978. Otis Air Force Base, Mass.

Sen PK (1968) On a class of aligned rank order tests in two-way layouts. Ann Math Stat **39**:1115-1124.

Snedecor GW, Cochran WG (1980) Statistical Methods, 7th ed. Iowa State University Press, Ames.

Sokal RR, Rohlf FJ (1969) Biometry. Freeman, San Francisco, pp 239-246.

Stanley JM, Webb JC, Wolf WW, Mitchell ER (1976) Electrocutor grid insect traps for research purposes. Trans Am Soc Agric Eng **20**:175-178.

Tilden PE, Bedard WD, Lindahl KQ, Wood DL (1983) Trapping *Dendroctonus brevicomis*: Changes in attractant release rate, dispersion of attractant and silhouette. J Chem Ecol **9**:311–321.

Timmons GM, Potter DA (1981) Influence of pheromone trap color on capture of lilac borer males. Environ Entomol **10**:756–759.

Wall C, Perry JN (1978) Interactions between pheromone traps for the pea moth, *Cydia nigricana* (F.). Entomol Exp Appl **24**:155–162.

Wall C, Perry JN (1980) Effect of spacing and trap numbers on interactions between pea moth pheromone traps. Entomol Exp Appl **28**:313–321.

Wall C, Perry JN (1981) Effects of dose and attractant on interactions between pheromone traps for the pea moth, *Cydia nigricana* (F.). Entomol Exp Appl **30**:26–30.

Wall C, Perry JN (1982) The behaviour of moths responding to pheromone sources in the field: A basis for discussion. Les Médiateurs chimiques INRA Coll **7**:171–188.

Wall C, Sturgeon DM, Greenway AR, Perry JN (1981) Contamination of vegetation with synthetic sex-attractant released from traps for the pea moth, *Cydia nigricana* (F.). Entomol Exp Appl **30**:111–115.

Webb RE (1982) Mass trapping of the gypsy moth. In: Insect Suppression with Controlled Release Pheromone Systems. Kydonieus AF, Beroza M (eds), CRC Press, Boca Raton, Florida, Vol 2, pp 27–56.

Webb JC, Agee HR, Stanley JM (1978) Electrocutor grid trap arcing sounds—Analysis and bollworm moth detection. Trans Amer Soc Agric Eng **21**:982–985.

Welch SM, Croft BA, Michels MF (1981) Validation of pest management models. Environ Entomol **10**:425–432.

Chapter 5

Electroantennogram Assays: Rapid and Convenient Screening Procedures for Pheromones

Wendell L. Roelofs[1]

I. Introduction

Although odor perception in general is poorly understood, the insect's peripheral sensing system—the antenna—is easily accessible and has been studied in great detail (Steinbrecht and Schneider, 1980; Kaissling and Thorson, 1980; Kaissling, 1971). Schneider (1957) pioneered the electroantennogram (EAG) technique and carried out the first electrophysiological experiments on olfaction in insects with *Bombyx mori.* Studies with *B. mori* revealed that slow olfactory receptor potentials could be recorded from an isolated antenna positioned between two glass capillary microelectrodes connected to an amplifier and a recording instrument. Schneider (1963) suggested that the "EAG is essentially the sum of many olfactory receptor potentials recorded more or less simultaneously by an electrode located in the sensory epithelium." He interpreted (Schneider, 1969) the negative potential of these slow electrical reactions of dendrites in the olfactory cells as a receptor membrane depolarization. The amplitude of the response, which correlates to the frequency of generated nerve impulses, was found to increase with increasing concentrations of the chemical stimulus until a saturation level was reached.

Kaissling and Thorson (1980) recently summarized information on the structural, chemical, and electrical aspects of olfactory sensilla related to EAGs. In general, male moth antennae possess thousands of long olfactory hairs, *sensilla trichodea,* which contain receptor cells that respond to the female pheromone

[1]Department of Entomology, New York State Agricultural Experiment Station, Geneva, New York.

components. The lumen of the hair contains receptor lymph and the fiber endings (dendrites) from several receptor cells. The hair walls are perforated with thousands of 100-Å pore tubules that are probably involved in the diffusion of odor molecules to the receptor-cell dendrite.

A working model of the olfactory receptors includes a transepithelial current path through the sense cells and another through the surrounding auxillary cells. An odor stimulus causes a negative deflection of the transepithelial potential, or receptor potential, which rises relatively fast and declines more slowly after the end of the stimulus.

Specificity of an olfactory cell to pheromone components and related compounds can be effected by several factors in the transduction process, including different kinds of ion-channel openings produced by different conformational changes of the receptor molecule as it interacts with various chemical stimuli. Each sense cell in the sensillum possesses its own chemical specificity for pheromone components and each can be distinguished by the size of the nerve impulse it generates.

Studies with *Bombyx mori* indicated that one molecule of pheromone could trigger a nerve impulse. Further studies on this phenomenon (Kaissling, 1979) showed that there are small potential fluctuations that precede each nerve impulse by 10–40 msec. These elementary receptor potentials were thought to be produced by the opening of single ion channels.

The EAG records a potential difference between the recording (distal end of antennae) and the indifferent electrodes. Nagai (1981) found a nearly linear relationship between the initial fast negative charge and the antennal length. The density of the sensilla trichodea was found to be uniform over the whole antenna and all responding sensillae located between the recording, and the indifferent electrode seemed to be involved in the initial fast negative charge. His data indicated that there is a gradient in the response potential along the whole length of the antenna with the elicited potential becoming more negative toward the distal end. He also suggested that the potential is effectively summated only in the proximal direction and is the result of an interaction of the initial fast negative charge due to excited sensory receptors and of slower developing electrotonic potentials that spread from stimulated cells to neighboring units through epithelial cell layers of high electrical resistance. Electrotonic spread in the proximal direction would increase the overall response amplitude.

The EAG amplitude and shape are produced by the processes described above and depend on the concentration and chemical structure of the stimulus. One of the pheromone components of male moths usually elicits the greatest EAG response at the saturation level of all compounds tested, although various less active derivatives can produce responses with a falling phase as slow or slower than that of the pheromone component.

The sensitivity and specificity of the male's antenna to its pheromone components make it a powerful tool in assaying for pheromone components and in predicting the structure of the pheromone components. This technique has been used very successfully in the Lepidoptera, but also has been used for assaying pheromone components in many other types of insects, such as Coleoptera (Payne, 1979, and references therein; Levinson et al., 1978; Angst and Lanier, 1979; Gutmann et al., 1981; Wadhams, 1984, Chapter 7, this volume), Hymenoptera (Jewett et al., 1977; Kraemer et al., 1981; Williams et al., 1982), Homoptera (Moore, 1981), and Orthoptera (Persoons et al., 1979; Nishino and Kimura, 1981, 1982; Nishino and Takayanagi, 1981).

II. Basic Apparatus

In general the EAG setup consists of electrodes, an amplifier with high input impedance (10^{12} ohm), and a recording device. There are many ways of putting together a useable system, but we have found that for most purposes, a simple system provides the best results (Roelofs, 1977).

2.1. Electrodes

Electroantennograms can be recorded by inserting tungsten electrodes into antennae, but more typically Ag/AgCl electrodes are used. In our system, the electrodes are short lengths of silver wire that have been scraped shiny, then dipped into molten silver chloride for several seconds. If the dip is too lengthy, the solder connections of the silver wires with the lead wires could melt. The silver chloride can be melted in a ceramic dish with a Bunsen burner, and can be done in the light.

The chloridized silver wire used for the indifferent (ground) electrode is simply submerged in a saline solution in a shallow glass dish containing a bed of Tackiwax (Fig. 1), and makes electrical contact with the antennal base through the saline solution. The recording electrode is made by inserting another chloridized silver wire into a glass disposable pipet until the end of the wire is about 1 cm from the pipet tip. The electrode is dipped into the saline solution so that it fills the tip by capillarity and contacts the chloridized silver wire within. Aluminum foil is wrapped around the outside of the pipet, particularly the portion containing the exposed Ag/AgCl wire, and connected to ground to shield the input electrode. In most cases, this provides adequate shielding and a Faraday cage is unnecessary. A short coaxial cable connects the recording electrode to the amplifier.

2.2. Saline Solution and Antenna Positioning

In the setup shown in Fig. 1, a plucked-off antenna or the head of an insect is pressed into a bed of Tackiwax. In most cases a few distal antennal segments are snipped off with microscissors, and then the recording electrode is micromanipulated over the antennal tip. In Lepidoptera we have found that almost

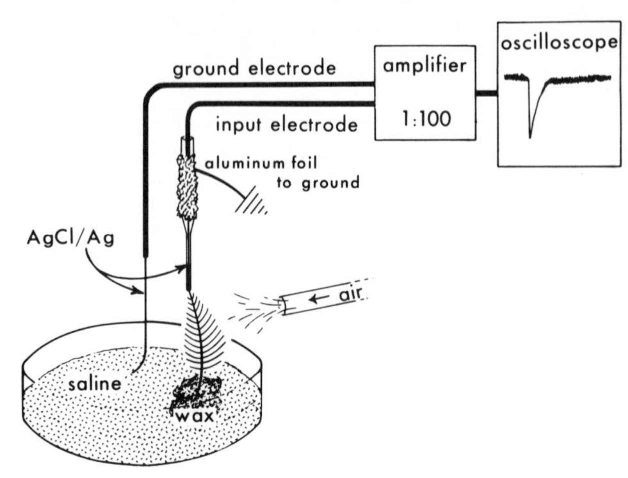

Figure 1. EAG setup with antenna positioned in wax. The antennal base makes contact with ground through a saline solution, and the distal antennal tip makes contact with saline solution in the input electrode tip. Test chemicals are puffed into an airstream that continuously passes over the antenna (from Roelofs, 1978).

any saline solution can be used in the dish and in the recording electrode for good EAG recordings. For years we have used a solution (suggested by K. Kaissling) consisting of the following: NaCl (7.5 g/liter); $CaCl_2$ (0.21 g/liter); KCl (0.35 g/liter); and $NaHCO_3$ (0.2 g/liter). Another commonly used solution is 3 *M* KCl.

If antennal preparations degenerate rapidly, they sometimes can be improved by adjusting the ion concentration, osmolarity, and pH of the solution to that of the insect's hemolymph. Kaissling and Thorson (1980) made these calculations (Table 1) for both hemolymph and receptor lymph from two *Antheraea* moth species. The osmolarity was adjusted with glucose. Electrophysiological studies were conducted using the hemolymph saline in the ground electrode and receptor lymph saline in the input electrode. This system produced recordings for several days if the antenna was not detached from the insect.

2.3. Amplifier

Commercial amplifiers with high-input impedance (10^{12} ohm) are available from numerous companies. The antennal responses usually are in the range of 1–20 mV and so we amplify the signal a hundredfold for monitoring on an oscilloscope. A simple inexpensive (U.S. $20.00) high-impedance ×100 amplifier can be constructed (Bjostad and Roelofs, 1980) specifically for the EAG technique. Amplifiers and many other accessories specially designed for the EAG technique by Jan van der Pers can be obtained from Murphy Developments, PO 199, NL-9750 RD Haren, The Netherlands.

Table 1. Hemolymph (H) and receptor lymph (RL) saline solutions (from Kaissling, 1980)

	H	RL
	Millimoles per liter	
KCl	6.4	171.9
KH_2PO_4	20.0	9.2
K_2HPO_4	–	10.8
$MgCl_2$	12.0	3.0
$CaCl_2$	1.0	1.0
KOH	9.6	–
HCl	–	1.5
Glucose	354.0	22.5
NaCl	12.0	25.0
	Milliosmoles per liter	
Osmotic pressure	450	475
pH	6.5	6.5

2.4. Recording Devices

The signals can be monitored with a high-speed recorder or an oscilloscope. Jewett et al. (1977) recorded EAGs of pine sawflies with a recording galvanometer. The advantage of a permanent recording is that the response amplitudes and shapes are available for later analyses and comparisons. A disadvantage is that data are taken from the recording paper after conducting the EAGs and this can be as time consuming as the EAG recording session.

In most cases involving the EAG technique as a tool for pheromone identification, the response amplitude is the only value that need be recorded. This can be obtained directly from a storage oscilloscope and recorded along with notes on responses with slow recovery rates. Recording values in this manner eliminates the second data acquisition session usually involved when using permanent recorders. We set our storage oscilloscope (Tetronix Model 564B) at 10 sec/sweep and each vertical division represents 1 mV of antennal response.

The process of recording EAG amplitudes can be made simpler by interfacing with an inexpensive digital voltmeter (Bjostad and Roelofs, 1980). A circuit has been designed to overcome problems with using a voltmeter directly. First, voltmeters are designed to read voltages that remain constant for at least several seconds, whereas an EAG peak voltage occurs over a few milliseconds. A peak detector circuit detects the maximum input voltage and holds the output voltage at that level–this digital value is read as the antennal response. The second problem when using a voltmeter is that the baseline voltage from an antenna drifts up or down at approximately 1 mV min^{-1}, presumably due to changing electrolyte concentrations and neuronal activity in the excised antenna. This baseline drift is eliminated with a high-pass filter with a cut-off frequency of about 1 Hz. After an antennal response is displayed, the voltmeter readout is reset to zero with a reset pushbutton switch just prior to puffing the next test sample across the antenna.

2.5. Sample Delivery System

A number of mechanically activated or solenoid-activated systems can be used successfully to blow specific concentrations of test material across an antennal setup for specified time intervals. For the type of information needed in evaluating relative activities of a series of synthetic chemicals at their saturation levels and in assaying gas–liquid chromatography (GLC) collections, we have found that a very simple "puff" technique has some advantages over other methods (Fig. 2).

Air (ca. 1 ml) from a 5-ml glass syringe is quickly (ca. 30 msec) puffed through a glass disposable pipet containing test material (see Section IV) or through a glass capillary tube used with GLC collections (see Section III) into an airstream that passes continuously over the antennal setup. A test cartridge or capillary tube is easily connected to the glass syringe (with the plunger already pulled back to the 1 ml mark) and inserted into a 0.5-cm-diameter hole located 2.5 cm from the terminal end of the airstream tubing through which filtered air is continuously blown over a positioned antenna at a rate of ca. 1 m/sec. Slight variations in the speed at which the syringe plunger is depressed and in the amount of puffed air do not seem to be critical for the resulting response amplitude. The same syringe can be used for the disposable pipets and the capillary tubes.

Not only can samples be tested in rapid succession with this method but also the short puff allows one to observe distinct differences in recovery rates among responses. Many test materials elicit very sharp responses with the "puff" technique, whereas some elicit a sharp depolarization response with a very slow recovery time to baseline. In many instances, this type of "sticky" response is associated with biologically active materials.

2.6. Biological Amplification

In most cases involving Lepidoptera, there are one or two pheromone components that are easily detected by the GLC analysis. However, other pheromone components present in small quantities or eliciting low EAG activity can easily be missed. Also, EAG responses from some very small insects are difficult to discern from background noise. To overcome these difficulties, Moore (1981) developed a simple technique to amplify the EAG signals based on in-series resistor and capacitor theory.

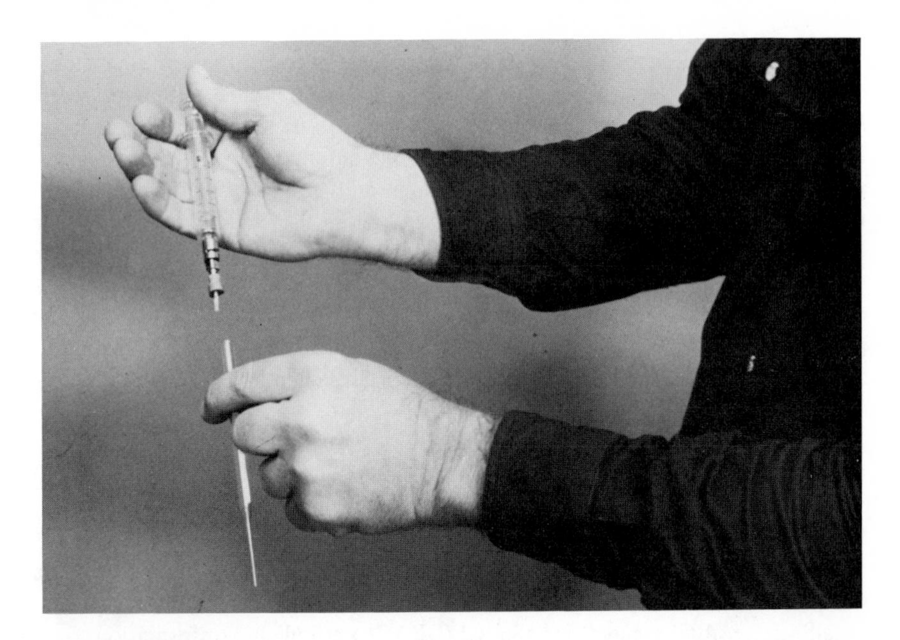

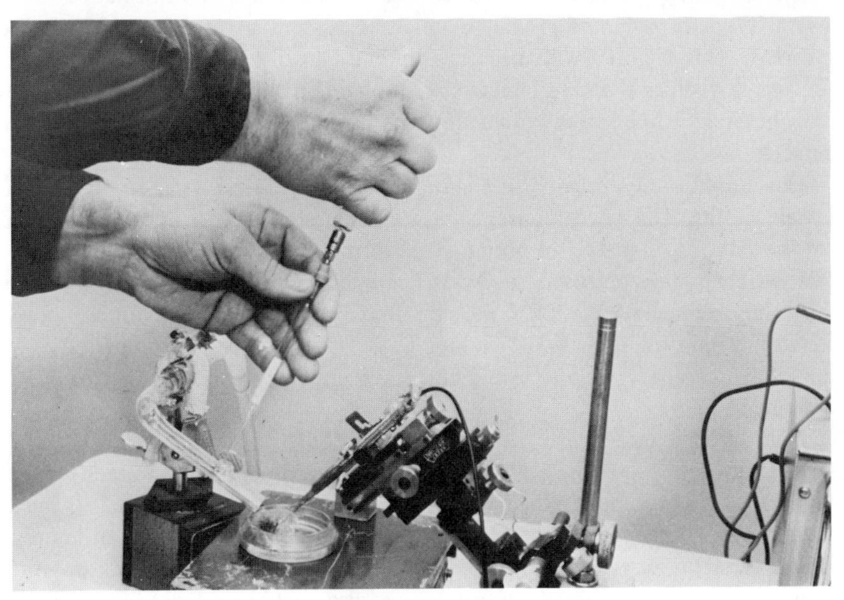

Figure 2. Technique for "puffing" test chemicals for EAG responses. The syringe plunger is withdrawn to 1–2 ml, a disposable pipet containing 10 μg of test chemical on a filter paper strip is connected to the syringe, the syringe-pipet unit is positioned so that the pipet tip can be inserted into the hole in the airstream tubing, the unit is lowered so the pipet tip protrudes into the airstream, the plunger is immediately depressed to "puff" test chemical across the antenna, and the unit is immediately withdrawn from the airstream.

With antennae connected in series, the voltage between the terminal should correspond to the sum of individual antennal voltages. Studies with *Spodoptera littoralis* showed that four antennae connected in series did indeed give larger EAG responses (ca. 6.7 mV) compared to one antenna (ca. 1.7 mV) when exposed to *Z,E*-9,11-14:0Ac. The four antenna series also gave significantly higher responses to the pheromone blend (1% *Z,E*-9,12-14:0Ac added in) compared to *Z,E*-9,11-14:0Ac alone, whereas a single antenna showed no differences. With a very small scale insect, *Matsucoccus josephi,* single antennae gave poor signal: noise ratios with airborne-collected pheromone, but the use of antennae in series produced an EAG response of 1.3 mV. This appears to be a powerful technique and could be used for all the assays described in this chapter.

III. Assay of GLC-Collected Material

When faced with the problem of defining a previously unknown moth sex pheromone, we find that an important first step is to obtain information on GLC retention times of the main EAG-active component. This component usually is easily detected by the GLC–EAG method and information from several types of columns provide a quick assessment of the type of structure involved. This chapter will discuss techniques involved in collecting material for assay, whereas the next chapter by Struble and Arn will discuss techniques for obtaining EAG activity data directly with a GLC electroantennographic detector.

Our initial study usually is conducted with crude material from 1 to 50 female moths. The pheromone glands are either stripped off or the abdominal tip is excised and placed in about 50 μl of methylene chloride. An aliquot of this extract is injected onto an OV-101 nonpolar GLC column usually set at a fairly high temperature (200°C) for the initial run. Eluted compounds are collected in 1-min fractions in 30-cm-long glass capillary tubes over a period of 30 min. Each capillary tube is tested for EAG activity by the "puff" technique described earlier.

One antenna usually can be used to go through the entire series of tubes several times. In most cases, this particular temperature program for elution will produce at least one or two adjacent fractions with EAG activity that is much greater than all other fractions. Initially, this represents the material of interest. Other areas of lesser activity could represent precursor compounds or other pheromone components and those fractions also should be saved for additional investigations.

The fractions containing the main EAG-active component are rinsed with 20 μl of methylene chloride and the solution is injected onto a polar GLC column (for example XF-1150, cyclohexanedimethanol succinate (CHDMS), or phenyldiethanolamine succinate). One-minute fractions are again collected throughout a time interval to cover structural possibilities taken from the OV-101 retention time. The fractions are assayed and the retention time of main activity is compared to standard compound retention times. An advantage of this method of collecting and reinjection is that the active retention time on the polar column can be associated directly with a particular active component retention time on the OV-101 column. Also, the material has been isolated from the GLC columns and can be used for further characterizations, such as capillary chromatographic analysis for *Z*/*E* isomers, mass spectral analysis, microozonolysis, microhydrogenation or hydrogenolysis, etc.

3.1. Unexpected GLC Retention Times

In Lepidoptera, many sex pheromone structures have been identified and a logical guess about the general type of pheromone structure can be made many times. The GLC–EAG method provides a fast determination of the accuracy of that guess. With the spruce budworm, *Choristoneura fumiferana* (Weatherston et al., 1971), we were perplexed with the retention times of the active material from nonpolar and polar columns until it became clear that they were producing a 14-carbon aldehyde instead of the 14-carbon acetate or alcohol we had expected from a tortricid moth in the subfamily tortricinae.

With the saltmarsh caterpillar moth, *Estigmene acrea* (Hill and Roelofs, 1981), we had no active fractions when we collected crude extract as usual from an OV-101 column. We increased the temperature to 200°C and collected for 25 min and got a large EAG response in the last 1-min fraction. The compound, a 21-carbon epoxydiene, had a larger molecular weight than we had anticipated from our previous studies.

With the peach twig borer, *Anarsia linestella* (Roelofs et al., 1975a), we also had an unexpected result. After receiving some pupae from Washington, we had conducted some preliminary studies with monounsaturated standards (see Section IV, 4.1) and found that the 12-carbon acetates elicited the largest response and that response to E5-12:0Ac was the best in that series. We set the OV-101 column at 160° (retention time 12:0Ac = 6.5 min), injected 20 female equivalents (FE) of crude female extract, carried out the 1-min collections, and analyzed the fractions by EAG. All tubes gave background responses of under 0.2 mV except for the 3- to 4-min collection. This retention time was obviously too short for a 12-carbon acetate, and we had to consider 10-carbon compounds or a 12-carbon alcohol or aldehyde.

The material was collected from both polar and nonpolar columns at lower temperatures and it correlated in each case to a monounsaturated 10-carbon acetate. The collected material was shown to be *E*5-10:0Ac by microozonolysis, mass spectral data, microhydrogenation, and further GLC and thin-layer chromatography (TLC) data. This compound was smaller than we had anticipated from all our previous studies.

3.2. Potato Tuberworm Example of Several GLC-Active Areas

The use of two different types of GLC columns sometimes can be quite enlightening. In the case of the potato tuberworm moth, *Phthorimaea operculella* (Roelofs et al., 1975b), crude pheromone extract collected from an OV-1 column (160°C) showed several areas of EAG activity (Fig. 3) with the greatest activity in the 7- to 9-min fractions and a second area of lesser activity in the 5- to 6-min fractions. The 7- to 9-min fractions were combined and collected from a polar column (CHDMS). EAG analysis of these 1-min fractions showed two distinct areas of high EAG activity. Retention times of these active areas relative to standards suggested an unsaturated 13-carbon acetate from OV-1 data and at least a conjugated or homoconjugated double-bond system from the polar column data. The 5- to 6-min OV-1 fraction with lesser activity was reacted with acetyl chloride and recollection from OV-1 and CHDMS with EAG analyses showed that it was converted to acetate with retention times similar to one of the pheromone acetates. In this case the alcohol was found in the gland extract, but was not shown to be a part of the pheromone system.

Collection of the two main EAG-active areas from both OV-1 and CHDMS made it possible to carry out further structural characterizations on the isolated compounds. The earlier compound was determined to be a Δ4,7-13:0Ac by (a) saponification and reacetylation reactions assayed by GLC–EAG techniques, (b) mass spectral data, and (c) microozonolysis. The standard EAG profile (Section IV, 4.1) was used to determine a *E*4,*Z*7 system in this compound. The standard profile was not useful for prediction of the third double bond found in the second EAG-active area from the CHDMS column, but it was subsequently

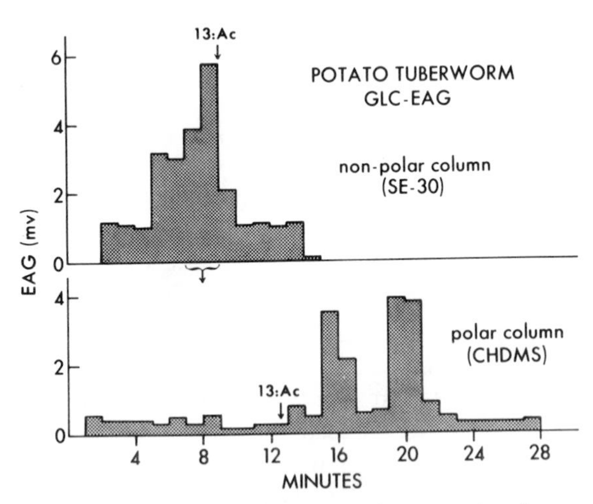

Figure 3. Male antennal responses of potato tuberworm to pheromone gland extract collected from nonpolar and polar GLC columns. Active fractions from the nonpolar column were injected and collected from the polar column to show the presence of two active components (from Roelofs, 1979).

identified to be (*E,Z,Z*)-4,7,10-tridecatrienyl acetate (Persoons et al., 1976; Yamaoka et al., 1976).

3.3. Winter Moth Example of GLC Data on a Unique Structure

There were no sex pheromone structures known for species in the family Geometridae at the time we analyzed the winter moth, *Operophthera brumata*, sex pheromone extract, and so there were no preconceived ideas on what structure to expect. Crude extract collected from an OV-101 column (180°) collected in 2-min fractions gave high EAG activity (5 mV) in the 8- to 10-min fraction. A 14-carbon acetate standard had a retention time of 6.1 min. On the polar XF-1150 column, however, the 14-carbon acetate standard came off in 11.5 min and the EAG-active material in 5.0 min. These data indicated an active component with little functional group polarity.

The retention times of a 19-carbon hydrocarbon (nonadecane) were 8.9 and 2.9 min on the OV-101 and CHDMS columns, respectively, which showed good correlation on the nonpolar column but indicated polarity of at least a conjugated double-bond system with the polar column. Visualization of the collected material produced carbon numbers of 18.95 and 20.9 on the nonpolar and polar columns, respectively. The monounsaturated standards in the EAG assay were ineffective with this species, again indicating a different type of pheromone structure. Further analyses with the GLC-collected material proved the pheromone to be the unique compound (*Z,Z,Z*)-1,3,6,9-nonadecatetraene (Roelofs et al., 1982).

IV. EAG Standard Profiles

The GLC retention times of the main EAG-active component are important in determining the general type of structure involved. The next important step is to define the specificity of the male's antennae to geometrical, positional, and enantiomeric isomers with standards in various series of different carbon lengths (e.g., 12-, 14-, and 16-carbon) and different functional groups (e.g., alcohol, acetate, and aldehyde). Comparative data with standard series can be generated with response amplitudes at the saturation level or with response spectra from dosage-response studies. We have routinely used the former method (Section IV, 4.1) for pheromone identifications, although the latter method (IV, 4.2) could be equally effective.

4.1. Amplitude Profiles

The absolute antennal response in mV varies among antennae and possibly can be affected by temperature, air flow rate, juxtaposition of the air tube to antenna, previous responses, etc. However, relative response amplitudes to a particular standard can be very reproducible in replications with several antennae. The whole series of compounds can be tested with one antennal setup, which eliminates any variations in temperature, flow rate, and the physical aspects of the setup. It generally was found that maximum responses could be generated with a 1 ml puff of air through a disposable pipet containing 10 μg of test chemical on a filter paper strip (2 × 1/2 cm). The test chemical pipets are stored in a refrigerator and used for 6 months to a year. Although the dosage decreases slightly throughout that period, these changes in concentration or differences in compound volatility at the 10-μg dosage level have little effect on the relative response profiles (Roelofs and Comeau, 1971; Minks et al., 1974).

Antennal responses are reduced after large amplitude responses to most active test chemicals, but relative responses to a standard compound are still reproducible. A standard is chosen for each species after preliminary screening has indicated a test compound eliciting 1–2 mV responses. In most cases with male moths this response level is much below that of the most active test chemicals and does not effect significant reduction of the subsequent test chemical response. A test chemical response is compared to a standard response obtained within a minute prior to the test chemical puff. The EAG responses of all the test chemicals then can be expressed relative to the standard given an arbitrary value of 10. Replication of the series of test chemicals will give mean values for each test chemical to be used to generate a "response profile."

A. Monounsaturated Pheromones

Many lepidopteran pheromones include a monounsaturated compound that elicits the greatest EAG response. In these cases, this component can readily be identified by obtaining GLC–EAG retention times and determining the response profile for the appropriate series of standards. For example, with *Grapholita prunivora* it was readily determined that the main EAG-active component was a 12-carbon acetate compound. Antennal responses to the monounsaturated standards (Fig. 4) showed that (a) the 12-carbon acetate series indeed did elicit greater responses than all other carbon length and functional group series, (b) the *Z* compounds produced greater responses than the corresponding *E* isomers, and (c) the *Z*8-12:0Ac compound elicited the greatest amplitude and "stickiest" response of all test compounds. A 98:2 *Z*/*E*-8-12:0Ac mixture subsequently was found to be a good attractant for *G. prunivora* males in the field (Roelofs and Cardé, 1974).

In some cases the response profiles are not as clear-cut as with the above case, but unusual response patterns can provide valuable information. The response profile for redbanded leafroller moths, *Argyrotaenia velutinana* (Fig. 5A) is normal for a *Z*11-14:0Ac pheromone component. Two leafrollers in Europe, however, have different EAG profiles (Minks et al., 1973) (Figs. 5B,C), reflecting the fact that *A. orana* uses mostly *Z*9-14:0Ac and 10–20% *Z*11-14:0Ac, whereas

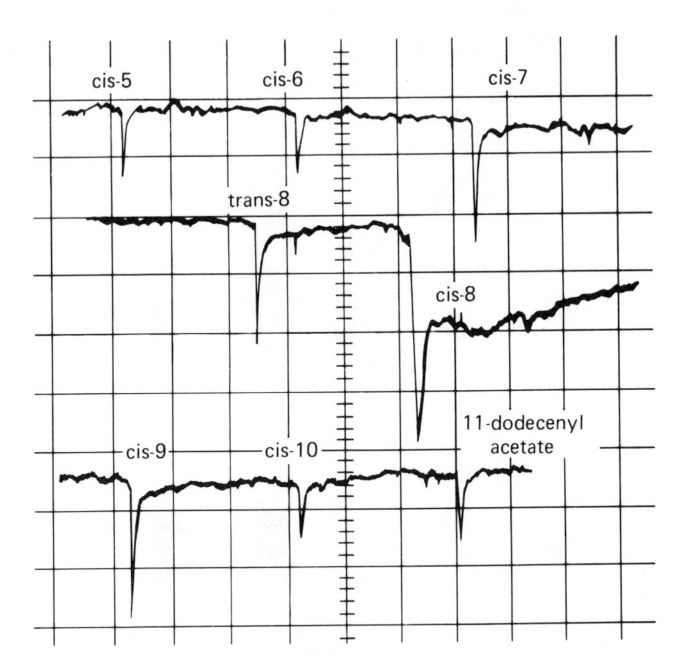

Figure 4. EAG responses of a male *Grapholita prunivora* antenna to monounsaturated 12-carbon chain standards (from Roelofs and Comeau, 1971).

C. spectrana uses mostly *Z*11-14:0Ac and ca. 10% *Z*9-14:0Ac. In our initial studies with the threelined leafroller, *Pandemis limitata,* we determined that the retention time of main EAG activity correlated with an unsaturated 14-carbon acetate. The response profile (Fig. 5D) was similar to that of *C. spectrana* (Fig. 5C) in that the *Z*11 isomer produced the greatest response, but the *Z*9 response was larger than the adjacent positional isomers. This could indicate a *Z,Z*-9,11 conjugated system or two compounds as found in *C. spectrana.* The retention time of activity from a polar column ruled out the polar conjugated double-bond system, and so a mixture of positional isomers was indicated. Further analyses confirmed the pheromone to be a 91:9 mixture of *Z*11- and *Z*9-14:0Ac's (Roelofs et al., 1976).

In another case, the response profiles gave an important clue for a missing component. GLC–EAG analysis of crude female extract for *Argyrotaenia citrana* had shown the presence of a monounsaturated 14-carbon acetate. The EAG response profile of this series was similar to *A. velutinana* (Fig. 5A) and indicated the pheromone to be *Z*11-14:0Ac. Further analysis showed no detectable *E* isomer, but unfortunately the *Z*11-14:0Ac compound was not effective in luring males to field traps. Response profiles with various series of standards showed that the 14-carbon aldehydes were more active than the acetates with *A. citrana,* whereas the aldehyde series had been quite inactive with *A. velu-*

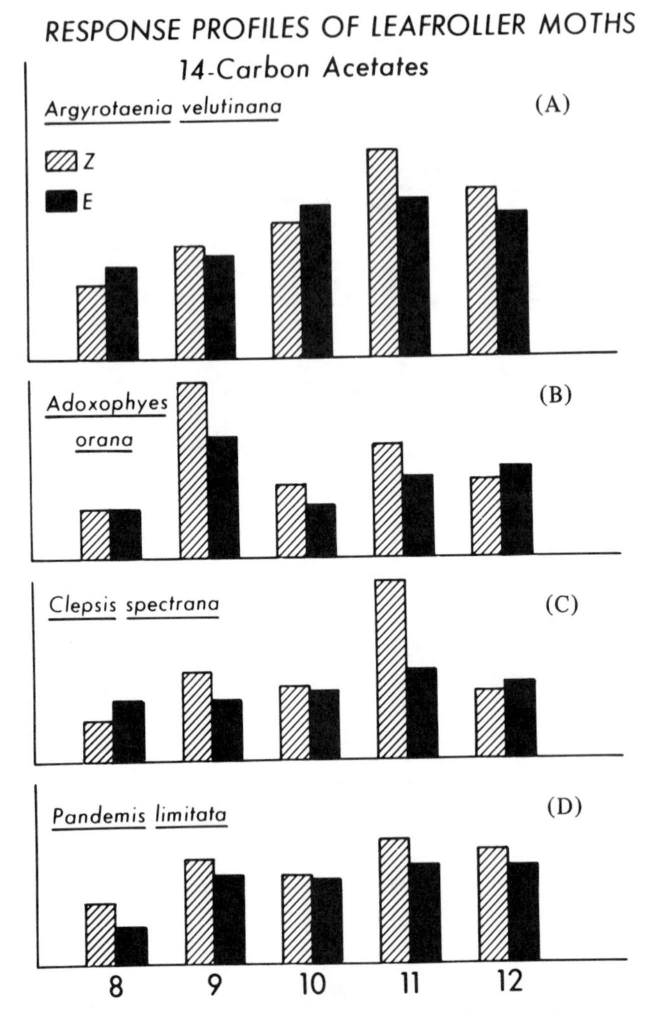

Figure 5. EAG profiles to monounsaturated 14-carbon chain compounds for a species, *A. velutinana,* that utilizes *Z*- and *E*-11-14:0Ac's as pheromone components and of three species, *A. orana, C. spectrana,* and *P. limitata,* that utilize a pheromone mixture of *Z*9- and *Z*11-14:0Ac's in ratios of 90:10, 25:75, and 9:91, respectively.

tinana. A search for the indicated *Z*11-14:Ald compound in gland extract was largely unsuccessful, but airborne collections of emitted pheromone showed that they released 15 times more aldehyde than acetate. Field studies proved that the aldehyde was a key component and that a wide range of aldehyde:acetate ratios were potent male moth lures for *A. citrana* (Hill et al., 1975).

B. Multiple Sites of Unsaturation

The library of monounsaturated standards not only is useful in determining the position and configuration of monounsaturated pheromone components but also can be used in many cases involving two double bonds in a pheromone component.

The first identification of this type was with the codling moth, *Cydia pomonella* (Roelofs et al., 1971). GLC–EAG runs had shown high EAG activity for a component that corresponded to a 12-carbon alcohol on OV-101, but with the polarity of a conjugated double-bond system on a polar column. An EAG response profile (Fig. 6A) immediately showed that the greatest activity in all the series of compounds tested was with the monounsaturated 12-carbon alcohols, and specifically with the *E*8, *E*9, and *E*10 isomers in that series. These data indi-

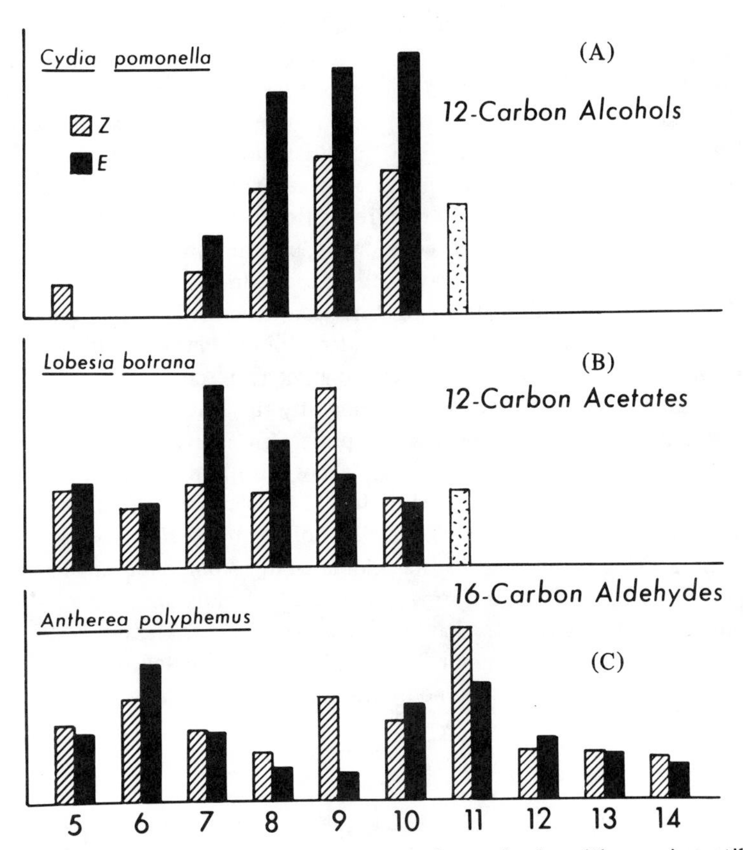

Figure 6. EAG profiles of monounsaturated standards with species utilizing doubly unsaturated pheromone components. From tip to bottom the pheromones involved are *E,E*-8,10-12:OH; *E,Z*-7,9-12:OAc; and *E,Z*-6,11-16:Ald.

cate an *E*8,*E*10 double-bond system. Synthesis and EAG analyses of all four 8,10 geometrical isomers showed that the *E,E* isomer produced the greatest antennal responses of all test compounds. Further research (Roelofs et al., 1971; McDonough and Moffit, 1974) proved it to be a sex pheromone component of the codling moth.

A conjugated double-bond system was also defined by the EAG technique with the European grapevine moth, *Lobesia botrana* (Roelofs et al., 1973). In this case a conjugated double-bond system in a 12-carbon acetate was indicated from the GLC–EAG analyses of crude female extract. The monounsaturated 12-carbon acetate series (Fig. 6B) elicited the greatest response of all test compounds, and the *E*7 and *Z*9 isomers were by far the most active isomers in that series. These data were used to define this EAG-active material to be *E,Z*-7,9-12:0Ac, which subsequently was proved to be a pheromone component of this insect (Roelofs et al., 1973; Buser et al., 1974).

In cases involving multiple double bonds that are not conjugated, monounsaturated standards sometimes can be used to predict the positions and configurations of all the double bonds, but in other cases not all sites of unsaturation will be reflected in the antennal response profile. With the potato tuberworm moth (see Section III) the GLC–EAG analyses showed the presence of di- and triunsaturated 13-carbon acetate pheromone components. Monounsaturated 13-carbon acetate standards showed high responses to the *E*4 and the *Z*7 isomers, which defined the diunsaturated component, but the additional *Z*10 position of the triunsaturated compound was not evidenced in the response profile. Presumably the triene configuration is distorted sufficiently by the *Z*7 double bond that a monounsaturated *Z*10-13:0Ac standard does not interact sufficiently with the active sites for the triene to generate an antennal response.

Two isolated double bonds were defined by the EAG response profile with *Antheraea polyphemus* (Fig. 6C), which utilizes an *E*6,*Z*11-16:0Ac pheromone component. The pheromone structure is not distorted too much by the *E*6 double bond, and so the monounsaturated 11-isomers were useful in predicting the second site of unsaturation. With pink bollworm, *Pectinophora gossypiella,* however, a *Z*7 double bond in the middle of the compound created enough change in the molecule's conformation so that 11-monounsaturated standards did not elicit larger responses than other inactive isomers. Therefore the *Z*7,*E*11 and *Z*7,*Z*11 pheromone components could not be predicted from an EAG response profile generated with monounsaturated standards.

4.2. Response Spectra

The evaluation of relative response activities of hundreds of test chemicals for many lepidopteran species, particularly with species in the Noctuidae, has been carried out by Priesner (Priesner, 1973, 1979a,b; Priesner et al., 1975, 1977). Instead of using a comparison of the highest response amplitude elicited by a saturation level of the test compounds (Section V), a dosage-response curve is generated for each test chemical and a single activity value assigned each chemical representing the half-log amount (μg) needed to elicit the same response amplitude as 0.001 μg of the most active EAG compound. The dosages ranged from 10^{-6} to 10^{2} μg and were tested by passing air (100 cm/sec) over the source for 1 sec.

The most effective chemical is initially determined by comparing response amplitudes of all test chemicals at the 0.1-μg dosage level. The most active EAG compound in concentrations of 0.001 to 0.1 μg is used as a reference stimulus. Test chemicals are compared to the reference stimuli used immediately preceding and following reference stimuli to correct for differences among antennal and time-dependent changes in antennal responsiveness. Comparisons between a test stimulus and reference stimuli usually are carried out at 3 decadic steps of stimulus dosages and replicated between 8 to 30 times. The "activity value" of the test chemical is assigned in half-decadic steps for the dosage of test chemical calculated to elicit the response amplitude equivalent to that of the 0.001 μg reference stimulus. Thus, if the test chemical dosage is 1.8 to 5.6 times more than 0.001 μg, an activity value of 0.003 is assigned to that chemical; a value of 0.01 is assigned if 5.6 to 18 times more chemical is required, etc.

A table of activity values for all the various test chemicals becomes the response spectrum for that species. Typical response spectra for three noctuid species are given in Table 2. The reference chemical almost always corresponds to a component of the female sex pheromone, and the position of the double bond relative to either the oxidized end or the hydrocarbon end is reflected in the test chemicals of different carbon lengths. An activity value can be compared with all others directly within the response spectrum for a species and with its relative effect in response spectra of other species.

Response spectra have been generated for many lepidopteran species and they show, in general, that a geometrical isomer of the reference chemical is between 3- and 10-fold less effective, that some positional isomers in the same carbon length series can be as much as 3000 times less effective, that increasing or decreasing the carbon chain by two methylene groups reduces the activity by 10- to 100-fold, and that the double-bond configuration of the reference chemical is more active than the corresponding geometrical isomer throughout the response spectrum.

The response spectra can be used in conjunction with GLC–EAG data in the same way as the amplitude profiles (Section V) for prediction of double-bond positions and configuration in a pheromone component. For species using several pheromone components, Priesner (1979a, b, 1980) also employed single-cell techniques to generate "activation profiles" in each of the two to five different receptor types associated with a particular sensillum. This method involves recording from the cut tip of a male's antennal trichoid hair to monitor the nerve impulse responses of the various receptor cells. The cells are distinguished

Table 2. Response spectra of three noctuid species each utilizing a pheromone component of a different carbon-chain length[1]

	Trichoplusia ni		*Hyssia cavernosa*		*Chersotis multangula*	
Δ	*Z*	*E*	*Z*	*E*	*Z*	*E*
	12-0Ac's		12-0Ac's		12-0Ac's	
3	0.03	0.1	0.3			
4	0.1	0.1	0.3		1	1
5	0.01	0.03	0.1	0.3	0.3	1
6	0.01	0.03	0.1	0.3	0.3	1
7	0.001	0.003	0.03	0.1	0.1	0.3
8	0.01	0.03	0.1	0.1	0.3	1
9	0.03		0.03		0.3	
	14-0Ac's		14-0Ac's		14-0Ac's	
4	0.3	0.3	0.1	0.3	1	1
5	0.1	0.3	0.03	0.1	0.3	1
6	0.1	0.3	0.03	0.1	0.3	1
7	0.03	0.1	0.1	0.03	0.1	0.3
8	0.03	0.1	0.01	0.03	0.1	0.3
9	0.01	0.03	0.001	0.003	0.01	0.03
10	0.1	0.3	0.01	0.03	0.03	0.1
11	0.1	0.3	0.03	0.1	0.1	0.3
12	0.3		0.1		0.3	
	16-0Ac's		16-0Ac's		16-0Ac's	
5	1	1	3	3	0.3	0.3
6	1	1	3	3	0.3	0.3
7	0.3	1	1	3	0.1	0.3
8	0.3	1	1	3	0.1	0.3
9	0.1	0.3	0.1	0.3	0.03	0.1
10	0.3	1	0.03	0.1	0.01	0.03
11	0.1	0.3	0.01	0.03	0.001	0.003
12	0.3	1	0.03	0.1	0.03	0.03
13	1		0.1		0.1	

[1] The numbers express amounts (μg) of chemical required to produce an EAG response of the same amplitude (taken from Priesner et al., 1975).

by the relative amplitudes of their nerve impulses and each can be sensitive to different key compounds that are either part of the pheromone or adversely affect the male's response to it. Details of the single-cell technique will be covered in Chapter 7 by Wadhams (this volume).

A comparison of the amplitude profile and the response spectrum methods on the same species is shown in the research on the sugar beet moth, *Scrobipalpa ocellatella* (Renou et al., 1980). A comparison of the amplitude profile method with both EAG and single sensillum is shown in Fig. 7. The profiles are almost identical, but the single sensillum recordings show that responses from two cells contribute to the EAG profile.

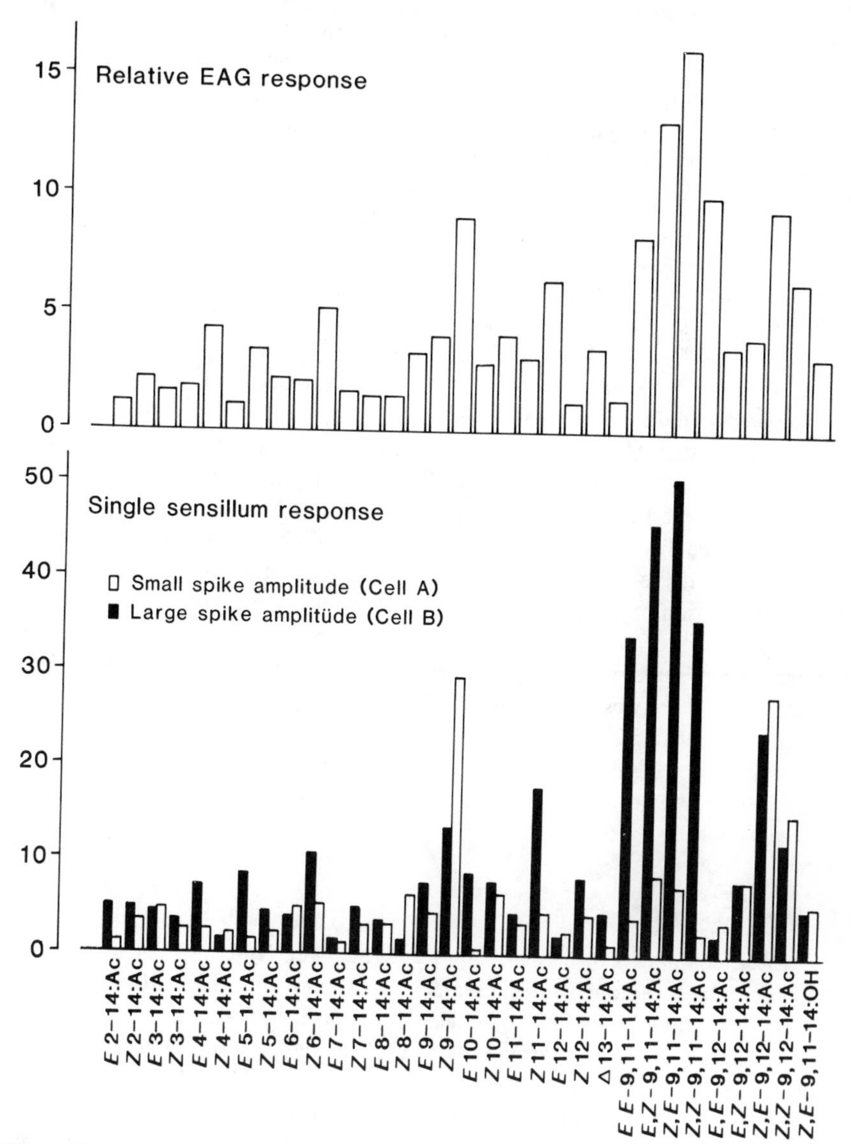

Figure 7. A comparison of an EAG and a single sensillum standard profile for *Dioryctria abietella* shows the two profiles to be similar (C. Löfstedt et al., 1983).

4.3. Stereochemical Predictions

A. Geometric Isomers

The EAG techniques described above have been used successfully in predicting the configuration of a major sex pheromone component, but, generally, the EAG has not been useful in determining blend ratios of geometric isomers. Although several different receptor cells could be activated by several pheromone components, the afferent signals are integrated in the central nervous system, and the overall pattern of response is affected by both blend ratios and concentration (Baker et al., 1981; Linn and Roelofs, 1981). With the European corn borer moth, *Ostrinia nubilalis,* populations exist that use opposite blends of geometric isomers, e.g., 96:4 *Z*/*E*-11-14:0Ac in Iowa (Klun et al., 1973) and 4:96 *Z*/*E* in New York (Kochansky et al., 1975). EAG recordings (Nagai et al., 1977) from these populations showed that (a) both isomers produced equal responses in the New York population, (b) the *Z* isomer was slightly better with the Iowa population, and (c) mixtures of the two isomers did not elicit responses any larger than did the *Z* isomer alone.

In beetles, the EAG technique also can be used to predict the most active geometrical isomer. A number of structurally related compounds have been found to elicit mating responses to varying degrees with the khapra beetle, *Trogoderma granarium.* The EAG responses to these individual components were found to correlate directly to the degree of attraction and copulatory activity of the compounds (Levinson et al., 1978). Thus (*Z*)-14-methyl-8-hexadecenal was the most active pheromone component behaviorally and by EAG, followed by the *E* isomer. The *Z* isomer produced a 1.0 mV response at 10^{-4} μg, whereas a concentration of 10^{-2} μg was needed for the *E* isomer to produce a 1.0 mV response. The corresponding alcohols were 10^4 times less active in stimulating copulation and required a 10^4-fold increase in concentration to elicit a 1.0 mV response. The corresponding esters were less effective by six to eight orders of magnitude both behaviorally and by EAG.

B. Optical Isomers

The EAG technique can play an important role in determining relative antennal response activity among optical isomers that cannot be characterized with the minute quantities of natural material. Although optical activity of the gypsy moth, *Lymantria dispar,* pheromone has not been defined from natural material, EAG analyses with the two isomers, (+)- and (−)-*cis*-7,8-epoxy-2-methyloctadecane (Fig. 8), immediately showed that the (+)-enantiomer produces a much greater response than the (−)-enantiomer (Yamada et al., 1976; also see Fig. 9, Section V). The (+)-compound was found to be very active in field trapping studies (Vité et al., 1976; Miller et al., 1977; Cardé et al., 1977) and the (−)-compound resulted in lower trap catches than unbaited traps.

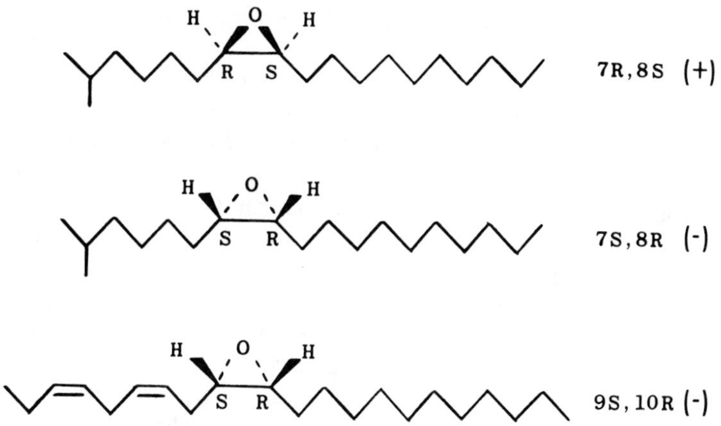

Figure 8. The top structure is the pheromone structure of the gypsy moth, the middle structure is its inactive enantiomer, and the bottom structure is a pheromone component of the salt marsh caterpillar. EAG analyses showed that male salt marsh caterpillar antennae gave greater responses to both (—) enantiomers compared to the (+) enantiomers.

Another chiral lepidopteran pheromone was found with the saltmarsh caterpillar moth, *Estigmene acrea* (Hill and Roelofs, 1981). This species produces a relatively large quantity (1 μg) of pheromone, (*Z,Z*)-3,6-*cis*-9,10-epoxyheneicosadiene, which was accumulated to show that this natural material exhibits negative ORD and CD curves (Mori and Ebata, 1981). An enantiomeric assignment could not be made until each optical isomer was synthesized, but EAG analyses using the disparlure epoxide optical isomers of the gypsy moth pheromone were used to predict the active isomer. With male saltmarsh caterpillar antennae, (—)-disparlure consistently produced EAG responses two to five times greater in amplitude than those produced with (+)-disparlure. A comparison of the structures involved (Fig. 8) indicates that the active enantiomer for saltmarsh caterpillar will possess a 9*S*,10*R* configuration similar to the 7*S*,8*R* configuration of (—)-disparlure. Synthesis and analyses of the two enantiomers (Mori and Ebata, 1981) showed that the (9*S*,10*R*)-isomer did exhibit negative ORD and CD curves, similar to the natural material, and that it elicited EAG responses with saltmarsh caterpillar antennae two to three times greater than the (9*R*,10*S*)-isomer (Hill et al., 1982). EAG analyses with the two optical isomers are being carried out to determine enantiomeric specificities among many arctiid species utilizing the same diene epoxide (W. Meyer and W. Roelofs, unpublished).

The value of the EAG technique in predicting acid moiety specificity and optical configuration was also shown in an interesting study involving several species of diprionid sawflies utilizing acetate or propionate esters of 3,7-dimethylpentadecan-2-ol (Jewett et al., 1976). The three asymmetric centers in these compounds allow for eight possible optical isomers with each ester. EAG analyses (Kraemer et al., 1981) with *Neodiprion lecontei* showed that male antennal responses were 20–50% more active with acetate isomers than with propionate isomers, and that the (*S,S,S*)-acetate elicited the greatest EAG response by far of all other isomers. The (*S,S,S*)-acetate was also the only compound to show activity in field tests with *N. lecontei.* Additionally, EAG responses from the other optical isomers correlated to their relationship to the (*S,S,S*)-isomer. Isomers containing two *S* configurations elicited stronger EAG response than those isomers with only one *S* configuration, and these in turn elicited greater responses than the (*R,R,R*)-isomer.

Most of the identified chiral pheromone components are in coleopteran species. In some cases EAG analyses have been used to predict or verify the active optical isomers, whereas in other cases single-cell recordings are required to differentiate responses. With the large European elm bark beetle, *Scolytus scolytus* (Wadhams et al., 1982), EAG analyses correlated directly with the known production and field activity of isomers. The active isomers, (−)-threo and (−)-erythro-4-methyl-3-heptanol elicited greater EAG responses with male and female beetles throughout a large portion of the dosage-response curves, and with male beetles the (−)-threo isomer was more active than the (−)-erythro isomer.

In *Dendroctonus frontalis* bark beetles (Payne, et al., 1982), an 85(−)/15(+) mixture of frontalin is produced by females boring in pine bolts and the (−)-isomer is more active than the (+)-isomer in field tests. EAG analyses also showed that the (−)-isomer elicited greater antennal responses than the (+)-isomer. Similar results (Payne, 1979) have been reported for the spruce bark beetle, *Ips typographus,* wherein EAG responses to the behaviorally active (*S*)-*cis*-verbenol were greater than to (*R*)-*cis*-verbenol.

EAG results (Angst and Lanier, 1979) with the pine engraver, *Ips pini,* appeared to be similar to the case of the European corn borer (Section 4.3 A). Two separate populations of beetles were shown to produce and behaviorally respond to different enantiomeric blends. In Idaho, *Ips pini* males produce only (−)-ipsdienol, and in New York the male beetles produce a mixture of 65% (+) and 35% (−)-ipsdienol. Although EAG responses were the greatest to ipsdienol compared to all other beetle-produced and host odors tested, responses to the two optical isomers of ipsdienol with beetles from both populations were similar. As with the two populations of European corn borer, there apparently are not major differences in the receptor systems of the two populations, but the afferent signals are interpreted differently.

V. Selective Adaptation

Single sensillum recordings are invaluable in determining receptor specificities, but the simple EAG apparatus can be used to provide some insights on this as well. A particular compound or mixture is blown continuously over an antennal preparation until the receptors become adapted and then other test materials are immediately puffed across the antennae. If the two test materials stimulate the same receptors, then adaptation of these receptors with one compound should greatly reduce peripheral response to the other.

5.1. Southern Pine Beetle

Payne (1975, 1979) showed that with the southern pine beetle, *Dendroctonus frontalis,* frontalin and *exo* brevicomin share the same receptors, but that the host terpenes, 3-carene and α-pinene, share some, but not all, of the same receptors. The adaptation studies showed that there are greater numbers of receptors for the pheromone frontalin than for host odors. These data were substantiated with single sensillum recordings (Dickens and Payne, 1977).

5.2. Redbanded Leafroller Moth

Selective adaptation studies with the three-component pheromone system of the redbanded leafroller moth (Baker and Roelofs, 1976) showed that at least two functionally different receptors are involved and that the close-range component, 12:0Ac, produces a temporal modulation of the neuronal response to one component, *Z*11-14:0Ac, but not to the *E*11-14:0Ac. In this study, airstream saturation was accomplished by applying 10 μg of a component to a 1.3-cm-diameter filter paper disc and placing it at the mouth of the glass air delivery tube. The tube was first tilted up so that the chemically laden air blew directly into an exhaust vent 20 cm behind the antennal preparation. Once the antennal baseline response was located on the oscilloscope, the airstream was lowered quickly to blow continuously across the antennae. After 5–6 sec a test chemical (10 μg dosage) was puffed into the saturated airstream and the response amplitude and recovery time were recorded either by tracings from the oscilloscope or with a Polaroid camera. Antennae were used for one response and then discarded. The glass air delivery tube was removed and rinsed thoroughly with acetone after every response.

Puffing the same chemical as used in the saturated airstream did not elicit responses greater than a blank puff, but a *Z*11-14:0Ac airstream did not eliminate a puff response to *E*11-14:0Ac, and vice versa. These data support single-cell studies (O'Connell, 1975) that gave evidence for two different, functionally independent receptor sites on redbanded antennae with these two pheromone components. Saturation with the close-range component, 12:0Ac, did not eliminate response to either *Z*11 or *E*11-14:0Ac, but it had a striking effect on recovery rate of the *Z*11 component. In the first one-third recovery phase after maximum depolorization, *Z*11-14:0Ac normally exhibits a recovery rate of 2.17 ± 0.85 mV/sec, whereas, its recovery rate when puffed in a 12:0Ac saturated airstream was only 0.72 ± 0.36 mV/sec. The recovery rate of the other pheromone component, *E*11-14:0Ac, was not affected. This suggested an altering of the temporal aspects of the peripheral responses. Data were consistent with the interaction of these two components found in single-cell studies (O'Connell, 1975) and behavioral studies (Baker et al., 1976).

5.3. Gypsy Moth

As described above (Section 4.3 B), (+)-disparlure was found to be the active pheromone component, whereas the addition of either (−)-disparlure or the corresponding olefin decreases trap catch, and, in fact, they both act as repellents in that their traps catch fewer moths than unbaited traps (Cardé et al., 1977; Miller et al., 1977). Selective adaptation studies were carried out with these three materials (Miller et al., 1977). In this case, one compound was blown continuously over a male gypsy moth antenna until an additional puff of that compound elicited no additional response. Then, an equal quantity of a second compound was puffed in the airstream and the EAG response was recorded. The results (Fig. 9) show that antennae exposed to (+)-disparlure still respond to both (−)-disparlure and the olefin.

Antennae preexposed either to (−)-disparlure or to olefin still respond to (+)-disparlure. Antennae exposed to olefin did not respond to (−)-disparlure, but antennae exposed to (−)-disparlure still responded to olefin, as well as to

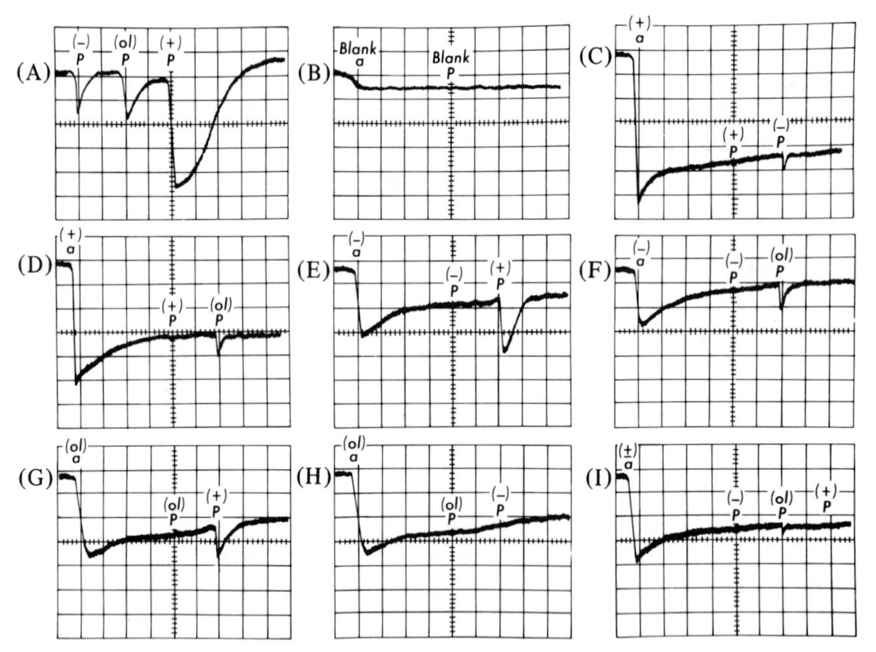

Figure 9. Male *Lymantria dispar* EAG responses to olefin (ol), (+)- and (−)-disparlure in a blank airstream (A), and after continuous application of each compound (C–I) introduced into the airstream by injection of a filter paper disc bearing the test compounds. (B) Represents the EAG response to a blank airstream and blank puff (P). Filter papers used in cartridges and in airstreams have 10 μg test compound. Each vertical division represents 1 mV and each horizontal division equals 2 sec in A and 5 sec in B–I (from Miller et al., 1977).

(+)-disparlure. These data support the existence of two receptor types, one sensitive to (+)-disparlure and one sensitive to olefin. The olefin receptor type has a greater affinity for (−)-disparlure than (+)-disparlure. These receptor types also were found in single sensillum studies (Schneider et al., 1977).

5.4 *Sparganothis pilleriana*

Selective adaptation also can be used in conjunction with the generation of response spectra (Section IV). A response spectrum is easy to interpret when it reflects a single receptor type. In many cases, one obtains a bioassay EAG spectrum composed of two, nonoverlapping response spectra. Antennal adaptation to one of the compounds allows one to generate a response spectrum indicative of the other receptor type. With *Sparganothis pilleriana,* both *E*9-12:0Ac and *E*9-12:OH elicit similar dosage-response curves in nonadapted antennae (Priesner, 1979a,b). Adaptation to *E*9-12:OH strongly reduced antennal responses to alcohol stimuli without affecting the responses to the acetate concentration series. Adaptation with the acetate produced the opposite effect and, in this case, the response spectrum to the alcohol receptor type could be generated.

VI. References

Angst ME, Lanier GN (1979) Electroantennogram responses of two populations of *Ips pini* (Coleoptera:Scolytidae) to insect-produced and host tree compounds. J Chem Ecol **5**:131–140.

Baker TC, Cardé RT, Roelofs WL (1976) Behavioral responses of male *Argyrotaenia velutinana* (Lepidoptera: Tortricidae) to components of its sex pheromone. J Chem Ecol **2**:333–352.

Baker TC, Meyer W, Roelofs WL (1981) Sex pheromone dosage and blend discrimination by Oriental fruit moth males. Entomol Exp Appl **30**:269–279.

Baker TC, Roelofs WL (1976) Electroantennogram responses of the male moth *Argyrotaenia velutinana* to mixtures of sex pheromone components of the female. J Insect Physiol **22**:1357–1364.

Bjostad LB, Roelofs WL (1980) An inexpensive electronic device for measuring electroantennogram responses to sex pheromone components with a voltmeter. Physiol Entomol **5**:309–314.

Buser HR, Rauscher S, Arn H (1974) Sex pheromone of the *Lobesia botrana*: (E,Z)-7,9-dodecadienyl acetate in the female grape vine moth. Z. Naturforsch **29**:781–783.

Cardé RT, Doane CC, Baker TC, Iwaki S, Marumo S (1977) Attractancy of optically active pheromone for male gypsy moths. Environ Entomol **6**: 768–772.

Dickens JC, Payne TL (1977) Bark beetle olfaction: pheromone receptor system in *Dendroctonus frontalis* Zimm. (Coleoptera: Scolytidae). J Insect Physiol **23**:481–489.

Gutmann A, Payne TL, Roberts EA, Schulte-Elte K-H, Giersch W, Ohloff G

(1981) Antennal olfactory response of boll weevil to grandlure and vicinal dimethyl analogs. J Chem Ecol **7**:919–926.

Hill AS, Cardé RT, Kido H, Roelofs WL (1975) Sex pheromones of the orange tortrix moth, *Argyrotaenia citrana.* J Chem Ecol **1**:215–224.

Hill AS, Kovalev BG, Nikolaeva LN, Roelofs WL (1982) Sex pheromone of the fall webworm moth *Hyphantria cunea.* J Chem Ecol **8**:383–396.

Hill AS, Roelofs WL (1981) Sex pheromone of the saltmarsh caterpillar moth, *Estigmene acrea.* J Chem Ecol **7**:655–668.

Jewett DM, Matsumura F, Coppel HC (1976) Sex pheromone specificity in the pine sawflies: Interchange of acid moieties in an ester. Science **192**:51–53.

Jewett DM, Matsumura F, Coppel HC (1977) A simple system for recording electroantennograms from male pine sawflies. J Electrophysiol Tech **5**: 23–28.

Kaissling KE (1971) Insect olfaction. In: Handbook of Sensory Physiology, Vol IV/1. Beidler LM (ed), Springer-Verlag, Berlin.

Kaissling KE (1979) Recognition of pheromones by moths, especially in *Saturniids* and *Bombyx mori.* In: Chemical Ecology: Odour Communication in Animals. Ritter FJ (ed), Elsevier/North-Holland Biomedical Press, Amsterdam.

Kaissling KE, Thorson J (1980) Insect olfactory sensilla: Structural, chemical and electrical aspects of the functional organization. In: Receptors for Neurotransmitters, Hormones and Pheromones in Insects. Sattelle DB, Hall, LM, Hildebrand JG (eds), Elsevier/North-Holland Biomedical Press, New York.

Klun JA, Chapman OL, Mattes KC, Wojtkowski PW, Beroza M, Sonnet PE (1973) Insect sex pheromones: Minor amount of opposite geometrical isomer critical to attaction. Science **181**:661–663.

Kochansky, J, Cardé RT, Liebherr J, Roelofs WL (1975) Sex pheromones of the European corn borer (*Ostrinia nubilalis*) in New York. J Chem Ecol **1**: 225–231.

Kraemer ME, Coppel HC, Matsumura F, Wilkinson RC, Kikukawa T (1981) Field and electroantennogram responses of the red-headed pine sawfly, *Neodiprion lecontei* (Fitch), to optical isomers of sawfly pheromones. J Chem Ecol **7**:1063–1072.

Levinson AR, Levinson HZ, Schwaiger H, Cassidy RF, Silverstein RM (1978) Olfactory behavior and receptor potentials of the khapra bettle *Trogoderma granarium* (Coleoptera: Dermestidae) induced by the major components of its sex pheromone, certain analogues and fatty acid esters. J Chem Ecol **4**: 95–108.

Linn CE, Roelofs WL (1981) Modification of sex pheromone blend discrimination in male Oriental fruit moths by pre-exposure to (E)-8-dodecenyl acetate. Physiol. Entomol **6**:421–429.

Löfstedt C, Van Der Pers JNC, Löfqvist J, Lanne BS (1983) Sex pheromone of the cone-pyralid *Dioryctria abietella.* Ent Exp Appl **34**:20–26.

McDonough LM, Moffitt HR (1974) Sex pheromone of the codling moth. Science **183**:978.

Miller JR, Mori K, Roelofs WL (1977) Gypsy moth field trapping and electroantennogram studies with pheromone enantiomers. J Insect Physiol **23**: 1447–1453.

Minks AK, Roelofs WL, Ritter FJ, Persoons CJ (1973) Reproductive isolation of two tortricid moth species by different ratios of a two-component sex attractant. Science **180**:1073–1074.

Minks AK, Roelofs WL, Schuurmans-van Kijk E, Persoons CJ, Ritter FJ (1974) Electroantennogram responses of two tortricid moths using two-component sex pheromones. J Insect Physiol **20**:1659–1665.

Moore I (1981) Biological amplification for increasing electroantennogram discrimination between two female sex pheromones of *Spodoptera littoralis* (Lepidoptera: Noctuidae). J Chem Ecol **7**:791–798.

Mori K, Ebata T (1981) Synthesis of optically active pheromones with an epoxy ring, (+)-disparlure and the saltmarsh caterpillar moth pheromone [*Z,Z*)-3,6-*cis*-9,10-epoxyheneicosadiene]. Tet Lett **22**:4281–4282.

Nagai T (1981) Electroantennogram response gradient on the antenna of the European corn borer, *Ostrinia nubilalis.* J Insect Physiol **27**:889–894.

Nagai T, Starratt AN, McLeod DGR, Driscoll GR (1977) Electroantennogram responses of the European corn borer, *Ostrinia nubilalis,* to (Z)-and (E)-11-tetradecenyl acetates. J Insect Physiol **23**:591–597.

Nishino C, Kimura R (1981) Isolation of sex pheromone mimic of the American cockroach by monitoring with male/female ratio in electroantennogram. J Insect Physiol **27**:305–311.

Nishino C, Kimura R (1982) Olfactory receptor responses of the nymphal American cockroach to sex pheromones and their mimics. Comp Biochem Physiol **72A**:237–242.

Nishino C, Takayanagi H (1981) Male/female ratio in electroantennogram responses of the American cockroach to (+)-*trans*-verbenyl acetate and related compounds: Relationships between the ratio and sex pheromonal activity. Comp Biochem Physiol **70A**:229–234.

O'Connell RJ (1975) Olfactory receptor responses to sex pheromone components in the redbanded leafroller moth. J Gen Physiol **65**:179–205.

Payne TL (1975) Bark beetle olfaction. III. Antennal olfactory responsiveness of *Dendroctonus frontalis* Zimmerman and *D. brevicomis* LeConte (Coleoptera: Scolytidae) to aggregation pheromones and host tree terpene hydrocarbons. J Chem Soc **1**:233–242.

Payne TL (1979) Pheromone and host odor perception in bark beetles. In: Neurotoxicology of Insecticides and Pheromones. Narahashi T (ed), Plenum, New York.

Payne TL, Richerson JV, Dickens JC, West JR, Mori K, Berisford CW, Hedden RL, Vire JP, Blum MS (1982) Southern pine beetle olfactory receptor and behavior discrimination of enantiomers of the attractant pheromone frontalin. J Chem Ecol **8**:873–881.

Persoons CJ, Verwiel PEJ, Talman E, Ritter FJ (1979) Sex pheromone of the American cockroach, *Periplaneta americana*–Isolation and structure elucidation of periplanone-B. J Chem Ecol **5**:221–236.

Persoons CJ, Voerman S, Verwiel PEJ, Ritter FJ, Nooyen WJ, Minks AK (1976) Sex pheromone of the potato tuberworm moth, *Phthorimaea operculella*: Isolation, identification and field evaluation. Entomol Exp Appl **20**:289–300.

Priesner E (1973) Artspezifität und Funktion einiger Insektenpheromone. Fortschr Zool **22**:49–135.

Priesner E (1979a) Specificity studies on pheromone receptors of noctuid and tortricid Lepidoptera. In: Chemical Ecology: Odour Communication in Animals. Ritter FJ (ed), Elsevier/North-Holland Biomedical Press, Amsterdam.

Priesner E (1979b) Progress in the analysis of pheromone receptor systems. Ann Zool Ecol Anim **11**:533-546.

Priesner E (1980) Sensory encoding of pheromone signals and related stimuli in male moths. In: Insect Neurobiology and Pesticide Action (Neurotox 79). Soc Chem Ind, London.

Priesner E, Bestmann HJ, Vostrowsky O, Rösel P (1977) Sensory efficacy of alkyl-branched pheromone analogues in noctuid and tortricid Lepidoptera. Z Naturforsch **32**:979-991.

Priesner E, Jacobson M, Bestmann HJ (1975) Structure-response relationships in noctuid sex pheromone reception. J Naturforsch **30**:283-293.

Renou M, Descoins C, Lallemand JY, Priesner E, Lettere M, Gallois M (1980) L'acétoxy-l dodécène 3E, composant principal de la phéromone sexuelle de la teigne de la betterave: *Scrobipalpa ocellatella* Boyd. (Lepidoptère: Gelechiidae). Z Angew Entomol **90**:275-289.

Roelofs WL (1977) The scope and limitations of the electroantennogram technique in identifying pheromone components. In Crop Protection Agents– Their Biological Evaluation. McFarlane NR (ed), Academic Press, New York.

Roelofs WL (1978) Chemical control of insects by pheromones. In: Biochemistry of Insects. Rockstein M (ed), Academic Press, New York.

Roelofs WL (1979) Electroantennograms. Chemtech **9**:222–227.

Roelofs WL, Cardé RT (1974) Oriental fruit moth and lesser appleworm attractant mixtures redefined. Environ Entomol **3**:586-588.

Roelofs WL, Cardé A, Hill A, Cardé R (1976) Sex pheromone of the threelined leafroller, *Pandemis limitata*. Environ Entomol **5**:649-652.

Roelofs W, Comeau A (1971) Sex attractants in Lepidoptera. In: Chemical Releasers in Insects. Tahori AS (ed), Gordon and Breach, New York.

Roelofs W, Comeau A, Hill A, Millicevic G (1971) Sex attractant of the codling moth: Characterization with electroantennogram technique. Science **174**: 297-299.

Roelofs WL, Hill AS, Linn CE, Meinwald J, Jain S, Herbert HJ, Smith RF (1982) Sex pheromone of the winter moth–A geometrid with unusually low-temperature precopulatory responses. Science **217**:657-658.

Roelofs W, Kochansky J, Anthon E, Rice R, Cardé R (1975a) Sex pheromone of the peach twig borer (*Anarsia lineatella*). Environ Entomol **4**:580-582.

Roelofs WL, Kochansky J, Cardé R, Arn H, Rauscher S (1973) Sex attractant of the grape vine moth, *Lobesia botrana*. Bull Soc Entomol Sci **46**:71-73.

Roelofs WL, Kochansky JP, Cardé RT, Kennedy GG, Henrick CA, Labovitz JN, Corbin VL (1975b) Sex pheromone of the potato tuberworm moth, *Phthorimaea operculella*. Life Sci **17**:699-706.

Schneider D (1957) Elektrophysiologische Untersuchungen von Chemo-und Mechanorezeptoren der Antenne des Seidenspinners *Bombyx mori* L. Z Vergl Physiol **40**:8-41.

Schneider D (1963) Electrophysiological investigation of insect olfaction. In: Olfaction and Taste, I. Zotterman Y (ed), Pergamon Press, Oxford.

Schneider D (1969) Insect olfaction: Deciphering system for chemical messages. Science **163**:1031–1037.

Schneider D, Kafka WA, Beroza M, Bierl BA (1977) Odor receptor responses of male gypsy and nun moths (Lepidoptera: Lymantriidae) to disparlure and its analogues. J Comp Physiol **113**:1–5.

Steinbrecht RA, Schneider D (1980) Pheromone communication in moths sensory physiology and behavior. In: Insect Biology in the Future. Locke M, Smith DS (eds), Academic Press, New York.

Vité JP, Klimetzek D, Loskant G, Hedden R, Mori K (1976) Chirality of insect pheromones: Response interruption by inactive antipodes. Naturwiss **63**: 582–583.

Wadhams LJ, Angst ME, Blight MM (1982) Responses of the olfactory receptors of *Scolytus scolytus* (F.) (Coleoptera: Scolytidae) to the stereoisomers of 4-methyl-3-heptanol. J Chem Ecol **8**:477–492.

Weatherston J, Roelofs W, Comeau A, Sanders CJ (1971) Studies of physiologically active arthropod secretions X. Sex pheromone of the Eastern spruce budworm, *Choristoneura fumiferana* (Lepidoptera: Tortricidae). Can Entomol **103**:1741–1747.

Williams IH, Pickett JA, Martin AP (1982) Nasonov pheromone of the honeybee, *Apis mellifera* L. (Hymenoptera, Apidae), IV. Comparative electroantennogram responses. J Chem Ecol **8**:567–574.

Yamada M, Saito T, Katagiri K, Iwaki S, Marumo S (1976) Electroantennogram and behavioral responses of the gypsy moth to enantiomers of disparlure and its *trans* analogues. J Insect Physiol **22**:755–761.

Yamaoka R, Fukami H, Ishii S (1976) Isolation and identification of the female sex pheromone of the potato tuberworm, *Phthorimaea operculella* (Zeller). Agric Biol Chem **40**:1971–1977.

Chapter 6

Combined Gas Chromatography and Electroantennogram Recording of Insect Olfactory Responses

Dean L. Struble[1] *and Heinrich Arn*[2]

I. Introduction

Electroantennogram (EAG)[3] recording of insect olfactory responses has been a great asset as a method of bioassay for chemical identifications of pheromones and for the development of synthetic attractants (Roelofs, Chapter 5). When EAG is used separately from the purification method, e.g., gas chromatography (GC), it is necessary to isolate and transfer small amounts (ng) of purified components to the EAG preparation. This does not take full advantage of the GC resolution capabilities and it may result in a loss of essential minor pheromone components. It is also time consuming.

Coupling GC and EAG has led to the development of extremely sensitive and specific detection systems for pheromone components, which fully utilize the tremendous analytical capabilities of these two techniques.

II. GC-Pulsed EAG

Moorhouse et al. (1969) reported the first coupling of GC and EAG. The GC-packed column effluent was split so that 75% went to a regular flame ionization detector (FID) and 25% was collected in a reservoir installed in the GC detector oven. Column effluent was allowed to accumulate in the reservoir for 15 sec and

[1]Research Station, Agriculture Canada, Lethbridge, Alberta, Canada.
[2]Swiss Federal Research Station, CH-8820 Wädenswil, Switzerland.
[3]Abbreviations used: EAG, electroantennogram; GC, gas chromatography; MS, mass spectrometry: EAD, electroantennographic detector; FID, flame ionization detector; SSR, single sensillum recording; FET, field effect transistor; FE, female equivalent.

then it was flushed with nitrogen for 3 sec over a standard EAG preparation. A second stream of nitrogen was used to cool the gases to 30°C before they reached the antenna. This system has been very useful as a bioassay technique during the identification of pheromone components of several insects (e.g., Beevor et al., 1975; Campion et al., 1979; Nesbitt et al., 1973, 1975a,b, 1977, 1979a,b, 1980; Rothschild et al., 1982).

Moorhouse et al. (1969) reported that passing the GC effluent continuously over the antenna was unsatisfactory because the EAG responses were masked by the baseline drift. One explanation for the drift may be that in packed column GC, peak widths are commonly 60 to 120 sec and exposure of the antenna to pheromone components for this length of time may cause the olfactory receptors to saturate. The reservoir collection system allowed components to be passed over the antenna in a 3-sec pulse, but this was also the limiting factor in the resolution of the components.

EAG normally involves a high-speed recording system such as an oscilloscope, oscillographic recorder, or a high-speed ink recorder. The antennal preparation requires either glass capillary electrodes or finely tapered metal electrodes, and the use of a micromanipulator to establish electrical contact on the antenna. Although these are common to electrophysiology, they are not common in most organic chemistry laboratories.

III. GC–Electroantennographic Detector (EAD)

The first direct coupling of GC–EAG with the GC effluent passed continuously over the antenna was reported by Arn et al. (1975). This system used high-resolution glass capillary GC columns where peak widths are only a few seconds, e.g., 15 sec or less, and the components reach the antenna with a sharp rise in concentration or as a pulse. Antennal olfactory receptors do not saturate under these conditions. This system takes full advantage of the high-resolution GC and the extreme sensitivity and selectivity of the antennal olfactory receptors for the detection of pheromone components. Because an insect antenna acts as a sensing element for the GC effluent, the authors suggested the term electroantennographic detector (EAD) for this system.

GC–EAD has a number of special features that will be considered under separate subheadings. These are: a simple amplifier and electronic filter that permit antennal responses to be recorded on a standard GC strip chart recorder; simple electrodes that do not require a micromanipulator for the antennal preparation; and detector antennae from selected species which provide extreme sensitivity and selectivity for detection and confirmation of pheromone components.

3.1. EAD Amplifier and Filter

Antennal signals are measured with a high-impedance amplifier circuit similar to those used for EAG. A basic satisfactory design is illustrated in Fig. 1. Any high-performance field effect transistor (FET) operational amplifier with an input

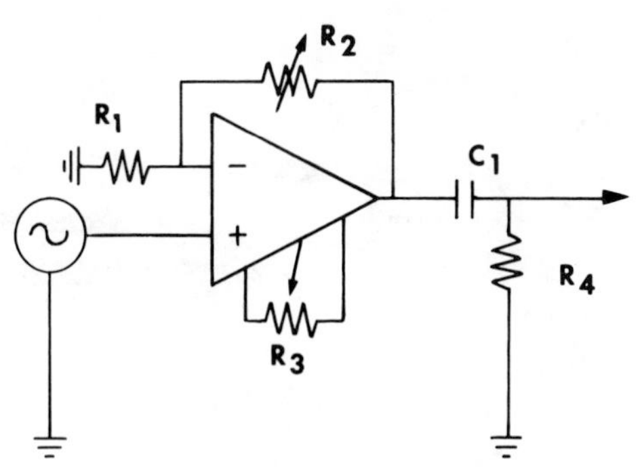

Figure 1. Circuit diagram of an electroantennographic detector (EAD) amplifier and passive high-pass filter; a high-performance field effect transistor operational amplifier, with resistors and a capacitor of the following values: R_1, 10 kΩ; R_2, 100 kΩ; R_3, will be specified by manufacturer of the operational amplifier; R_4, 130 Ω; and C_1, 0.01 F.

impedance of 10^{10} to 10^{12} ohm such as Burr-Brown (Tucson, Ariz.) No. 3523K, Intersil (Cupertino, Calif.) No. 8007, or their equivalent is suitable. Antennal signals may be up to 12 mV, so an amplifier gain, $(R_1 + R_2)/R_1$, of one to two is adequate. The offset trim resistor R_3 and a power supply will be specified by the manufacturer of the operational amplifier.

Baseline drift is eliminated from the amplifier output signal with a passive high-pass filter, e.g., C_1, 0.01 F and R_4, 130 ohm. This filters frequencies (f) below 0.12 Hz, i.e., $f = 1/(2\pi \cdot C_1 \cdot R_4)$. Filtering is reduced by increasing R_4 or by using a series of resistors, but for practical purposes this is not necessary.

Filtered antennal signals are recorded on any regular GC strip chart recorder with a high-input impedance (e.g., 10^6 ohm) and a pen response of 0.3 sec/full-scale deflection. A variable input, e.g., 1 to 20 mV, is required on the recorder. It is advantageous to use a dual-channel recorder so that FID and EAD signals can be recorded simultaneously on the same chart paper, which permits accurate comparison of retention times.

3.2. EAD Electrodes

A simple, adequate electrode system is illustrated in Fig. 2. Electrodes are made from silver wire, 1.2 mm diameter, with the tips flattened, rolled into a spoon shape, and electrolytically chloridized from 1% NaCl solution. A drop of 0.75% NaCl solution placed in each chloridized spoon tip holds the excised antenna in place by surface tension. One or two segments of the antenna tip must be excised to permit electrical contact. A low-power microscope is beneficial, but not essential, for connecting antennae. This method has been used successfully with antennae ranging from 0.4 cm to several centimeters in length. The entire antennal preparation is inserted into the end of a glass tube which transfers the GC effluent (Figs. 2,3).

Figure 2. Antennal preparation on chloridized silver electrodes, placed inside a glass transfer line for effluent from the gas chromatograph (GC).

The silver electrodes can be soldered directly to the operational amplifier leads or connected by gold-plated microbanana plugs (Type LS2-F, Multi Contact AG, CH-4123, Allschwil, Switzerland, or Interlok, Burlingame, Calif., 94010). These connectors do not add to the background noise of the antennal signal, and the electrodes can be conveniently removed for chloridizing, e.g., once per week. No external electronic shielding is required with this electrode and amplifier system.

Glass capillary electrodes used for EAG may also be used for EAD, but they do not fit easily inside the GC effluent transfer tube.

3.3. GC Effluent Splitter and Transfer System

A GC effluent splitter and transfer system is illustrated in Fig. 3. The GC column effluent is diluted with N_2 make-up gas (2) at 20 to 30 ml/min, which is then split to the EAD–FID in a ratio of 1:3 to 1:30. Make-up gas (1) at 30 ml/min is added to carry the effluent to the glass transfer tube. Flow rates of the make-up gases can be adjusted to alter the split ratio over a narrow range. It is necessary to have an auxillary heater at the exit of the GC to prevent the effluent from condensing in the tip of the splitter capillary. The effluent can be delivered to the glass transfer tube directly from the splitter capillary or via a modified FID quartz flame tip.

The splitter may be made from stainless steel, Pt-Ir, fused silica, or other materials commonly used to transfer GC effluent in GC–mass spectrometry (GC–MS). Stainless-steel capillaries may cause losses of polar compounds such as alcohols and aldehydes, but losses can be minimized by coating the capillaries with the liquid phase that is used on the GC column. Coated capillaries are satisfactory provided a flow of nitrogen (make-up gas 1) is maintained when the capillaries are above ambient temperature, even though the GC is not in use.

The glass effluent transfer tube (0.7 to 1.4 cm inside diameter at the antennal

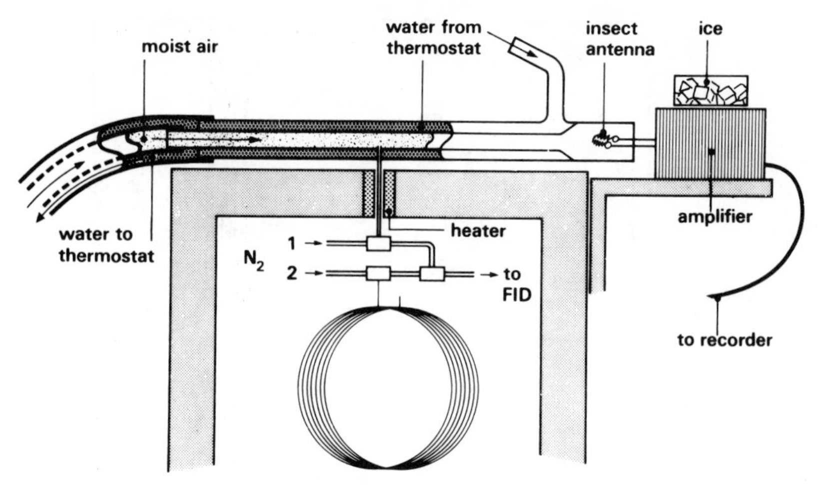

Figure 3. Drawing of a GC effluent splitter and glass transfer system.

preparation) and the EAD amplifier are supported on the GC frame in any convenient location. The GC effluent enters the transfer tube 10 to 15 cm from the antennal preparation. The temperatures of the water jacket and the humidified air (medical grade or synthetic air, 200 to 300 ml/min) are controlled between 18 and 26°C by a circulating water bath. The air is humidified by bubbling through a water bottle that is contained in the water bath. Humidified air cools the GC effluent and prevents the antennal preparation from becoming dry. In practice the air temperature is increased to about 26°C when a fresh antennal preparation is installed and it is gradually (20 min) reduced to 20 to 22°C as condensation accumulates on the electrodes (Fig. 2). At high ambient temperatures (28 to 30°C) it is beneficial to cool the electrodes by placing a beaker of ice water on the amplifier (Fig. 3), which allows the humidified air temperature to be maintained at 20 to 22°C without drying the antennal preparation. Excised antennae of most lepidopterous species will survive for 4 to 8 hr under these conditions.

3.4. Directly Coupled GC-EAD

The first application of capillary GC-EAD was for the analysis of the sex pheromone of the European grapevine moth, *Lobesia botrana* (Arn et al., 1975). The GC-EAD and GC-FID responses were recorded from successive injections, without the aid of a GC effluent splitter (Fig. 4).

EAG data indicated that the male antennae had strong responses to (*E*)-7, (*Z*)-9-dodecadienyl acetate (abbrev. *E*7,*Z*9-12:OAc), which was confirmed by analyzing the four 7,9-12:OAc isomers by GC-EAD. A minor EAD response was recorded to the *Z,E* isomer and a major response to the *E,Z* isomer (Fig. 4). Injection of a female abdomen tip extract under identical conditions showed an

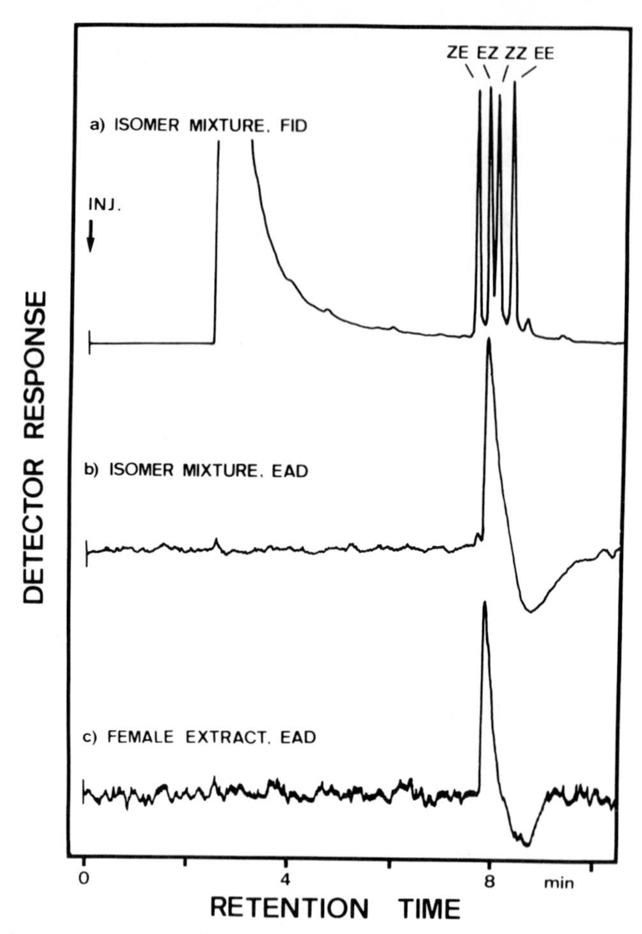

Figure 4. Separately recorded male *Lobesia botrana* EAD and flame ionization detector (FID) responses to *L. botrana* female extract and an isomer mixture. Conditions: Ucon 50 HB 5100, 50 m × 0.3 mm glass capillary column, injector split ratio was 1:12.5. (A) Mixture of four geometrical isomers of 7,9-dodecadienyl acetate, 250 ng of each injected. (B) Same mixture, 250 pg of each injected. (C) *L. botrana* female extract 0.1 female equivalent (FE).

antennal response to only one component, which had the same retention time and also cochromatographed with *E*7,*Z*9-12:OAc. There were no other responses to the pheromone extract, which indicated that the pheromone may be a single component. The structure of the pheromone component was confirmed by GC-MS and the synthetic compound attracted males under field conditions.

An EAD dose-response curve for the *L. botrana* male detector antenna is shown in Fig. 5. In general, signal intensities varied considerably from one antenna to another. This may be partially due to variations within antennae, but a large part was due to the antennal preparation. For example, stimulation of *L.*

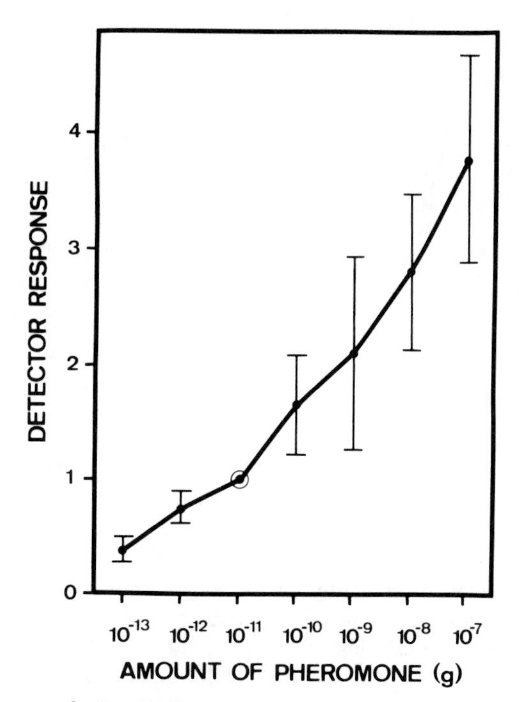

Figure 5. Response of the EAD with *L. botrana* male antenna to (*E*)-7,(*Z*)-9-dodecadienyl acetate (abbrev. *E*7,*Z*9-12:OAc). Values were actual amounts reaching the antennal preparation under the conditions given in Fig. 4. Responses (± standard deviation) were normalized for 1.0 at 10^{-11} g.

botrana with 10 pg of *E*7,*Z*9-12:OAc gave responses from 1.0 to 11.4 mV, with an average of 3.5 mV. Ratios of the responses with different amounts of stimulus were fairly reproducible, and each response was normalized by dividing its value by the response to a preceding 10 pg standard. A 3-min disadaptation period was allowed after each standard. Signal intensities were approximately linear with the logarithm of the stimulant between 0.1 pg and 100 ng. Stimulation with more than 1 ng reaching the antenna, however, seemed to cause irreversible adaptation of the receptors. In general, antennae of most species will function for 4 to 8 hr with little deterioration if the stimuli are kept at a low concentration (e.g., splitless injection of 1 to 2 ng and simultaneous recording of EAD-FID with a split ratio of 1:3). This permits the recording of several chromatograms with the same antenna.

3.5. Simultaneous Recording of EAD and FID Responses

A good example of recording EAD and FID responses simultaneously was in the detection of pheromone components in *Tortrix viridana* (Arn et al., 1979). EAD of the female wash showed a major response at a retention time of *Z*11-14:OAc, which coincided with a small peak in the FID chromatogram (Fig. 6). The

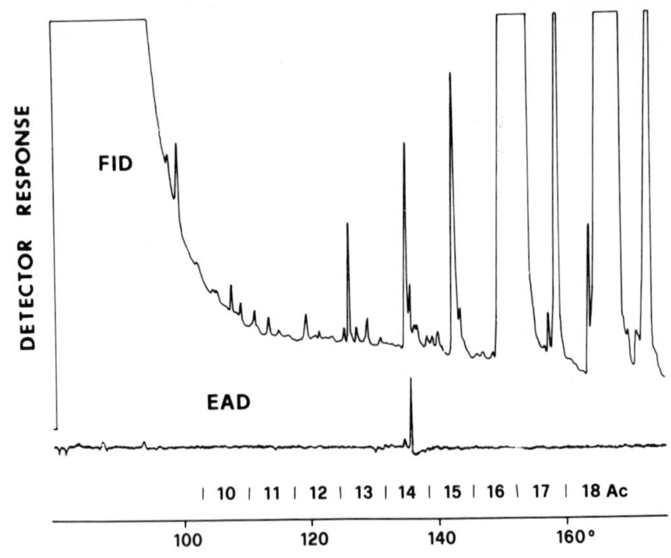

Figure 6. Simultaneously recorded EAD–FID chromatograms of a *Tortrix viridana* female wash. Retention times of saturated acetates and elution temperatures are indicated. Conditions: Silar 10C, 22 m × 0.3 mm glass capillary column, 1 FE injected splitless at 70°C for 2 min, 30°/min to 90°C, and 6°/min to 160°C, EAD–FID split ratio 1:10.

pheromone component cochromatographed with *Z*11-14:OAc and the synthetic compound gave an EAD response of similar intensity. The structure of the pheromone component was confirmed by GC–MS, and *Z*11-14:OAc attracted males under field conditions.

The EAD chromatogram showed a minor response just before *Z*11-14:OAc, which corresponded to the retention time of *E*11-14:OAc or *Z*9-14:OAc. There was insufficient quantity of this component(s) to be detected by GC–MS, and combinations of *E*11-14:OAc and *Z*9-14:OAc with *Z*11-14:OAc reduced the catches of moths under field conditions. The minor pheromone component(s) may be an attractant inhibitor or it may be involved with some other aspect of the mating behavior.

Two other small EAD responses were observed during the elution of the solvent, but the corresponding components were not isolated.

Combined EAD–FID has been an important technique for the detection of major and minor pheromone components in other lepidopterous species, for example, *Eupoecilia ambiguella* (Arn et al., 1976, 1982); *Agrotis segetum* (Bestmann et al., 1978; Arn et al., 1980; Löfstedt et al., 1982); *Amathes c-nigrum* (Bestmann et al., 1979); *Scotia exclamationis* (Bestmann et al., 1980); *Adoxophyes orana* (Guerin et al., 1982); *Euxoa ochrogaster* (Struble et al., 1980b); *Mamestra brassicae* (Struble et al., 1980a); *Euxoa drewseni* (Struble, 1983); and *Vitula edmandsae serratilineella* (Struble and Richards, 1983).

GC-EAD also provides a rapid method of measuring relative antennal responses to synthetic chemicals. High-resolution GC provides on-line purification of the compounds, which eliminates any contribution of impurities to the antennal responses. Mixtures of compounds can be analyzed by one injection on the GC, provided the compounds can be completely resolved on the GC column. These are considerable advantages over EAG.

3.6. Selected Detector Species for Increasing Selectivity and Sensitivity

Male antennae may be highly sensitive to the major sex pheromone component(s) produced by females of their own species, but relatively less sensitive to the minor components. Therefore, an EAD with male antennae from the same species as the female being studied may provide little evidence, if any, for the presence of minor pheromone components.

Male detector antennae can be selected from other species that are known to have strong responses to the major female pheromone component(s) of their own species. Recent reviews on sex pheromones and attractants have been reported by Tamaki (1983) and Steck et al. (1982). Selected detector species can be used to confirm the presence or absence of suspected major or minor component(s) in the female extracts of the species being studied. The use of selected detector antennae increases the selectivity and sensitivity of the EAD system, and the following examples illustrate the value of this technique.

EAD of *Euxoa ochrogaster* female abdomen tip wash with detector antennae from the same species showed a major response to the main pheromone component, *Z*5-12:OAc (Struble et al., 1980b). There was a weak response to a minor pheromone component at the retention time of *Z*7-12:OAc. Male antennae of two other species *Agrotis ipsilon* and *Grapholitha funebrana* were used on the EAD to help confirm the structure of this minor component. These species were known to respond specifically to *Z*7-12:OAc (Hill et al., 1977) and *Z*8-12:OAc (Granges and Baggiolini, 1971), respectively.

A. ipsilon antennal responses to *E. ochrogaster* pheromone wash and to synthetic *Z*5-12:OAc and *Z*7-12:OAc are shown in Figs. 7 and 8. The strong antennal response to the minor pheromone component, *Z*7-12:OAc, provided confirmatory evidence for its presence in the *E. ochrogaster* pheromone (Fig. 7). *A. ipsilon* antennae had no response to *Z*8-12:OAc.

G. funebrana antenna gave a 0.5 mV response to 1 ng of *Z*8-12:OAc by splitless injection on the GC–EAD, but it did not respond to 1 FE of *E. ochrogaster* pheromone wash under the same conditions. It was concluded that *Z*8-12:OAc was not present in the pheromone and it was not detected by any other methods.

The structures of *Z*5-12:OAc and *Z*7-12:OAc were confirmed by GC–MS and the synthetic compounds were biologically active in a four-component blend under field conditions (Struble, 1981).

A second example of the selected detector species technique was in the identification of the *E. drewseni* pheromone components (Struble, 1983). *E. drewseni* male antennae gave consistent responses to female pheromone components that were identified by GC–MS and attractancy tests as 12:OAc, *Z*5-12:OAc, *Z*5-14:OAc, *Z*7-14:OAc, and *Z*7-14:OH. Antennal responses of similar intensities were obtained with these synthetic compounds.

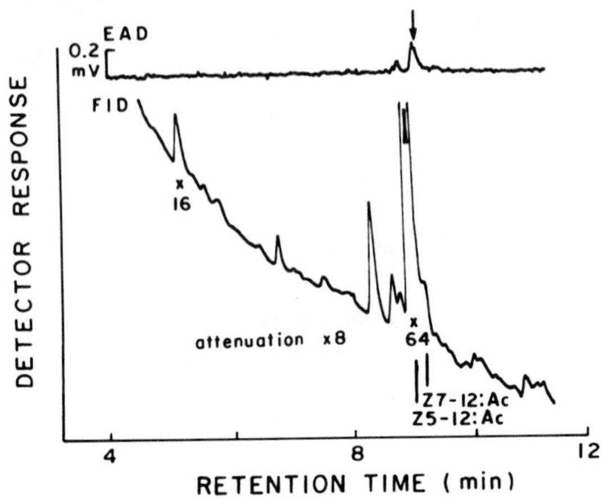

Figure 7. EAD–FID chromatograms with responses of *Agrotis ipsilon* male antenna to *Euxoa ochrogaster* female wash. Conditions: Silar 10C, 22 m × 0.3 mm glass capillary column; 1 FE injected splitless at 70°C for 2 min, 30°/min to 90°C, 3°/min to 190°C; EAD–FID split ratio 1:10.

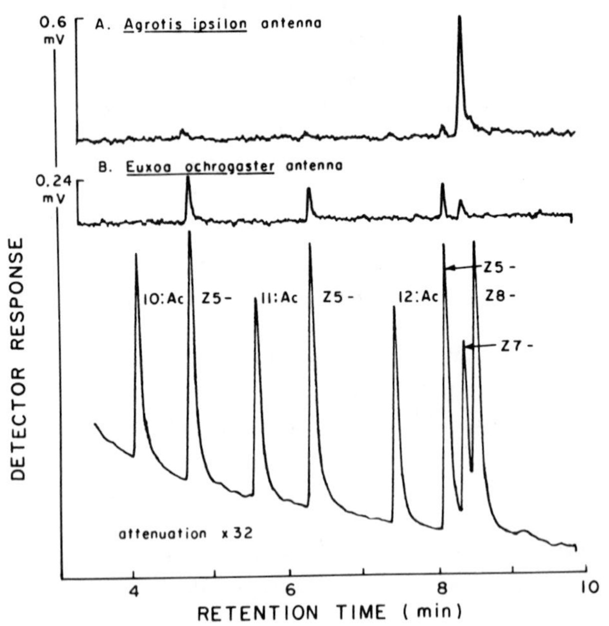

Figure 8. EAD–FID chromatograms of (A) *Agrotis ipsilon* and (B) *Euxoa ochrogaster* male antennae with FID responses to a mixture of synthetic compounds, 6 ng of each injected under conditions as in Fig. 7.

Three other detector species, *E. ochrogaster, Autographa californica,* and *E. auxiliaris,* were used to confirm the presence of these components. *E. ochrogaster* antennae were known to respond strongly to *Z*5-10:OAc and *Z*5-12:OAc (Fig. 8) and, as detector antennae for the *E. drewseni* pheromone extract, responded strongly (0.7 mV) for only *Z*5-12:OAc. The lack of any other response indicated that *Z*5-10:OAc was not present in the extract.

A. californica male antennae were known to respond specifically to *Z*7-12:OAc and *Z*7-14:OAc (Struble, unpublished data; Steck et al., 1979) and splitless injection of 1 ng of each compound gave EAD responses of 0.8 and 0.14 mV, respectively. Responses of 0.1 and 0.13 mV were recorded with *E. drewseni* pheromone extract at retention times that corresponded to *Z*7-12:OAc and *Z*7-14:OAc. This suggested the presence of a trace amount, ca. $<$10 pg/FE, of *Z*7-12:OAc and confirmed the presence of *Z*7-14:OAc in *E. drewseni* extract. *Z*7-12:OAc was not detectable by any other methods; however, it was found to be a potent attractant inhibitor under field conditions.

E. auxiliaris was known to respond to *Z*5-14:OAc (Struble, unpublished data) and as a detector antenna it responded to this suspected *E. drewseni* pheromone component. *Z*5-14:OAc had no obvious effect on the attraction of males, but it may be involved with other aspects of their mating behavior or reproductive isolation.

This selected detector species technique provides extreme sensitivity and specificity for the detection and, in part, the structural confirmation of major and minor pheromone components. Some species that have been used for the detection of various compounds by GC–EAG or GC–EAD are listed in Table 1.

Table 1. Species of which male antennae were used for GC–EAG or GC–EAD detection of pheromone components in female extracts of the same (A) or other (B) species.

Compound	Species	A[1]	B[1]
*Z*5-10:OAc	*A. segetum*	1, 2	
	E. ochrogaster	3	4
*Z*5-12:OAc	*E. ochrogaster*	3	4
	E. drewseni	4	
*Z*7-12:OAc	*A. californica*		4
	A. ipsilon		2, 3
*E*7,*Z*9-12:OAc	*L. botrana*	5	
*Z*8-12:OAc	*G. funebrana*		2, 3
*Z*9-12:OAc	*E. ambiguella*	6	2
*E*9-12:OAc	*D. castanea*	7	
	Z. diniana (pine form)	8	
*E*9,11-12:OAc	*D. castanea*	7	
*Z*5-14:OAc	*S. exclamationis*	9	
	E. auxiliaris		4

Table 1. (continued)

Compound	Species	A[1]	B[1]
*Z*7-14:OAc	*A. c-nigrum*	10	
	E. drewseni	4	
*Z*7-14:Ald	*P. citri*	11	
	P. oleae	12	
*Z*8-14:OAc	*S. ocellana*	5	
*Z*9-14:OAc	*S. exempta*	13	
	S. littoralis	14	
*Z*9,*E*11-14:OAc	*S. littoralis*	14	
*Z*9,*E*12-14:OAc	*S. exempta*	13	
*Z*9-14:OH	*V. edmandsae*	15	
	H. electellum		15
*Z*9,*E*12-14:OH	*V. edmandsae*	15	
	H. electellum		15
*Z*9-14:Ald	*V. edmandsae*	15	
*Z*9,*E*12-14:Ald	*V. edmandsae*	15	
*Z*11-14:OAc	*T. viridana*	16	
*E*11-14:OAc	*S. littoralis*	14	
	Z. diniana (larch form)	8	17
*Z*9-16:Ald	*H. armigera*	18	
*Z*11-16:OAc	*M. brassicae*	19	
	S. inferens	20	
	H. punctiger	21	
*Z*11-16:OH	*H. armigera*	18	
	H. punctiger	21	
*Z*11-16:Ald	*C. partellus*	22	
	C. suppressalis	23	
	H. armigera	18	
	H. punctiger	21	
*Z*13-18:OAc	*C. sacchariphagus*	24	
*Z*13-18:OH	*C. sacchariphagus*	24	
*Z*13-18:Ald	*C. suppressalis*	23	

[1] Numbers are for the following references.

1. Bestmann et al. (1978)
2. Arn et al. (1980)
3. Struble et al. (1980b)
4. Struble (1983)
5. Arn et al. (1975)
6. Arn et al. (1976)
7. Nesbitt et al. (1975a)
8. Arn et al. (1982)
9. Bestmann et al. (1980)
10. Bestmann et al. (1979)
11. Nesbitt et al. (1977)
12. Campion et al. (1979)
13. Beevor et al. (1975)
14. Nesbitt et al. (1973)
15. Struble and Richards (1983)
16. Arn et al. (1979)
17. Guerin et al. (1982)
18. Nesbitt et al. (1979a)
19. Struble et al. (1980a)
20. Nesbitt et al. (1976)
21. Rothschild et al. (1982)
22. Nesbitt et al. (1979b)
23. Nesbitt et al. (1975b)
24. Nesbitt et al. (1980)

3.7. Pheromone and Parapheromone Receptor Cells

Relative structure-activity relationships of a homologous series of compounds can be readily determined by GC-EAD. The GC is an excellent quantitative delivery system and antennal responses can be recorded under nearly identical conditions. This type of measurement appears to provide information on the type of receptor cells that are present on antennae and it allows inferences on the receptor sites with which pheromones and parapheromones may be reacting. Recordings with male antennae of *E. drewseni* serve as a good example (Struble, 1983).

Male *E. drewseni* olfactory responses were determined for saturated and most monounsaturated C_{10} to C_{16} acetates and alcohols where the double bond was beyond carbon atom number 4. The strongest responses were to the saturated acetates and the *Z*5- and *Z*7-acetates of this homologous series, which reached a maximum with the C_{12} acetates (Fig. 9). *Z*7-14:OAc also showed a relatively strong response. Similar structure-activity relationships have been reported for single sensilla recordings of other noctuids (Priesner, 1979).

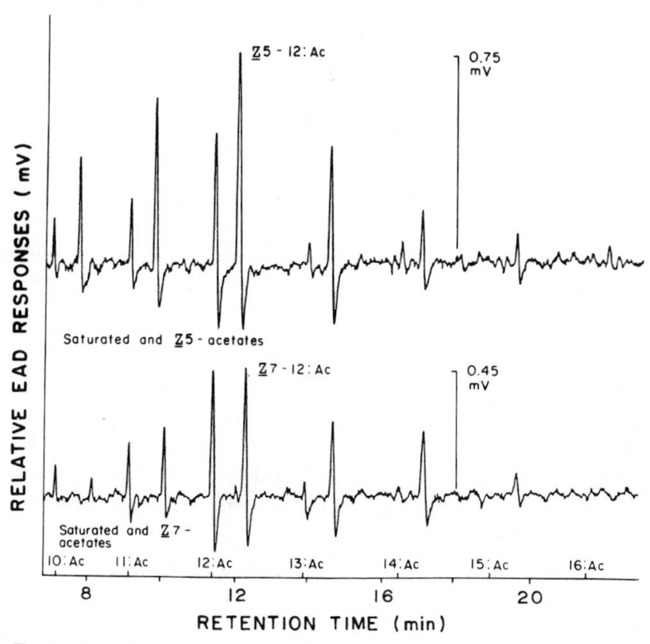

Figure 9. EAD chromatograms with *Euxoa drewseni* male antennal responses to C_{10} to C_{16} saturated and monounsaturated *Z*5 and *Z*7 acetates. Retention times of saturated acetates are indicated and conditions were: Carbowax 20M TPA, 28 m × 0.25 mm glass capillary column; splitless injection at 50°C for 0.3 min, 25°/min to 110°C, 4°/min to 205°C; EAD–FID split ratio 1:3. Compounds (1 ng each) were in two blends: saturated and *Z*5 acetates, and saturated and *Z*7 acetates; FID responses were omitted.

Although single sensilla recordings were not made for *E. drewseni*, it is assumed that the main receptor type was for Z5-12:OAc, which gave the strongest EAD response and was the major pheromone component. The other Z5-acetates of the homologous series presumably react on the receptor cells for Z5-12:OAc and the decrease in their relative responses with an increase or decrease in their carbon chain length is an indication of a decrease in their "fit" on these cells.

The strongest response of the homologous series of Z7-acetates was to Z7-12:OAc, which was an attractant inhibitor. There are presumably specific receptor cells for Z7-12:OAc, which may be an evolutionary "hang over" (Priesner, 1979) to provide some interspecific responses such as reproductive isolation. There are likely specific receptor cells for Z7-14:OAc since it is an essential pheromone component and the response to Z7-14:OAc was greater than would be expected if it were reacting only at the Z7-12:OAc receptor (Fig. 9).

The odd-numbered carbon compounds, Z7-13:OAc and Z7-15:OAc, were not detected in the female extract, so there are probably no specific receptor cells for them. These compounds did function as parapheromones, however, as substitution of Z7-14:OAc by them in the three-component blend of 12:OAc, Z5-12:OAc, and Z7-14:OAc attracted males under field conditions. The blend involving Z7-13:OAc was the least effective attractant. These compounds must have some "fit" on the receptor cells for Z7-14:OAc; however, since Z7-13:OAc was the least effective parapheromone it may have some reaction on the receptor cells for the attractant inhibitor Z7-12:OAc.

EAD responses are from the total antenna rather than from single sensilla, so it is not possible to locate and record the responses of specific receptor cells.

IV. GC–Single Sensillum Recording (GC–SSR)

The first coupling of GC and a single sensillum recording system was independently reported by Van Der Pers and Löfstedt (1983) working with the moth *Agrotis segetum* and Wadhams (1982; see also Wadhams, Chapter 7, this volume) working with bark beetles of the genus *Scolytus.* In either version, the GC effluent was transferred to the antennal preparation as in EAD. Recordings were made from single sensilla trichodea on excised male antennae by the tip recording technique (Van Der Pers and Den Otter, 1978). Action potentials were continuously recorded on a tape recorder and transferred to paper using an ink jet recorder. The system was tested with two species that have known pheromones, *A. segetum* and *Adoxophyes orana.*

The literature is not consistent on the composition of *A. segetum* pheromone components, but the presence of Z5-10:OAc, Z7-12:OAc and Z9-14:OAc have been reported (Löfstedt et al., 1982, and references therein). Cells that responded to Z5-10:OAc and Z7-12:OAc were found in different sensilla trichodea that were morphologically indistinguishable. One cell was activated by Z5-10:OAc and one or two cells were activated by Z7-12:OAc. These cells were highly sensitive as only 10 and 170 pg of these compounds reached the antenna. This is well below the detection threshold of the FID.

*Z*9-14:OAc was known to be in the *A. segetum* female extract, but no GC–SSR responses were recorded at the corresponding retention time. In single sensilla recording it is essential to locate sensilla that have receptors of interest, and specialized receptor cells may occur in relatively low numbers. For example, Löfstedt et al. (1982) recorded by EAG from a male antennal sensillum that contained a cell highly sensitive to *Z*9-14:OAc; however, this cell occurred in only 2 of 100 sensilla tested. This explained why activation of receptor cells at the retention time of *Z*9-14:OAc was not found by GC–SSR of the female extract.

In comparison to EAD, SSR has the advantage of differentiating among responses of different receptor cells. On the other hand, a single SSR analysis can provide information on only one type of sensillum, whereas EAD provides information on several types of olfactory receptor cells on an antenna. For this reason Van Der Pers and Löfstedt (1983) suggested that GC–SSR should be considered a valuable supplement, but not a substitute for GC–EAD, an opinion shared by Wadhams (chapter 7, this volume).

V. GC–EAD Analyses of Allelochemics

A recent review of semiochemicals (Nordlund et al., 1981) provided a comprehensive and well-referenced account of allelochemic-induced interactions of phytophagous and entomophagous insects. Relatively few allelochemics have been chemically identified even though they have great potential for pest management. Many of the analytical techniques developed for pheromones, particularly GC–EAD and GC–SSR, are applicable for their identification. A good example was the application of GC–EAD with a detector antenna of the female carrot fly, *Psila rosae*, to detect a kairomone for this species in carrot foliage (Guerin et al., 1983).

VI. Conclusion

Direct coupling of high-resolution GC and EAG has added a new dimension to electrophysiological measurements. The EAD system is readily added to any GC, and it does not require any specialized equipment. GC–EAD provides a rapid bioassay of pheromone components and synthetic compounds that can be done in a chemistry laboratory. It takes full advantage of high-resolution GC and the selected detector species technique provides extreme sensitivity and selectivity for the detection of pheromone components and for confirming their structures.

The direct coupling of high-resolution GC and SSR should be a valuable supplement to GC–EAD. GC–SSR does require specialized antennal preparation and recording techniques.

GC-EAD has been used to detect a kairomone in plant material and it should continue to find general application for the detection and identification of semiochemicals in general.

VII. References

Arn H, Baltensweiler W, Bues R, Buser HR, Esbjerg P, Guerin P, Mani E, Rauscher S, Szőcs, G, Tóth M (1982) Refining lepidopteran sex attractants. Les Médiateurs chimiques INRA, Versailles 261–265.

Arn H, Priesner E, Bogenschütz H, Buser HR, Struble DL, Rauscher S, Voerman S (1979) Sex pheromone of *Tortrix viridana* (*Z*)-11-tetradecenyl acetate as the main component. Z Naturforsch **34c**:1281–1284.

Arn H, Rauscher S, Buser HR, Roelofs WL (1976) Sex pheromone of *Eupoecilia ambiguella*: *cis*-9-Dodecenyl acetate as a major component. Z Naturforsch **31c**:499–503.

Arn H, Städler E, Rauscher S (1975) The electroantennographic detector—A selective and sensitive tool in the gas chromatographic analysis of insect pheromones. Z Naturforsch **30c**:722–725.

Arn H, Städler E, Rauscher S, Buser HR, Mustaparta H, Esbjerg P, Philipsen H, Zethner O, Struble DL, Bues R (1980) Multicomponent sex pheromone in *Agrotis segetum*: Preliminary analysis and field evaluation. Z Naturforsch **35c**:986–989.

Beevor PS, Hall DR, Lester R, Poppi RG, Read JS, Nesbitt BF (1975) Sex pheromones of the armyworm moth, *Spodoptera exempta* (Wlk). Experientia **31**: 22–23.

Bestmann HJ, Brosche T, Koschatzky KH, Michaelis K, Platz H, Vostrowsky O, Knauf W (1980) Pheromone xxx. Identifizierung eines neuartigen Pheromonkomplexes aus der Graseule *Scotia exclamationis* (Lepidoptera). Tetrahedron Lett **21**:747–750.

Bestmann HJ, Vostrowsky O, Koschatzky KH, Platz H, Brosche T, Kantardjiew I, Rheinwald M, Knauf W (1978) (*Z*)-5-Decenyl-acetat, ein Sexuallockstoff für Männchen der Saateule *Agrotis segetum* (Lepidoptera). Angew Chem **90**:815–816.

Bestmann HJ, Vostrowsky O, Platz H, Brosche T, Koschatzky KH, Knauf W (1979) (*Z*)-7-Tetradecenylacetat, ein Sexuallockstoff für Männchen von *Amathes c-nigrum* (Noctuidae, Lepidoptera). Tetrahedron Lett 497–500.

Campion DG, McVeigh LJ, Polyrakis J, Michelakis S, Stavrakis GN, Beevor PS, Hall DR, Nesbitt BF (1979) Laboratory and field studies of the female sex pheromone of the olive moth, *Prays oleae*. Experientia **35**:1146–1147.

Granges J, Baggiolini M (1971) Une phéromone sexuelle synthétique attractive pour le carpocapse des prunes (*Grapholitha funebrana* Tr.) (Lep. Tortricidae). Rev Suisse Vitic Arboric **3**:93–94.

Guerin PM, Arn H, Buser HR, Charmillot PJ (1982) The sex pheromone of *Adoxophyes orana*: Preliminary findings from a reinvestigation. Les Médiateurs chimiques INRA, Versailles 267–269.

Guerin P, Städler E, Buser HR (1983) Identification of host plant attractants for the carrot fly, *Psila rosae*. J Chem Ecol **9**:843–861.

Hill AS, Roelofs WL, Rings RW, Swier SR (1977) Sex pheromone of the black cutworm moth, *Agrotis ipsilon* (Hufnagel) (Lepidoptera:Noctuidae). J NY Entomol Soc **85**:179–180.

Löfstedt C, Van Der Pers JNC, Löfqvist J, Lanne BS, Appelgren M, Bergström G, Thelin B (1982) Sex pheromone components of the turnip moth, *Agrotis segetum*: Chemical identification, electrophysiological evaluation and behavioral activity. J. Chem Ecol **8**:1305–1321.

Moorhouse JE, Yeadon R, Beevor PS, Nesbitt BF (1969) Method for use in studies of insect chemical communication. Nature (London) **223**:1174–1175.

Nesbitt BF, Beevor PS, Cole RA, Lester R, Poppi RG (1973) Sex pheromones of two noctuid moths. Nature New Biol **244**:208–209.

Nesbitt BF, Beevor PS, Cole RA, Lester R, Poppi RG (1975a) The isolation and identification of the female sex pheromones of the red bollworm moth, *Diparopsis castanea.* J Insect Physiol **21**:1091–1096.

Nesbitt BF, Beevor PS, Hall DR, Lester R (1979a) Female sex pheromone components of the cotton bollworm, *Heliothis armigera.* J Insect Physiol **25**: 535–541.

Nesbitt BF, Beevor PS, Hall DR, Lester R, Davies JC, Seshu Reddy KV (1979b) Components of the sex pheromone of the female spotted stalk borer, *Chilo partellus* (Swinhoe) (Lepidoptera:Pyralidae): Identification and preliminary field trials. J Chem Ecol **5**:153–163.

Nesbitt BF, Beevor PS, Hall DR, Lester R, Dyck VA (1975b) Identification of the female sex pheromones of the moth, *Chilo suppressalis.* J Insect Physiol **21**:1883–1886.

Nesbitt BF, Beevor PS, Hall DR, Lester R, Dyck VA (1976) Identification of the female sex pheromone of the purple stem borer moth, *Sesamia inferens.* Insect Biochem **6**:105–107.

Nesbitt BF, Beevor PS, Hall DR, Lester R, Sternlicht M, Goldenberg S (1977) Identification and synthesis of the female sex pheromone of the citrus flower moth, *Prays citri.* Insect Biochem **7**:355–359.

Nesbitt BF, Beevor PS, Hall DR, Lester R, Williams JR (1980) Components of the sex pheromone of the female sugar cane borer, *Chilo sacchariphagus* (Bojer) (Lepidoptera:Pyralidae). Identification and field trials. J Chem Ecol **6**:385–394.

Nordlund DA, Jones RL, Lewis WJ (1981) Semiochemicals, Their Role in Pest Control. Wiley, New York.

Priesner E (1979) Specificity studies on pheromone receptors of noctuid and tortricid Lepidoptera. In: Chemical Ecology: Odour Communication in Animals. Ritter J (ed), Elsevier/North-Holland Biomedical Press, Amsterdam, pp 57–71.

Rothschild GHL, Nesbitt BF, Beevor PS, Cork A, Hall DR, Vickers RA (1982) Studies of the female sex pheromone of the native budworm, *Heliothis punctiger.* Entomol Exp Appl **31**:395–401.

Steck W, Underhill EW, Chisholm MD (1982) Structure-activity relationships in sex attractants for North American noctuid moths. J Chem Ecol **8**:731–754.

Steck WF, Underhill EW, Chisholm MD, Gerber HS (1979) Sex attractant of male alfalfa looper moths, *Autographa californica*. Environ Entomol **8**: 373–375.

Struble DL (1981) A four-component pheromone blend for optimum attraction of redbacked cutworm males, *Euxoa ochrogaster* (Guenée). J Chem Ecol **7**: 615–625.

Struble DL (1983) Sex pheromone components of *Euxoa drewseni*: Chemical identification, electrophysiological evaluation and field attractancy tests. J Chem Ecol **9**:335–346.

Struble DL, Arn H, Buser HR, Städler E, Freuler J (1980a) Identification of 4 sex pheromone components isolated from calling females of *Mamestra brassicae*. Z Naturforsch **35c**:45–48.

Struble DL, Buser HR, Arn H, Swailes GE (1980b) Identification of sex pheromone components of redbacked cutworm, *Euxoa ochrogaster*, and modification of sex attractant blend for adult males. J Chem Ecol **6**:573–584.

Struble DL, Richards KW (1983) Identification of sex pheromone components of the female driedfruit moth, *Vitula edmandsae serratilineella*, and a blend for attraction of male moths. J Chem Ecol **9**:785–801.

Tamaki Y In press. Sex pheromones, Chapter 10. In: Comprehensive Insect Physiology, Biochemistry and Pharmacology. Kerkut GA, Gilbert LI (eds), Pergamon Press, Oxford.

Wadhams LJ (1982) Coupled gas chromatography-single cell recording: A new technique for use in the analysis of insect pheromones. Z Naturforsh **37C**: 947–952.

Van Der Pers JNC, Den Otter CJ (1978) Single cell responses from olfactory receptors of small ermine moths (Lepidoptera:Yponomeutidae) to sex attractants. J Insect Physiol **24**:337–343.

Van Der Pers JNC, Löfstedt C (1983) Continuous single sensillum recordings as a detection method for moth pheromone components in the effluent of a gas chromatograph. J Chem Ecol **9**:000–000.

Chapter 7

The Coupled Gas Chromatography–Single Cell Recording Technique

Lester J. Wadhams[1]

I. Introduction

The first identification of a pheromone in 1961 by Butenandt et al. (1961) was achieved without the aid of a gas chromatograph–a tool which subsequent workers have found essential. Although gas chromatography (GC) is a powerful technique for the separation and isolation of pheromones, additional techniques are required to establish which components in the mixture are involved in the behavior of the insect.

The perception of olfactory stimuli in insects is mediated largely through their antennal receptors. The electroantennogram (EAG) technique is commonly used to monitor the activity of fractions collected from the gas chromatograph (cf. Roelofs, Chapter 5, this volume). Several systems have been described which directly combine the EAG technique with gas chromatography (Struble and Arn, Chapter 6, this volume); those in which the column effluent is monitored simultaneously by the EAG and gas chromatographic detectors provide a most powerful tool for pheromone studies.

Although the EAG gives some information on the specificity of insect olfactory receptors, a more detailed understanding of the perception of odor molecules by insects can be obtained by recording the response of individual olfactory cells. Multicomponent pheromone systems are common in insect communication (Silverstein and Young, 1976) and receptor cells specialized to different components of the mixture are being increasingly implicated in pheromone

[1] Department of Insecticides and Fungicides, Rothamsted Experimental Station, Harpenden, Herts. A15 2JQ.

perception (Kaissling, 1979; Priesner, 1979; Mustaparta, 1979). A system has therefore been developed in which the GC effluent is monitored continuously by recording the responses of a single olfactory cell (GC–SCR) (Wadhams, 1982).

II. The GC–SCR Technique

The coupled GC–SCR system is shown in Fig. 1. Components eluted from the GC column are detected simultaneously by the flame ionization detector (FID) of the gas chromatograph and the olfactory cell of the insect. Both signals are continuously monitored by loudspeaker and chart recorder and are recorded on magnetic tape for subsequent analysis.

The effluent from the fused silica capillary column is split about equally be-

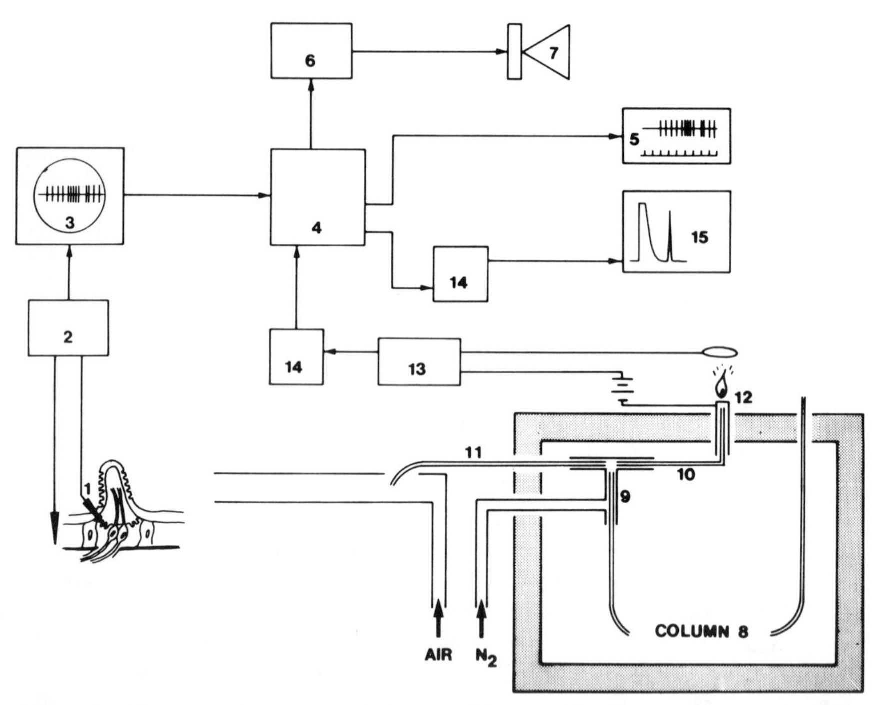

Figure 1. Diagram of the coupled GC–SCR system. 1, Recording electrode; 2, AC coupled amplifier; 3, oscilloscope; 4, tape recorder; 5, mingograph 34T ink jet recorder; 6, audio amplifier; 7, loudspeaker; 8, capillary column; 9, glass-lined tubing "T" piece; 10, deactivated fused silica transfer line to FID; 11, deactivated fused silica transfer line to insect preparation; 12, flame ionization detector (FID); 13, FID amplifier; 14, phase lock loop; 15, chart recorder.

tween the FID and the transfer line to the insect preparation by a low volume splitter constructed from glass-lined stainless-steel tubing (GLT) (Scientific Glass Engineering Limited) and deactivated fused silica tubing (Phase Separations Limited). The GC column is inserted into the end of a GLT "T" piece (0.7 mm inside diameter) and two equal lengths of 0.3 mm inside diameter fused silica tubing are inserted into the other two arms. One of these lengths of tubing leads to the FID while the other is taken through the oven wall and directs the remainder of the column effluent into a purified and humidified airstream (300 ml/min) which passes continuously over the preparation. This latter fused silica transfer line is supported in a length of thin walled 1/16-in. stainless-steel tubing, one end of which is securely clamped to the body of the gas chromatograph.

This arrangement provides some protection and rigidity to the transfer line while retaining a degree of flexibility. The 1/16-in. tubing is wound with heating tape and the transfer line temperature, controlled with a variable transformer, is generally maintained 10–15°C above that of the column oven. The split ratio between the FID and the antennal preparation can be altered by varying the lengths or inside diameter of the fused silica transfer lines, or by using an adjustable microneedle valve effluent splitter (Scientific Glass Engineering Limited).

To maximize the FID response and to diminish residence time in the transfer lines, N_2 (60 ml/min) is added as a make-up gas via a Swagelok 1/16-in. "T" coupling mounted at the base of the splitter.

Recordings of the action potentials generated by the receptor cells associated with individual olfactory sensilla are made using tungsten microelectrodes (Boeckh, 1962). The indifferent electrode is positioned in the mouthparts of the insect and the recording electrode is manipulated into the surface of the antennal club until impulses are recorded. The signals are amplified with an AC coupled amplifier, displayed on an oscilloscope (Devices 3131, Dynamic Electronics Limited) and stored on magnetic tape. Permanent copies of the recordings are made on a Mingograph 34T ink jet recorder (Siemens Limited).

The olfactory specificity of the cell is determined by stimulation with a series of synthetic reference compounds and/or natural product extracts. When the required cell type is located the fused silica transfer line from the GC column is inserted into the glass tube (4 mm inside diameter) which directs the airstream over the antenna. The sample is injected onto the gas chromatograph and the responses elicited from the FID and the olfactory cells are recorded. The signals from the FID amplifier are fed through a phase lock loop to the tape recorder (Teac A3440) the output from which is monitored on a chart recorder. This system upgrades a relatively inexpensive AC tape recorder into a DC recorder of limited bandwidth (0–1500 Hz).

The impulses given by the individual olfactory cells are summed over 3-sec intervals and then plotted against the FID response (e.g., Figs. 2,3). This manual analysis of the data is time consuming and the results will be manipulated by computer in the future.

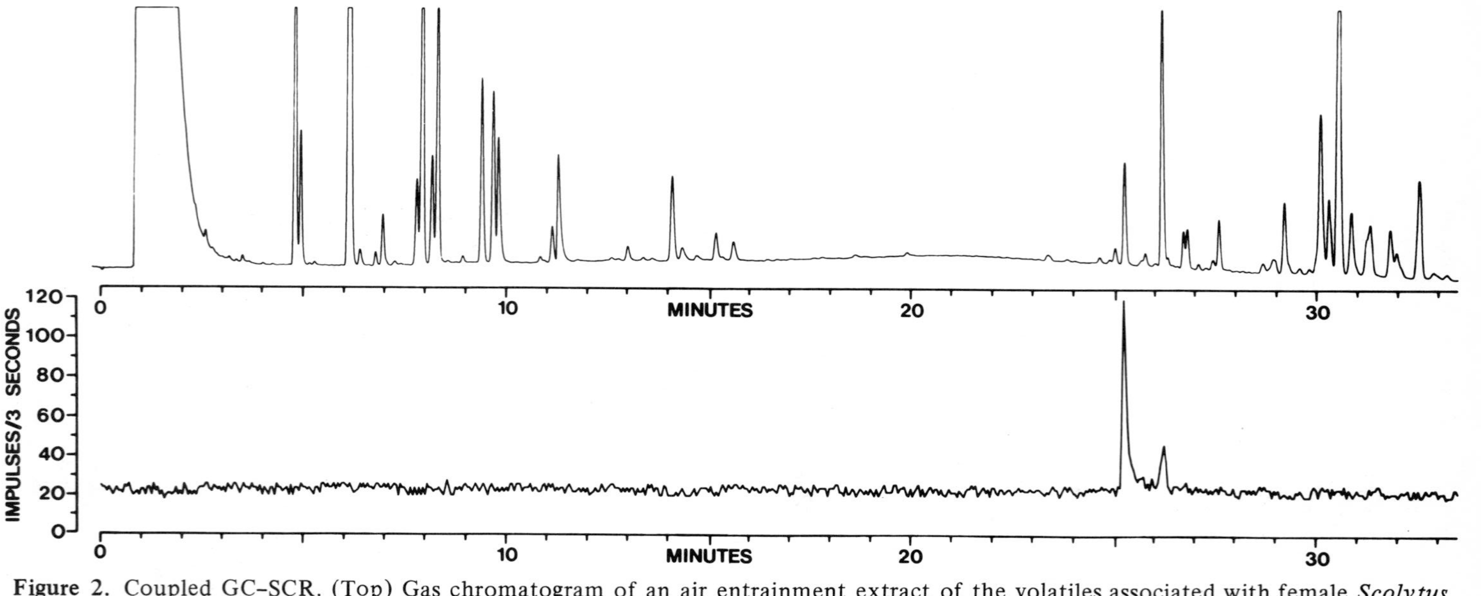

Figure 2. Coupled GC–SCR. (Top) Gas chromatogram of an air entrainment extract of the volatiles associated with female *Scolytus scolytus* boring into English elm. GC conditions: Carlo Erba 2151 AC gas chromatograph fitted with a 21 m × 0.3 mm inside diameter SE-30 fused silica column programmed from 40 to 75°C at 4°/min with hydrogen as the carrier gas, inlet pressure 0.5 kg/cm^2 (Bottom) Response of an olfactory cell specialized to α-cubebene, to stimulation with the air entrainment extract. The cell was recorded from an *S. scolytus* female preparation. The impulses elicited from the cell were summed over 3-sec intervals. The increase in activity of the cell at 25.2 min corresponded to the elution from the capillary column of α-cubebene. The weak response shown by the cell at 26.2 min corresponded to the elution of α-ylangene from the column.

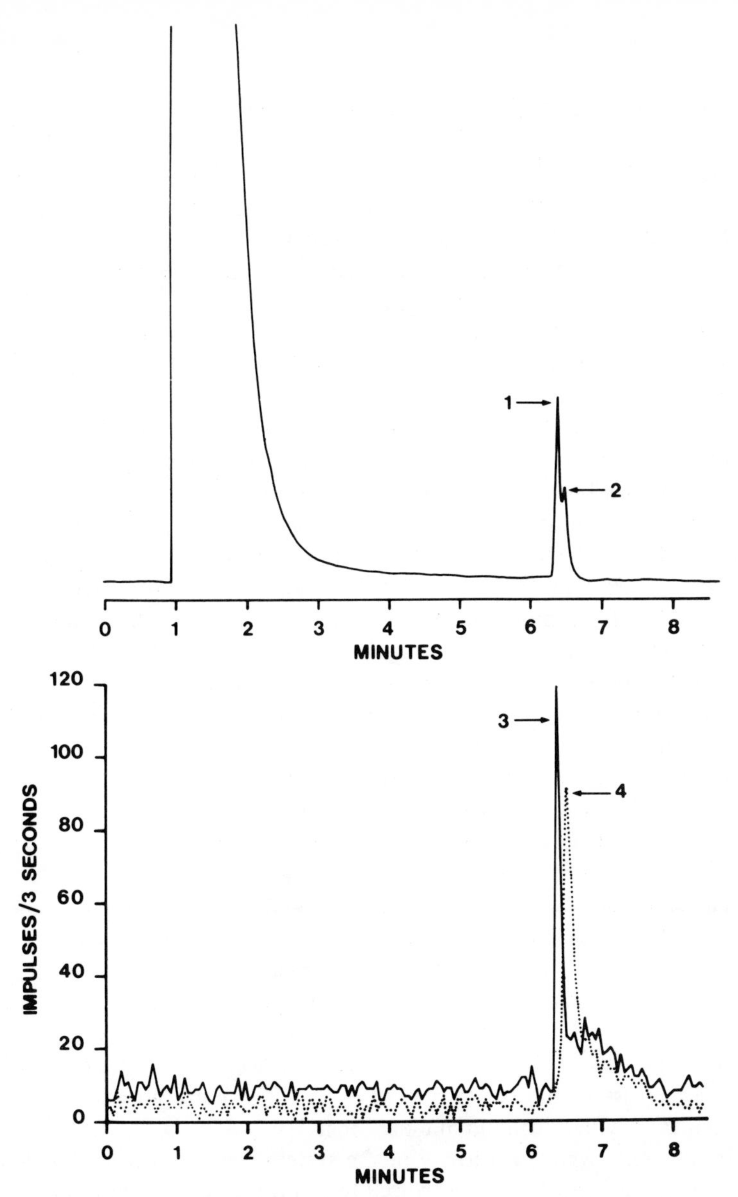

Figure 3. Response of (top) flame ionization detector and (bottom) two olfactory cells from a female *S. scolytus* preparation to (1) (±)-*threo*- and (2) (±)-*erythro*-4-methyl-3-heptanol. The cell (3) was specialized to the (—)-*threo* stereoisomer while the cell (4) was specialized to (—)-*erythro*-4-methyl-3-heptanol. The impulses recorded from each cell were summed over 3-sec intervals. Total sample injected onto the GC was 6.7 ng (±)-*threo*- and 3.3 ng (±)-*erythro*-4-methyl-3-heptanol. The inlet split ratio was 5:1 and the column effluent was split about equally between the FID and the insect preparation. GC conditions: 21 m × 0.3 mm inside diameter SE-30 fused silica column at 50°C with hydrogen as the carrier gas, inlet pressure 0.5 kg/cm^2.

III. Applications of the GC–SCR Technique

The use of the coupled GC–SCR technique in analysis of insect pheromones is illustrated by our work (Grove, 1983) on the aggregation pheromones of the bark beetles, *Scolytus scolytus* and *S. multistriatus.* Its primary application has been to provide rapid and precise information on the retention times of physiologically active compounds in the complex natural product extracts. In these *Scolytus* species, most if not all the individual components of the aggregation pheromone are perceived by separate groups of specialized olfactory cells. The proportions of these different cell types present on the antenna vary considerably. In *S. scolytus,* 20–25% of the cells recorded from the whole antenna respond specifically to each of the key beetle metabolites, (—)-*threo* and (—)-*erythro*-4-methyl-3-heptanol (Blight et al., 1979), while only 2–3% of the cells are specialized to a host synergist such as α-cubebene (Wadhams, 1982).

The EAG and SCR threshold concentrations for the 4-methyl-3-heptanol stereoisomers are similar, but the EAG threshold concentration for α-cubebene is two orders of magnitude higher than that estimated by the SCR technique. This difference in the EAG sensitivity between α-cubebene and the 4-methyl-3-heptanol stereoisomers presumably reflects differences in the proportions of the cell types on the antenna, since the amplitude of the EAG response is thought to be directly related to the numbers of receptors responding (Boeckh et al., 1965). Thus, although significant responses to the 4-methyl-3-heptanol stereoisomers are observed with the GC–EAG technique, none were obtained with α-cubebene since the concentrations required to elicit a response are attained only at concentrations which overload the capillary column.

Such limitations do not apply to the GC–SCR system. Threshold concentrations of olfactory cells in *S. scolytus* are similar, irrespective of their individual specificities. These concentrations, of the order of 10^9–10^{10} molecules/ml of air are readily attainable under high-resolution GC conditions. For example, Fig. 2 shows the response of a specialized α-cubebene cell to stimulation with an extract of the volatiles associated with female *S. scolytus* infestations in English elm. The increase in activity of the cell at 25.2 min corresponded with the elution from the capillary column of α-cubebene.

Although the overall sensitivity of the system has not been fully examined, stimulation of a pair of cells specialized to (—)-*threo* and (—)-*erythro*-4-methyl-3-heptanol with 200 pg of (±)-*threo* and (±)-*erythro* isomers resulted in a signal to noise ratio of 6:1. The amount of active material reaching the antennal preparation, 67 pg (—)-*threo* and 33 pg (—)-*erythro*-4-methyl-3-heptanol, was at the detection limits of the FID. This suggests that the present 1:1 split ratio between the FID and the antenna could be altered in favor of the FID.

The selectivity of the coupled GC–SCR system is shown by the responses of a pair of cells to stimulation with a mixture of (±)-*threo* and (±)-*erythro*-4-methyl-3-heptanol (Fig. 3). Fig. 4 shows the actual responses of the cells over the period of stimulation by the isomers. The cell with the large amplitude

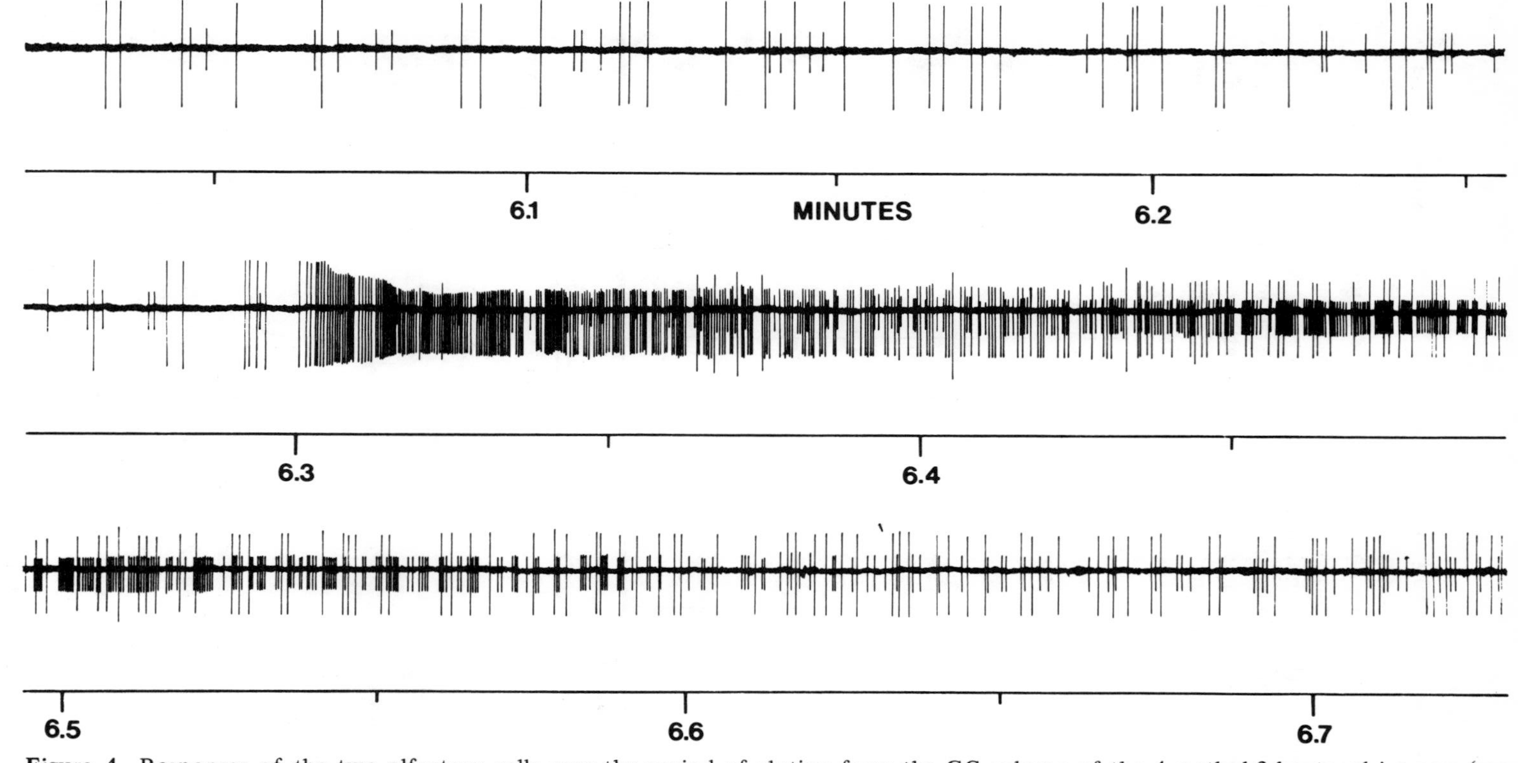

Figure 4. Responses of the two olfactory cells over the period of elution from the GC column of the 4-methyl-3-heptanol isomers (see Fig. 3). The cell with the larger amplitude was specialized to (—)-*threo*-4-methyl-3-heptanol while the cell with the smaller amplitude was specialized to the (—)-*erythro* stereoisomer.

spikes was specialized to (−)-*threo*-4-methyl-3-heptanol. A sharp increase in impulse frequency was recorded from the cell 6.3 min after injection of the sample and this corresponded with the elution from the column of the (±)-*threo* isomer. The cell giving the small amplitude spikes was specialized to the (−)-*erythro* stereoisomer. The maximum impuse frequency of this cell occurred between 6.45 and 6.5 min which corresponded with the elution of the (±)-*erythro* isomer from the column. Despite incomplete separation on the GC column of the 4-methyl-3-heptanol isomers, the peaks corresponding to the key compounds for each cell can be readily distinguished. In contrast, the GC–EAG technique re-

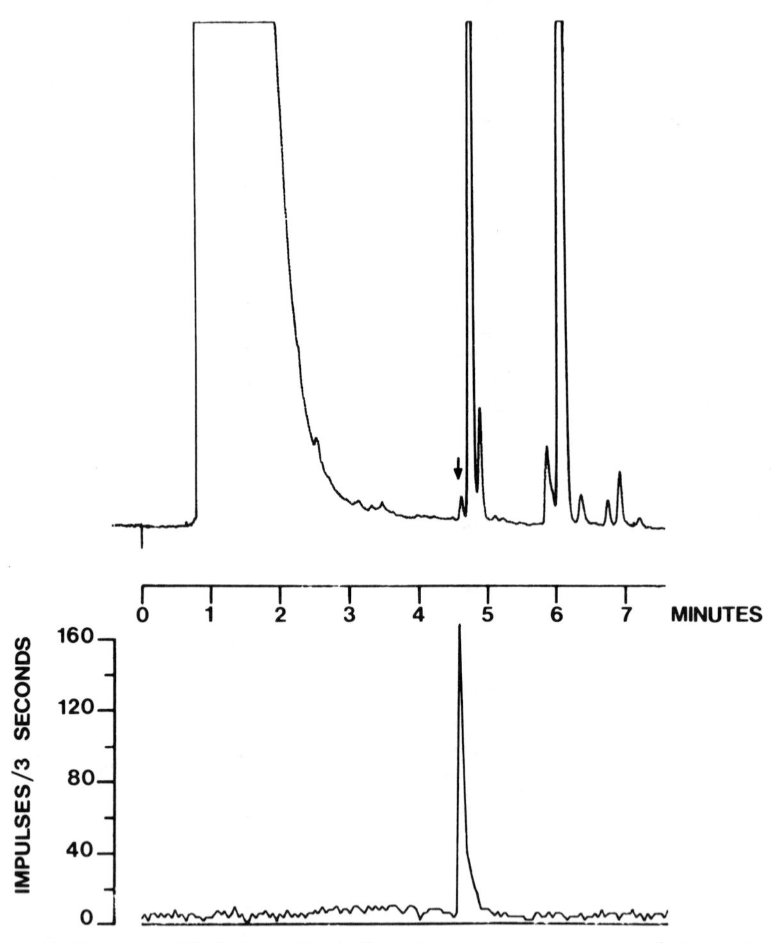

Figure 5. Coupled GC–SCR: (Top) Partial gas chromatogram of the volatiles associated with infestations of male *S. scolytus* boring in English elm. GC conditions: Injection via Grob injector, split ratio 1:5, on a 21 m × 0.3 mm inside diameter SE-30 fused silica column programmed from 40°C at 2°C/min with hydrogen as the carrier gas. (Bottom) Response of an olfactory cell, specialized to 4-methyl-3-heptanone, to stimulation with the extract. The cell was recorded from a female *S. scolytus*.

quired almost baseline separation of the two peaks before unambiguous evidence for the electrophysiological activity of the two stereoisomers was obtained (Blight et al., 1979).

The SCR preparations are stable and, provided stimulus concentrations remain low, recordings may be continued for several hours allowing many samples to be tested against each preparation. This, together with the specificity of the olfactory cells, has enabled us to use the system as a specific detector to confirm the presence of trace components in complex gas chromatograms. *S. scolytus* male beetles produce significant amounts of 4-methyl-3-heptanone which is perceived by a small group of olfactory cells (Blight et al., 1983). Fig. 5 shows the

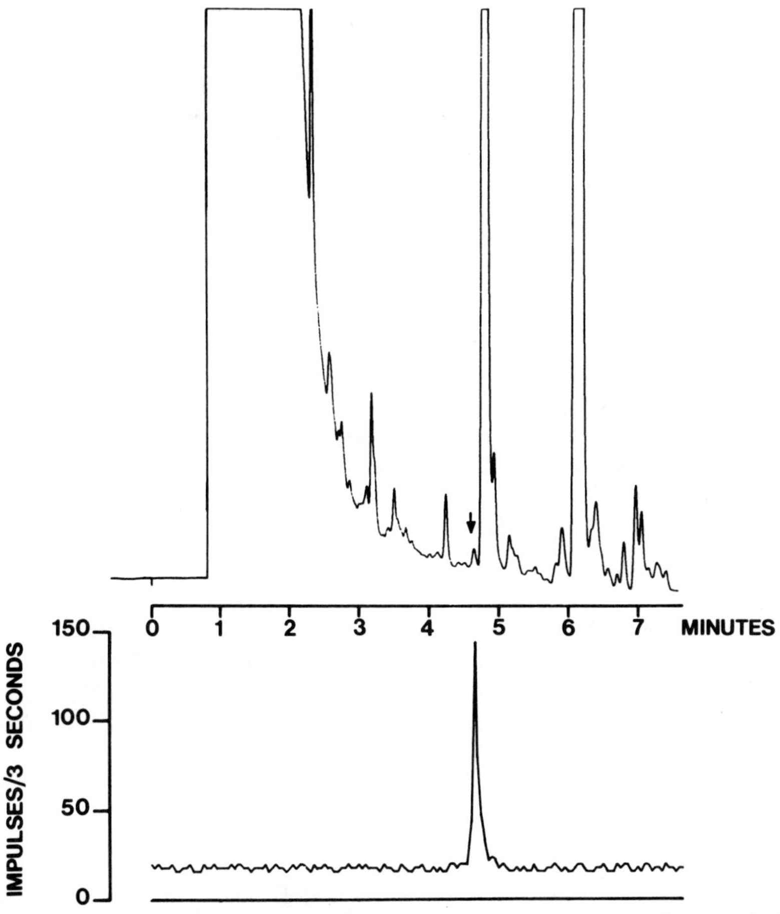

Figure 6. Coupled GC–SCR: (Top) Partial gas chromatogram of the volatiles associated with female *S. multistriatus* boring into English elm. The GC conditions were the same as those for Fig. 5. (Bottom) Response of the same olfactory cell shown in Fig. 5 to stimulation with the extract. The cell responded to the component, retention time 4.62 min, which had been found to coinject with an authentic sample of 4-methyl-3-heptanone.

response of a female *S. scolytus* cell specialized to this substance, to stimulation with part of a male *S. scolytus* extract. The cell responded to the component, retention time 4.65 min, unambiguously identified by GC–MS and coinjection with authentic material as 4-methyl-3-heptanone. The same cell was then used to confirm the presence of 4-methyl-3-heptanone, retention time 4.62 min, in the extract of volatiles from females of the other Scolytid, *S. multistriatus* (Fig. 6). It should be noted (see Figs. 5,6) that equivalent amounts of 4-methyl-3-heptanone injected onto the column from the two extracts elicited about the same number of impulses from the cell.

In general each cell responds strongly to one specific component. However, other components may elicit weak responses. Where these secondary compounds have been identified they have been found to bear a close structural relationship to the cells' key compound. Thus the α-cubebene cell shown in Fig. 2 also responded weakly to a component eluting at 26.2 min. This compound, which is either α-ylangene or its stereoisomer α-copaene, is structurally related to α-cubebene.

IV. Discussion

The examples of the use of GC-SCR technique presented above have all related to our study of bark beetle aggregation pheromones. However, the technique should be useful with other species which utilize complex multicomponent pheromone systems (see Löfsted et al., 1982; Struble and Arn, Chapter 6, this volume). The system is designed to complement rather than to replace the GC–EAG technique. While the latter can provide information from many types of olfactory cells, the GC–SCR system functions as a more selective monitor of the GC effluent. The choice of technique will depend ultimately on the nature of the insect's olfactory system.

Acknowledgments. The author thanks Mr. P. Buchan and Dr. R. Moreton, University of Cambridge, for the design and construction of the AC coupled amplifier and the loan of the phase lock loop system, and Margaret M. Blight and Dr. J. Pickett for reviewing the manuscript.

V. References

Blight MM, Henderson NC, Wadhams LJ (1983) The identification of 4-methyl-3-heptanone from *Scolytus scolytus* (F.) and *S. multistriatus* (Marsham). Absolute configuration, laboratory bioassay and electrophysiological studies on *S. scolytus*. Insect Biochem **13**:27–38.

Blight MM, Wadhams LJ, Wenham MJ (1979) The stereoisomeric composition of the 4-methyl-3-heptanol produced by *Scolytus scolytus* and the preparation and biological activity of the four synthetic stereoisomers. Insect Biochem **9**:525–533.

Boeckh J (1962) Electrophysiologische Untersuchungen an einzelnen Geruchs-Rezeptoren auf den Antennen des Totengräbers (Necrophorus:Coleoptera). J Vergl Physiol **46**:212–248.

Boeckh J, Kaissling KE, Schneider D (1965) Insect olfactory receptors. Cold Spring Harbor Symp Quant Biol **30**:263–280.

Butenandt A, Beckmann R, Stamm D (1961) Über den Sexuallockstoff des Seidenspinners. Konstitution und Konfiguration des Bombykols. Z Physiol Chem **32**:84.

Grove JF (1983) Biochemical investigations related to Dutch Elm Disease carried out at the Agricultural Research Council, Unit of Invertebrate Chemistry and Physiology, University of Sussex, 1973–1982. In: Research on Dutch Elm Disease in Europe. Burdekin D (ed), Forestry Commission Bull 60, HMSO.

Kaissling KE (1979) Recognition of pheromones by moths, especially in Saturniids and *Bombyx mori*. In: Chemical Ecology: Odour Communication in Animals. Ritter FJ (ed), Elsevier/North-Holland, Amsterdam, pp 43–56.

Löfstedt C, Van Der Pers JNC, Löfqvist J, Lanne BS, Appelgren M, Bergström G, Thelin B (1982) Sex pheromone components of the Turnip Moth, *Agrotis segetum*: Chemical identification, electrophysiological evaluation and behavioral activity. J Chem Ecol **8**:1305–1322.

Mustaparta H (1979) Chemoreception in bark beetles of the genus Ips: Synergism, inhibition and discrimination of enantiomers. In: Chemical Ecology: Odour Communication in Animals. Ritter FJ (ed), Elsevier/North-Holland, Amsterdam, pp 147–158.

Priesner E (1979) Specificity studies on pheromone receptors of Noctuid and Tortricid Lepidoptera. In: Chemical Ecology: Odour Communication in Animals. Ritter FJ (ed), Elsevier/North-Holland, Amsterdam, pp 57–71.

Silverstein RM, Young JC (1976) Insects generally use multicomponent pheromones. In: Pest Management with Insect Sex Attractants. Beroza M (ed), ACS Symposium Series, American Chemical Society, Washington, D.C., Vol 23, pp 1–29.

Wadhams LJ (1982) Coupled gas chromatography–single cell recording: A new technique for use in the analysis of insect pheromones. Z Naturforsch **37C**: 947–952.

Chapter 8

The Tandem Gas Chromatography—Behavior Bioassay[1]

Hans E. Hummell[2]

I. Introduction

Genuinely new detection methods like the flame ionization detector (FID)[3] for gas chromatography (McWilliams and Dewar, 1958; Harley et al., 1958), mass spectrometry for structure determination (Biemann, 1962), and the use of whole insects (Flaschenträger et al., 1957) or isolated insect antennae (Schneider, 1957; Roelofs, 1984, Chapter 5, this volume) as biological detectors significantly increased the number of techniques available to the pioneers in the field 25 years ago (see Hecker and Butenandt, 1984, Chapter 1, this volume). Equally important and numerous are the examples in which a novel combination of already existing techniques opened up additional analytical avenues. Examples are coupled gas–liquid chromatography–mass spectrometry (GLC–MS) (Gohlke, 1959; Ryhage, 1964), tandem GLC–EAD (Moorhouse et al., 1969; Arn et al., 1975; Beevor et al., 1975), tandem GLC single-cell recordings (Wadhams, 1982; Wadhams, 1984, Chapter 7, this volume; Löfsted et al., 1982; Struble and Arn, Chapter 6, this volume), and the tandem GLC–behavior bioassays (t-GLC–BB) (Table 1).

---

[1] Dedicated to Dr. G. Eger-Hummel for her unfailing support. Contribution No. 3 in the series titled: "Invertebrate Chemical Communication." For contribution No. 2, see Hummel HE, Andersen JF (1982). In: Proc. 5th Int. Symp. Insect-Plant Relationships, Wageningen, Netherlands, March 1–4, 1982, Visser JH and Minks AK, (eds), Pudoc Press, Wageningen, pp 163–167.

[2] 2011 Cureton, Urbana, Illinois 61801 and Harbor Branch Foundation, Ft. Pierce, Florida 33450, contribution no. 377.

[3] Abbreviations used: FID, Flame ionization detector; GLC–MS, gas–liquid chromatography–mass spectrometry; EAG, electroantennogram; t-GLC–BB, tandem GLC behavioral bioassays; INJ, injector; COL, column; DET, detector; SPL, splitter; ACT, actograph; REC, dual-channel recorder.

Table 1. Tandem techniques in pheromone research in their order of increasing complexity

Level of biological organization	Method	Measurement of	Reference
Organ			
Isolated antenna	GLC–EAD	Combined generator potentials of many sensilla	Moorhouse et al. (1969) Arn et al. (1975)
Specialized sensilla on isolated antenna	GLC–SCR	Generator potential of one defined sensillum	Beevor et al. (1975) Löfsted et al. (1982) Wadhams (1982) Wadhams (Chapter 7, this volume) (1984)
Organism			
Single individual or a group of individuals	t-GLC–BB	Activation of behavioral displays stimulated by exposure to pheromones and their analogs	Anders and Bayer (1959) Bayer and Anders (1959) Feeny (1962) Gaston et al. (1966) Bierl et al. (1970) Hummel et al. (1973) Hummel and Kucera (1977) Hummel and Sternburg (1979) Hummel (1979)
	t-GLC–flight tunnel	Orientation and upwind flight toward pheromone-emitting source	So far undescribed

In much the same way as the linkup of gas chromatography with mass spectrometry combines the virtues of both methods, the t-GLC–BB technique couples the sensitivity and specificity of insect behavior with the analytical accuracy and reproducibility of gas chromatography. Thus, the t-GLC–BB can describe to what extent and degree a specific chemical message is translated into behavior of the test insect. Moreover, it can make determinations of the number of carbon atoms and double bonds in pheromones before the isolation and chemical identification of pure material have been achieved. The low cost, easy construction of the hardware and its compatibility with many existing GLC systems and the general simplicity of the method can provide as incentive for adopting it on a wider scale. Given this range of possibilities, this chapter will describe the advantages and the versatility of the method and thus hopes to introduce it to a wider circle of users.

Historically, Anders and Bayer (1959) described the first very crude system of this kind. The FID was not commonly available then, and the thermal conductivity cell was too insensitive for work at the behaviorally relevant nanogram level. For their analysis of crude female gland extracts of *Bombyx mori,* Bayer and Anders (1959) therefore used male silkmoths as biological detectors, a principle already described 2 years earlier by Flaschenträger et al. (1957) for black cutworm moths. Anders and Bayer (1959) could define three areas of activity in their *Bombyx* extracts. Of those, the second area of activity was subsequently equated with bombykol (Butenandt et al., 1959; Hecker and Butenandt, Chapter 1, this volume). It took 19 years until the first area of activity was chemically characterized as bombykal by Kasang et al. (1978), whereas the third component remains unknown. A few years after the discovery of Anders and Bayer, Feeny (1962) reported his principally similar, but differently designed, combination of GLC with a behavioral assay. He used it for the characterization of pheromones from the garden tiger moth, *Arctias virgo.* His unpublished work, however, was never followed up by chemical investigations.

Similar approaches on several economically important insect species originated from laboratories in the United States, again without reference to Anders and Bayer (1959) whose work was published in German. Gaston et al. (1966) described the most advanced apparatus for t-GLC–BB work. It allowed the accurate chromatographic characterization of the sex pheromone of *Trichoplusia ni* before the chemical identification as *Z*7-12:OAc by Berger (1966) via microdegradation and spectrometric studies was published. In this work, Gaston et al. (1966) describe the retention characteristics of several other species in the Noctuid family. Four years later, Bierl et al. (1970) used a laboratory bioassay coupled to GLC for pinpointing disparlure, the pheromone of female gypsy moths. This work made use of a bioassay described 10 years earlier by Block (1960). Within the last decade, the technique was repeatedly used by Hummel et al. (1973) for the purification and isolation of gossyplure, the *Pectinophora gossypiella* sex pheromone; by Hummel (1976, unpublished) for the chromatographic characterization of bombykal; by Hummel and Kucera (1977) for the chromatographic investigation of sex pheromone components of *Diabrotica undecimpunctata howardi*, the southern corn rootworm beetlee, by Hummel and Sternburg (1979) for showing a multiplicity of sex pheromones in *Hyalophora cecropia*, and by Hummel (1979) for investigations of the mechanism of reproductive isolation based on differences of pheromones in species of the Saturniid moth family. The refined version of the apparatus used in these studies and some of the results obtained are described in the following sections.

II. General Description of the Apparatus

The t-GLC–BB arrangement, schematically drawn in Fig. 1, consists of a GLC equipped with injector (INJ), column (COL), and detector (DET) compartments, a gas stream splitter (SPL), the behavior bioassay (BB) chamber attached to the side of the GLC, an actograph (ACT), and a dual-channel recorder (REC). These individual parts are connected with lines for the transfer of gases and electrical signals.

In brief, nitrogen gas N_2 enters the injector port at point B and carries the sample injected through syringe A onto the column head C of a glass capillary or a packed column. Individual components of the samples, separated according to their physical and chemical properties, leave the column and are split into two portions as described by Brownlee and Silverstein (1968). One portion leaves at D and enters the FID where it generates the electrical signal S fed to the REC and is written on a strip chart T. The second portion of the column effluent leaves the DET compartment at E and is introduced into the transfer line F that connects the GLC with the BB part via the ball and socket joint H. The effluent is, at point G, diluted with a stream of prepurified and rehumidified air. Upon

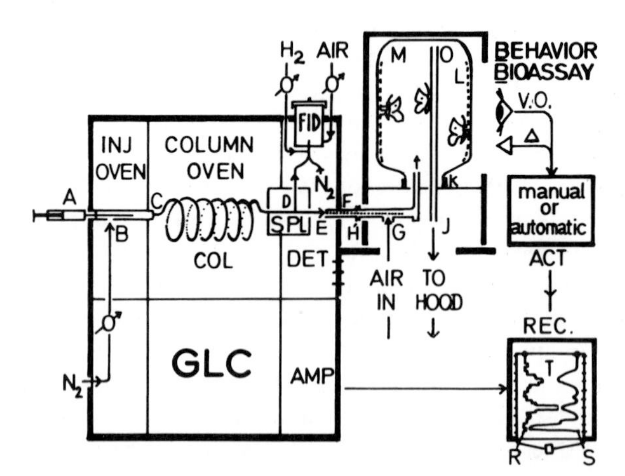

Figure 1. Schematic view of the t-GLC–BB unit assembled with transfer lines and connections to the actograph ACT and recorder REC.

Table 2. Operating parameters for the GLC part of the apparatus

Part of the apparatus considered	Symbol used in Fig. 1	Specifications and remarks	Ranges of operations: For gas streams (ml/min)	Ranges of operations: For temperatures (°C)
Injector	INJ	Splitless, glass lined; 1–10 μl injection volume	High purity, low O_2, gas 10–30 N_2	150–300
Column[1]	COL	(a) Silanized, packed glass columns	10–30	150–250
		(b) Support coated open tubular (SCOT) columns	5–20	Depending on specifications of manufacturers
		(c) Wall-coated open tubular (WCOT) columns	1–5	
		N_2 Make-up gas required for operation of FID with WCOT	20	
Detector	DET	FID	20–30 H_2	200–250
Stream splitter	SPL	Operated in range of 10:1 to 1:10 split ratio depending on the required sensitivity, sample size, and type	150–200 air	Same as detector oven
Recorder	REC	Dual-channel, dual-pen instrument with 1–10 mV sensitivity and paper speeds adjustable between 0.5 and 10 cm/min		

[1]Common liquid phases include: Apiezon L (Ap L), Dexsil, SE-30 (unpolar); SF-96, QF-1, NPGA (medium polarity); Carbowax 20M, DEGS (highly polar).

entering the observation flask M, the airstream charged with biologically active substances activates the test insects. The level and temporal patterns of their activity are recorded simultaneously as trace R on the same strip chart as the FID signal S. Care is taken that FID signal and insect response indeed occur in synchrony, a condition which is verified by injecting a chemically defined test substance of known quantity and biological activity, for example, the cabbage looper sex pheromone *Z*7-12:OAc.

The range of operation parameters for all parts of the GLC are listed in Table 2. Sizes and specifications of the transfer line, the BB chamber and the REC are contained in Figs. 1-4 and in the part list attached as Appendix 1.

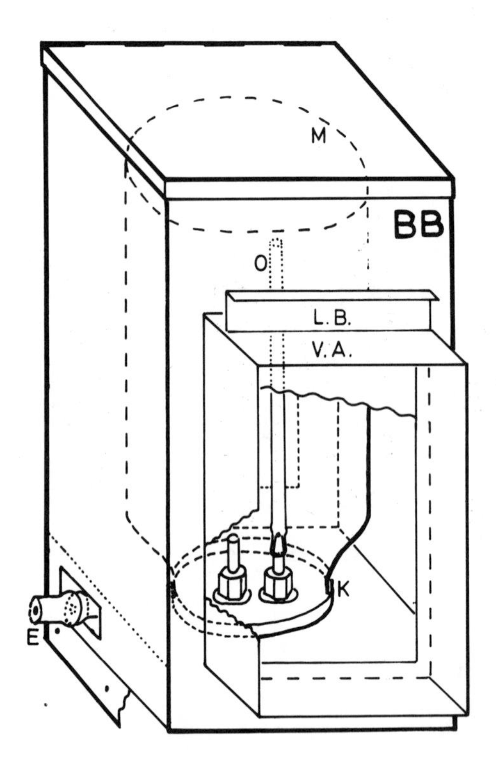

Figure 2. Behavior bioassay chamber seen from the upper left. Part of the viewing aperture (V.A.) and the light baffle (L.B.) used for control of the level of illumination inside the chamber are removed. The upper rear and the lower right edge of the chamber are hinged for ease of changing bioassay flasks containing the test insects. To the lower left, part of the transfer line connecting gas chromatograph and bioassay chamber is visible. For additional explanation, see Sections II–IV. Drawn to scale: the upper front edge is 220 mm long.

III. Operation of the t-GLC–BB Apparatus

By heating the bioassay flask in a drying oven at 150°C for an hour, any residual pheromone traces from previous runs are removed. After cooling the flask to ambient temperature, care is taken to handle it only on the outside while it is being loaded with 10-30 test insects. The proper light conditions and correct circadian phase for the particular insect species have to be observed. While its cap is being loosened, the flask M is inverted and quickly attached to the base K of the BB chamber. Proper light conditions [for nocturnal moths usually 0.3 lumens/m^2 (Gaston et al., 1964), for diurnal insects daylight or incandescent light] and an airflow of 2 liters/min are adjusted while the insects are monitored for background activity for at least 15 min. A background of 10-15% is considered satisfactory. Higher activity may indicate pheromone contamination of the glassware, the GLC column, or the transfer line; or the particular batch of insects may be unsuitable for the test.

If the background level is acceptable, a small aliquot of the pheromone solution spiked with an internal standard is injected into the GLC. To minimize memory effects, typically less than 1 μg of pheromone is being used. Temporal reversal of the air stream by commutation of joints G and J (Figs. 1,3) flushes out solvent vapors (e.g., CS_2) that are irritative to the insects. Typically, the REC trace S from the FID will be used as an indicator. When the REC pen is returning to the starting level, connections G and J will be reversed to the original setting. At this point, recording of insect activity can start. Depending on the nature of the sample, its complexity, and the chromatographic conditions used, data acquisition may take 20 to 120 min. During this period, food and water are withheld from the insects. For convenience of operation and for ensuring full behavioral viability of the insects, the peaks of insect activity should appear between 5 and 60 min after injection of the sample, as shown in Gaston et al. (1966) and in Figs. 5-8 of this chapter. After completion of the test, the insects are returned to their holding cage and are allowed access to food and water. The glass flask M is rinsed with solvent, detergent, and distilled water, and heated to 150°C for 1 hr. The COL is temperature programmed to the limit recommended by the manufacturer and kept under N_2 flow for 1 hr to remove remaining traces of biological activity. Subsequently, the GLC is equilibrated at the starting temperature for the next bioassay. Periodic checks of gas flows and retention times of test mixtures ensure proper operating conditions of the GLC part (see Table 2). More details of the apparatus can be seen in the cutaway drawing of Fig. 2. Measurements of all dimensions and connections can be taken from Fig. 3.

GLC columns are of high-performance quality: Solid support in the packed columns is acid washed and pretreated with dichlorodimethylsilane. The glass columns are internally silanized before they are being packed. Conditioning of columns occurs under moderate N_2 flow rates over a period of 24 hr while

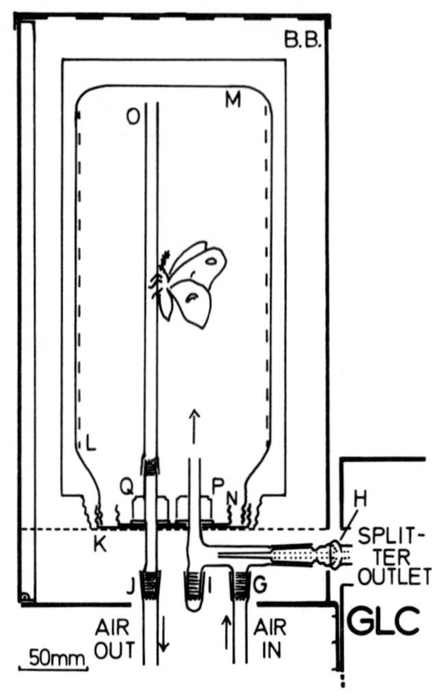

Figure 3. Rear view of behavior bioassay chamber showing all connections. Paper layers covering the rear and part of the frame supporting the paper are omitted for clarity. For details see the text in Sections II–IV. Drawn to scale: the mark in the lower left corner is 50 mm long.

the columns are disconnected from the detector. Temperatures are increased in steps of 10°C and maintained for 1 hr per step until a temperature 10°C above the maximum operating level is reached. At this point the column is held for another 8 hr. Readers unfamiliar with operating conditions and theory of GLC may consult the books by Jennings (1980) and Perry (1981) as reference sources.

IV. Brief Description of the Behavior-Bioassay Procedure

The bioassay chamber BB rests on a platform attached to the detector side of the GLC. The connection via a ball and socket joint is designed in such a manner that one person can hook up the chamber to the GLC outlet within a few seconds. Care is taken to avoid condensation of volatiles in the transfer line F. An electrically heated tape can be used to cover the line, the joint H and the glass adapter up to point G. The temperature of the gas mixture entering the observation chamber M should, however, not exceed 40°C. A brass screen L covering the side walls of the flask M (volume, 3.8 liters) provides a foothold for the insects. For small organisms with body length of less than 3 mm, the screen

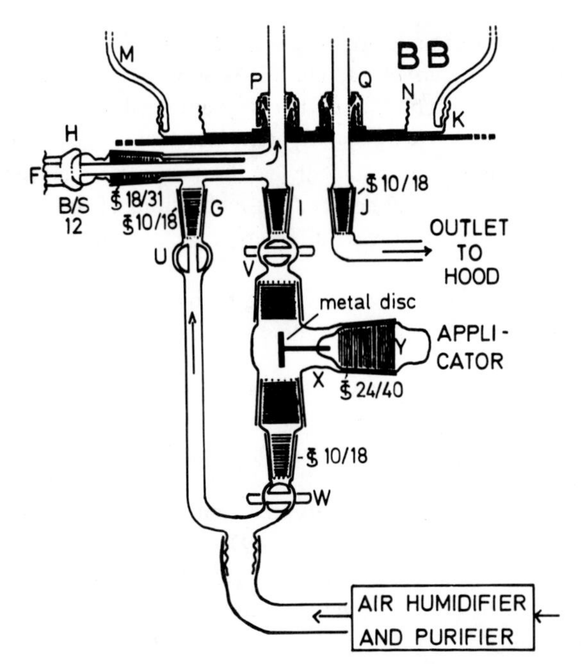

Figure 4. Detailed front view of the transfer line leading from the gas chromatograph GLC to the behavior bioassay chamber BB. The alternative inlet system X with applicator Y and valves U, V, and W allows external testing of compounds, with or without simultaneous GLC input.

is omitted for better visibility. Small insects can also be assayed in a flask of half the volume fitted to the concentrically arranged base N. An exhaust tube O sealed to J by a Luer type joint extends almost to the top of the observation flask. This arrangement forces air to flush the entire volume of the flask and to reach all insects before another component enters from the GLC column. The 6 mm inner diameter Swagelok nuts P and Q are tightened by hand. A conical graphite or lead ferrule replaces the customary stainless-steel rings and provides the necessary seal between glass tube and metal parts (see part list in Appendix 1 for details).

V. Evaluation of Insect Responses

By observing and monitoring male insect behavior through a viewing aperture (VA) containing a light baffle (LB)(Fig. 2), the duration and intensity of their biological activity can be correlated with the electrical response of the FID as shown by Gaston et al. (1966). The observer, after some training, is well capable of registering the simultaneous motion of five to seven insects or, instead, the few nonresponders in a field of constant motion created by the active members of the population. Simple subtraction from the total number of insects subsequently produces the number of responders. However, this approach is only valid for observing unspecified "activity." As soon as closer examination becomes essential, e.g., a discrimination between antennal movement, clasper extension, and copulation attempts, only one insect at a time can be assayed. Several independent observers could divide the task after marking the insects with different colors; or several consecutive replications of the experiment with fresh insects could be performed in order to obtain statistically significant results.

In the order of increasing activation, the following behavioral sequence has been described for *Bombyx mori* by Anders and Bayer (1959):

1. attentive posture (raising of head, slight antennal movements);
2. wing vibration with small, and
3. wing vibration with increasingly large amplitude;
4. undirected searching runs;
5. copulation attempts while claspers are extended and abdomens are curved.

This sequence for *B. mori* may serve as a guide for the assay of many other Lepidoptera species. In the following sections, a few examples from the orders Lepidoptera (see Figs. 5-8) and Coleoptera (Fig. 9) will illustrate the wide applicability of the method to insects of both purely scientific and practical importance. Other insect orders with clear-cut behavioral responses to pheromones (e.g., Orthoptera, Hemiptera, and Hymenoptera) should be accessible as well although the literature to date lists no examples.

VI. The t-GLC–BB Technique and Its Value for Structural Assignments of Pheromones

6.1. Examples from the Order of Lepidoptera

A. *Pectinophora gossypiella*

In the early 1970's, a great deal of confusion existed about the biological activities and correct structural assignments of the sex pheromones, parapheromones and putative pheromones of the pink bollworm moth. Relative retention data based on the t-GLC–BB technique helped to clarify the situation in this economically highly important species (Hummel et al., 1973). Table 3 lists the chromatographic results and supports the following three conclusions:

Nonidentity of gossyplure and propylure. The relative retention data of propylure, the putative sex pheromone of *P. gossypiella* (Jones et al., 1966) and DEET, its putative activator (Jones and Jacobson, 1968) are decidedly different from gossyplure (Hummel et al., 1973) isolated from female moths by a classical purification procedure. Since gossyplure and the putative pheromones differ on

Table 3. Relative retention data for gossyplure, hexalure, propylure, and DEET as determined by the t-GLC–BB technique

GLC glass columns with liquid phase	Relative retention data				
	Determined by the FID detector response				Determined by the behavior bioassay detector
	n-hexadecyl-acetate	Hexalure	Propylure	DEET	Extracted pink bollworm sex pheromone = gossyplure
Apiezon L	Standard = 1.00	0.83	0.51	0.15	0.80
Dexsil		–	–	–	0.81
SF-96		0.86	0.62	0.15	0.78
QF-1		0.91	0.63	0.60	0.86
NPGA		1.01	0.71	0.44	1.03
Carbowax 20M		1.05	0.83	0.87	1.24
DEGS		1.26	1.04	0.42	1.46

six columns of low and high polarity, a negative statement of nonidentity is justified. One mismatch in retention time rules out identity.

Not only on the basis of relative retention data but also by reading the behavior bioassay detector, propylure, and DEET can be ruled out as sex pheromone candidates. Milligram doses of both compounds could not convincingly excite the male pink bollworm moths, even when eluted from the column simultaneously in a fashion similar to Section VII. Hexalure, under similar conditions, showed moderate activity at the microgram level whereas gossyplure proved to be active at the nanogram level. Gossyplure alone induced copulation attempts, the highest response level, in the males exposed to the column effluent.

Determination of the number of carbon atoms and the number of C–C double bonds in gossyplure at the submicrogram level. The relative retention data of hexalure Z7-16:OAc, a parapheromone of *P. gossypiella,* and gossyplure, the genuine sex pheromone, are consistent with acetate esters of 16 carbon atoms in the alcohol moiety (Figs. 5-7). On the nonpolar Ap L phase and on the polar phases Carbowax 20M (a polyethyleneglycol) and DEGS (a diethyleneglycolsuccinate polyester), the carbon–carbon double bonds are indicated by charac-

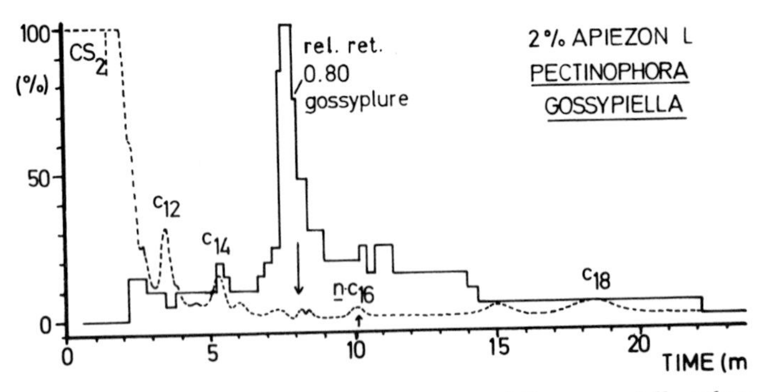

Figure 5. t-GLC–BB of 100 female equivalents of *P. gossypiella* ether extract dissolved in CS_2 and gas chromatographed on an apolar glass column of 50 cm × 0.2 cm size and packed with 2% Apiezon L on Chromosorb W, HMDS treated, at 150°C with 27 ml N_2/min. For better chromatographic results, the crude extract is partially purified by (1) cold temperature precipitation (−20°C, 1 week), (2) high-vacuum distillation in a falling film distillation apparatus at 190°C/0.35 mm Hg, and (3) urea inclusion of excess saturated, straight-chain hydrocarbons. The relative retention of the biologically active peak is 0.80. *n*-16:OAc, actually contained in the female extract (symbol *n*-C_{16}), was chosen as internal standard (1.00 = retention time 10.1 min). At highest amplification, other peaks appear at 3.5 min (12:OAc), 5.5 min (14:OAc), and 18.5 min (18:OAc). Behavioral activities shown are averaged from two different experiments. For further explanation see the text. (—) Solid line: percentage male moths response (N=10 males). (- - -) dashed line: percentage FID response (100% = 10 mV recorder sensitivity).

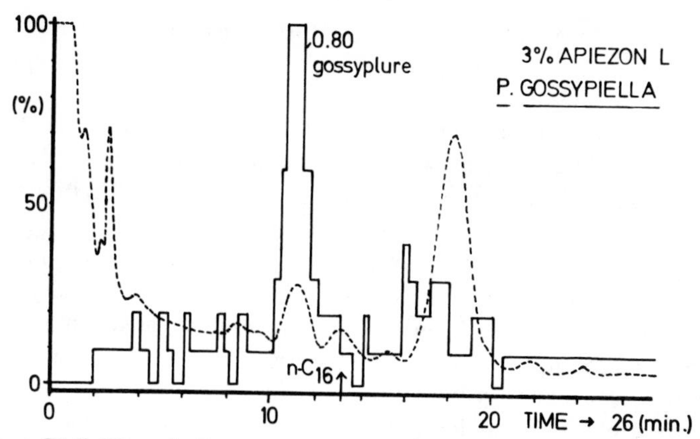

Figure 6. t-GLC–BB of *P. gossypiella* extract on a preparative apolar 3% Apiezon L column, 190°C, 80 ml N_2/min. The relative retention of the biologically active peak is 0.80 as in Fig. 5. Inactive peaks appearing after 27 min are not shown.

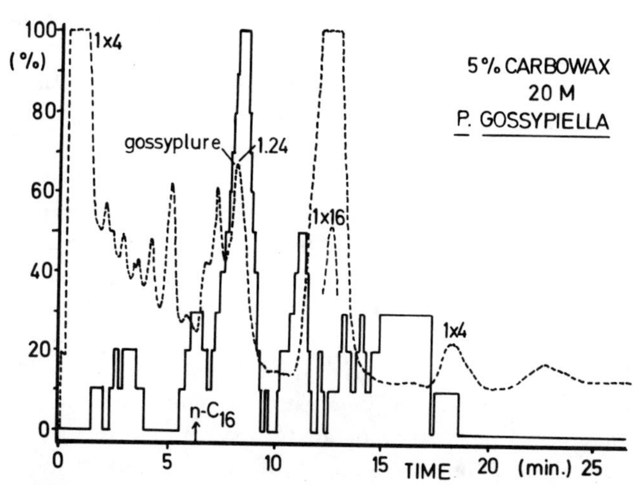

Figure 7. t-GLC–BB of *P. gossypiella* extract on a polar column: 5% Carbowax 20 M, size 50 cm × 0.2 cm, at 190°C, 50 ml N_2/min. The extract assayed underwent, in addition to the three purification steps listed in Fig. 5, two subsequent column chromatographies on florisil impregnated with 20% $AgNO_3$, and one preparative gas chromatographic purification on 3% Apiezon L under conditions shown in Fig. 6. The sample, at this point, contains about 10% gossyplure in the biologically active peak appearing at relative retention 1.24. Note also a peak of lesser, but still significant activity at relative retention 1.77. The arrow indicates the retention time of the internal standard, *n*-16:OAc (relative retention 1.00). Two more biologically inactive peaks, not shown, appear after 45 min.

teristic shifts to lower (non-polar phase)(Figs. 5,6) and higher (polar phase) (Fig. 7) retention data. These shifts are in all cases stronger for gossyplure than for hexalure. On the apolar Ap L phase, for example, gossyplure consistently shows (Table 3) a relative retention of 0.80 and hexalure of 0.83. More strikingly, on the highly polar Carbowax 20M phase, gossyplure has a relative retention of 1.24 and hexalure of 1.05. The polyester phase DEGS offers the best distinction between gossyplure and hexalure with 1.46 and 1.24, respectively. Note that negative increments apply for phases with relative retentions smaller than 1, positive increments for phases with relative retentions larger than 1. Since the increments per C–C double bond are additive, gossyplure is likely to have two C–C double bonds. This result was obtained with a submicrogram quantity of a partially purified sample.

Other spectrometric techniques such as mass spectrometry, nuclear magnetic resonance and infrared absorption, and microozonolysis required a highly purified sample of 99% or better purity. The latter method unequivocally confirmed the presence of two carbon double bonds by three oxidative cleavage products described in more detail in Hummel et al. (1973). Bierl et al. (1974) independently arrived at the same conclusions. Structural assignments for unsaturated acetate esters of this type would preferably be carried out on the florosilicon phase SF-96, on Carbowax 20M, and on DEGS where the relative retention differences are largest.

Gossyplure can be isolated at the 100-μg scale by preparative gas chromatographic fractionation based on exactly known retention data. On the basis of relative retention data first established by painstaking work on analytical columns, nonpolar and polar preparative GLC columns can be used in sequence for effective purification and isolation of gossyplure (Figs. 5–7). The success of this approach is guaranteed as long as the fractionation conditions are held highly constant. In conjunction with the micropreparative fraction collector by Brownlee and Silverstein (1968), retention data provided an avenue for the isolation of 100 μg of gossyplure. Only two steps were required for purification of gossyplure from the 1% (see Fig. 5) to the 10% level (see Fig. 7). Final purification to 99% and isolation of the pure natural product were accomplished on a DEGS column. The overall recovery rate for the last four steps was 80%. More information on purification and isolation of microsamples can be found in Chapter 11 by Heath and Tumlinson (this volume).

In addition to gossyplure, a third component of the pheromone mixture, probably an aldehyde, was characterized by its retention value on Carbowax 20M (1.77 relative to *n*-16:OAc. But it was not chemically identified.

Predictable relative retention characteristics may well be established for other functional groups such as alcohols, aldehydes, or carboxylic acid esters. Required are a set of high-performance analytical columns, a partially purified pheromone sample, 10–20 test insects, and a few synthetic reference compounds with zero, one, two, and three C–C double bonds.

B. *Bombyx mori*

In solvent extracts of freshly emerged *B. mori* females the t-GLC–BB technique reveals three major well-resolved peaks of activity. This observation is in support of the early work on Anders and Bayer (1959). Peaks labeled 1, 2, and 3 in Fig. 8 most closely resemble the chromatographic properties of bombykal (relative retention 0.29), bombykol (0.92), and a major unidentified component (1.67) in the range of 18 carbon atoms. The latter component has an unknown satellite at relative retention of 2.00. *n*-16:OAc was used as a standard whose relative retention was set as 1.00.

Peaks 2 and 3 induce male dances, curved abdomens, clasper extension, and copulation attempts. Other peaks below 30% response represent walking, running, wing vibration, and antennal flicking as described by Bayer and Anders (1959) and listed in Section V. Bombykal was chemically identified by Kasang et al. (1978). Interestingly, it also serves as a sex pheromone for female *Manduca sexta* in the family Sphingidae (Starratt et al., 1979).

Work with the more refined t-GLC–BB technique on several different columns now can provide a straightforward avenue towards the isolation and chemical identification of the unknown compounds. Further testing of *B. mori* extracts on the moth families Saturniidae and Sphingidae also may provide information on the cross-specificity of their pheromone signals and on factors contributing to reproductive isolation.

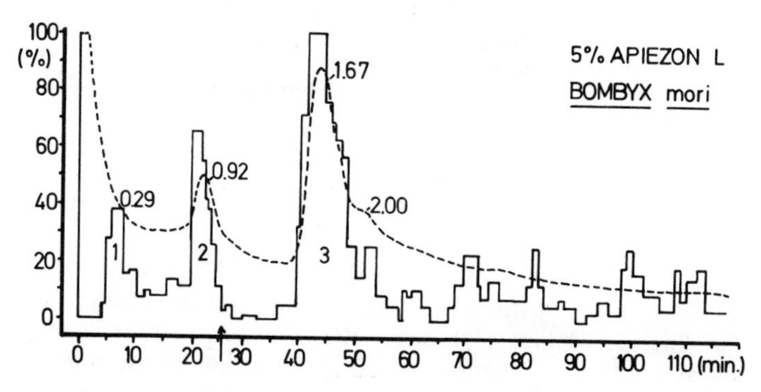

Figure 8. t-GLC–BB on a 5% Apiezon L packed glass column of *Bombyx mori* female sex pheromone gland extract (1 female equivalent in CS_2) tested on male *B. mori* moths. The arrow indicates the relative retention of *n*-16:OAc (1.00) at 26 min. For comparison, hexalure (*Z*-7-16:OAc) has a relative retention of 0.80. Other chromatographic conditions: column dimensions 180 cm × 0.2 cm inside diameter; temperatures of injector, column and detector (°C): 250; 190; 225. N_2 carrier gas flow rate: 24 ml/min. Behavioral activity and FID data were averaged from two separate experiments. (—) Percentage behavioral activity of males (*N*=20). (---) FID response (100% = 1 mV recorder sensitivity).

6.2. Examples from the Order of Coleoptera

A. *Diabrotica undecimpunctata howardi*

The t-GLC–BB technique is not only applicable to the analysis of Lepidoptera sex pheromones, but also can handle some of the notoriously complicated Coleoptera pheromone systems. Figs. 9A and B show results obtained for *D. undecimpunctate howardi,* the economically important southern corn rootworm or cucumber beetle. On the apolar Ap L column, five peaks with behavioral activity above 30% appear. Their relative retentions (*n*-16:OAc = 1.00) are

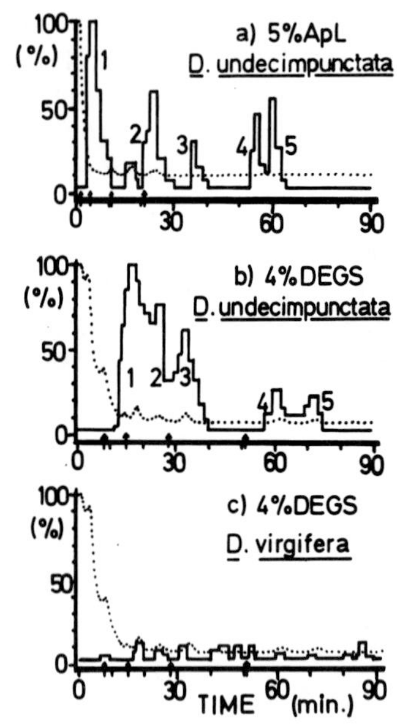

Figure 9. t-GLC–BB of virgin female beetle extract of *Diabrotica undecimpunctata howardi* collected on the glass surface of a jar and washed off with CS_2. (—) Solid line: percentage of male beetles activated to running, flights, and copulation attempts. (---) dashed line: percentage of FID response (100% = 1 mV recorder sensitivity). The arrows 1, 2, 3, and 4 in each chromatogram indicate the retention times of *n*-decyl-, *n*-dodecyl-, *n*-tetradecyl-, and *n*-hexadecyl acetate on (A) 5% Ap L column, 180 cm × 0.2 cm, 190°C, 22 ml N_2/min; 350 female beetle hour equivalents injected and tested on 25 male *D. undecimpunctata howardi*. Solid line: average from two replications. (B) 4% DEGS column, 560 cm × 0.6 cm, 160°C, 24 ml N_2/min, 580 female beetle hour equivalents injected and tested on 13 male *D. undecimpunctata howardi.* Solid line: average from three replications. (C) Four percent DEGS column, same conditions as under A and B, with 12 male *D. virgifera* as test insects. Solid line: average from two replications.

0.27, 1.21, 1.80, 2.75, and 2.96. Peak 1 is probably identical with 10-methyltridecane-2-one. Peaks 2 to 5 are chemically unknown. Their retention characteristics place them in the range of 18–20 carbon atoms. Fig. 9B, obtained in 1976 on a DEGS polyester column, correctly predicted the carbon number of the first pheromone component. Peaks 1–5 have relative retentions of 0.33, 0.50, 0.65, 1.18, and 1.41.

The chromatograms (Figs. 9A,B) may be helpful in the chemical characterization of further components.

B. *Diabrotica virgifera*

Evidence for the nonidentity of the *D. virgifera* and the *D. undecimpunctata howardi* pheromones also was obtained by Hummel and Kucera (1977). At a time when both pheromones were chemically unknown, they analyzed extracts of both species. Evidence for the nonidentity followed from the lack of cross-reactivities in the t-GLC–BB system (see Fig. 9C). Under conditions when *D. undecimpunctata howardi* males would become wildly excited (Figs. 9A,B), *D. virgifera* males showed only listless responses statistically inseparable from the biological background activity. If 8-methyl-2-decanol propanoate, one of the *D. virgifera* sex pheromones (Guss et al., 1982) were coinjected onto the column, its activity peak would appear near the second arrow.

VII. Behavior Tests with Multiple Chemical Stimuli

Several different chemical substances can be simultaneously applied to the test insects. Three different ways are available for operating the apparatus:

1. the multiple injection of samples into the GLC column in such a sequence that they are eluted simultaneously from the column (Hummel and Dolce, 1978) (internal application mode)
2. the stimulation of insects by behaviorally active compounds from a pheromone applicator similar to Bartell and Shorey (1969) arranged in parallel to the GLC column (external application mode)
3. the simultaneous introduction of chemical stimuli from GLC *and* the external applicator.

7.1. Multiple Injection Technique

An example of this versatile technique is given in Table 4. Under the condition of low column loading (which is met in most pheromone work) peaks emerge from the column without measurable interference. By proper timing of injections, many different combinations of pheromone components and disruptants can be tested. In Table 4 the disruptive effect of the terpene ester *l*-bornylace-

Table 4. Reduction of the sex pheromone responses in male *D. undecimpunctata howardi* by the behavioral disruptant *l*-bornylacetate[1]

Peak No.[2]	Retention time[3] (min)	Control: pheromone alone[4] (% response)	Treatment with pheromone extract and disruptant[5] (% response)
1	1.0	96	33
2	8.5	50	16
3	17.5	35	3

[1]From Hummel and Dolce (1978).

[2]Additional minor peaks of activity emerging between 24 and 90 min are not shown.

[3]Chromatographic conditions: glass column 5% Ap L on Chrom. W, AW, DMCS, temperatures of injector, column, and detector (°C): 250; 140; 200. N_2 flow rate 43 ml/min. Air flow through BB chamber: 2 liters/min.

[4]Injection at time 0 min of 90 virgin female beetle hour equivalents in 10 μl of pentane/ether (1/1) solution. The solvent itself does not excite nor inhibit the male beetles. Only running, flight, and copulation attempts, but not antennal movements and walking were counted as positive responses. The male beetles (N=30) were collected in the field 8 days before testing.

[5]Retention time of *l*-bornylacetate: 2 min. 0.1 mg of the disruptant was used per injection: at min: −1.5 (peak 1); at +6.0 (peak 2); and at +15.0 (peak 3). For additional explanations, see text.

tate on the behavioral response of male southern corn rootworm beetles, *Diabrotica undecimpunctata howardi,* is recorded. For all three components of the female sex pheromone, a reduction by a factor of 3 and more was observed when 100 μg of *l*-bornylacetate were applied simultaneously with 90 virgin female beetle hour equivalents of the pheromone. Thus this test, when performed on 30 test insects within 3 hr, provided a quick preliminary evaluation of the disruptive effect of *l*-bornylacetate before an expensive field study with several weeks of data acquisition was undertaken.

In addition, the experiment provided evidence for the transitory, reversible nature of the disruptive effect: Half an hour after the simultaneous test of pheromone extract and *l*-bornylacetate, the same male beetles responded to pheromone extract alone with the same sensitivity as they did in a control experiment preceding the test.

VIII. External Introduction of Chemical Stimuli

Substances released from an external applicator can be tested simultaneously with substances eluting from the GLC column. Fig. 4 shows the arrangement of the applicator and the gas streams merging between points I and P with the gas stream from the column. All parts are made from glass and can be easily cleaned and reassembled before the test. Known amounts of substances can be mixed with the airstream without interrupting the GLC analysis or without changing the total gas flow the test insects are exposed to.

The applicator Y is inserted into the glass adapter X while stopcocks W and V are closed, stopcock U is open, and while the test insects are being observed. At a predetermined time, W and V can be opened for a precisely measured time interval while the change in reactions of the insects is being recorded. If a quantitatively known chemical stimulus is necessary, it could be introduced internally via the second channel of the same gas chromatograph, measured with the FID and introduced through a branch in the transfer line. Most modern GLC instruments now have dual column operation capability in their standard outfit.

8.1. The Mixed Operation Mode

Simultaneous GLC and external introduction of samples may be of advantage in the assay of multiple pheromone mixtures as they occur, e.g., in the summer fruit tortrix moth *Adoxophyes orana* (Tamaki, 1979), the cotton bollworm *Heliothis virescens* (Tumlinson et al., 1982), and in most other of the over 800 insect pheromone systems thoroughly investigated to date (Silverstein and Young, 1976; Inscoe, 1982).

IX. Estimation of Pheromone Release by Females Based on t-GLC–BB Results

In Section VI of this chapter the sex pheromone contained in 580 virgin female beetle hour equivalents of *D. undecimpunctata howardi* was analyzed (Fig. 9A). While the males responded strongly to a component eluting from the column at 5 min, the FID did not show a corresponding signal. Since the maximal FID sensitivity in our system is 1 ng, the biologically active component must be present at the subnanogram level. Consequently, 1 virgin female beetle will release less than 0.01 ng/hr in an aeration experiment. This estimated value allows extrapolations of how many beetles will have to be aerated for how many hours in order to collect quantities sufficient for chemical identification.

X. Coupling of the t-GLC–BB with Devices for Automatic Registration and Evaluation of Insect Activity

In the system described by Gaston et al. (1966), visual observations (V.O.) have been manually translated into pencil marks on the strip chart of the recorder, or into electrical signals generated by a step switch Δ (for symbols see Fig. 1). As an alternative, the bioassay chamber can be fitted with an electronic insect activity detector as described by Haskell (1954), Hunsaker and Lalonde (1975), or Pinniger and Collins (1976). For nocturnal insects, e.g., *Trichoplusia ni* or *Pectinophora gossypiella,* records of activity can be based on IR light-emitting diodes as described by Cohen and Denenberg (1973). Diurnal insect can be monitored with visible light. Miller (1979) lists many additional references describing insect activity detectors.

XI. Adaptation of the Apparatus to Other GLC Systems with Different Configuration

Some GLC systems, e.g., older Hewlett-Packard models, are equipped with splitter outlets on top rather than at one side of the GLC assembly. A configuration of the latter case is shown in Fig. 1 for the Varian 2700 gas chromatograph. In the former case, one of two alternative steps can be taken: (1) Either a horizontal outlet through the detector oven is established or (2) the transfer line F to the bioassay chamber can be modified: The joint G, for example, could be used as the sample inlet and the joint H as the inlet for air (and external stimuli), while the entire bioassay cage would be placed on top (rather than at one side) of the GLC.

XII. Potential Problems Inherent in the Method

The t-GLC–BB technique so far has been tested with insect sex pheromones only. Other semiochemicals like allomones can elicit (or inhibit) insect activity as well. Future experimentation would have to show the usefulness of the method for substances other than sex pheromones.

Problems may arise from the use of highly efficient capillary (WCOT) columns: They can be charged with minute quantities of substances only. During the short time intervals in which the typically narrow peaks emerge from such columns, the test insects may not attain full activation.

Potentially more serious might be the loss of activity due to the high separation power of capillary columns. They can distinguish geometrical or positional isomers. On the other hand, full activation of test insects may be linked to the simultaneous presence of several compounds or of a certain ratio of geometrical (*P. gossypiella,* Hummel et al., 1973) or positional (*A. orana,* Tamaki, 1979) pheromone isomers. Distortion of this ratio may result in the partial or complete loss of activity. Such unfavorable cases should reveal themselves in control tests of unfractionated mixtures. For this purpose, the additional input port I in the transfer line F is provided. Through it, unfractionated samples may be introduced and comparisons with fractionated samples may be made. Other highly efficient coupled methods like t-GLC–EAG or t-GLC–SSR may potentially suffer from the same problems. Failures of the method, however, so far have not been reported and are unlikely to occur with the use of the less efficient packed columns.

XIII. Advantages of the Method

The apparatus described is an improvement over the qualitative olfactometers of all kinds and shapes previously reviewed by Young and Silverstein (1975). The t-GLC–BB gives quantifiable, on-line readings for the pheromones and synthetic chemicals tested and for the behavioral reactions they induce. Time-consuming transfers of volatile material that are prone to evaporation losses and external contamination are no longer required. The method also offers an inexpensive addition to the standard GLC instrument available to most laboratories involved in pheromone research and olfactory physiology. The hardware can be manufactured in any machine shop equipped for sheet metal work. Glass parts can be provided by any glass blower shop. Costs for material including metal and glass, but excluding the actograph and recorder, are of the order of 100–200 U. S. dollars. Many GLC systems have a built-in splitter system. Otherwise, the system described by Brownlee and Silverstein (1968) can be recommended. Decontamination, if necessary, can be accomplished by taking the apparatus apart and by baking the glass parts overnight in a drying oven. Reassembly of the ground glass joints and the metal caps holding the transfer lines in place takes only a few minutes. Should contamination of the entire apparatus occur, it can be heated in its entirety since it consists only of metal, glass, and a few temperature-resistant fittings.

XIV. Extension of the Method to Related Fields

14.1. The GLC–Flight Tunnel Combination

For the highest level of behavioral complexity in laboratory assays, the flight tunnel, its combination with GLC is listed in Table 1. A prototype has never been described in the literature (Drs. B. A. Leonhardt and M. Inscoe, Beltsville, Md., personal communication March 8, 1983). The reason may be that chemical stimuli emerging from GLC columns, especially from capillary columns, are available only for too short a time interval for sustaining a sufficient sample concentration necessary for covering a convoluted insect flight path over many meters of length. However, for tests close to the source the possibility may gain appeal in the future.

14.2. Olfactometers Based on the t-GLC–BB Principle

Dravnieks (1975) described a similar device for delivering small measured quantities of odorants to a human olfactometer. In his applications, the separation power of GLC combined with the sensitivity and specificity of human olfaction was superior to all other available analytical methods.

14.3 Work in Aquatic Systems

In analogy to the t-GLC–BB in air, the same principle is applicable to aquatic systems by combining a high pressure liquid chromatograph with a water-filled aquarium chamber which contains the test organisms.

XV. Appendix 1: Part List for Behavior Bioassay Chamber and Accessories

1. Behavior Bioassay Chamber

Sheet metal, stainless steel, 2 mm strong, for base of chamber
Sheet metal, stainless steel, 1 mm strong, for chamber
Sheet metal, stainless steel, 0.5 mm strong, for light baffle
Two hinges, each 300 mm long, for movable top and side wall of chamber
Five to ten legal-size typewriter paper sheets for adjusting the light intensity inside the bioassay chamber to appropriate levels
Duct tape for attaching paper sheets
Two stainless-steel Swagelok nuts, 6 mm inner diameter, part No. 402-1
Two stainless-steel Swagelok unions, with one threading sawed off in each, and prepared for silver soldering to the base plate of the chamber as depicted in Fig. 4; part No. 400-6.
Two graphite or lead ferrules, 6 mm inside diameter, fitting the Swagelok nuts
Three male and female 18/10 standard taper glass joints (see Fig. 4)
One male and female Luer type glass joint (see Fig. 3)
One glass male and female 31/10 standard taper joint with glass extension (Fig. 4)
One male and female ball and socket standard taper glass joint, ST 12 (Fig. 4)
One glass flask 3.9 liters, with screw cap top (Fig. 3)
One glass flask 1.9 liters, with screw cap top
One heating tape, 120 V, with variable transformer

2. Accessories

A. Optional Unit for External Input of Chemical Stimuli

Three stopcocks ST 2, 3-mm-inside diameter bore width
Three male/female standard taper 24/40 glass joints
One male/female standard taper 10/18 glass joint
One metal disc mounted on applicator similar to Bartell and Shorey (1969)

B. Carbon Filter and Air Rehumidifier

One activated carbon pellet filter, length 20 cm, width 2 cm, particle size 1 mm

One glass wash bottle, 0.5 liter, with distilled water, hose adapters, and Gilmont ball flow meter with scale of 0 to 2.5 liters/min air flow

C. Insect Activity Detector

See part list cited in specific papers mentioned in the text

D. Recorder

Dual-channel strip chart recorder, e.g., Cole-Parmer SD-8376-30, 1-mV input for each channel; 20-cm paper width, 1–30 cm/min chart speed

Acknowledgments. Dr. L. K. Gaston, Riverside, California, first introduced the author to a prototype of the t-GLC–BB in 1970. Over the years, experiences with this simple design focused his interest on searching for technical improvements and applications to other insect species. The author also thanks Dr. K. -E. Kaissling, Seewiesen, for bringing to his attention the references of Anders and Bayer (1959). Dr. P. Feeny, Ithaca, New York, provided a copy of, and permission to quote results from, his unpublished B.S. thesis. Dr. Bierl-Leonhardt and Dr. Inscoe of Beltsville, Maryland, kindly scanned their extensive literature base for published examples of the GLC–wind tunnel combination. T. -F. Hsueh assisted with the rearing and bioassays of *B. mori*; R. A. Kucera ran some of the *Diabrotica* bioassays, and Sigma Xi Research Society provided financial support.

XVI. References

Anders F, Bayer E (1959) Versuche mit dem Sexuallockstoff aus den Sacculi laterales vom Seidenspinner (*Bombyx mori* L.). Biol Zentralbl **78**:584–589.

Arn H, Städler E, Rauscher S (1975) The electroantennographic detector—A selective and sensitive tool in the gas chromatographic analysis of insect pheromones. Z Naturforsch **30c**:722–725.

Bartell RJ, Shorey HH (1969) A quantitative bioassay for the sex pheromone of *Epiphyas postvittana* (Lepidoptera) and factors limiting male responsiveness. J Insect Physiol **15**:33–40.

Bayer E, Anders F (1959) Biologische Objekte als Detektoren für Gaschromatographie. Naturwissenschaften **46**:380.

Beevor PS, Hall DR, Lester R, Poppi RG, Read JS, Nesbitt BF (1975) Sex pheromones of the army worm moth, *Spodoptera exempta* (Wlk.). Experientia **31**:22–23.

Berger RS (1966) Isolation, identification, and synthesis of the sex attractant of the cabbage looper, *Trichoplusia ni.* Ann Ent Soc Amer **59**:767–771.

Biemann K (1962) Mass Spectrometry: Organic Chemical Applications. McGraw Hill, New York.

Bierl BA, Beroza M, Collier CW (1970) Potent sex attractant of the gypsy moth: Its isolation, identification and synthesis. Science **170**:87–89.

Bierl BA, Beroza M, Staten RT, Sonnet PE, Adler VE (1974) The pink bollworm sex attractant. J Econ Entomol **67**:211-216.

Block BC (1960) Laboratory method for screening compounds as attractants to gypsy moth males. J Econ Entomol **53**:172-173.

Brownlee RG, Silverstein RM (1968) A micropreparative gas chromatograph and a modified carbon skeleton determinator. Anal Chem **49**:2077-2079.

Butenandt A, Beckmann R, Stamm D, Hecker E (1959) Über den Sexual-Lockstoff des Seidenspinners, *Bombyx mori.* Reindarstellung und Konstitution. Z Naturforsch **14b**:283-284.

Cohen A, Denenberg VH (1973) A small-animal activity-analysing system for behavioral studies. Med Bio Engin **11**:490-498.

Dravnieks A (1975) Instrumental aspects of olfactometry. In: Methods in Olfactory Research. Moulton DG, Turk A, Johnston Jr. JW (eds), Academic Press, London/New York/San Francisco, pp 1-61.

Feeny PP (1962) Some Studies in Gas Chromatography with Particular Reference to Biological Problems. B.S. thesis, Merton College, Oxford.

Flaschenträger B, Amin ES, Jarczyk HJ (1957) Ein Lockstoffanalysator (Odour-analyzer) für Insekten. Mikrochim Acta 385-389.

Gaston LK, Fukuto TR, Shorey HH (1966) Sex pheromones of noctuid moths. IX. Isolation techniques and quantitative analysis for the pheromones, with special reference to that of *Trichoplusia ni* (Lepidoptera: Noctuidae). Ann Ent Soc Amer **59**:1062-1066.

Gaston LK, Shorey HH (1964) Sex pheromones of noctuid moths. IV. An apparatus for bioassaying the pheromones of six species. Ann Ent Soc Amer **57**:779-780.

Gohlke RS (1959) Time-of-flight mass spectrometer and gas-liquid partition chromatography. Anal Chem **31**:535-541.

Guss PL, Tumlinson JH, Sonnet PE, Proveaux AT (1982) Identification of a female-produced sex pheromone of the Western Corn Rootworm. J Chem Ecol **8**:545-556.

Harley J, Nel W, Pretorius V (1958) Flame ionization detector for gas chromatography. Nature (London) **181**:177-178.

Haskell PT (1954) An automatic recording maze for insect behavior studies. Brit J Anim Behav **2**:153-158.

Hummel HE (1979) Chemical communication in the reproductive strategies of some nearctic Saturniid moths. Proceedings of the EUCHEM Conference on Insect Chemistry: Chemical Communication and Interaction. Borgholm, Sweden, August 13-17, 1979, p 17.

Hummel HE, Dolce GJ (1978) Disruption of sex pheromone communication in Spotted Cucumber Beetles (Chrysomelidae) Demonstrated in Various Test Systems. Paper No. 487, Annual National Meeting Entomol. Society of America, Houston, Texas, November 29, 1978.

Hummel HE, Gaston LK, Shorey HH, Kaae RS, Byrne KJ, Silverstein RM (1973) Clarification of the chemical status of the pink bollworm sex pheromone. Science **181**:873-875.

Hummel HE, Kucera RA (1977) Sex pheromone communication in the southern corn rootworm. Proc N Centr Br Ent Soc Amer **32**:55-56.

Hummel HE, Sternburg JG (1979) *Hyalophora cecropia* mating communication:

Purification and isolation of several sex pheromone components. 11th International Congress of Biochemistry, Toronto, Canada, July 8–13, 1979, No. 13-9-R65.

Hunsaker WG, Lalonde JM (1975) Photo-electric detector for monitoring behavioural activities in animals. Lab Pract **24**:526–527.

Inscoe MN (1982) Insect attractants, attractant pheromones and related compounds. In: Insect Suppression with Controlled Release Pheromone Systems. Kydonieus AF, Beroza M (eds), CRC Press, Boca Raton, Florida, Vol 2, pp 201–295.

Jennings W (1980) Gas Chromatography with Glass Capillary Columns. 2nd ed. Academic Press, New York.

Jones WA, Jacobson M (1968) Isolation of N,N-diethyl-*m*-toluamide (DEET) from female pink bollworm moths. Science **159**:99–100.

Jones WA, Jacobson M, Martin DF (1966) Sex attractant of the pink bollworm moth: Isolation, identification, and synthesis. Science **152**:1516–1517.

Kasang G, Kaissling KE, Vostrowsky O, Bestmann HJ (1978) Bombykal, a second pheromone component of the silkworm moth *Bombyx mori* L. Angew Chem Int Ed Engl **17**:60.

Löfsted C, Van der Pers JNC, Lofquist J, Lanne BS, Appergren M, Bergström G, Thelin B (1982) Sex pheromone components of the turnip moth, *Agrotis segetum*: Chemical identification, electrophysiological evaluation and behavioral activity. J Chem Ecol **8**:1305–1321.

McWilliams IG, Dewar RA (1958) Flame ionization detector for gas chromatography. Nature (London) **181**:760–761.

Miller TA (1979) Insect Neurophysiological Techniques. Springer-Verlag, New York, pp 59–117.

Moorhouse JE, Yeadon R, Beevor PS, Nesbitt BF (1969) Method for use in studies of insect chemical communication. Nature (London) **223**:1174–1175.

Perry JA (1981) Introduction to Analytical Gas Chromatography. Dekker, New York.

Pinniger DB, Collins LG (1976) Two insect activity detector systems using light dependent resistors. Lab Pract **25**:523–524.

Ryhage R (1964) Use of a mass spectrometer as a detector and analyzer for effluents emerging from high-temperature gas-liquid chromatography columns. Anal Chem **36**:759–764.

Schneider D (1957) Elektrophysiologische Untersuchungen von Chemo-und Mechanorezeptoren der Antenne des Seidenspinners *Bombyx mori* L. J Comp Physiol **40**:8–41.

Silverstein RM, Young JC (1976) Insects Generally Use Multicomponent Pheromones. ACS Symposium Series No. 23:1–29.

Starratt AN, Dahm KH, Allen N, Hildebrand JG, Payne TL, Röller H (1979) Bombykal, a sex pheromone of the Sphinx moth *Manduca sexta.* Z Naturforsch **34c**:9–12.

Tamaki Y (1979) Multicomponent sex pheromones of Lepidoptera with special reference to *Adoxophyes sp.* In: Chemical Ecology: Odour Communication in Animals. Ritter FJ (ed), Elsevier/North-Holland, Amsterdam, pp 169–180.

Tumlinson JH, Heath RR, Teal PEA (1982) Analysis of Chemical Communication Systems in Lepidoptera. ACS Symposium Series No. 190:1–25.

Wadhams LJ (1982) Coupled gas chromatography–single cell recording: A new technique for use in the analysis of insect pheromones. Z Naturforsch **37c**: 947–952.

Young JC, Silverstein RM (1975) Biological and chemical methodology in the study of insect communication. In: Methods in Olfactory Research. Moulton DG, Turk A, Johnston Jr JW (eds), Academic Press, New York, pp 75–160.

Chapter 9

Technique and Equipment for Collection of Volatile Chemicals from Individual, Natural, or Artificial Sources

Lyle K. Gaston [1]

I. Introduction

The major problem in the identification of semiochemicals has been the small amount of biologically active material in a large amount of chemically similar inactive material. Thus for identification purposes, there is a great need for a general technique and procedure that can easily yield pure, natural material on a submicrogram level. Rational application of these materials in insect control programs requires a knowledge of the rate at which the organism releases the material and then designing an artificial substrate which duplicates the composition of the natural system at a total rate of evaporation anticipated to achieve the desired effect, i.e., population density surveys or control by disruption of mating communication.

II. Applications

Equipment and techniques are described that make it possible to collect, analyze, and quantitate nanogram quantities of chemicals volatilizing from submillimeter to multimillimeter diameter sources. The source can range from an artificially extruded, individual female pheromone gland or male scent brush to artificial substrates such as filter paper, glass or metal surfaces, plastic reservoirs, hollow fibers, or droplets of microdispersed chemicals.

[1] Division of Toxicology and Physiology, Department of Entomology, University of California, Riverside, California.

The ability of this technique to provide nearly pure unknown compounds makes it possible to identify them tentatively by comparison with standards using gas–liquid chromatography (GLC) retention times or mass spectral analysis. The rate of evaporation of each individual component can be quantified to give both total rate of evaporation of the pheromone and its composition on an individual organism basis. Individual data can be summed to give a population average and individual variation from the average. Since individual collections can be completed in a few minutes, emitter age, time of day, and previous emission history can be investigated. In favorable cases, after collection, the gland can be excised, extracted, and analyzed for the amount of pheromone remaining.

Rates of evaporation from artificial evaporation substrates can be measured and adjusted to specific values according to species and use. In addition, the time for the rate of evaporation to reach a predetermined level can be determined.

III. Apparatus

3.1. Flow System

A totally enclosed flow system was used to collect the volatile chemicals (Fig. 1). Prepurified nitrogen was metered through a 4A molecular sieve filter using a two-stage pressure regulator. A capillary tube provided a pressure drop such that 40 psig gave a volumetric flow rate of 2 ml/sec, which corresponds to a linear velocity by the evaporation substrate of about 11 cm/sec. This system can maintain a constant flow rate with any combination of two collectors and one concentration system. The lowest flow rate possible is desirable because it prevents or minimizes loss of highly volatile material from the collector.

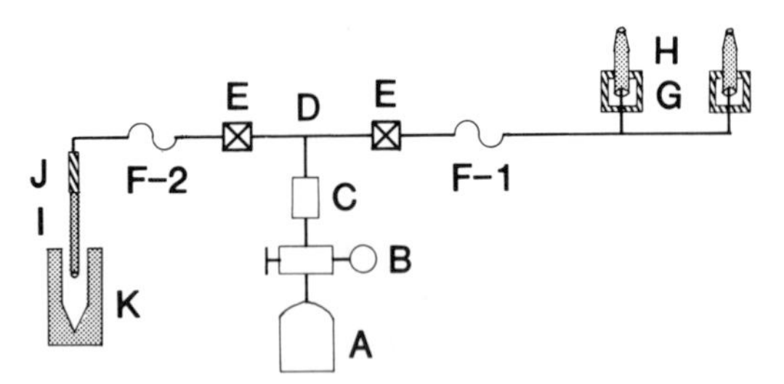

Figure 1. Block diagram for the nitrogen flow system for collection of volatile chemicals and concentration of the collected materials. (A) High-pressure nitrogen cylinder. (B) Two-stage pressure regulator. (C) Molecular sieve filter. (D) 1/8-in. copper tubing. (E) On–off valve. (F) Capillary flow restrictor. (1) 2 ml/sec ca. 40 psi g. (2) 4.5 ml/sec ca. 40 psi g. (G) Teflon union, (H) Male 14/20 ₮ glass joint for collector. (I) 3-mm-outside-diameter glass tube. (J) No. 9 Teflon tubing. (K) Conical evaporation tube.

3.2. Collectors

Three different collectors (Baker et al., 1981; Pope et al., 1982) have been designed to collect volatile chemicals. All collectors have employed nonsilanized glass wool as the main absorbing medium and were miniaturized to the size of a medium-sized moth. Attachment of the collector to the nitrogen manifold was through a 14/20 Ŧ female joint.[2] The evaporation substrate, biological or artificial, was placed in the nitrogen stream at right angle to the direction of flow. This opening was either a 10/18 Ŧ female joint which provided a gas-tight seal or an 11-mm-OD (9-mm-ID) glass tube sealed with a Teflon-backed GLC septum. The volatile chemical was collected on 4–10 mg of Pyrex glass wool. For systems with a low rate of evaporation, up to four evaporation ports can be put on one collector body. In this way, collections can be obtained from four sources simultaneously reducing breakthough and wash solvent contamination.

3.3. Holders

A. Biological

Various sizes and shapes of glass holders (10/30 Ŧ male joint or glass tubing) were designed to minimum dimensions of the abdomen of a particular species. These glass holders were constricted at one end such that, when the releasing structure was extruded, it would just pass through the hole. This dimension is critical because it regulates the amount of extraneous material that is in the nitrogen stream and which may be a source of volatile, nonpheromone compounds. These extraneous compounds may interfere with the quantitation of the pheromone or be mistaken as a component of the pheromone.

B. Artificial

Holders for man-made substrates are easier to design and fabricate because their dimensions can be specified and held constant. One substrate was a 1-mm-diameter glass bead fused to the end of a glass rod. This was used as a pseudo gland to establish the maximum time and flow rate that could be used for various different volatility compounds before they started to break through the glass wool. This pseudo gland was also used to make a material balance between the initial amount of material put on the glass bead, the amount collected on the glass wool, and the amount remaining on the glass bead at the end of the collection. Other substrates such as hollow fibers or stainless-steel discs were pushed through a small hole in the septum. Plastic laminates were impaled on the end of a pin stuck through the septum.

---

[2] Ŧ: Notation for standard taper ground glass joint. 10/18 Ŧ means a diameter of 10 mm at the small end, and 18 mm is the length of the ground joint. The taper is 1 mm per 10 mm length.

IV. Preparation

The procedures and techniques for preparing an insect for collection must be tailored to fit each species. The most favorable case, generally small female moths, requires only that the entire moth be placed abdomen first in the holder and then a slight vacuum be applied to the holder. The moth will be pulled down the tube and frequently the gland can be extruded at the same time. If the gland does not extrude or stay extruded, a screw-type pusher or a pipe cleaner can be forced against the moth to create and keep hydrostatic pressure on the abdomen. The most difficult case is where large amounts of other body parts such as scales or hairs surround the gland. In this case, scales or hairs can be vacuumed off with a small glass pipet. The abdomens of large female moths frequently are too fragile to withstand the necessary pressure to extrude the gland. For these cases, the abdomen tip can be ligated with dental floss, excised anteriorly and then put in the holder for gland extrusion. Sometimes the gland retractor muscles are still functional and will withdraw the gland part way through the collection. One solution to this is to cut off the head and inject 1 μl of 50% Na_4EDTA near the abdomen tip by going through the thorax into the abdomen. After 1 min, the abdomen can be ligated as above.

The gland should be observed periodically during the collection to verify that it is still fully extruded. If not, the thumb screw can be slightly tightened.

V. Collection, Concentration, Analysis

Collection times ranged from 1 to 30 min depending on the rate of evaporation and break-through volume of the material being collected. In the case of biological material, the estimated time that a particular glandular source could maintain a constant evaporation rate was also considered. For most lepidoptera sex pheromones, a 10-min collection is a reasonable starting point. At the end of the collection, the holder, and septum if used, is removed and a known amount of internal standard is added to the glass wool. The volume of internal standard added should be between 3 and 8 μl and is best measured with a 10-μl syringe. The internal standard should be pure, stable, and preferably have the same type of functional groups as the unknown and be separable from the unknown compound(s) by the GLC column. The collector is washed with six ca. 30-μl volumes of freshly distilled carbon disulfide. Each aliquot, after the first one, is forced out of the collector by closing both inlets with one's fingers and gently squeezing. The first aliquot is added directly to the glass wool. Succeeding aliquots are used to wash the sides of the collector around the evaporation substrate. The total wash, ca. 200 μl, is collected in a 1-ml conical, microreaction vial. The solvent is removed by directing a slow stream of nitrogen from a 3-mm-outside-diameter glass tube held above the surface. This operation is the single-most critical part of the entire procedure. The velocity of the nitrogen at the surface of the solvent must be low enough so that no splattering occurs. At least four times during the evaporation, the vial must be removed from the nitrogen stream and gently tapped with a fingernail to cause the meniscus to rise 2–3 mm. This washes any material on the sides of the vial back into the solution. Evaporation is continued until a volume of ca. 5 μl for packed column GLC or ca. 1 μl for capillary GLC is obtained. With good technique and total concentration by the operator, a recovery of internal standard of greater than 80% at 5 μl and 40% at 1 μl can be obtained. For GLC identification and quantification, a 10-μl syringe is filled to 0.5 μl with freshly distilled carbon disulfide, and then the entire contents of the concentration flask is pulled into the barrel of the syringe. The sample is analyzed by conventional GLC techniques.

VI. Special Precautions

The only solvent that we have had success with concentrating 100-fold and not causing a huge problem with a hydrogen flame detector is carbon disulfide. Carbon disulfide is unstable and for this type of work must be freshly distilled every morning.

Collectors and holders must be thoroughly washed after each use with distilled hexane and then dried for 20 min at 130°C in a beaker. Thus each complete assembly is segregated, making it easier to locate any problems such as deterioration of the glass wool or contamination. The beaker also makes a convenient handle so that no fingerprints get on the inside surfaces of the holder or on the outside of the tip of the collector.

Syringes should be cleaned by rinsing the plunger with distilled acetone and pulling acetone through the needle and barrel with a vacuum. A vacuum seal with the back of the barrel can be made with a Teflon-backed silicone rubber ring.

Concentration flasks should be washed with distilled acetone and dried for 20 min at 130°C. The glass nitrogen inlet tube for solvent removal should be removed and cleaned daily or anytime that it may have become contaminated.

VII. References

Baker TC, Gaston LK, Mistrot Pope M, Kuenen LPS, Vetter RS (1981) A high-efficiency collection device for quantifying sex pheromone volatilized from female glands and synthetic sources. J Chem Ecol **7**:961–968.

Mistrot Pope M, Gaston LK, Baker TC (1982) Composition, quantification, and periodicity of sex pheromone gland volatiles from individual *Heliothis virescens* females. J Chem Ecol **8**:1043–1055.

Chapter 10

Techniques for Extracting and Collecting Sex Pheromones from Live Insects and from Artificial Sources

Mitzi A. Golub[1] *and Iain Weatherston*[2]

I. Introduction

As the field of pheromone research evolved, and it became clear that pheromones possessed great potential as components of pest management strategies, it became necessary (a) to define precisely the pheromonal blend emitted by the insect, (b) to determine the rates of production and release of the blends by the insect, and (c) to develop controlled release systems for use in monitoring, mass trapping, and aerial dissemination control programs; such development, as noted by Roelofs (1979), requiring a knowledge of the useful longevity of the formulated materials under normal use conditions.

Prior to 1972 most efforts at pheromone collection were directed at structural elucidation. Collection usually involved laborious solvent extraction of either whole insects or excised parts (Jacobson, 1972). In the last decade with the advances in physicochemical analysis techniques, particularly in capillary gas–liquid chromatography, selective ion mass spectrometry and high-performance liquid chromatography, detection methods have achieved high levels of sensitivity and resolution without the necessity to collect or culture thousands of insects.

[1]973 Furnace Brook Parkway, Quincy, Massachusetts 02169.

[2]Département de Biologie, Université Laval, Quebec, Canada G1K 7P4.

II. Methods of Collection

There are basically only two methods to collect and quantify a pheromone, namely, solvent extraction (residue analysis) and effluvial collection. The recent work by Silk et al. (1980) on the chemistry of the sex pheromone of the spruce budworm *Choristoneura fumiferana* exemplifies both methods and shows that the contents of the pheromone producing gland and the composition of its emissions are not necessarily identical.

In the determination of pheromone release rates from artificial sources, ideally, both methods should be applied to show mass balance. This has been achieved in the measurements of release rates of gossyplure from both hollow fibers and rubber septa (Weatherston et al., 1982; Golub et al., 1983).

2.1. Collection from Insect Sources

A. Extraction

(*a*) *Selection of Solvent.* Jacobson (1972) reported that dichloromethane, hexane, and diethyl ether are the preferred solvents since they are sufficiently volatile for extracts to be concentrated without exposing them to high temperatures, and dichloromethane is nonflammable. Working with *Plodia interpunctella,* Brady and Smithwick (1968) tested several solvents for extraction efficiency as measured by a laboratory bioassay. They concluded that 95% ethanol was superior to dichloromethane, diethyl ether, hexane, chloroform, benzene, and acetone. Since 95% ethanol was the poorest lipid solvent tested, they hypothesized that this resulted in the extraction of fewer contaminants.

Benzene has also been used as an extraction solvent, for example, in the original isolation of 2,6-dichlorophenol as a tick pheromone (Berger, 1972). The flammability and toxicity of carbon disulfide have probably restricted its use, although its advantages in terms of ease of concentration and nonresponse to the flame ionization detector have resulted in limited use (e.g., Hirai, 1980; Sower et al., 1972); however, its use should increase as the added sensitivity of newer methods reduces the necessity to use larger volumes of solvent followed by concentration.

(*b*) *Method of Extraction.* Again this can depend on several factors, such as whether or not the site of production are known and the size of the insect. Recently, Bartelt et al. (1982) isolated a multicomponent sex pheromone from the yellow-headed spruce sawfly *Pikonema alaskensis* (Rowher). Prior to eclosion they placed sawfly cocoons individually in gelatin capsules; females 3 to 5 days postemergence were killed by freezing, and the insects, their cocoons, and capsules were separately washed with hexane.

Whole insects have also been used in the isolation of the sex pheromone of the lone star tick, *Amblyomma americanum* (L.) (Berger, 1972); the velvet bean caterpillar, *Anticarsia gemmatalis* (Hübner) (Johnson et al., 1981); the Indian meal moth, *Plodia interpunctella* (Hübner) (Brady and Smithwick, 1968); the tobacco budworm, *Heliothis virescens* (F.) (Mitchell et al., 1974) and the olive fly, *Dacus oleae* (Gmelin) (Haniotakis et al., 1977).

Although it is the female of the lone star tick which produces the pheromone, Berger (1972) found it more efficient to extract both sexes; lots of 50,000 two- to six-week-old adults were homogenized in 1 liter of benzene, which after filtering, drying, and removal of the solvent, yielded 1.5 g of residue including much *non*pheromonal material from which the active compound(s) had to be separated. In this case since the tick pheromone is phenolic, the initial purification step involved suspending the residue in alkali, removal of neutral lipids by partitioning with hexane and acidification of the alkaline solution with carbon dioxide, followed by benzene extraction.

The procedure used by Haniotakis et al. (1977) to isolate the olive fly pheromone involved aspirating 7- to 10-day-old female flies, during their period of sexual activity, into flasks containing ether and storing at −4 to −5°C for 48 hr. After removal of the bodies by filtration, the filtrate was returned to the refrigerator for a further 24 hr, then filtered and concentrated.

The extraction of whole bodies for short periods of time characterized the isolation procedures of the pheromones of the velvet bean caterpillar and the tobacco budworm. With the former, calling females were vacuumed into a 1-liter flask containing sufficient ether to cover the insects. After soaking for 30 min, the bodies were removed by filtration and the filtrate plus solvent rinses of the flask were concentrated. For *H. virescens,* Mitchell et al. (1974) compared the biological activity of crude extracts obtained by various methods. Three of their methods used whole insects. Two 5-min soaks in ether produced the most active extract, superior to either soaking the whole insects in ether for 2 to 4 hr or grinding them in ether.

The classical method of preparing pheromone extracts from female Lepidoptera is the excision of the pheromone gland, usually located in the abdominal tip, or the entire abdominal tip (Table 1).

In another comparison the levels of pheromone extracted from *Cadra cautella* (Walker), as measured by laboratory bioassay, were obtained by subjecting excised female abdominal tips to three extraction methods (Daley et al., 1978). These authors determined that there was no difference in the amount of pheromone recovered by (a) excising the abdominal tips from 25 two-day-old females during their optimal calling period and soaking them in 1 ml of hexane:ether (1:1) at −22°C for exactly 24 hr, (b) homogenizing them in a glass tissue grinder after the 24-hr period at −22°C, and similarly extracting the residue after centrifugation for 30 sec with further 1- and 0.5-ml portions of solvent, or (c) sonication of the tips after the 24-hr period at −22°C with a Bronwill Biosonik III fitted with a microtip.

Table 1. Summary of extraction methodology for insect sex pheromones, 1977–1982

Insect	Sex	Part of insect or source processed	Solvent	Method	Reference
ORTHOPTERA					
Periplaneta americana	F	Frass and rearing container linings	Aqueous lead nitrate and hexane	The centrifuged aqueous extract reduced to half-volume, then extracted with hexane	Persoons et al. (1979)
COLEOPTERA					
Anthonomus grandis	F	Frass	Steam, 20% ether in pentane or dichloromethane	Frass collected over 2 weeks steam distilled for 1 hr; distillate extracted with organic solvent	McKibben et al. (1977) Hedin et al. (1979a)
Attagenus elongatulus	F	Whole insect plus rearing container linings	Hexane	Insects and solvent plus glass beads vibrated until completely macerated	Barak and Burkholder (1977) Fukui et al. (1977)
Curculio caryae	M,F	Whole insect	Pentane	Field trapped (cone traps) insects (1000) immersed in solvent at −20°C, then the extracts concentrated	Hedin et al. (1979a,b)

Dendroctonus jeffreyi	F	Hindgut	None	Insects frozen at −70°C, then dissected and volatiles analyzed by headspace GC–analysis	Renwick and Pitman (1979)
			Ether	Hindgut extracts examined by GC/MS	
Dermestes maculatus	F	Whole insect	Hexane	Insects immersed in solvent at 0°C for 24 hr	Abdel-Kader and Barak (1979)
	M	Producing gland from 2- to 3-week-old adults	Hexane	Glands immersed in hexane	Levinson et al. (1978)
Diabrotica virgifera virgifera	F	Rearing container linings	25% ether in hexane	Filter paper eluted in solvent	Guss et al. (1982)
Ips accuminatus	M	Hindgut	—	—	Bakke (1978)
Pityogenes chalcographus	M	Whole insects	Pentane	Insects treated with juvenile hormone analogue and extracted	Francke et al. (1977)
Platypus flavicomis	M	Hindgut	Ether	Hindguts immersed in solvent	Renwick et al. (1977)
Popillia japonica	F	Container	Benzene	Container rinsed 3X with solvent	Tumlinson et al. (1977)

Table 1. (continued)

Insect	Sex	Part of insect or source processed	Solvent	Method	Reference
Scolytus multistriatus	M,F	Whole insects	Hexane	Insects homogenized in Waring blender	Gore et al. (1977)
	M&F F	Head, prothorax, mesothorax, abdomen, abdominal tip, hindgut, accessory gland	Hexane	Individual parts from 125 insects crushed in 50 μl of solvent, the mixture frozen, thawed, and the supernatant removed	
Trypodendron lineatum	F	Frass	Benzene	Extraction carried out in cooled Waring blender	MacConnell et al. (1977)
LEPIDOPTERA					
Agrotis ipsilon	F	Abdominal tip of 3- to 6-day-old adults	Dichloromethane	The glandular area was snipped off into the solvent 3–5 hr into the scotophase	Hill et al. (1979)
Agrotis segetum	F	Producing gland	Pentane or hexane	The gland from 3- to 5-day-old insects immersed in solvent for 24 hr at −20°C	Löfstedt et al. (1982)

Amathes c-nigrum	F	Abdominal tip	Hexane	Tips extracted with hexane	Bestmann et al. (1979)
Amorbia cuneana	F	Abdominal tip	Dichloromethane	2- to 4-day-old females, 7–10 hr into scotophase frozen and the tips excised into solvent for 0.5–1.5 hr	McDonough et al. (1982a)
Amyelois transitella	F	Producing gland	Ether	Excised glands dipped in 200 μl of solvent for 1–2 sec	Coffelt et al. (1979)
Archips argyrospilus	F	Abdominal tip	Dichloromethane	Excised tips of 2- to 3-day-old females put in solvent	Cardé et al. (1977)
Archips mortuanus	F	Abdominal tip	Dichloromethane	Excised tips of 2- to 3-day-old females immersed in solvent	Cardé et al. (1977)
Atrophaneura alcinous alcinous	M&F	Wings	Dichloromethane	Wings were extracted for 24 hr at room temperature with solvent	Honda (1980)
Bombyx mori	F	Abdominal tip	Benzene	Excised tips from newly emerged insects extracted	Kasang et al. (1978)

Table 1. (continued)

Insect	Sex	Part of insect or source processed	Solvent	Method	Reference
Chilo partellus	F	Abdominal tip	Dichloromethane	Insects, 1 day old, placed in dark for 2 hr, frozen at −20°C for 10 min and the excised tips soaked in solvent at room temperature for 15 min	Nesbitt et al. (1979b)
Chilo sacchariphagus	F	Abdominal tip	Dichloromethane	Insects, 1 day old, placed in dark for 2 hr, frozen at −20°C for 10 min and the excised tips soaked in solvent at room temperature for 15 min	Nesbitt et al. (1980)
Choristoneura fumiferana	F	Producing gland	(Not stated)	Gland of 2- to 4-day-old insects everted, dipped in liquid nitrogen, and snipped off into solvent	Wiesner et al. (1979)
Choristoneura occidentalis	F	Abdominal tips	Hexane	Tips excised in final 1–2 hr of photophase and dipped in solvent for 10–20 sec	Cory et al. (1982)

	F	Producing gland	Hexane	Gland excised in final 2 hr of photophase and soaked in solvent	Silk et al. (1982)
Choristoneura parallela	F	Ovipositor	Heptane	Ovipositor excised, blotted, and transferred to ampoule containing internal standard and washed with 3 μl solvent	Neal et al. (1982)
Choristoneura rosaceana	F	Abdominal tips	Dichloromethane	Abdominal tips extracted with solvent	Hill and Roelofs (1979)
Chrysoteuchia topiaria	F	Abdominal tips	Dichloromethane	Excised tips steeped in solvent 0.5–1.5 hr	McDonough and Kamm (1979)
Colias eurytheme	M	Wings	Ether	Wings washed for 5 min at 5°C with magnetic stirring	Grula et al. (1980)
Colias philodice	M	Wings	Ether	Wings washed for 5 min at 5°C with magnetic stirring	Grula et al. (1980)
Cryptophlebia leucotreta	F	Whole insects	Dichloromethane	Insects 2–5 days old immersed and then homogenized in solvent. The residue was further Soxhlet extracted for 8 hr	Persoons et al. (1977)

Table 1. (continued)

Insect	Sex	Part of insect or source processed	Solvent	Method	Reference
Dioryctria disclusa	F	9/10 abdominal segments	Dichloromethane	Insects 3–4 days old, 4 hr into the scotophase were excised and the tips soaked in solvent overnight	Meyer et al. (1982)
Erias insulana	F	Abdominal tips	Ether	Tips cut 3 hr into scotophase of 1- to 2-day-old insects and soaked in ether at room temperature for 15 min	Hall et al. (1980)
Estigmene acrea	F	Abdominal tips	Dichloromethane	Insects, 1–3 days old, in first part of scotophase, excised and extracted with solvent	Hill and Roelofs (1981)
Euxoa ochrogaster	F	Producing glands	Dichloromethane	Insects, 3–19 days old, that exhibited "calling" in the first 4 hr of scotophase had their glands everted and washed with solvent	Struble et al. (1980b)

Grapholitha molesta	F	Abdominal tip	None	Tips (2) put into a capsule for direct GC injection	Biwer et al. (1979)
	F	Rearing containers	Hexane	Insects (50), 2–6 days old, left in stoppered flasks during calling time, the flasks cooled at −5°C, the insects removed, and the flasks washed with solvent	Cardé et al. (1979)
	M	Hair pencils and claspers	Skellysolve B	The organs from 5-day-old insects were excised and dipped in solvent for several minutes	Nishida et al. (1982)
Hedya nubiferana	F	Producing gland	None	Glands from 4-day-old insects encapsulated for direct GC injection	Frerot et al. (1979b)
Heliothis armigera	F	Abdominal tip	Dichloromethane	Tips from 2- to 3-day-old insects 3.5–5 hr into the scotophase were macerated with solvent	Piccardi et al. (1977)

Table 1. (continued)

Insect	Sex	Part of insect or source processed	Solvent	Method	Reference
Heliothis subflexa	F	Abdominal tip	Ether, then *iso*-octane	Tips (5) from 2- to 5-day-old insects excised, blotted, and put in 250 μl ether for 2–3 minutes. The ether was removed and *iso*-octane added.	Teal et al. (1981)
	F	Ovipositor	Heptane	The ovipositors of 2- to 4-day-old insects, 2–5 hr into scotophase were excised, blotted, and washed with 3 μl heptane.	Klun et al. (1982)
Heliothis virescens	F	Ovipositor	Heptane	Individual ovipositors were excised, the cut surface was blotted and transferred to an ampoule containing internal standard and washed with 3 μl of solvent	Klun et al. (1980a)

Heliothis zea	F	Ovipositor	Heptane	Indiviudal ovipositors were excised, the cut surface was blotted and transferred to an ampoule containing internal standard and washed with 3 μl of solvent	Klun et al. (1980c)
Homeosoma electellum	F	Producing gland	Dichloromethane	Tips of insects, 3–5 days old, excised and steeped in solvent for 30 min	Underhill et al. (1979)
		Containers	Pentane	Groups of 25–50 insects held in glass jars for 3 days, then the jars rinsed with pentane	
Homona coffeania	F	Abdominal tips	Dichloromethane	The tips from 3-day-old insects ground with chilled solvent in a tissue grinder	Kochansky et al. (1978)
Hyphantria cunea	F	Abdominal tip	Ethyl acetate: benzene (1:1) or dichloromethane	Tips excised from 1-day-old insects Tips excised from 1- to 3-day-old insects	Hill et al. (1982)

Table 1. (continued)

Insect	Sex	Part of insect or source processed	Solvent	Method	Reference
Mamestra brassicae	F	Abdominal tip	Hexane	Tips excised and extracted with solvent	Bestmann et al. (1978)
	F	Abdominal tip	Dichloromethane	Extended tip of 3- to 9-day-old insects, 3–7 hr into scotophase washed with a few drops of solvent	Struble et al. (1980a)
	F	Abdominal tip	None	The tip from 2-day-old insects clipped 2 hr before end of scotophase and encapsulated for direct GC injection	Descoins et al. (1978)
Mamestra configurata	F	Abdominal tip	Dichloromethane	Tips extracted with solvent	Underhill et al. (1977)
Mamestra oleracea	F	Abdominal tip	None	Tips from 4-day-old insects clipped 2 hr before end of scotophase and encapsulated for direct GC injection	Descoins et al. (1978)
Manduca sexta	F	Abdominal tips	Ether	Tips clipped 11 P.M.–4 A.M. and homogenized in ether containing sodium sulfate	Starratt et al. (1979)

Ostrinia furnicales	F	8/9 abdominal segments	Dichloromethane	1- to 2-day-old insects had the segments excised and stored in solvent at 0°C	Cheng et al. (1981)
	F	Ovipositors	Heptane	The ovipositors of 1- to 4-day-old insects excised and chopped in 3 μl heptane for 10 sec	Klun et al. (1980b)
Ostrinia nubilalis	F	Abdominal tips	Ether	Insects, 2 days old, chilled to 6°C, the tips excised and homogenized in ether with a tissue grinder	Klun and Junk (1977)
Pandemis heparana	F	Producing gland	Hexane	Gland excised into hexane.	Frerot et al. (1979a, 1982)
			None	Gland encapsulated for direct injection into GC	
Pandemis pyrusana	F	Producing gland	Dichloromethane	Glands from 2- to 3-day-old insects excised into solvent	Roelofs et al. (1977)

Table 1. (continued)

Insect	Sex	Part of insect or source processed	Solvent	Method	Reference
Platynota flavedana	F	Producing gland	Dichloromethane	Glands excised and soaked in solvent at less than $-5°C$ for 24 hr	Hill et al. (1977)
Plusia chalcites	F	Producing gland	Hexane and Heptane	Glands from 4- to 6-day-old females 1–3 hr into scotophase extracted with 250 μl hexane, a few drops of heptane added and the hexane removed	Dunkelblum et al. (1981)
Prays citri	F	Abdominal tips	Dichloromethane or ether	Day following emergence 2 hr before "calling" the tip excised and extracted in solvent for 15 min	Nesbitt et al. (1977)
Prays oleae	F	Abdominal tip	None	Tips excised and encapsulated for direct injection into GC	Renou et al. (1979)
Rhyacionia frustrana	F	Abdominal tip	None	Tips of 1-day-old insects excised	Hill et al. (1981)

Rhyacionia subtropica	F	Abdominal tip	Dichloromethane	Tips of 2- to 3-day-old moths 4 hr into scotophase snipped into solvent	Roelofs et al. (1979)
Scotia exclamationis	F	Abdominal tip	Hexane	Tips extracted with hexane	Bestmann et al. (1980)
Spaelotis clandestina	F	Abdominal tip	Dichloromethane	Tips from light trap collected insects extracted with solvent for 1/2 hr	Steck et al. (1982)
Sparganothis directana	F	Ovipositors	Dichloromethane	Ovipositors (200) were extracted with 1 ml solvent for 24 hr	Bjostad et al. (1980c)
Sparganothis spp.	F	Producing gland	Dichloromethane	Ovipositors (200) from 2-day-old insects extracted with 1 ml solvent for 24 hr	Bjostad et al. (1980d)
Trichoplusia ni	F	Producing gland	Carbon disulfide	Glands (2000) extracted with 50 ml solvent or individual glands soaked overnight	Bjostad et al. (1980a)
Xylomyges curialis	F	Producing gland	Dichloromethane	Glands excised from 2- to 6-day-old insects 5 hr into scotophase and extracted for 0.5–1.5 hr	McDonough et al. (1982b)

Table 1. (continued)

Insect	Sex	Part of insect or source processed	Solvent	Method	Reference
DIPTERA					
Dacus neohumeralis	M	Producing glands	Ether	Glands dissected from mature adults, contents collected in micro-pipets and placed in ether	Bellas and Fletcher (1979)
Dacus oleae	F	Rectal glands	Ether	Glands dissected out and extracted with ether over at least 24 hr	Mazomenos and Haniotakis (1981)
Dacus tryoni	M	Producing glands	Ether	Glands dissected from mature adults, the contents collected in micropipets and stored in ether	Bellas and Fletcher (1979)
Glossina morsitans morsitans	F	Whole insects	Ether	Insects immobilized at 4°C and rinsed with ether	Carlson et al. (1978)

Lycoriella mali	F	Whole insects	Hexane	Whole bodies ground in a tissue grinder with solvent	Kostelc et al. (1980)
Sarcophaga bullata	M	Whole insects	Ethyl acetate	Insect immobilized with CO_2 and ground in a Waring blender with solvent	Girard et al. (1979)
HYMENOPTERA					
Itoplectis conquisitor	F	Whole insects	Dichloromethane	Newly emerged insects washed with 100 μl solvent	Robacker and Hendry (1977)
Pikonema alaskensis	F	Rearing capsules, cocoons, and whole insects	Hexane	The sources separately washed with solvent	Bartelt et al. (1982)

[1] Abbreviations used: GC, gas chromatography; MS, mass spectrometry.

Coffelt et al. (1978) and Sower et al. (1972, 1973) were among the first researchers to appreciate the advantages of using small volumes of solvent for short durations of extraction. The use of 0.2 to 0.5 ml of ether for 1–2 sec yielded measurable quantities of pheromone clean enough for quantitative analysis by gas–liquid chromatography without further clean-up.

Working with various stored products pests, females were anesthetized by ether or by chilling, glands were everted by finger pressure on the abdomen, and the tip of the abdomen was dipped in ether for 1–2 sec. After drying over $MgSO_4$, the solution was concentrated to 3 to 5 μl in a gentle stream of nitrogen. When known quantities of (*Z,E*)-9,12-tetradecadien-1-yl acetate, the corresponding alcohol, and (*Z*)-9-tetradecen-1-yl acetate were concentrated to 5 μl and introduced into the syringe prior to gas chromatographic analysis a 50 to 90% loss of material was observed. This variable loss was corrected for by adding known amounts of internal standards of similar volatility prior to concentration.

This brief dip method has been shown to yield about 5% of the pheromone contained in the glands of *Sitotroga cerealella* (Oliver), *Trichoplusia ni* (Hübner), and *Pseudaletia includens* (Walker); the value for *P. interpunctella* and *C. cautella* being about 10%. The amount of pheromone obtained from *T. ni* by gland surface wash (dipping in carbon disulfide for 10 sec) was 7.6% of the amount obtained by whole gland extraction (61 abdominal tips in 2–3 ml CS_2 for 24 hr). Nesbitt et al. (1979a, b), used the abdominal tip washing technique as a means of avoiding possible enzymatic degradation of *Heliothis armigera* pheromone components during isolation.

The collection of pheromone by extraction from male Lepidoptera has been simpler than that from females, and almost always involves the excision of the producing or storage gland and immersion in the solvent of choice. A typical procedure was that of Grant et al. (1972), who anesthetized male armyworm *Pseudaletia unipuncta* with CO_2 or by cooling, pulled the scent brushes from their pockets, cut them off at the base, and placed them in chilled pentane or dichloromethane. Weatherston and Percy (1976) everted the hair pencils of male bertha armyworm *Mamestra configurata* by injecting air into their abdomens.

Shorter-duration extractions have also recently been reported. Nishida et al. (1982) extracted the excised hair pencils and claspers of male *Grapholitha molesta* by immersion for a few minutes in Skellysolve B, while Grula et al. (1980) used two 5-min ether washes of the wings of 50–100 male butterflies when isolating the aphrodisiac pheromones of *Colias* species. The wings were stirred with 100 ml of anhydrous ether at 5°C; the solvent was filtered through glass wool and concentrated to 0.5 μl at 0°C.

A summary of extraction methodology employed for the isolation of insect sex pheromones over the last five years is presented in Table 1.

2.2. Effluvial Collection

In the collection of airborne volatiles from live insects, one must choose whether or not still or moving air should be used; the trapping method, i.e., cryogenic or adsorptive techniques, and if the latter, which type of adsorbent and which method of desorption should be used; and the dimensions, configuration, and construction materials of the collection device. Recently Ma et al. (1980) briefly reviewed the relative merits of the various methods used to collect airborne pheromones from live insects.

A. Passive Airflow

Findlay and Macdonald (1966) had obtained active material by passing air over virgin female *Choristoneura fumiferana* contained in plastic bags or by rinsing out with ether plastic bags which had contained females. To avoid the extraction of contaminants and plasticizers, Weatherston et al. (1971) placed newly emerged females (100/jar) in glass Mason jars containing a piece of cheesecloth on which the moths could rest. After 2 days the moths were removed and the jars and the cheesecloth were well rinsed with ether.

Sower and Fish (1975) used a similar technique when determining the pheromone release rate from Indian meal moths. Female *P. interpunctella* in groups of 20 were confined in 500-ml glass stoppered flasks at about 25°C for 1 hr. After quickly shaking out the moths, the flasks were rinsed with 5 or 6 ml of anhydrous ether, the solution was filtered and reduced to 3–5 μl with dry N_2 at 30°C, drawn into a syringe, and analyzed. The efficiency of pheromone recovery was determined by placing known nanogram amounts of (*Z,E*)-9,12-tetradecadien-1-yl acetate and the corresponding alcohol on stainless-steel spatula tips suspended in clean sealed flasks for 1 hr. It was found that 80% of the acetate and 40% of the alcohol were recoverable from the rinses.

Additional and more variable losses occurred during the concentration and introduction of the small volume of solution into the syringe. These losses were individually calibrated by adding a known amount of internal standard ((*Z*)-7-hexadecen-1-yl acetate) to each sample before concentration and assuming that the losses of the standard and the two pheromone components were about equal. By this method the authors determined that calling female *P. interpunctella* less than 5 days old released at all times of the day a mean of 3 ± 1 ng/hr of the acetate and 6 ± 2 ng/hr of the alcohol.

For the identification of the complete pheromone blend of *Grapholitha molesta* Cardé et al. (1979) used a similar technique; 50 two- to six-day-old females were placed in 250-ml stoppered round-bottom flasks just prior to the initiation of calling. After about 3 hr the flasks were cooled to −5°C for 5 min and the females were removed; rinsing with 3 × 30 ml of hexane gave a solution which was filtered through glass wool, concentrated to 1 ml, transferred to an ampoule, and taken to dryness with a stream of nitrogen. The residue was dissolved in 100 μl of dichloromethane and this solution was stored at −5°C until analyzed.

Using a method similar to that of Weatherston et al. (1971), Baker et al. (1980) were the first to attempt quantification of pheromone emission rates from both live females and synthetic sources by glass adsorption. They studied the effects of the number of rinses and cooling of the flasks on the efficiency of recovery before turning their attention to the rate of emission from live females.

Since this is one of the most important reports on the measurement of emission rates, the method is described here in detail.

The pheromone source (microscope slide coverslip, rubber septum, or live moth) was placed in a 250-ml round-bottom flask with a 24/40 ground glass joint and sealed with a glass stopper. After a selected period of time, the flask was opened, the source was removed, and the flask was rinsed with small quantities of hexane. (*Z*)-11-Tetradecen-1-yl acetate was added as internal standard and the pheromone components, (*Z*)-8-dodecen-1-yl acetate (DDA) and (*Z*)-8-dodecen-1-ol (DDOL), two major components of the *Grapholitha molesta* sex pheromone, were quantified by gas-liquid chromatography.

Stock solutions of the internal standard (1 μg/μl) and DDA (10 μg/μl) in hexane were formulated gravimetrically. DDOL was added to the DDA solution to make a 1:1 mixture. A stock mixture of DDA and the internal standard was made by dissolving 8.9 and 8.8 mg, respectively, in 1 ml of hexane.

To test the effect on the recovery of chilling the flask, and the number of hexane rinses, 1 μl of the DDA/DDOL solution was applied to each of six coverslips with a 10-μl syringe; each coverslip was carefully placed in a 250-ml flask after 10 sec and the flasks were stoppered. After 12 hr, three of the flasks were chilled for about 5 min at −20°C while the remaining three were held at ambient temperature. Ten milliliters of hexane were added to each flask, it was restoppered, and the solvent was swirled for 30 sec and then removed by pipet. Two further separate identical rinses were performed, internal standard solution (5 μl) was added to each rinse which was then evaporated to less than 1 ml. Fifty-microliter aliquots were evaporated to dryness and a few microliters of carbon disulfide added prior to analysis.

Results indicated that chilling the flasks did not effect the recovery, and that at least 99% of the recovered pheromone components were recovered in the first two rinses. Only 60% of the DDA and 50% of the DDOL applied to the coverslips were recovered; however, it was shown that, in fact, the small volume (1 μl) and the concentration of the solution used to prepare the coverslips were contributing factors to the low recovery due to loss in transfer with the syringe needle.

To minimize variations in the recovery, data on collection efficiency were obtained by using the DDA/DDOL mixture diluted to 1 μg/μl. Five microliters of the solution was applied to five coverslips, which after 10 sec were placed in closed flasks at 25°C for 3-5 hr. The coverslips were then removed and allowed to stand in hexane (10 ml) for 5 min; the flasks were twice rinsed with 10-ml aliquots of hexane. Internal standard (5 μg) was added to both the coverslip wash and the first flask rinse. The flask rinses were combined. To measure how much DDA/DDOL was actually applied, 5 μl of the dilute mixture was applied to 10 coverslips, and after 10 sec they were rinsed in hexane (10 ml) containing 5 μg of internal standard. Results revealed that 4.4 μg of DDA and 3.9 μg of DDOL were applied to the coverslips with a 100 and 97.4% recovery, respectively, from the flask walls (about 2% DDA and 2.5% DDOL were recovered from the coverslips). Septa impregnated with either 1000, 100, or 10 μg of the two pheromone components were aged in a fume hood for 12 hr at 22°C and then introduced into the flasks at 25°C for 3 hr (1000 μg septa) or 16 hr (100 and 10 μg septa). The work up was as described above except that only 50 ng of internal standard was used.

In another series of tests 50 noncalling Oriental fruit moth females were placed in each flask with a septum (1000 and 100 μg loading only). For both DDA and DDOL, the decade increase in loading on the septa increased the emission rate by at least 10 times, and the alcohol was emitted between 2 and 3 times faster at each loading. The presence of insects reduced the amount of DDA recovered by 90% at both loadings, and the amount of DDOL recovered by 75 and 83% for the 1000- and 100-μg loadings, respectively.

The rate of emission from the live insects was determined by placing groups of 50 *G. molesta* females, 3-4 days old, in the flasks about 1 hr after the onset of calling. At the end of the calling period, the flasks were cooled to –20°C for 5 min and then the immobilized females were quickly removed. The mean amounts of DDA and DDOL emitted by the insects were, respectively, 3.2 and 0.7 ng/hr/female (N=11); these values take into account the recovery efficiency, the losses due to the adsorption onto the insect bodies, and the presence of the other pheromonal components.

Another method, passive airflow, was used regularly in the early work on nonlepidopteran pheromones; for example, the isolation studies of Haniotakis et al. (1977) working with the olive fly *D. oleae,* those of Persoons et al. (1979) on the isolation and elucidation of periplanone B from the American cockroach *Periplaneta americana,* and the release rate and production studies of the pheromone of *P. interpunctella* by Nordlund and Brady (1974).

Nordlund and Brady placed 50 virgin female meal moths in small wire cages which were then put inside a 1-gal- glass jar completely lined with filter paper (Curtin Qualitative Paper No. #7760). After exposure to the moths, the filter paper was extracted with diethyl ether at –22°C for a minimum of 7 days. A stock solution was made from the extract and serially diluted since quantification was by laboratory bioassay. For comparison, dilutions were obtained by evaporating 10, 100, and 1000 ng of (*Z,E*)-9,12-tetradecadien-1-yl acetate from coverslips suspended in the cages and extracting the filter paper and the coverslips separately with 40 ml of ether. Control dilutions were made by dissolving 10, 100, and 1000 ng of neat (*Z,E*)-9,12-tetradecadien-1-yl acetate in 40 ml of ether and diluting.

Haniotakis et al. (1977) placed female olive flies in a filter-paper-lined airtight Plexiglas cylinder, which they had previously used in a total condensation trapping airflow method. After the period of sexual activity, crude pheromone was obtained by ether extraction of the filter paper in a Soxhlet for 8 hr.

For studies on periplanone B (Persoons et al., 1979) the female cockroaches were allowed to inhabit boxes lined with filter paper which was changed every 3-4 weeks. After storage at –15°C the paper was ground up in water containing lead nitrate; the resultant slurry was centrifuged and the supernatant distilled. Hexane extraction of the aqueous distillate yielded crude active material. In an alternate method, accumulated frass from the cockroaches was similarly treated.

B. Moving Airflow

(*a*) *Cryogenic trapping.* This method has been developed to aid in environmental sampling since, at the temperature of liquid nitrogen (−196°C), the usual refrigerant, all trace organics which comprise air pollution constituents can be trapped out (Rasmussen, 1972; Rasmussen and Hutton, 1972). A quantitative cryogenic sampler was designed by Rasmussen (1972). In the system of cryogenic trapping, air intake is achieved by creating a vacuum in the trap by condensing all of the air which enters. The air intake rate is dependent on three factors: the surface area for condensing air, the temperature differences between the refrigerant and the liquifaction temperature of air, and the size of the intake orifice.

The use of the air condensation trap to collect insect pheromones was first reported by Browne et al. in 1974. These authors developed three cryogenic trapping techniques for the collection of volatile pheromones from bark beetles. As can be seen in Fig. 1, logs containing producing beetles are held in a containment chamber through which air flows once the collection device, in this instance a 750-ml Erlenmeyer flask, is placed in the wide-mouthed Dewar flask containing liquid nitrogen. In order to avoid ice build-up and blockage of the inlet tube, it is necessary that the neck of the collection flask protrudes from the Dewar.

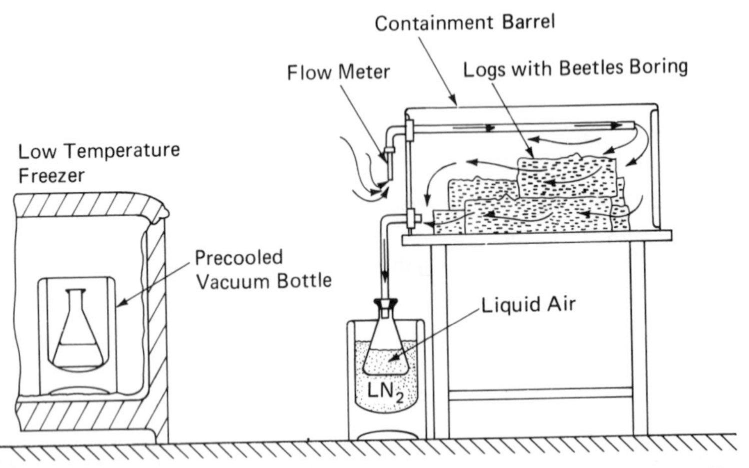

Figure 1. Condensation apparatus for recovery of airborne pheromones, Browne et al. (1974).

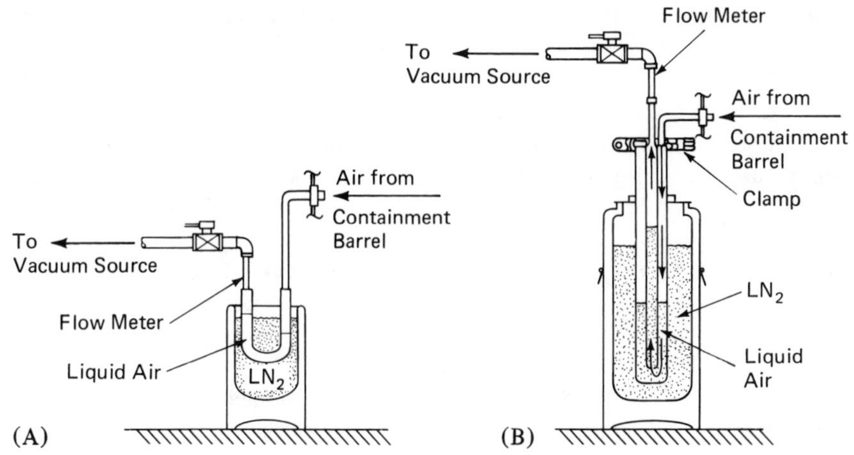

Figure 2. Condensation trap for continuous collection of airborne pheromones. (A) Glass U-tube apparatus. (B) Stainless-steel tube-within-a-tube apparatus, Browne et al. (1974).

As the Erlenmeyer flask fills up with liquid air containing the beetle volatiles, the surface area available for heat exchange is reduced and hence the flow rate of air through the containment vessel is also reduced. When the flow drops below a desired rate, the collection flask inside a prechilled Dewar is transferred to a low-temperature freezer (–50°C) where the air slowly evaporates but the residue of beetle volatiles is retained. To increase the operation time of 4-6 hr to continuous operation for 48-96 hr (or until blocked by ice formation) the Erlenmeyer flask was replaced by a 5-cm-diameter U tube (Fig. 2A) connected on the out-flow side to an adjustable vacuum source.

The vacuum is applied only when the liquid air partially fills both arms of the U tube, and then in such a way that the levels of the liquid air are as shown in Fig. 2A. This gives a greater heat-exchange capability on the in-flow side while allowing slow evaporation of the air on the out-flow side. A flow meter positioned on the out-flow side is used to adjust the vacuum so that the "liquid lock" is always maintained. The modification to the stainless-steel "tube within a tube" design (Fig. 2B) was undertaken to conserve liquid nitrogen.

This technique has been used to collect pheromone from the olive fly (Haniotakis et al., 1977) where 2000 sexually mature females inside a 55 cm × 27 cm wire cage were placed inside an airtight Plexiglas cylinder. Using the cryogenic trap shown in Fig. 1, 200 ml of liquid air was collected in 3 hr. The collection flask was allowed to come to room temperature and then rinsed with diethyl ether.

Working with the western pine beetle, Browne et al. (1979) collected pheromone from cut bolts and extended the cryogenic trapping technique to obtaining quantitative estimates of *exo*-brevicomin, frontalin, and myrcene from a natural population of colonizing beetles. Portions of trees being colonized were wrapped in an airtight aluminum sheet to form an airtight cylinder which was sealed to the tree. Inlet and airflow holes were cut in the aluminum, and the outflow was connected to a cryogenic trap.

(*b*) *Cold Trapping.* Cold trapping was used during isolation and identification studies of pheromones from insects of several orders, e.g., *Periplaneta americana* (Yamomoto, 1963) (Orthoptera); *Ips confusus* (Vite et al., 1963); *Anthonomus grandis* (Keller et al., 1964) and *Tenebrio molitor* (Happ and Wheeler, 1969) (Coleoptera); *Cochliomyia hominivorax* (Fletcher et al., 1966) (Diptera); *Diparopsis castanea* (Tunstall, 1965) and *Lasiocampa quercus* (Kettlewell, 1961) (Lepidoptera). The first attempt to use this method as a quantitative measurement is the report by Sower et al. (1971).

Nitrogen at a flow of 150 ml/min for 10 min was passed either over live individual female *Trichoplusia ni* whose pheromone gland had been forcibly extruded by means of a dental matrix retainer, or over microgram amounts of synthetic pheromone on a copper disc. The effluent gas stream passed through a small volume of carbon disulfide contained in a U tube held at −70°C in a dry ice-acetone bath. After the collection period, the glass surfaces of the collection tube were washed with solvent, and the combined carbon disulfide solution (ca. 3 ml) was reduced to 100 μl by distillation at water bath temperatures. A 10-μl aliquot was then analyzed by gas–liquid chromatography and quantitatively compared to a standard curve of detector response to various concentrations of (*Z*)-7-dodecen-1-yl acetate.

Calibration of the system using 0.1 to 0.3 μg of the pheromone on copper discs resulted in an average recovery of 66%; 13% was recovered from the collection tube, 35% from washes of the apparatus, and 18% remained as a residue on the copper discs. Apparently there was no breakthrough as evidenced by the inability to recover pheromone from a second collection tube placed in series to the first, and the authors (Sower et al., 1971) concluded that the 34% loss occurred during transfer and concentration procedures.

Another source of error may have been in the loading of the copper discs, since the volume of the solvent used was not reported. In all subsequent release rate calculation based on this method a correction factor of 1.515 was used.

A similar method was reported by Richerson and Cameron (1974) in their comparison of pheromone release rates from laboratory-reared and feral gypsy moths. The female moth was allowed to acclimate in the holding chamber (a 9 cm × 2 cm diameter glass tube); once calling began, air was passed through the chamber at 10 ml/min for 30 min and then through 1 ml of technical grade hexane contained in the capillary section of a U tube immersed in a dry ice-acetone bath. After aeration, the U tube was rinsed with two 1-ml aliquots of solvent, and the combined hexane solution (3 ml) was reduced to 300 μl under nitrogen without, the authors claim, the loss of pheromone. Calibration of the efficiency over the 30-min collection period was undertaken using 1 μg of disparlure on copper discs, the average recovery being 61 ± 3%. Based on the residue on the disc, 819 ± 30 ng evaporated; of this 46% was recovered from the solvent in the collection tube, 11% from collection tube washes, and 4% from solvent washes of the holding chamber. The 39% unaccounted for is presumed to have been lost from the open end of the holding chamber since once again a second U tube in series failed to recover any pheromone; however, it could have been lost in the loading, handling, and concentration procedures. A correction factor of 1.754 was applied to the release rates calculated from this study.

Several collection vials in series containing triple-distilled pentane or hexane over a 2% aqueous sodium hydroxide solution cooled in an ice bath were used to collect 2,6-dichlorophenol from *Dermacentor andersoni* (Sonenshine et al., 1977). Charcoal-filtered compressed air at 50–100 ml/min was passed for 5-min periods over female ticks contained in a stainless-steel holding chamber and connected to the collecting vials via a stainless-steel tube. The efficiency of the recovery procedure was tested by placing known weights (0.658, 0.94, 8.8, and 35.2 μg) of 2,6-dichlorophenol on filter paper in the holding chamber and carrying out the procedure outlined above. Inexplicably only 0.5 μg of the phenol was recovered from the 8.8 μg control, none being recovered from the other three amounts, indicating that this aeration system is not reliable and requires refinement before being used for the quantification of the rates of emission.

(*c*) *Resin Adsorbents.* For the collection of insect pheromones, resin adsorption has evolved as an extension of their use in other fields, e.g., flavor research, water pollution, aroma analysis, and the study of body fluid metabolites. The adsorbent capacity of chromatographic packing materials has been extensively studied, usually with regard to a particular type of compound, but nevertheless such materials could be used as general adsorbents for organic volatiles. Zlatkis et al. (1973) investigated three such adsorbents with a view to their use in trapping volatile materials from human breath and urine.

The materials were heat desorbed and analyzed by capillary gas chromatography. Porapak P, a porous polymer of styrene and divinylbenzene, from Waters Associates, Framingham, Massachusetts, appeared to suffice as a general adsorbent but was limited by an upper temperature limit for heat desorption of 200°C which could prevent efficient desorption of high-molecular-weight volatiles. Carbosieve from Supelco Inc., Bellefonte, Pennsylvania, prepared by thermal cracking of polyvinylidene chloride has exceedingly great surface area (1000 m^3/g) and a high temperature stability; however, these assets combined with high reactivity to ambient volatiles may possibly be a serious limiting factor, since temperatures in excess of 400°C required to desorb the volatiles can cause pyrolysis.

The third chromatography packing material tested by Zlatkis et al. (1973), and which best appeared to fulfill their requirements of efficient adsorptivity and desorptivity, was Tenax GC, a porous polymer of 2,6-diphenyl-*p*-phenylene oxide obtained from Applied Science Laboratories, State College, Pennsylvania. It should be noted that Zlatkis et al. (1973) did not undertake solvent desorption studies with these adsorbents.

The first report of adsorbents being used to collect a pheromone from live insects came from the work of Rudlinsky et al. (1973), who isolated 3-methyl-2-cyclohexene-2-one from female *Dendroctonus pseudotsugae* by pulling air over them and then through a 100 × 6 mm column of Porapak contained in a stainless-steel tube. The volatiles were heat desorbed to a gas chromatograph for analysis. As will be discussed later, many workers have reported erroneous peaks from Porapak Q traps, and care must be exercised in the preconditioning of the adsorbent prior to use. Rudinsky et al. (1973) conditioned their collection tubes by baking them out in a stream of nitrogen for 30 min at 200°C followed by 24 hr at 100°C.

Porapak Q, the resin adsorbent most frequently used today, is an ethylvinylbenzene-divinylbenzene copolymer. The trap used by Byrne et al. (1975) consisted of 25 g of Porapak Q contained in a 20 cm × 2 cm glass tube constricted to 6 mm at both ends, and with a 24/40 joint near one end to permit access. The Porapak, which was preconditioned in a stream of nitrogen (3 liters/min) for 24 hr at 180°C followed by a second 24 hr at 110°C, was held in place between two plugs of glass wool.

Tests with aliphatic alcohols, injected into the inlet side of the trap and flushed for 168 hr with an airflow of 2 liters/min which on exiting the trap passed through a 3-mm glass U tube immersed in liquid air, indicated that propanol had broken through by the fourth hour and was totally eliminated from the trap by the eighth hour. Butanol which also broke through by the fourth hour was eliminated by hour 16. Hexanol, octanol, and decanol did not appear to break through during the course of the test.

The recovery of model compounds at 40-mg and 500-μg levels was studied by impregnating a hexane solution onto filter paper contained in a 16 cm × 6 mm evaporation chamber and aerating for 1 week. The airflow rate for the 40-mg samples was 2 liters/min, and, in an alcohol series, no butanol mass balance was achieved. With octane, tetradecane, and tetracosane total recoveries were, respectively, 9.5, 76.7, and 101%; total recovery figures for pentyl acetate, methyl caproate, and methyl oleate were 65, 66, and 87%; the authors did not comment on either the lack of mass balance or the ratio of the percentages recovered from the adsorbent versus those obtained from the evaporation chamber.

Of 40-mg samples of the behavior-modifying chemicals frontalin and myrcene, 85 and 75%, respectively, were recovered from the adsorbent. This also was the total recovery. Studies at the 500-μg level using an airflow of 1 liter/min for 1 week with pentyl acetate, dodecyl acetate, hexalure, sulcatol, ipsenol, and ipsdienol yielded recoveries ranging from 58% for the pentyl acetate to 89% for the ipsenol; in all cases no material was recovered from the evaporation chamber. These recoveries were calculated from gas chromatographic analysis subsequent to solvent desorption by Soxhlet extraction with pentane over 24 hr and concentration.

The advantages of this general method for the collection of pheromones, from live insects or artificial sources, for structural elucidation, behavioral studies, release rate studies, etc., as stated by Byrne et al. (1975) and Peacock et al. (1975) were obviously also perceived by many workers in the field as is attested to by the use data presented in Table 2. There are, however, several improvements and modifications to the Porapak entrainment technique which are of interest; these are discussed below.

The Porapak method was successfully applied to the collection of pheromones from four dermestid beetle species by Cross et al. (1976).

The apparatus (Fig. 3, see page 256) consisted of a glass tube into which were placed batches of 2000 virgin female beetles maintained on filter paper strips. The closures at each end of the tube reduced the diameter of the holding chamber. Air drawn through the system by vacuum was first purified by filtration through charcoal (and in the case of *Trogoderma granarium* dried over calcium chloride). The air exiting the insect holding chamber passed through Porapak Q (approximately 6 g/2000 insects) which had been conditioned by Soxhlet extraction over 24 hr with redistilled hexane.

The aeration took place for 14 or 28–30 days, and the beetles were maintained at 20–28°C on a 16:8 L:D cycle. It should be noted that the activity of the Porapak extract showed a dependence on the flow rate through the aeration chamber. The beetles appeared to call normally in a flow of up to 3 liters/min, and the extract was most active at this flow rate. The number of beetles calling was observed to decrease at a flow rate of 5 liters/min, hence an aeration rate of 2–3 liters/min was used in the subsequent studies.

The insect volatiles were desorbed by solvent extraction (Soxhlet) with hexane over 24 hr, and concentrated by distillation through a 20-cm glass bead packed column.

By miniaturizing the system, Ma et al. (1980) were able to estimate the pheromone release from a single furniture carpet beetle. Working under well-defined conditions of temperature, humidity, light phase, and intensity, these authors entrained the pheromone on Tenax in the apparatus shown in Fig. 4 (see page 256). The insect holding chamber comprises one-sixth of a 120-mm glass tube 20 mm in diameter and is separated from the activated charcoal air scrubber by a 2-mm coarse glass frit.

A single female furniture carpet beetle (*Anthrenus flavipes*) was placed in the holding chamber with a small piece of pleated filter paper to provide a calling surface. The glass tube was sealed with a foil-lined plastic cap through which is inserted the collection trap, 80 × 4 mm glass tube containing 10 mg of Tenax. The airflow at a rate of 2 liters/min was provided by vacuum. The Tenax was conditioned by eluting 5 g of the resin with 100 ml of ether followed by 3 × 100 ml double-distilled pentane and then baked out at 180°C for 24 hr under a stream of nitrogen.

The authors (Ma et al., 1980) tested the system for pheromone contamination of the holding chamber, and the possibility of breakthrough. They concluded that pheromone loss to glass adsorption was <10 ng, an acceptable experimental margin, and that there was minimum pheromone bleed as evi-

Table 2. Summary of effluent collection methodology for insect sex pheromones, 1977–1982

Insect	Sex	Adsorbent	Flow rate (l/min)	Duration	Method of desorption	Reference
COLEOPTERA						
Callosobruchus maculatus	F	Porapak Q Tenax	3	7–8 days	Solvent–hexane	Qi and Burkholder (1982)
Diabrotica virgifera virgifera	F	Porapak Q	2.5	Several weeks	Solvent–back flushing with 25% ether:hexane or 25% ether:pentane	Guss et al. (1982)
Gnathotrichus retusus	M	Porapak Q	1–2	230 hr		Borden et al. (1980)
Rhyzopertha dominica	M	Porapak Q	2	<16 days	Solvent–Soxhlet extraction with pentane	Williams et al. (1981)
Scolytus multistriatus	F& M+F	Porapak Q	9–11	3 days	Solvent–Soxhlet extracted with pentane	Gore et al. (1977)
Scolytus scolytus	M+F	Porapak Q	1.5–2.0	14 days	Solvent–Soxhlet extraction for 24 hr with pentane	Blight et al. (1978)
Sitophilus granarius	M	Tenax	0.5	7 days	Solvent–3 × 50 ml rinses	Faustini et al. (1982)
Trogoderma inclusum	F	Porapak Q	2	3 days	Solvent–pentane or hexane in Soxhlet over 24 hr	Greenblatt et al. (1977)

Trogoderma glabrum	F	Porapak Q	2	3 days	Solvent–pentane or hexane in Soxhlet over 24 hr	Greenblatt et al. (1977)
Trogoderma variabile	F	Porapak Q	2	3 days	Solvent–pentane or hexane in Soxhlet over 24 hr	Greenblatt et al. (1977)
	F	Porapak Q	2	15 days	Solvent–pentane in Soxhlet over 24 hr	Cross et al. (1977)
Trogoderma granarium	F	Porapak Q	3	28 days	Solvent–pentane or hexane in Soxhlet over 24 hr	Greenblatt et al. (1977)
HOMOPTERA						
Aonidiella aurantii	F	Porapak Q	–	14 days	Solvent–200 μl pentane	Roelofs et al. (1978)
Aonidiella citrina	F	Porapak Q	–	–	Solvent–Skellysolve	Gieselmann et al. (1979a)
Planococcus citri	F	Porapak Q	–	–	Solvent–pentane	Bierl-Leonhardt et al. (1981)
Pseudaulacaspis pentagona	F	Porapak Q	–	2 weeks	Solvent–Back-flushed with 2 ml 50% ether in hexane using an LC pump with flow at 1 ml/min	Heath et al. (1979)
Pseudococcus comstocki	F	Porapak Q	–	4 weeks	Solvent–pentane	Bierl-Leonhardt et al. (1980)
		Porapak Q	0.7–4.0	14 days	Solvent–pentane	Bierl-Leonhardt et al. (1982)

Table 2. (continued)

Insect	Sex	Adsorbent	Flow rate (l/min)	Duration	Method of desorption	Reference
Quadraspidiotus perniciosus	F	Porapak Q	1–3	14 days	Solvent–pentane	Anderson et al. (1981) Gieselmann et al. (1979b)
LEPIDOPTERA						
Agrotis ipsilon	F	Porapak Q	1–2	–	Solvent–(not specified)	Hill et al. (1979)
Agrotis segetum	F	Charcoal	1.3	5 hr	Solvent–carbon disulfide	Löfstedt et al. (1982)
Chilo partellus	F	Charcoal	2	–	Solvent–10 μl carbon disulfide	Nesbitt et al. (1979b)
Choristoneura occidentalis	F	Porapak Q	1.5	3–4 nights during calling period	Solvent–3–4 ml pentane	Silk et al. (1980) Silk et al. (1982)
	F	Porapak Q	0.25	4–8 hr	Solvent–hexane	Cory et al. (1982)

Chrysoteuchia topiaria	F	Porapak Q	1	–	Solvent–dichloromethane	McDonough and Kamm (1979)
Earias insulana	F	Charcoal	–	–	Solvent	Hall et al. (1980)
Heliothis virescens	F	Glass wool	0.12	10 min	Solvent–carbon disulfide	Pope et al. (1982)
Hyphantria cunea	F	Porapak Q	1–2	8 hr	Solvent–Skellysolve	Hill et al. (1982)
Rhyacionia frustrana	F	Porapak Q	1–2	–	Solvent–not specified	Hill et al. (1981)
Rhyacionia subtropica	F	Porapak Q	1–2	–	Solvent–not specified	Roeloefs et al. (1979)
DIPTERA						
Dacus oleae	F	Cold trapping	–	–	–	Mazomenos and Haniotakis (1981)
ACARINA						
Dermacentur andersoni	F	Cold trapped in pentane or hexane	0.05–0.1	5 min over 2–3 hr	–	Sonenshine et al. (1977)

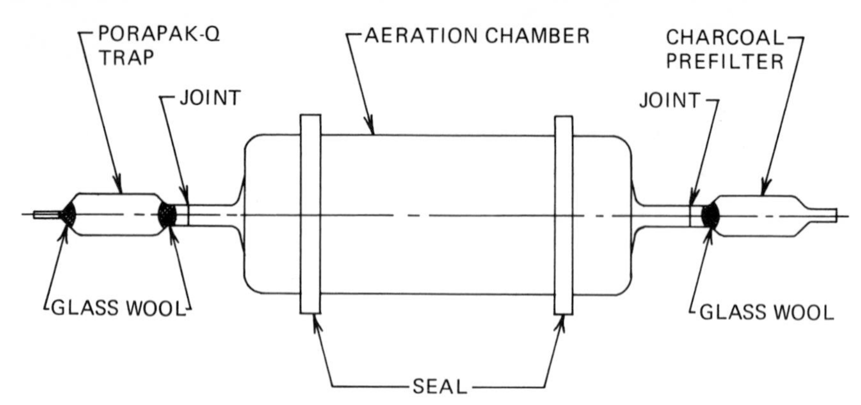

Figure 3. Collection apparatus for aeration of insects and recovery of pheromone from Porapak Q adapted from Cross et al. (1976). Airflow is from the charcoal tube to the Porapak Q.

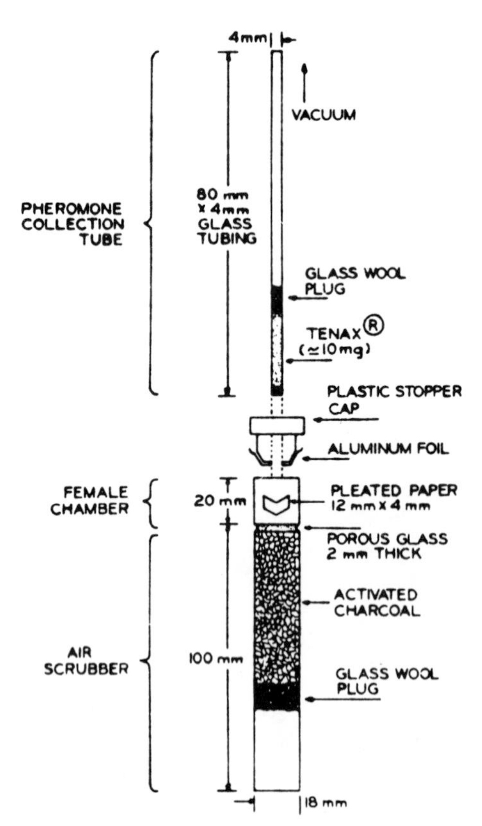

Figure 4. Single female aeration apparatus of Ma et al. (1980).

denced by the failure of a back-up Tenax trap to yield biologically active extract after 2 days of pheromone collection.

To analyze individual gypsy moth sex pheromone production, Ma and Schnee (1983) modified the device used in the furniture carpet beetle study, using an 80 X 23 mm glass tube containing a 1-mm mesh screen cylinder to hold the moth. Both ends were sealed with No. #8 rubber stoppers. Inserted through the outlet and stopper are two pheromone collection tubes, each 50 X 5 mm and packed with 25 mg of either Porapak or Tenax held between glass wool plugs. The recovery efficiency of the system was determined to be 89% by aerating 5 μg of synthetic disparlure, impregnated on filter paper, from the holding chamber over 24 hr at an airflow of 2 liters/min. The resins were pretreated by washing with pentane and then methanol followed by baking at 150°C for 24 hr under a constant nitrogen stream.

Cane and Jonsson (1982) reported the development of a portable collection system which may be used in the laboratory or in the field (Fig. 5). The entrainment air flow is generated by passing a compressed gas through a suction valve (AGA suction ejector No. #323190101 AGA AB Stockholm, Sweden), controlling the flow to the desired rate by valves at the outlet side of the collection tube. The application of the compressed gas as a high velocity jet across the suction valve's chamber orifice reduces the air pressure in the chamber to less than ambient and generates a vacuum; thus, the suction valve has no moving parts.

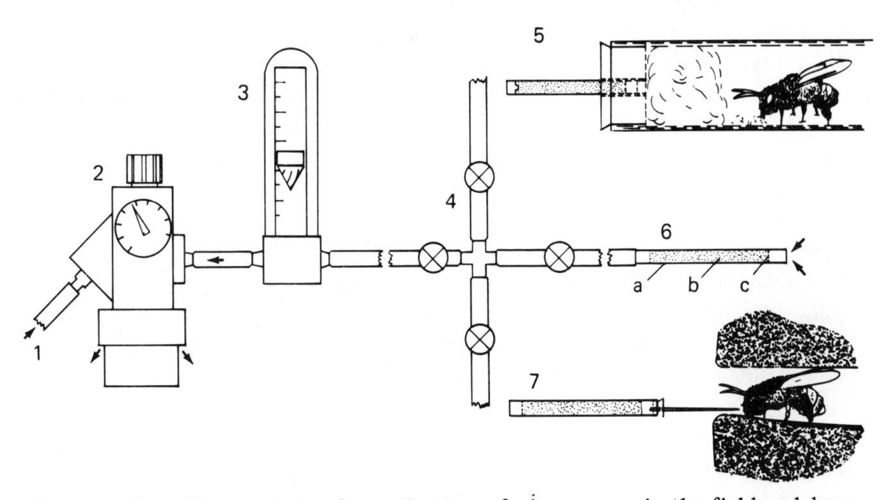

Figure 5. Portable apparatus for collection of pheromone in the field or laboratory. Air or nitrogen from a compressed gas cylinder (1) enters the suction valve (2) to generate a vacuum. Airflow from the sample tubes is monitored by an in-line gas-flow meter (3) and equalized using C clamps or valves (4). The glass sample tube (6a) is packed with Tenax (6b) and plugged with glass filter paper discs (6c). Samples were collected from bees either digging (5) or exhibiting defensive behaviors (7). Cane and Jonsson (1982).

A 5-liter nitrogen tank weighing only 8 kg and operated at a pressure of 250-400 mm Hg will provide a vacuum source for 8 hr of continuous collection. The adsorption tubes (5 cm × 4 mm inside diameter) were packed with Tenax. By attaching the inlet end of the adsorption tube to glass tubes containing shredded glass wool to simulate nest materials, Cane and Tengö (1981) isolated pheromones responsible for digging behavior in male *Colletes cunicularius.* For the collection of alarm and defense volatiles the collection tube is fitted with a 5-μl micropipet which is used to probe and disturb individual insects.

A most complete study of the use of Porapak Q to collect airborne pheromone is that of Bjostad et al. (1980b). Many workers have experienced difficulties in eliminating extraneous materials from this adsorbent. Bjostad et al. (1980b) extracted 100-mg samples of unconditioned Porapak with carbon disulfide and found it to contain microgram amounts of several contaminants which they showed could only be completely removed by baking the resin at 200°C for 24 hr in a stream of nitrogen, followed by a 24-hr Soxhlet extraction with pentane. Adsorbent tubes prepared in this manner could be stored for several days before use. Contaminants acquired from ambient air could be removed by elution with a small volume of pentane. Glass adsorption tubes (100 mm × 7 mm diameter) containing 100 mg of preconditioned Porapak Q were positioned as shown in Fig. 6 to collect pheromone from individual *Trichoplusia ni* females.

Glass cylinders (85 × 50 mm) with screen-centered plastic lids were used as holding chambers. A cotton dental wick, soaked in a 10% sucrose solution, covered with a glass vial was set up as shown on top of the screen portion of the lid. During collection, undertaken in the normal calling period of this insect, the Porapak Q tube was positioned about 1 mm from the everted gland, and the vacuum was activated. Each collection tube was connected to a manifold which in turn was connected through a flow meter to the vacuum source (house vacuum) producing a flow of approximately 2 liters/min.

Because females could select any location in the holding chamber from which to call, a device made from steel discs and button magnets was used to allow positioning of the collection tube near to the insect's abdominal tip. At the conclusion of each particular collection period the resin was desorbed with 5 ml of carbon disulfide, a known amount of internal standard, (*E*)-9-tetradecen-1-yl acetate, added and, after concentration under nitrogen to 4 μl, the solution was analyzed by gas chromatography.

Recovery efficiency was calculated for different quantities of (*Z*)-7-dodecen-1-yl acetate added to collection tubes aerated for different periods of time; 100-ng samples aerated for 5 min had a recovery efficiency of 89 ± 1%, while 85 ± 1% of 500-ng samples were recovered after 4 hr. The loss of 11% of the pheromone was not discussed by the authors and could possibly be a result of the method of introduction into the collection tube or the concentration and introduction of the small volume into the syringe prior to analysis. The loss of (*E*)-9-tetradecen-1-yl acetate, the internal standard, during concentration and transfer was not reported.

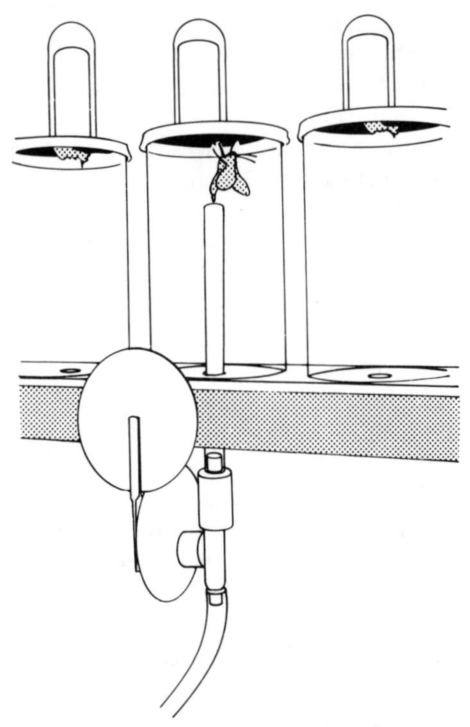

Figure 6. Female *T. ni* with pheromone gland everted is positioned near the orifice of a Porapak Q tube in the apparatus used by Bjostad et al. (1980b). The apparatus is attached to the house vacuum through the tygon tubing at the bottom.

(*d*) *Miscellaneous Adsorbents.* Discussed in this section are charcoal and glass. Several other adsorbents which have previously been used include cellulose powder, lanolin, and a tallow-lard mixture. The first was used to trap volatile-laden air from *Diparopsis castanea* (Tunstall, 1965), while cheesecloth coated with lanolin was used to collect pheromone from *Plodia interpunctella.* The lanolin impregnated cheesecloth was then Soxhlet extracted for 2 hr with ether, the ethereal solution taken to dryness at 40°C under vacuum and the residue distilled for 2 hr at 130°C (0.1 mm Hg). The biologically active volatile fraction was collected in a dry ice trap (Brady and Smithwick, 1968).

In the study which identified undecanal as the sex pheromone of *Galleria mellonella,* Röller et al. (1968) adsorbed the volatiles from air exiting from a container of male moths on glass plates coated with a 2:1 mixture of retropenitaneal tallow and mesenteric lard stabilized with hydroquinone. The active material was isolated by molecular distillation of the fat followed by preparative gas chromatography.

As early as 1964, Keller et al. used charcoal to adsorb pheromone from male boll weevils, the charcoal being subsequently desorbed with chloroform.

Studies of trace contamination by organics in water have been well documented by Grob and his colleagues in Switzerland, and several of the techniques evolved by this group are directly applicable to pheromone collection (Grob, 1973; Grob and Zürcher, 1976). Several researchers have used the charcoal entrainment filters described by Grob and Zürcher (1976) to entrain pheromone from individual females in structural elucidation studies of lepidopteran sex pheromones (Nesbitt et al., 1979a, b; Löfstedt et al., 1982; Tumlinson et al., 1982).

In a recent review, Tumlinson et al. (1982) described the apparatus (Fig. 7) and method which they used for collecting airborne volatiles. Charcoal (3–5 mg) is sealed between two 325-mesh stainless-steel frits in a 6-mm outside diameter X 3.7-mm inside diameter Pyrex tube approximately 25 mm long and tapered as shown in the figure. This entrainment filter is placed at the exit of the insect-holding chamber and purified air is drawn through the holding chamber and through the filter at a rate of 2.5 liters/min. Among the many factors which affect the collection efficiency is the size of the holding chamber; hence it should be constructed as small as possible to accommodate the target insect.

On completion of aeration, the filter is removed and the charcoal desorbed with six aliquots of 15–20 μl of distilled dichloromethane. The combined eluate is concentrated to 5–10 μl by gentle warming; *iso*-octane or another solvent of choice is added prior to capillary gas chromatographic analysis. Recoveries vary with conditions and the apparatus should be calibrated with standards under the conditions to be used with the live insects. Evaluation of the method using *Z* aldehydes, acetates and alcohols with 14 and 16 carbon atoms in the chain gave good recoveries (57–86%) when the standards (0.5 μg) were evaporated from stainless-steel planchets.

Evolving from the studies (Baker et al., 1980) of emission rates by glass adsorption in still air Baker et al. (1981) developed a device for the collection of pheromones volatilized from the forcibly extruded glands of individual females by gas flow; the device is shown in Fig. 8. Purified nitrogen is introduced at the desired rate, flows over the extruded gland or synthetic pheromone source, and

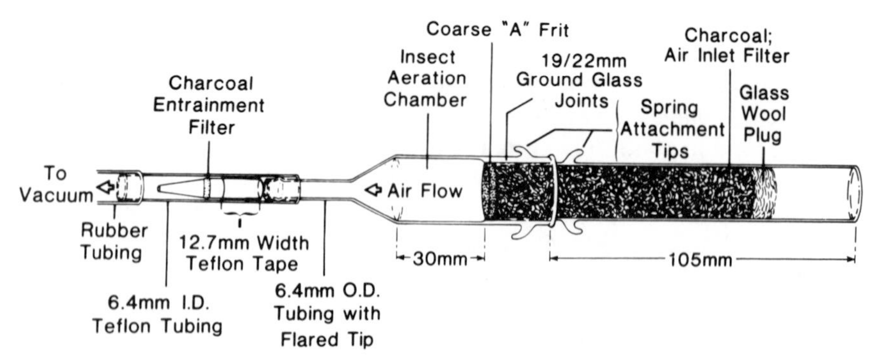

Figure 7. Vapor collection apparatus of Tumlinson et al. (1982) used to collect pheromone from live insects. Reprinted with permission.

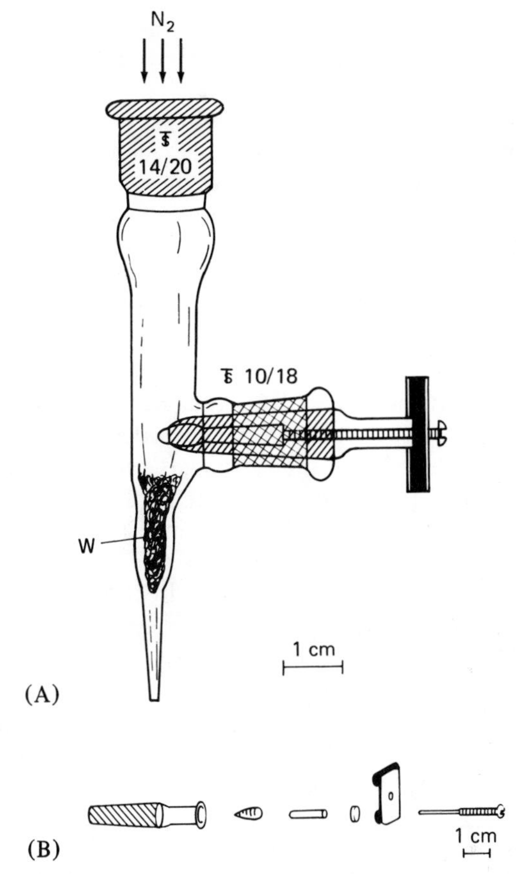

Figure 8. (A) Pheromone emitted from the forcibly extruded gland of a female is collected from the prefiltered nitrogen stream on glass wool (w). (B) Expanded view of the extrusion tube components shows (from left to right) the extrusion tube, ligated gland, Teflon plunger, GLC-conditioned silicon rubber gasket, steel clip and screw for exerting pressure (Baker et al., 1981).

the volatiles are adsorbed on the glass wool. Desorption with several small aliquots of carbon disulfide is followed by addition of the internal standard, concentration and gas chromatographic analysis. At 21–23°C and with a flow rate of 30 ml/min the authors showed that for seven compounds (decyl acetate, decyl alcohol, dodecyl acetate, (*Z*)-9-tetradecenal, tetradecyl acetate, (*Z*)-11-hexadecenal, and hexadecyl acetate) at the 90-ng level recovery efficiency was 100% and that the device could be operated for up to 2 hr without appreciable breakthrough with the exception of decyl acetate.

As the chain length and hence the molecular weight increased, with the resultant decrease in volatility, less material was recovered from the glass wool and more from the synthetic source. For a 2-hr collection the recoveries from the glass wool ranged from 89 ± 10% for dodecyl acetate to 26 ± 4% for hexadecyl acetate. The concomitant recoveries from the sources ranged from 0% for dodecyl acetate to 72 ± 6% for hexadecyl acetate.

Using a 1-hr collection time and substrate loadings of 90 ng and 9 μg, and two additional compounds, hexadecanol and (*Z*)-11-hexadecen-1-ol, total recoveries were again excellent; 73–96% at the 90-ng level and 69–105% with loadings of 9 μg. The recovery range from the glass wool at the 90-ng level was spread from 0% for hexadecanol and (*Z*)-11-hexadcen-1-ol to 85 ± 44% for decyl acetate and 94 ± 13% for decanol. At the heavier loading no hexadecyl alcohol, acetate or (*Z*)-11-hexadecen-1-ol were recovered from the glass wool, and the best recovery was for decanol which was only 16 ± 3%.

To collect the pheromone from individual *Trichoplusia ni,* females 3–4 days old were cooled for 5–10 min at −20°C. After removal of some abdominal scales, the females were ligated at the thoracic end of the abdomen and the body was severed immediately anterior to the ligature. The abdomen was then placed in the device and the gland forcibly extruded with the Teflon plunger. A modification (Pope et al., 1982) used 1 μl of tetrasodium EDTA solution (500 μg) injected to the vicinity of the gland to prevent retraction. At the end of the 5-min collection periods, the glass wool was desorbed with carbon disulfide, 50 ng of internal standard was added, and the solution was analyzed.

By this method the emission rate from *T. ni* females was calculated to be 2.40 ± 0.65 ng/min for the (*Z*)-7-dodecen-1-yl acetate and 0.25 ± 0.07 ng/min for dodecyl acetate (cf. Bjostad et al., 1980a,b). Pope et al. (1982) have used the same method to collect and quantify the sex pheromone components of *Heliothis virescens* and study the effect of photoperiod on their ratio and emission rates.

Fig. 9 illustrates the pheromone collection apparatus used by Charlton and Cardé (1982) for the collection of disparlure for release rate and diel periodicity studies. Incorporating several features of the "miniflow" device reported by Weatherston et al. (1981) (see below) Charlton and Cardé carried out the most complete study of pheromone collection from live whole individual insects.

The collection device consists of two modified glass joints forming the insect holding chamber which incorporates a 2 × 3 cm screen hung such that it does not touch the sides of the chamber. The lower half of the holding chamber is inserted into the pheromone trap which has a volume of approximately 3 ml and contains 1-mm-diameter glass beads. The airflow used was approximately 3 liters/min, and at that rate it was shown by titanium tetrachloride "smoke" that pheromone emission from a moth clinging to the screen would be unlikely to come in contact with or be adsorbed by her body.

Prior to using the device for collection from individual female gypsy moths, its performance in regard to breakthrough time and collection efficiency was evaluated by emitting 1 μg of disparlure from a filter paper disc attached to the bottom of the screen to which an acetone-washed female moth was also attached. No breakthrough was observed over a 3-hr aeration period as evi-

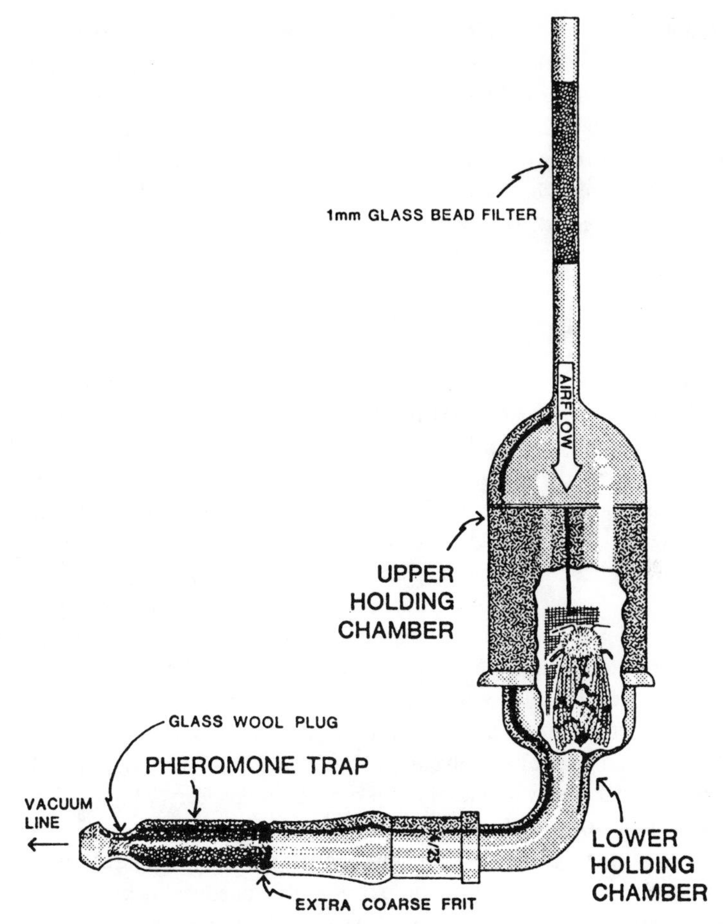

Figure 9. Apparatus used by Charlton and Cardé (1982) to collect airborne pheromone from *L. dispar* females on glass beads.

denced by the absence of pheromone desorbed from a second trap placed in series with the first.

Of the 92 ± 5% (N=5) recovered from the synthetic disparlure released from the 1-μg sample, 74 ± 4% was obtained from the pheromone trap, an additional 18 ± 3% from the lower portion of the holding chamber, 3% from the screen and upper portion of the chamber, leaving 5% unaccounted for. To collect from live insects, the female was placed in the device about 45 min after eclosion and once the insect became acquiescent and before calling started, the airflow was initiated.

Analysis of the adsorbed pheromone was by gas chromatography following extensive sample preparation. The pheromone was desorbed from the glass beads and lower holding chamber with three 2-ml aliquots of hexane; 50 ng of an internal standard was added to the combined hexane rinses, the solution was concentrated in a gas flow to about 6 μl and injected onto a packed column. The column effluent corresponding to disparlure and the internal standard were collected in chilled capillary tubes. The compounds were eluted from the capillaries with five 3-μl portions of carbon disulfide; the combined eluate was reduced to about 1 μl and analyzed by capillary gas chromatography. Table 2 summarizes the use of moving airflow methods during the last 5 years in the collection of pheromone from insects.

2.3. Collection from Artificial Sources

The methodology for the collection and analysis of behavior-modifying chemicals from artificial sources has been the subject of several recent reviews (Weatherston et al., 1981, 1982; Bierl-Leonhardt, 1982; Taylor, 1982). The current interest in the collection and quantification of pheromones from artificial sources is a consequence of their increased use in monitoring and control strategies against pest insects. The use of several controlled release devices for this purpose has been reviewed by Campion et al. (1978).

The only two ways to obtain the rate of release of a volatile compound from a controlled release device are to measure, as a function of time, the amount of active material remaining in the device or the amount of material being released by the device. Ideally both methods should be used wherever possible.

The first method, residue analysis, is applicable to both laboratory and field studies and usually involves solvent extraction of the source device followed by chromatography. The variables in the extraction process which affect the recovery are choice of solvent, extraction time, and temperature.

Because of the presumed simplicity and familiarity with the method, few authors have reported details of the extraction procedure or give details regarding the efficiency of the extraction. Butler and McDonough (1979, 1981) and Flint et al. (1978), e.g., simply report that extraction was carried out in 50 ml of hexane:dichloromethane (1:1) over 1 hr with stirring. Maitlen et al. (1976), working with mass trapping and population survey baits for the codling moth and testing rubber bands, septa, and polyethylene vial caps as release substrates, used the following method: The substrate was placed in a 50-ml conical flask and shaken on a mechanical shaker for 15 min with 25 ml of a hexane:dichloromethane (1:1) mixture. The extract was then chromatographed through a column of a 15 g alumina (Baker 0536) washed with 50 ml hexane before use. For transfer to the column 20 ml of hexane was used followed, at the point of dryness, with a further 75 ml hexane. The synthetic pheromone was eluted from the column with 125 ml of hexane:dichloromethane (85:15) which was evaporated to dryness at 30°C on a rotary evaporator and then analyzed by gas chromatography.

Working with both laminated plastic and cigarette filter dispensers, Shaver et al. (1981), in the case of the laminate, took a 1.6-cm subsample and soaked it over 1 hr with 40 ml of dichloromethane, decanted the solution, and added another 30 ml of solvent for 30 min. This solution was added to the first and the residue was washed with two 1-ml portions of solvent. The combined extracts and rinses were adjusted to 10 ml and analyzed. The cigarette filters (8 × 30 mm) had been aged in glass shell vials for extraction. The filter was removed from the shell vial and placed in a 20-ml screw cap vial. The shell vial was rinsed with approximately 5 ml dichloromethane. After shaking for 1 hr, the solution was decanted and the filter and vial were washed with 3 × 1 ml portions of solvent. The extract and washes were combined and the volume was adjusted to 10 ml prior to analysis. Several recent reports, namely those of Hall et al. (1982) and Leonhardt and Moreno (1982), also report solvent extraction methods.

The methodology for the trapping of effluent from release devices closely parallels that described above in the section dealing with collection from live insects, i.e., the effluent may be trapped in cold traps (Browne et al., 1974) or solvent traps (Beroza et al., 1975; Bierl and DeVilbiss, 1975; Bierl-Leonhardt et al., 1979) or on adsorbents (Vick et al., 1978; Baker et al., 1980; Weatherston et al., 1981). Two examples of the entrainment methods currently in use are contained in the reports of Cross (1980a,b) and Golub et al. (1983).

The apparatus used by Cross is shown in Fig. 10 and consisted of two modified glass joints held together with a pinch clamp. Air filtered through charcoal at a flow rate of 1 liter/min was pulled through the dispenser holding chamber and then through a 7.5-cm long tube slipped inside the bottom exit of the apparatus and held in place with a Swagelok fitting. Each collector contained 150 mg Tenax. After aeration the trapped material was heat desorbed onto a gas chromatography column.

The miniairflow device developed by Weatherston et al. (1981) is illustrated in Fig. 11. The one-piece apparatus consists of a female $\overline{\$}$ 14/20 joint separated from the adsorbent holder by a 5-mm-thick coarse frit. At the exit end of the device is a Luer joint. The apparatus was silanized with hexamethyldisilazane and the collecting chamber filled with glass beads (ca. 8 g: 1–1.05 mm diameter) held in place with silanized glass wool. The artificial source was placed in the holding chamber, and the prefilter (molecular sieves and Amberlite X-AD resin) was attached to the front end of the miniairflow. The Luer joint was attached to a vacuum source through a flow meter, and air was pulled through the apparatus at a rate of 1 liter/min. Every 24 hr the apparatus was detached from the vacuum and disassembled. The source was removed, the apparatus was held in a vertical position and, with a syringe attached to the Luer joint, the entrained pheromone was solvent desorbed. Previously Weatherston et al. (1981) had shown that solvent desorption of the glass beads was quantitative.

This section would be incomplete without reference to the collection of pheromone vapor from air permeation field trials. The methodology used is given below although the present authors doubt the usefulness of the data obtained for the design of control formulations. Such doubts are based on the lack of knowledge as to the mechanisms of pheromone communication between the

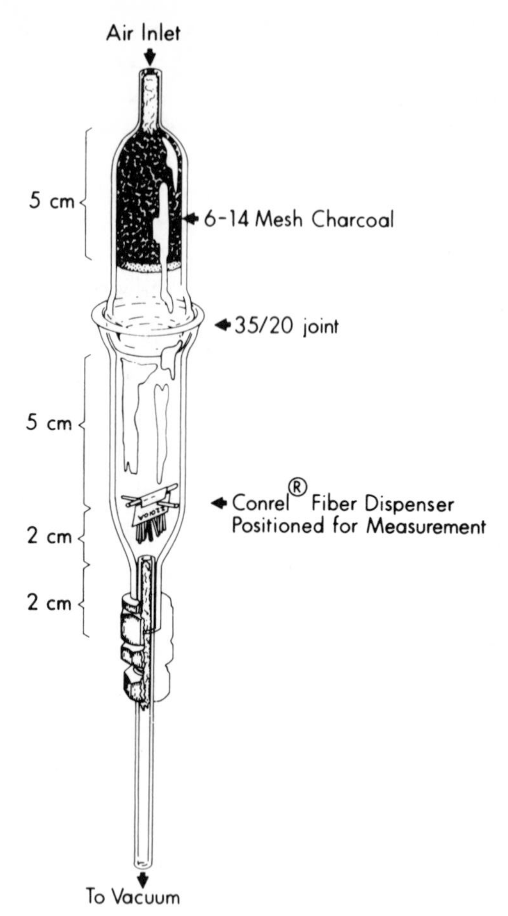

Figure 10. Apparatus showing collection of airborne volatiles from hollow fiber dispensers used by Cross (1980a).

sexes which are, in general terms, restricted to (i) sensory adaptation, (ii) competition, and (iii) trail camouflage. Hence time averaging of the pheromone concentration by using high-volume samplers suspended in or above the crop and drawing air through an adsorbent for several hours [e.g., Caro et al. (1979) reported an airflow of 1.2–1.5 m^3/hr for sampling periods of up to 4 hr] may give data which could be related to habituation concentrations. But sampling is grossly inadequate in relating the amount of pheromone required if the strategy is either trail camouflage or competition because the effects of plume size, shape, direction, and speed have been negated.

Another cause for concern in the interpretation of data obtained from such experiments is that due to the lack of sensitivity of the analytical methods, the application rates of the applied formulations are in excess of 100 times that used

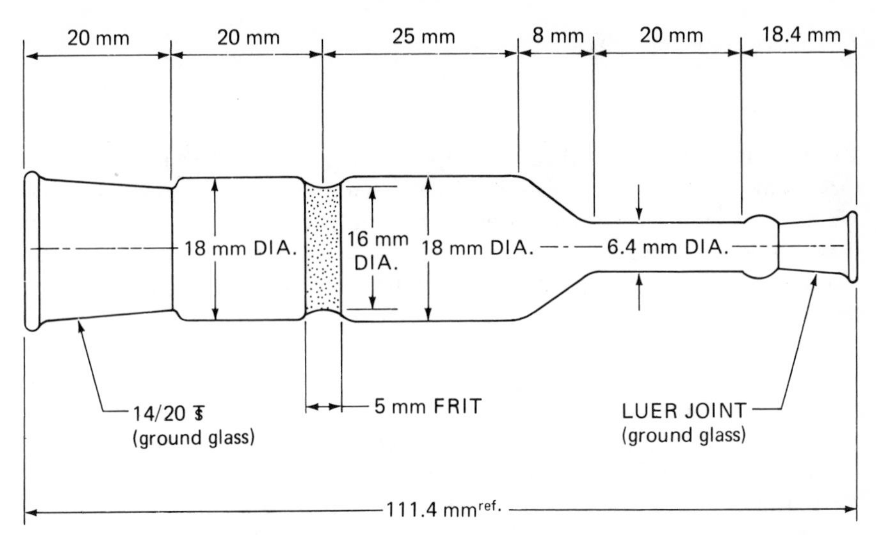

Figure 11. Mini-airflow apparatus used by Weatherston et al. (1981) to collect pheromone from artificial sources.

in control programs. The data obtained are then extrapolated for the lower rates; however, the effects of the biological sinks (saturation levels and foliage adsorption and rerelease) are usually ignored.

The methodology for collecting pheromone vapor under field conditions has been developed primarily by the Beltsville laboratory of the USDA (Plimmer et al., 1978; Caro et al., 1978-1981; Wiesner et al., 1980; Taylor, 1982; and Wiesner and Silk, 1982). Typically the method is as described by Caro et al. (1978) for disparlure. Air is pumped through an adsorbent bed of about 8 g of molecular sieves at a rate of 2-3 m^3/hr. Molecular sieves are the preferred adsorbent because they are inexpensive, easily desorbed and, since they are pelletized, they allow high airflow while minimizing the pressure drop through the system. Taylor (1982) reported the comparison of three commercial formulations of disparlure at four different heights on four hectare plots by this method. Prior to analysis the pheromone is solvent desorbed from the molecular sieves.

III. Methods of Analysis

In addition to Chapter 11 on analytical methodology in this volume (Heath and Tumlinson, 1984), three recent reviews are also relevant to the area of collecting and extracting pheromones from live insects and artificial sources, namely, those by Weatherston et al. (1981), Tumlinson et al. (1982), and Bierl-Leonhardt (1982). The analysis methods included here are restricted to those pertinent to the collection and extraction methodology reported above.

3.1. Gas Chromatography

A. Direct or Solid Sample Injection

This was first used by Weatherston and Maclean (1974) for the analysis of volatiles from the pheromone glands of the eastern spruce budworm. Using the MS-41, a solvent free gas chromatographic sampler developed by Perkin-Elmer, these authors excised the abdominal tips from two virgin female moths and placed them directly into an aluminum capsule of the sampler system which was then cold-weld sealed and introduced, by means of a probe, into the heated injection block of the gas chromatograph. Once the desired injection temperature was attained, the probe was fully inserted into the instrument causing a hollow needle on the front end of the column to pierce the sealed capsule allowing the accumulated volatiles to be flushed onto the column with carrier gas.

An alternate technique was to grind up the abdominal tips from six insects in 3 ml of hexane; after removal of the tissue the solvent was carefully evaporated to a small volume with a stream of nitrogen and introduced into the capsule dropwise from a syringe. The material in the capsule was taken to dryness with nitrogen, the capsule sealed and introduced into the chromatograph as described above. This method has also been reported by Descoins and Gallois (1979), who summarized its use in structural elucidation studies of the pheromones of 19 moth species.

One drawback of the MS-41 system is that it can only be used in conjunction with packed columns; however, Buser and Widmer (1979) developed a capsule introduction system based on the MS-41 but incorporating significant modification to allow it to be used with capillary columns. They reduced the capsule volume to 20 μl and, in order to achieve the high efficiencies obtainable with capillary columns, modified the introduction system such that the hollow piercing needle had a total volume of less than 1 μl and the maximum dead volume at the column–needle interface was less than 0.15 μl. Using hydrocarbons (C_8–C_{19}), an alcohol and an ester, Buser and Widmer claim that for qualitative analysis their capsule-insertion system is comparable to a sophisticated split system. Previously the insertion of insect material onto capillary columns via a precolumn system had been described by Stallberg-Stenhagen (1972) and Bergström (1973).

3.2. Electron-Capture Chromatography

This has been used for both the quantitative analysis of insect released material and that from artificial sources. After extracting the pheromone from phycitid moth abdominal tips, the trichloroacetate was either prepared directly or after hydrolysis using trichloroacetyl chloride at 25°C over 18 hr in order to assay the relative amounts of (*Z,E*)-9,12-tetradecadienyl acetate and the corresponding alcohol present. The resulting trichloroacetates were analyzed on packed columns with a ^{63}Ni detector, and the peak heights were transformed into amount of pheromone based on a standard curve. The authors (Kuwahara and Casida, 1973) reported that the amounts of pheromones detected in the four species analyzed are greater than alternate methods using large numbers of insects.

The determination of the release rate of (*Z*)-3-decenoic acid by *Anthrenus flavipes* was achieved by solvent desorption of the effluent from a single beetle from Tenax with 1 ml of hexane. To this solution was added 1 ml of 0.1 *M* tetrabutylammonium hydroxide, 10 μl pentafluorobenzyl bromide (6.8×10^{-2} *M*), and 9-bromophenanthrene as internal standard. The data obtained using a ^{63}Ni detector were quantified from a standard response curve of the pentafluorobenzyl ester of (*Z*)-3-decenoic acid (Ma et al., 1980).

Electron-capture chromatography has also been used in the analysis of atmospheric concentrations of synthetic pheromones and their analogs. Disparlure was desorbed from molecular sieves (8 g) by shaking for 1 min with 10 ml of a 1:1 mixture of hexane and ether (Caro et al., 1978). After standing for 1 hr at room temperature, an aliquot was carefully evaporated to 1 ml in a Kuderna-Danish apparatus using a Snyder column on a steam bath. The volume was adjusted to 0.2 ml in a gas stream, 0.5 ml triphenylphosphine dibromide (0.25 *M*) was added, and the mixture was heated at 37–40°C for 1 hr. On cooling the volume was adjusted to 0.5 ml with dichloromethane. The resultant mixture of diastereoisomeric 7,8-dibromo-2-methyloctadecanes was eluted from a pretreated Florisil column with 50 ml of hexane. Concentration was followed by analysis using a ^{63}Ni detector with a reported sensitivity as low as 0.2 ng/ml for disparlure. This method was also used by Caro et al. (1981) for the comparison of three commercial formulations of disparlure.

Caro et al. (1979, 1980) have also reported on the analysis of aerially disseminated (*Z*)-9-tetradecen-1-yl formate as its dibromide by electron-capture gas chromatography. The method as described by Caro et al. (1979) is as follows: The hexane-desorbed formate was concentrated in a Kuderna-Danish apparatus to 10 ml and briefly extracted with two 80-ml volumes of water. The hexane solution was next concentrated to 1 ml with purified nitrogen, and using a Vortex mixer, brominated with 0.5 ml bromine solution in carbon disulfide. After washing with water the organic solution of the dibromide was concentrated to 0.5 ml.

The dibromide was first subjected to clean-up on a microcolumn of 0.4 g silica gel (Woelm activity IV) topped with approximately 0.1 g sodium sulfate. The benzene eluate, after concentration to 0.2 ml, was transferred to a watch glass, allowed to evaporate to dryness, and then exposed to a germicidal UV lamp (254-nm radiation, 17 μW/cm^2 intensity at 1 m) for 30 min at a distance of 10 cm. The residue was then taken up in benzene and aliquots were analyzed using a ^{63}Ni electron-capture detector. The authors noted a consistent 40% loss of the dibromide during the UV clean up, however, after correction the recovery reported is about 89% with a quantification limit of less than 3 ng/m^3. Similar analytical methodology for aldehyde pheromones by derivatization to the *o*-(pentafluorobenzyl)oxime has been reported by Wiesner et al. (1980).

3.3. Sample Concentration Chromatography

The sole published application of this technique in pheromone research was reported by Ma and Schnee (1983) for the quantification of disparlure released by individual female gypsy moths. The pheromone, after elution from the Porapak or Tenax with 500 μl of methanol, was concentrated using a UNACON concentrator to enrich the solute. Incorporating a two-stage adsorbent trapping system, the removal of the solvent may be monitored by a shunt from the first trap to the flame ionization detector. Desorption from both traps is by heat and gas purge, the solute from the second trap going directly on to the capillary column. The amount of disparlure released by female gypsy moths, as determined by this method, varied between 2.4 and 4.0 μg/day in relation to a temperature range of 15–30°C.

3.4. High-Performance Liquid Chromatography

A fast and efficient method of measuring the release rate of 3-methyl-2-cyclohexen-1-one, the aggregation pheromone of the Douglas fir beetle, is the use of high-pressure liquid chromatography with a UV detector (Look, 1976). The pheromone was desorbed from Porapak into a 10-ml volumetric flask with trimethylpentane. Aliquots of the eluate were analyzed on a 120 cm × 0.3 cm column of Corasil 1 at ambient temperature with a mobile phase of dioxane: trimethylpentane (3:7) at a flow rate of 2 ml/min. The peak height of the sample was compared to standard curves. The usual sample size was 10 μl; however, in larger samples (50–100 μl) dilution with dioxane was undertaken to cancel the refractive index effects in the UV detector. Release rates obtained were of the order of 5–400 μg/hr/g depending on the formulation.

3.5. Bioluminescence Assay

The use of aldehyde-dependent bacterial luciferases to quantify nanogram amounts of aldehydic pheromones was first reported by Meighen et al. (1981). Briefly, the assay consists of rapidly mixing an aqueous solution of the aldehyde pheromone with a luciferase and reduced flavin nucleotide. The reaction is carried out in a light-tight cylinder so that the resultant emission of light can be detected with a photomultiplier tube and graphically displayed. Although the maximum response is obtained with tetradecanal, the hypothesized *in vivo* substrate for the enzyme, the chain length specificity of the enzymes brackets that of almost all the aldehydic pheromones.

The bioluminescence assay can be used to quantify pheromones from insects, lures, control formulations, and the atmosphere (Meighen et al., 1981, 1982; Grant et al., 1982; Szittner et al., 1982). Two types of assays, the standard assay and the dithionite assay, have been reported (Meighen et al., 1982) and are detailed below. In the standard assay, the flavin nucleotide is injected into a solution of aldehyde and luciferase (purified from either *Beneckea harveyi* or *Photobacterium phosphoreum* NCB844). One milliliter of 50 μM reduced nucleotide is injected into 1 ml of 0.05 M phosphate:0.001 M mercaptoethanol containing luciferase (5 μg) and the aqueous aldehyde solution. The light emission at 490 nm was detected by a photomultiplier tube and graphically recorded in light units (1 LU = 6×10^9 quanta/sec) with a pen recorder.

In the dithionite assay the aqueous aldehyde is injected into a solution of the luciferase and the nucleotide maintained in the reduced state with excess dithionite. One milliliter of an aqueous aldehyde solution is injected into 1.0 ml of 0.001 M ammonium hydroxide, 0.05 M mercaptoethanol, 0.05 M phosphate (pH 7) containing 10 μl luciferase (approximately 5 μg), and 50 μM flavin nucleotide reduced prior to use with approximately 0.5 mg solid sodium dithionite.

The dithionite assay produces a much lower background than the standard assay and hence is more sensitive enabling detection of lower aldehyde concentrations. The response is linear over the range 20 pg–600 ng, such amounts of pheromones having often been obtained from insects. With regard to artificial sources which often release much higher amounts of pheromones, the collection time may be adjusted or the aldehyde solution diluted to put the concentration into the above range. Szittner et al. (1982) claim that the bioluminescence assay can also be applied to alcohol and acetate pheromones, since, at low concentrations, these compounds can be enzymatically converted into aldehydes with high efficiency.

3.6. Miscellaneous Methods

The three analytical methods which fall into this category are weight loss, meniscus regression, and scintillation counting; they have recently been reviewed by Weatherston et al. (1981, 1982).

A. Weight Loss

This method has been widely used in the quantification of release rates from lures and formulations (Gaston et al., 1971; Fitzgerald et al., 1978), the most extensive report being that of Rothschild (1979).

B. Meniscus Regression

The meniscus regresson method is restricted to use with transparent or translucent capillaries and hollow fibers. The current methodology is to follow the regression of the pheromone meniscus with time by using a Wilder Varibeam optical comparitor fitted with IKL digital positioners connected to IKL Microcode digital readout systems. Release rates can be measured to ±0.015 μg translating the difference in length, internal diameter of the capillary, and the density of the liquid charge into weight of pheromone emitted.

C. Liquid Scintillation Counting

This method was first used by Kuhr et al. (1972), who determined the residual amount of radioactive pheromone analogs remaining in polyethylene vial caps which had been field aged in an orchard. Weatherston et al. (1981, 1982) and Golub et al. (1983), using gossyplure labeled with ^{14}C in the acetate moiety, have been able to demonstrate mass balance for hollow fibers and rubber septa release devices.

IV. Discussion

It is not surprising that the methodology for collecting pheromones from both natural and artificial sources is so similar. Indeed, it would be surprising if there were any significant differences. Certainly, in the area of solvent extraction it is reasonable that the same solvents are used since the materials being extracted are the same for a given insect and/or lure. And in the area of effluvial collection, it is also reasonable that the techniques and apparatus should be alike. This similarity can be seen by comparing Fig. 11, used by Weatherston et al. (1981, 1982) to collect gossyplure from hollow fibers and rubber septa (Golub et al., 1983), to Fig. 7 used by Tumlinson et al. (1982) to collect pheromone from *Heliothis virescens.* In fact, Tumlinson et al. (1982) did indeed collect from artificial sources (stainless-steel planchets) in their calibration studies.

Only in the area of actual analysis do the two methodologies exhibit any substantial differences and these are directly related to the intrinsic differences in the two sources. It is not reasonable to weigh an insect to determine pheromone production, but it is reasonable to consider weight loss as a means of determining pheromone emission from an artificial source. The same is true of liquid scintillation counting since synthetic pheromones can readily be labeled with a radioactive element while natural pheromones certainly are not so labeled. However, it is entirely within the realm of possibility that laboratory-reared insects could be fed a diet containing radiolabeled components which would eventually be used by such insects in the biosynthesis of pheromones. Such a procedure could yield information not only about the identity, amount, and release rate of the pheromone but also, where biosynthetic information is lacking, the method could be used to obtain such information.

Even as the technology progresses, it is likely that the similarity in methodology will persist. A comparison of the contents of Tables 1 and 2 easily indicates that far more effort has been expended in the area of extraction of pheromones from natural sources than in the area of effluvial collection. The problem has been, both with natural and artificial sources, in the development of analytical procedures sufficiently sensitive to detect the minute quantities being emitted from these sources. However, the necessary technology has been advancing very rapidly. In only 4 years it has become possible to reduce the number of insects needed to provide an adequate sample from 2000 beetles (Cross et al., 1976) to a single beetle (Ma et al., 1980). The present-day application of selective analytical techniques, such as electron-capture chromatography or bioluminescence assay procedures, have made the detection of subnanogram quantities much easier today than it was in the past, although these procedures are still far from routine.

The design of collection equipment is also a critical factor. Several researchers, e.g., Baker et al. (1981), Weatherston et al. (1981), Charlton and Cardé (1982), and Wiesner and Silk (1982), have already noted that the size and design of the apparatus, the air speed, and the collection conditions may be critical in determining the amount of pheromone emitted. Indeed, Tumlinson et al. (1982), as already stated, found that their apparatus needed to be calibrated for the specific conditions to be used.

The need for devices such as those used by Cane and Jonsson (1982) for collecting pheromone from *Colletes cunicularius* will grow as the need to obtain more accurate information about the amount and identity of pheromones present in nature increases. This information is of vital importance in gaining a better understanding of insect behavior and is also important in pheromone-based insect pest management strategies since, whatever the mechanism responsible for control, whether it is false trail following or habituation, it is the pheromone which is released not that which is contained in the producing gland or artificial source which ultimately affects the behavior of other insects. Since the development of pheromone-based pest control strategies is already becoming important in agriculture, the associated developments in collection techniques can be expected to continue in the future.

V. References

Abdel-Kader MM, Barak AV (1979) Evidence for a sex pheromone in the hide beetle, *Dermestes maculatus* (DeGeer) (Coleoptera:Dermestidae). J Chem Ecol **5**:805–813.

Anderson RJ, Gieselmann MJ, Chinn HR, Adams KG, Henrick CA, Rice RE, Roelofs WL (1981) Synthesis and identification of a third component of the San Jose scale sex pheromone. J Chem Ecol **7**:695–706.

Baker TC, Cardé RT, Miller JR (1980) Oriental fruit moth pheromone emission rates measured after collection by glass-surface adsorption. J Chem Ecol **6**:749–758.

Baker TC, Gaston LK, Pope MM, Kuenen LPS, Vetter RS (1981) A high efficiency collection device for quantifying sex pheromones volatilized from female glands and synthetic sources. J Chem Ecol **7**:961–968.

Bakke A (1978) Aggregation pheromone components of the bark beetle *Ips accuminatus*. Oikos **31**:184–188.

Barak AV, Burkholder WE (1977) Behavior and pheromone studies with *Attagenus elongatulus* Casey (Coleoptera:Dermestidae). J Chem Ecol **3**:219–237.

Bartelt RJ, Jones RJ, Kulman HM (1982) Evidence for a multicomponent sex pheromone in the yellow headed spruce sawfly. J Chem Ecol **8**:83–94.

Bellas TE, Fletcher BS (1979) Identification of the major components in the secretion from the rectal pheromone glands of the Queensland fruit flies *Dacus tryoni* and *Dacus neohumeralis* (Diptera:Tephritidae). J Chem Ecol **5**:795–803.

Berger RS (1972) 2,6-Dichlorophenol, sex pheromone of the lone star tick. Science **177**:704–705.

Bergström G (1973) Studies on natural odoriferous compounds (VI.). Use of a pre-column tube for the quantitative isolation of natural volatile compounds for gas chromatography-mass spectrometry. Chemica Scripta **4**: 135–138.

Beroza M, Bierl BA, James P, DeVilbiss ED (1975) Measuring emission rates of pheromones from their formulations. J Econ Entomol **68**:369–372.

Bestmann HJ, Brosche T, Koschatzky, KH, Michallis K, Platz H, Vostrowsky O, Knauf W (1980) Pheromone XXX Identifizierung eines Pheromonkomplexes aus der Graseule *Scotia exclamationis* (Lepidoptera). Tetrahedron Lett **21**:747–750.

Bestmann HJ, Vostrowsky O, Koschatzky KH, Platz H, Szymanska A (1978) Pheromone XVII. (Z)-11-Hexadecenylacetat, ein Sexuallockstoff des Pheromonsystems der Kohleule *Mamestra brassicae.* Tetrahedron Lett **19**: 605–608.

Bestmann HJ, Vostrowsky O, Platz H, Brosche T, Koschatzky KH, Knauf W (1979) Pheromone XXIII. (Z)-7-Tetradecenylacetat, ein Sexuallockstoff für Männchen von *Amathes c-nigrum* L (Noctuidae,Lepidoptera). Tetrahedron Lett **20**:497–500.

Bierl-Leonhardt BA (1982) Release rates from formulations and quality control methods. In: Insect Suppression Using Controlled Release Pheromone Systems. Kydonieus AF, Beroza M (eds), CRC Press, Boca Raton, Florida.

Bierl BA, DeVilbiss ED (1975) Insect sex attractants in controlled release formulations: Measurements and applications. In: Proceedings of the International Controlled Release Pesticide Symposium, Dayton, Ohio.

Bierl-Leonhardt BA, DeVilbiss ED, Plimmer JR (1979) Rate of release of disparlure from laminated plastic dispensers. J Econ Entomol **72**:319–321.

Bierl-Leonhardt BA, Moreno DS, Schwartz M, Fagerlund JA, Plimmer JR (1981) Isolation, identification and synthesis of the sex pheromone of the citrus mealy bug, *Planococcus citri* (Risso). Tetrahedron Lett **22**:389–392.

Bierl-Leonhardt BA, Moreno DS, Schwarz M, Forster HS, Plimmer JR (1980) Identification of the pheromone of the Comstock mealy bug. Life Sci **27**: 399–402.

Bierl-Leonhardt BA, Moreno DS, Schwarz M, Forster HS, Plimmer JR, DeVilbiss ED (1982) Isolation, identification, synthesis and bioassay of the pheromone of the Comstock mealybug and some analogs. Chem Ecol **8**:689–699.

Biwer G, Descoins C, Gallois M (1979) Etude des constituants volatils présents dans la glande productrice de phéromone de la femelle vierge de *Grapholitha molesta* Busck. (Lepidortère:Tortricidae:Olethreutinae). CR Acad Sci Ser D **288**:413–416.

Bjostad LB, Gaston LK, Noble LL, Moyer JH, Shorey HH (1980a) Dodecyl ace-

tate, a second pheromone component of the cabbage looper moth, *Trichoplusia ni.* J Chem Ecol **6**:727-734.
Bjostad LB, Gaston LK, Shorey HH (1980b) Temporal pattern of sex pheromone release by female *Trichoplusia ni.* J Insect Physiol **26**:493-498.
Bjostad LB, Taschenberg EF, Roelofs WL (1980c) Sex pheromone of the choke cherry leafroller moth *Sparganothis directana.* J Chem Ecol **6**:487-498.
Bjostad LB, Taschenberg EF, Roelofs WL (1980d) Sex pheromone of the Woodbine leaf roller moth, *Sparganothis sp.* J Chem Ecol **6**:797-804.
Blight MM, Wadhams LJ, Wenham MJ (1978) Volatiles associated with unmated *Scolytus scolytus* beetles on English elm: Differential production of multistriatins and 4-methyl-3-heptanol, and their activities in laboratory bioassay. Insect Biochem **8**:135-142.
Borden JH, Handley JR, McLean JA, Silverstein RM, Chang L, Slessor KN, Johnston BD, Schuler HR (1980) Enantiomer-based specificity in pheromone communication by two sympatric *Gnathotrichus* species (Coleoptera: Scolytidae). J Chem Ecol **6**:445-456.
Brady UE, Smithwick EB (1968) Production and release of sex attractant by the female Indian meal moth, *Plodia interpunctella.* Ann Entomol Soc Amer **61**:1260-1265.
Browne L, Birch MC, Wood DL (1974) Novel trapping and delivery system for airborne insect pheromones. J Insect Physiol **20**:183-193.
Browne LE, Wood DL, Bedard WD, Silverstein RM, West JR (1979) Quantitative estimates of the western pine beetle attractive pheromone components, *exo*-brevicomin, frontalin and myrcene in nature. J Chem Ecol **5**:397-414.
Buser HV, Widmer HM (1979) Capillary gas chromatography based on a capsule-insertion technique, general aspects and comparison with split injection systems. JHRC&CC **2**:177-183.
Butler LI, McDonough LM (1979) Insect sex pheromones: Evaporation rates of acetates from natural rubber septa. J Chem Ecol **5**:825-837.
Butler LI, McDonough LM (1981) Insect sex pheromones: Evaporation rates of alcohols and acetates from natural rubber septa. J Chem Ecol **7**:627-633.
Byrne KJ, Gore WE, Pearce GT, Silverstein RM (1975) Porapak-Q collection of airborne organic compounds serving as models for insect pheromones. J Chem Ecol **1**:1-7.
Campion DG, Lester R, Nesbitt BF (1978) Controlled release of pheromones. Pestic Sci **9**:434-440.
Cane JH, Jonsson T (1982) Field method for sampling chemicals released by active insects. J Chem Ecol **8**:15-21.
Cane JH, Tengö JO (1981) Pheromonal cues direct mate-seeking behavior of male *Colletes cunicularius* (Hymenoptera:Colletidae). J Chem Ecol **7**: 427-436.
Cardé AM, Baker TC, Cardé RT (1979) Identification of a four component sex pheromone of the female oriental fruit moth *Grapholitha molesta* (Lepidoptera:Tortricidae). J Chem Ecol **5**:423-427.
Cardé RT, Cardé AM, Hill AS Roelofs WL (1977) Sex pheromone specificity as a reproductive isolating mechanism among the sibling species *Archips argyrospilus* and *A. montuanus* and other sympatric tortricine moths (Lepidoptera:Tortricidae). J Chem Ecol **3**:71-84.
Carlson DA, Langley PA, Huyton P (1978) Sex pheromone of the tsetse fly:

Isolation, identification and synthesis of contact aphrodisiacs. Science **201**: 750–753.

Caro JH, Bierl BA, Freeman HP, Sonnet PE (1978) A method for trapping disparlure from air and its determination by electron capture gas chromatography. J Agric Food Chem **26**:461–463.

Caro JH, Freeman HP, Bierl-Leonhardt BA (1979) Determination of (Z)-9-tetradecen-1-ol formate, a *Heliothis* spp. mating disruptant, in air by electron-capture gas chromatography following photolytic clean up. J Agric Food Chem **27**:1211–1215.

Caro JH, Freeman HP, Brower DL, Bierl-Leonhardt BA (1981) Comparative distribution and persistence of disparlure in woodland air after aerial applications of three controlled-release formulations. J Chem Ecol **7**:867–880.

Caro JH, Glotfelty DE, Freeman HP (1980) (Z)-9-Tetradecen-1-ol formate. Distribution and dissipation in air within a corn crop after emission from a controlled-release formulation. J Chem Ecol **6**:229–239.

Charlton RE, Cardé RT (1982) Rate and diel periodicity of pheromone emission from female gypsy moths, (*Lymantria dispar*) determined with a glass-adsorption collection system. J Insect Physiol **28**:423–430.

Cheng Z-Q, Xiao J-C, Huang X-T, Chen D-L, Li J-Q, He Y-S, Huang S-R, Luo Q-C, Yang C-M, Yang T-H (1981) Sex pheromone components of the China corn borer, *Ostrinia furnacales* Greene (Lepidoptera:Pyralidae), (*E*)- and (*Z*)-12 tetradecenyl acetates. J Chem Ecol **7**:841–851.

Coffelt JA, Sower LL, Vick KW (1978) Quantitative analysis of identified compounds in pheromone gland rinses of *Plodia interpunctella* and *Ephestia cautella* at different times of the day. Environ Entomol **7**:502–505.

Coffelt JA, Vick KW, Sonnet PE, Doolittle RE (1979) Isolation, identification and synthesis of a female sex pheromone of the naval orangeworm, *Amyelois transitella* (Lepidoptera:Pyralidae). J Chem Ecol **5**:955–966.

Cory HT, Daterman GE, Davies Jr, GD, Sower LL, Shepherd RF, Sanders CJ (1982) Chemistry and field evaluation of the sex pheromone of western spruce budworm *Choristoneura occidentalis* Freeman. J. Chem Ecol **8**:339–350.

Cross JH (1980a) A vapor collection and thermal desorption method to measure semiochemical release rates from controlled release formulations. J Chem Ecol **6**:781–789.

Cross JH (1980b) Interpretation of release rate and extraction measurements made on two types of controlled-release formulation dispensers containing (*Z,Z*)-3,13-octadecadien-1-ol acetate. J Chem Ecol **6**:789–795.

Cross JH, Byler RC, Silverstein RM, Greenblatt RE, Gorman JE, Burkholder WE (1977) Sex pheromone components and calling behavior of the female dermestid beetle, *Trogoderma variabile* Ballian (Coleoptera:Dermestidae). J Chem Ecol **3**:115–125.

Cross JH, Byler RC, Cassidy Jr, RF, Silverstein RM, Greenblatt RE, Burkholder WE, Levinson AR, Levinson HZ (1976) Porapak-Q collection of pheromone components and isolation of (*Z*) and (*E*)-14-methyl-8-hexadecenal, sex pheromone components of four species of *Trogoderma* (Coleoptera:Dermestidae). J Chem Ecol **2**:457–468.

Cross JH, Tumlinson JH, Heath RR, Burnett DE (1980) Apparatus and procedure for measuring release rates from formulations of lepidopteran semiochemicals. J Chem Ecol **6**:759–770.

Daley RC, Kislow CJ, Brady UE (1978) Comparison of the levels of sex pheromone extracted from female *Cadra cautella* (Walker) by three methods. J Georgia Entomol Soc **13**:161–163.

Daterman GE, Sower LL, Sartwell C (1982) Challenges in the use of pheromones for managing western forest Lepidoptera. In: Insect Pheromone Technology: Chemistry and Applications, ACS Symposium Series 190. Leonhardt BA, Beroza M (eds), American Chemical Society, Washington, pp 243–254.

Descoins C, Gallois M (1979) Analyse directe par chromatographie en phase gazeuse des constituents volatils présent dans les glandes à phéromones des femelles de lépidoptères. Ann Zool Ecol Anim **11**:521–532.

Descoins C, Priesner E, Gallois M, Arn H, Martin G (1978) Sur la sécrétion phéromonale des femelles vierges de *Mamestra brassicae* L et de *Mamestra oleracea* L (Lépidoptères, Noctuidae:Hadeninae). CR Acad Sci Ser D **286**: 77–80.

Dunkelblum E, Gothilf S, Kehat M (1981) Sex pheromone of the tomato looper, *Plusia chalcites* (ESP). J Chem Ecol **7**:1081–1088.

Faustini DL, Giese WL, Phillips JK, Burkholder WE (1982) Aggregation pheromone of the male granary weevil *Sitophilus granarius* (L). J Chem Ecol **8**: 679–687.

Findlay JA, Macdonald DR (1966) Investigation of the sex attractant of the spruce budworm moth. Chem Can, Sept. 3–4.

Fitzgerald TD, St. Clair AD, Daterman GE, Smith RG (1978) Slow release plastic formulations of the cabbage looper pheromone *cis*-7-dodecenyl acetate: Release rate and biological activity. Environ Entomol **2**:607–610.

Fletcher LW, O'Grady Jr, JJ, Claborn HV, Graham OH (1966) A pheromone from male screw-worms. J Econ Entomol **63**:1611–1612.

Flint, HM, Butler L, McDonough LM, Smith RL, Forey DL (1978) Pink bollworm: Response to various emission rates of gossyplure in the field. Environ Entomol **7**:57–61.

Francke W, Heeman V, Gerken B, Renwick JAA, Vité JP (1977) 2-Ethyl-1, 6-dioxospiro[4,4]nonane, principle aggregation pheromone of *Pityogenes chalcographus* (L.) Naturwissenschaften **64**:590–591.

Frerot B, Gallois M, Einhorn J (1979a) La phéromone sexuelle produite par la femelle vierge de *Pandemis heparana* (Den et Schiff) (Lépidoptère:Tortricidae:Tortricinae). CR Acad Sci Ser D **288**:1611–1614.

Frerot B, Gallois M, Lettere M, Einhorn J, Michelot D, Descoins C (1982) Sex pheromone of *Pandemis heparana* (Den & Schiff) (Lepidoptera:Tortricidae). J Chem Ecol **8**:663–670.

Frerot B, Priesner E, Gallois M (1979b) A sex attractant for the green budworm moth *Hedya nubiferana*. Z Naturforsch **34C**:1248–1252.

Fukui H, Matsumura F, Barak AV, Burkholder W (1977) Isolation and identification of a major sex-attracting component of *Attagenus elongatulus* (Casey) (Coleoptera:Dermestidae). J Chem Ecol **3**:539–548.

Gaston LK, Shorey HH, Saario CA (1971) Sex pheromones of noctuid moths XVIII. Rate of evaporation of model compound of *Trichoplusia ni* sex pheromone from different substrates at various temperatures and its application to insect orientation. Ann Ent Soc Am **64**:381–384.

Gieselmann MJ, Moreno DS, Fagerlund J, Tashiro H, Roelofs WL (1979a) Identification of the sex pheromone of the yellow scale. J Chem Ecol **5**: 27–33.

Gieselmann MJ, Rice RE, Jones RA, Roelofs WL (1979b) Sex pheromone of the San José scale. J. Chem Ecol **5**:891–900.

Girard JE, Germino FJ, Budnis JP, Vita RA, Garrity MP (1979) Pheromone of the male flesh fly *Sarcophaga bullata*. J Chem Ecol **5**:125–130.

Golub M, Weatherston J, Benn MH (1983) The measurement of release rates of gossyplure from controlled release formulations by the mini-airflow method. J Chem Ecol **9**:323–333.

Gore WE, Pearce GT, Lanier GN, Simeone JB, Silverstein RM, Peacock JW, Cuthbert RA (1977) Aggregation attractant of the European elm bark beetle *Scolytus multistriatus*: Production of individual components and related aggregation behavior. J Chem Ecol **3**:429–446.

Grant GG, Brady UE, Brand JM (1972) Male armyworm scent brush secretion: Identification and electroantennogram study of major compounds. Ann Ent Soc Am **65**:1224–1227.

Grant GG, Slessor KN, Szittner RB, Morse D, Meighen EA (1982) Development of a bioluminescence assay for aldehyde pheromones of insects II. Analysis of pheromone glands. J Chem Ecol **8**:923–933.

Greenblatt RE, Burkholder WE, Cross JH, Cassidy Jr, RF, Silverstein RM, Levinson AR, Levinson HZ (1977) Chemical basis for interspecific responses to sex pheromones of *Trogoderma* species (Coleoptera:Dermestidae). J Chem Ecol **3**:337–347.

Grob K (1973) Organic substances in potable water and in its precursor. Part I. Methods for their determination by gas-liquid chromatography. J Chromatogr **84**:225–273.

Grob K, Zürcher F (1976) Stripping of trace organic substances from water—Equipment and procedure. J Chromatogr **117**:285–294.

Grula JW, McChesney JD, Taylor Jr, OR (1980) Aphrodisiac pheromones of the sulfur butterflies *Colias eurytheme* and *C. philodice* (Lepidoptera:Pieridae). J Chem Ecol **6**:241–256.

Guss PL, Tumlinson JH, Sonnet PE, Proveaux AT (1982) Identification of a female-produced sex pheromone for the western corn rootworm. J Chem Ecol **8**:545–556.

Hall DR, Beevor PS, Lester R, Nesbitt BF (1980) (*E,E*)-10,12-Hexadecadienal: A component of the female sex pheromone of the spiny bollworm, *Earias insulana* (Boisd.) (Lepidoptera:Noctuidae). Experientia **36**:152–154.

Hall DR, Nesbitt BF, Marrs GJ, Green A StJ, Campion DG, Critchley BR (1982) Development of microencapsulated pheromone formulations. In: Insect Pheromone Technology: Chemistry and Applications. Leonhardt BA, Beroza M (eds), ACS Symposium Series 190, American Chemical Society, Washington, pp 131–143.

Haniotakis GE, Mazomenos BE, Tumlinson JH (1977) A sex attractant of the olive fly *Dacus oleae* and its biological activity under laboratory and field conditions. Entomol Exp Appl **21**:81–87.

Happ GM, Wheeler J (1969) Bioassay, preliminary purification, and effect of age, crowding and mating on the release of sex pheromone by female *Tenebrio molitor.* Ann Entomol Soc Am **62**:846–851.

Heath RR, McLaughlin JR, Tumlinson JH, Ashley TR, Doolittle RE (1979) Identification of the white peach scale sex pheromone: An illustration of micro techniques. J Chem Ecol **5**:941–953.

Hedin PA, McKibben GH, Mitchell EB, Johnson WL (1979a) Identification and field evaluation of the compounds comprising the sex pheromone of the female boll weevil. J Chem Ecol **5**:617–627.

Hedin PA, Payne JA, Carpenter TL, Neel W (1979b) Sex pheromones of the male and female pecan weevil *Curculio caryae*: Behavioral and chemical studies. Environ Entomol **8**:521–523.

Hill AS, Berisford CW, Brady UE, Roelofs WL (1981) Nantucket pine tip moth, *Rhyacionia frustrana*: Identification of two sex pheromone components. J Chem Ecol **7**:517–528.

Hill AS, Cardé RT, Bode WM, Roelofs WL (1977) Sex pheromone components of the variegated leafroller moth *Platynota flavedana.* J Chem Ecol **3**:369–379.

Hill AS, Kovalev BG, Nikolaeva LN, Roelofs WL (1982) Sex pheromone of the fall webworm moth, *Hyphantria cunea.* J Chem Ecol **8**:383–396.

Hill AS, Rings RW, Swier SR, Roelofs WL (1979) Sex pheromone of the black cutworm moth *Agrotis ipsilon.* J Chem Ecol **5**:439–457.

Hill AS, Roelofs WL (1979) Sex pheromone components of the oblique banded leafroller moth, *Choristoneura rosaceana.* J Chem Ecol **5**:3–11.

Hill AS, Roelofs WL (1981) Sex pheromone of the salt marsh caterpiller moth *Estigmene acrea.* J Chem Ecol **7**:655–668.

Hirai K (1980) Male scent emitted by armyworms, *Pseudaletia unipuncta* and *P. separata* (Lepidoptera:Noctuidae). Appl. Ent Zool **15**:310–315.

Honda K (1980) Odor of a papilionid butterfly: Odoriferous substances emitted by *Atrophaneura alcinous alcinous.* J Chem Ecol **6**:867–873.

Jacobson M (1972) Insect Sex Attractants, Academic Press, New York.

Johnson DW, Mitchell ER, Tumlinson JH, Allen GE (1981) Velvet bean caterpillar: Response of males to virgin females and pheromone in the laboratory and field. Florida Entomologist **64**:528–533.

Kasang G, Kaissling KE, Vostrowsky O, Bestmann HJ (1978) Bombykal, a second pheromone component of the silkworm moth *Bombyx mori* L. Angew Chemie **90**:74–75.

Keller JC, Mitchell EB, McKibben G, Davich TB (1964) A sex attractant for female boll weevils from males. J Econ Entomol **57**:609–610.

Kettlewell HBD (1961) The radiation theory of female assembling in the lepidoptera. Entomologist **94**:59–65.

Klun JA, Bierl-Leonhardt BA, Plimmer JR, Sparks AN, Primiani M, Chapman OL, Lepone G, Lee GH (1980a) Sex pheromone chemistry of the female tobacco budworm moth, *Heliothis virescens.* J Chem Ecol **6**:177–183.

Klun JA, Bierl-Leonhardt BA, Schwarz M, Litsinger JA, Barrian AT, Chiang HC, Jiang Z (1980b) Sex pheromone of the Asian corn borer moth. Life Sci **27**: 1603–1606.

Klun JA, Junk GA (1977) Iowa European corn borer sex pheromone: Isolation and identification of four C_{14} esters. J Chem Ecol **3**:447–459.

Klun JA, Leonhardt BA, Lopez Jr, JD, LaChance LE (1982) Female *Heliothis subflexa* (Lepidoptera:Noctuidae) sex pheromone: Chemistry and congeneric comparisons. Environ Entomol **11**:1084–1090.

Klun JA, Plimmer JR, Bierl-Leonhardt BA, Sparks AN, Primiani M, Chapman OL, Lee GH, Lepone G (1980c) Sex pheromone chemistry of female corn earworm moth, *Heliothis zea.* J Chem Ecol **6**:165–175.

Kochansky JP, Roelofs WL, Sivapalan P (1978) Sex pheromone of the tea tortrix moth (*Homona coffearia* Neitner). J Chem Ecol **4**:623–631.

Kostelc JG, Girard JE, Hendry LB (1980) Isolation and identification of a sex attractant of a mushroom-infesting sciarid fly. J Chem Ecol **6**:1–11.

Kuhr RJ, Comeau A, Roelofs WL (1972) Measuring release rates of pheromone analogues and synergists from polyethylene caps. Environ Entomol **1**:625–627.

Kuwahara Y, Casida JE (1973) Quantitative analysis of the sex pheromone of several phycitid moths by electron-capture gas chromatography. Agr Biol Chem **37**:681–684.

Leonhardt BA, Moreno DS (1982) Evaluation of controlled release laminate dispensers for pheromones of several insect species. In: Insect Pheromone Technology: Chemistry and Applications. Leonhardt BA, Beroza M (eds), ACS Symposium Series 190, American Chemical Society, Washington, pp 159–173.

Levinson HZ, Levinson AR, Jen TI, Williams JLD, Kahn G, Francke W (1978) Production site, partial composition and olfactory perception of a pheromone in the male hide beetle. Naturwissenschaften **65**:543–545.

Löfstedt C, Van Der Pers JNC, Lofqvist J, Lanne BS, Appelgren M, Bergström G, Thelin B (1982) Sex pheromonal components of the turnip moth *Agrotis segetum*: Chemical identification, electrophysiological evaluation and behavioral activity. J Chem Ecol **8**:1305–1321.

Look M (1976) Determining release rates of 3-methyl-2-cyclohexene-1-one antiaggregation pheromone of *Dendroctonus pseudotsugae* (Coleoptera:Scolytidae). J Chem Ecol **2**:481–486.

Ma M, Hummel HE, Burkholder WE (1980) Estimation of single furniture carpet beetle (*Anthrenus flavipes* LeConte) sex pheromone release by dose response curve and chromatographic analysis of pentafluorobenzyl derivative of (*Z*)-3-decenoic acid. J Chem Ecol **6**:597–607.

Ma M, Schnee ME (1983) Analysis of individual gypsy moth pheromone production by sample concentrating gas chromatography. Can Entomol **115**:251–255.

MacConnell JG, Borden JH, Silverstein RM, Stokkink E (1977) Isolation and tentative identification of lineatin, a pheromone from the frass of *Trypodendron lineatum* (Coleoptera:Scolytidae). J Chem Ecol **3**:549–562.

Maitlen JC, McDonough LM, Moffitt HR, George DA (1976) Codling moth sex pheromone: Baits for mass trapping and population survey. Environ Entomol **5**:199–202.

Mazomenos BE, Haniotakis GE (1981) A multicomponent female sex pheromone of *Dacus oleae* Gmelin: Isolation and bioassay. J Chem Ecol **7**:437–444.

McDonough LM, Kamm JA (1979) Sex pheromone of the cranberry girdler, *Chrysoteuchia topiaria* (Zeller) (Lepidoptera:Pyralidae) J Chem Ecol **5**:211–219.

McDonough LM, Hoffman MP, Bierl-Leonhardt BA, Smithhisler CL, Bailey JB, Davis HG (1982a) Sex pheromone of the avocado pest, *Amorbia cuneana* (Walsingham) (Lepidoptera:Tortricidae): Structure and synthesis. J Chem Ecol **8**:255–265.

McDonough LM, Moreno DS, Bierl-Leonhardt BA, Butt BA (1982b) Isolation and identification of a component of the female sex pheromonal gland attractive to male *Xylomyges curialis*. Environ Entomol **11**:660–662.

McKibben GH, Hedin PA, McGovern WL, Wilson NM, Mitchell EB (1977) A sex pheromone for male boll weevils from females. J Chem Ecol **3**:331–335.

Meighen EA, Slessor KN, Grant GG (1981) A bioluminescent assay for aldehyde sex pheromones of insects. Experientia **37**:555–556.

Meighen EA, Slessor KN, Grant GG (1982) Development of a bioluminescence assay for aldehyde pheromones of insects. I. Sensitivity and specificity. J Chem Ecol **8**:911–921.

Meyer WL, Debarr GL, Berisford CW, Barber LR, Roelofs WL (1982) Identification of the sex pheromone of the webbing caneworm moth *Dioryctria disclusa* (Lepidoptera:Pyralidae). Environ Entomol **11**:986–988.

Mitchell ER, Tumlinson JH, Copeland WW, Hines RW, Brennan MM (1974) Tobacco budworm: Production, collection and use of natural pheromone in field traps. Environ Entomol **3**:711–714.

Neal Jr, JW, Klun JA, Bierl-Leonhardt BA, Schwarz M (1982) Female sex pheromone of *Choristoneura parallela* (Lepidoptera:Torticidae) Environ Entomol **11**:893–896.

Nesbitt BF, Beevor PS, Hall DR, Lester R (1979a) Female sex pheromone components of the cotton bollworm *Heliothis armigera*. J Insect Physiol **25**: 535–541.

Nesbitt BF, Beevor PS, Hall DR, Lester R, Davies JC, Seshu Reddy DV (1979b) Components of the sex pheromone of the female spotted stalk borer, *Chilo partellus* (Swinhoe) (Lepidoptera:Pyralidae): Identification and preliminary field trials. J Chem Ecol **5**:153–163.

Nesbitt BF, Beevor PS, Hall DR, Lester R, Sternlicht M, Goldenberg S (1977) Identification and synthesis of the female sex pheromone of the citrus flower moth, *Prays citri*. Insect Biochem **7**:355–359.

Nesbitt BF, Beevor PS, Hall DR, Lester R, Williams JR (1980) Components of the sex pheromone of the female sugar cane borer, *Chilo sacchariphagus* (Bojer) (Lepidoptera:Pyralidae): Identification and field trials. J Chem Ecol **6**:385–394.

Nishida R, Baker TC, Roelofs WL (1982) Hairpencil pheromone components of male oriental fruit moths, *Grapholitha molesta*. J Chem Ecol **8**:947–959.

Nordlund DA, Brady UE (1974) Factors affecting release rate and production of sex pheromone by female *Plodia interpunctella* (Hübner) (Lepidoptera: Pyralidae). Environ Entomol **3**:797–802.

Peacock JW, Cuthbert RA, Gore WE, Lanier GN, Pearce GT, Silverstein RM (1975) Collection on Porapak Q of the aggregation pheromone of *Scolytus multistriatus* (Coleoptera:Scolytidae). J Chem Ecol **1**:149–160.

Persoons CJ, Ritter FJ, Nooyen WJ (1977) Sex pheromone of the false codling moth *Cryptophlebia leucotreta* (Lepidoptera:Tortricidae): Evidence for a two component system. J Chem Ecol **3**:717–722.

Persoons CJ, Verwiel PEJ, Talman E, Ritter FJ (1979) Sex pheromone of the american cockroach *Periplaneta americana*: Isolation and structure elucidation of periplanone B. J Chem Ecol **5**:221–236.

Piccardi P, Capizzi A, Cassani G, Spenelli P, Arsura E, Massardo P (1977) A sex pheromone component of the old world bollworm *Heliothis armigera.* J Insect Physiol **23**:1443–1445.

Plimmer JR, Caro JH, Freeman HP (1978) Distribution and dissipation of aerially-applied disparlure under a woodland canopy. J Econ Entomol **71**: 155–157.

Pope MM, Gaston LK, Baker TC (1982) Composition, quantification and periodicity of sex pheromone gland volatiles from individual *Heliothis virescens* females. J Chem Ecol **8**:1043–1055.

Qi Y-T, Burkholder WE (1982) Sex pheromone biology and behavior of the cowpea weevil *Callosobruchus maculatus* (Coleoptera:Bruchidae). J Chem Ecol **8**:527–534.

Rasmussen RA (1972) A quantitative cryogenic sampler: Design and operation. Am Lab **4**:19–27.

Rasmussen RA, Hutton RS (1972) A freeze out system for trace organic collections in remote areas. Bioscience **22**:294–298.

Renou M, Descoins C, Priesner E, Gallois M, Lettere M (1979) Le tétradécène-7(*Z*)-al-1, constituant principal de la sécrétion phéromonale de la teigne de l'olivier: *Prays oleae* Bern. (Lepidoptere:Hypnomeutidae). CR Acad Sci Ser D **288**:1559–1562.

Renwick JAA, Pitman GB (1979) An attractant isolated from female Jeffrey pine beetles, *Dendroctonus jeffreyi.* Environ Entomol **8**:40–41.

Renwick JAA, Vite JP, Billings RR (1977) Aggregation pheromones in the ambrosia beetle *Platypis flavicomis.* Naturwissenschaften **64**:266.

Richerson JV, Cameron EA (1974) Differences in pheromone release and sexual behavior between laboratory-reared and wild gypsy moth adults. Environ Entomol **3**:475–481.

Robacker DC, Hendry LB (1977) Neral and geranial: Components of the sex pheromone of the parasitic wasp, *Itoplectis conquisitor.* J Chem Ecol **3**: 563–577.

Roelofs WL (ed) (1979) Establishing Efficacy of Sex Attractants and Disruptants for Insect Control. Entomological Society of America, Baltimore.

Roelofs W, Gieselmann M, Cardé A, Tashiro H, Moreno DS, Henrick CA, Anderson RJ (1978) Identification of the California red scale sex pheromone. J Chem Ecol **4**:211–224.

Roelofs WL, Hill AS, Berisford CW, Godbee JF (1979) Sex pheromone of the subtropical pine tip moth *Rhyacionia subtropica.* Environ Entomol **8**:894–895.

Roelofs WL, Lagier RF, Hoyt SC (1977) Sex pheromone of the moth *Pandemis pyrusana.* Environ Entomol **6**:353–354.

Röller H, Biemann K, Bjerke JS, Norgard DW, McShan WH (1968) Sex pheromones of pyralid moths. I. Isolation and identification of the sex attractant of *Galleria mellonella* (Greater Waxmoth). Acta Entomol Bohemoslov **65**: 208–211.

Rothschild GHL (1979) A comparison of methods of dispensing synthetic sex pheromone for the control of oriental fruit moth, *Cydia molesta* (Busck) (Lepidoptera:Tortricidae), in Australia. Bull Ent Res **69**:115–127.

Rudinsky JH, Morgan M, Libbey LM, Michael RR (1973) Sound production in *Scolytidae*: 3-Methyl-2-cyclohexen-1-one released by the female Douglas fir beetle in response to male sonic signal. Environ Entomol 2:505–509.

Shaver TN, Hendricks DE, Hartstack Jr, AW (1981) Dissipation of virelure, a synthetic pheromone of the tobacco budworm from laminated plastic and cigarette filter dispensers. Southern Entomol **6**:205–210.

Silk PJ, Tan SH, Wiesner CJ, Ross RJ, Lonergan GC (1980) Sex pheromone chemistry of the eastern spruce budworm, *Choristoneura fumiferana*. Environ Entomol **9**:640–644.

Silk PJ, Wiesner CJ, Tan SH, Ross RJ, Grant GG (1982) Sex pheromone chemistry of the western spruce budworm *Choristoneura occidentalis* Free. J Chem Ecol 8:351–362.

Silverstein RM, Rodin JO (1966) Insect pheromone collection with absorption columns. I. Studies on model organic compounds. J Econ Entomol **59**: 1152–1154.

Sonenshine DE, Silverstein RM, Collins LA, Saunders M, Flynt C, Hamsher PJ (1977) Foveal glands, source of sex pheromone production in the ixodid tick *Dermacentor andersoni* Stiles. J Chem Ecol **3**:695–706.

Sower LL, Coffelt JA, Vick KW (1973) Sex pheromone: A single method of obtaining relatively pure material from females of five species of moths. J Econ Entomol **66**:1220–1222.

Sower LL, Fish JC (1975) Rate of release of the sex pheromone of the female Indian meal moth. Environ Entomol **4**:168–169.

Sower LL, Gaston LK, Shorey HH (1971) Sex pheromones of noctuid moths. XXVI. Female release rate, male response threshold, and communication distance for *Trichoplusia ni*. Ann Ent Soc Am **64**:1448–1456.

Sower LL, Shorey HH, Gaston LK (1972) Sex pheromones of Lepidoptera. XXVII. Factors modifying the release rate of extractable quantity of pheromones from females of *Trichoplusia ni* (Noctuidae). Ann Ent Soc Am **65**:954–957.

Ställberg-Stenhagen S (1972) Studies on natural odoriferous compounds (V.). Splitter-free all glass intake system for glass capillary chromatography of volatile compounds from biological material. Chemica Scripta **2**:97–100.

Starratt AN, Dahm KH, Allen N, Hildebrand JG, Payne TL, Röller H (1979) Bombykal, a sex pheromone of the sphinx moth *Manduca sexta*. Z Naturforsch **34C**:9–12.

Steck WF, Underhill EW, Bailey BK, Chisholm MD (1982) (*Z*)-7-Tetradecenal, a seasonally dependent sex pheromone of the W-marked cutworm, *Spaelotis clandestina* (Harris) (Lepidoptera:Noctuidae). Environ Entomol **11**:1119–1122.

Struble DL, Arn H, Buser HR, Städler E, Freuler J (1980a) Identification of four sex pheromone components isolated from calling females of *Mamestra brassicae*. J Naturforsch **35C**:45–48.

Struble DL, Buser HR, Arn H, Swailer GE (1980b) Identification of the sex pheromone components of the red backed cutworm *Euxoa ochrogaster*, modification of sex attractant blend for adult male. J Chem Ecol **6**:573–584.

Szittner RB, Morse D, Grant GG, Meighen EA (1982) Development of a bioluminescence assay for aldehyde pheromones of insects. III. Analysis of airborne pheromones. J Chem Ecol **8**:935–945.

Taylor AW (1982) Field measurements of pheromone vapor distribution. In: Insect Pheromone Technology: Chemistry and Applications. Leonhardt BA, Beroza M (eds), ACS Symposium Series 190, American Chemical Society, Washington, pp 193–207.

Teal, PEA, Heath RR, Tumlinson JH, McLaughlin JR (1981) Identification of a sex pheromone of *Heliothis subflexa* (GN) (Lepidoptera:Noctuidae) and field trapping studies using different blends of components. J Chem Ecol **7**:1011–1022.

Tumlinson JH, Heath RR, Teal PEA (1982) Analysis of chemical communications systems of Lepidoptera. In: Insect Pheromone Technology: Chemistry and Applications. Leonhardt BA, Beroza M (eds), ACS Symposium Series 190, American Chemical Society, Washington, pp 1–25.

Tumlinson JH, Klein MG, Doolittle RE, Ladd TL, Proveaux AT (1977) Identification of the female Japanese beetle sex pheromone: Inhibition of male response by an enantiomer. Science **197**:789–792.

Tunstall JP (1965) Sex attractant studies in *Diparopsis.* Pest Ant New Summ Series **A11**:212.

Underhill EW, Arthur AP, Chisholm MD, Steck WR (1979) Sex pheromone components of the sunflower moth *Homoeosoma electellum,* (*Z*)-9-(*E*)-12-tetradecadienol and (*Z*)-9-tetradecenol. J Chem Ecol **8**:740–743.

Underhill EW, Steck WR, Chisholm MD (1977) A sex pheromone mixture for the bertha armyworm moth *Mamestra configurata*: (*Z*)-9-Tetradecen-1-ol acetate and (*Z*)-hexadecen-1-ol acetate. Can Entomol **109**:1335–1340.

Vick KW, Coffelt JA, Sullivan MA (1978) Disruption of pheromone communication in the *Angoumois* grain moth with synthetic female sex pheromone. Environ Entomol **7**:528–531.

Vité JP, Gara RI, Klieforth RA (1963) Collection and bioassay of a volatile fraction attractive to *Ips confusus* (Lee) (Coleoptera:Scolytidae). Contrib Boyce Thompson Inst **22**:39–50.

Weatherston I, Descoins C, Grant GG (1975) Adsorption sur Porapak Q des effluves émis par la femelle vierge de *Choristoneura fumiferana* (Clem.) (Lépidoptère Tortricidae). Mis en évidence du tetradecene-11-(*E*)al-1, principale phéromone sexuelle de cette espèce. CR Acad Sci Paris **281D**: 1111–1114.

Weatherston J, Golub MA, Benn MH (1982) Release rates of pheromones from hollow fibers. In: Insect Pheromone Technology: Chemistry and Applications. Leonhardt BA, Beroza M (eds), ACS Symposium Series 190, American Chemical Society, Washington, pp 145–157.

Weatherston J, Golub MA, Brooks TW, Huang YY, Benn MH (1981) Methodology for determining the release rates of pheromones from hollow fibers. In: Management of Insect Pests with Semiochemicals. Mitchell ER (ed), Plenum, New York, pp 425–443.

Weatherston J, Roelofs W, Comeau A, Sanders CJ (1971) Studies of physiologically active anthropod secretions. X. Sex pheromone of the eastern spruce budworm, *Choristoneura fumiferana* (Lepidoptera:Tortricidae). Can Entomol **103**:1741–1747.

Weatherston J, MacLean W (1974) The occurrence of (*E*)-11-tetradecen-1-ol, a known sex attractant inhibitor, in the abdominal tips of virgin female eastern spruce budworm *Choristoneura fumiferana* (Lepidoptera: Tortricidae). Can Entomol **106**:281–284.

Weatherston J, Percy JE (1976) The biosynthesis of phenethyl alcohol in the male bertha armyworm *Mamestra configurata.* Insect Biochem **6**:413–417.

Wiesner CJ, Silk PJ (1982) Monitoring the performance of eastern spruce budworm formulations. In: Insect Pheromone Technology: Chemistry and Applications. Leonhardt BA, Beroza M (eds), ACS Symposium Series 190, American Chemical Society, Washington, pp 209–217.

Wiesner CJ, Silk PJ, Tan SH, Fullarton S (1980) Monitoring of atmospheric concentrations of the sex pheromone of the spruce budworm, *Choristoneura fumiferana* (Lepidoptera:Tortricidae). Can Entomol **112**:333–334.

Wiesner CJ, Silk P, Tan SH, Palaniswamy P, Schmidt JO (1979) Components of the sex pheromone gland of the eastern spruce budworm, *Choristoneura fumiferana* (Lepidoptera:Tortricidae). Can Entomol **111**:1311.

Williams HJ, Silverstein RM, Burkholder WE, Khowamshahi A (1981) Dominicalure 1 and 2: Components of the aggregation pheromone from male lesser grain borer *Rhyzopertha dominica* (F) (Coleoptera:Bosichidae). J Chem Ecol **7**:759–780.

Yamomoto R (1963) Collection of the sex attractant from female American cockroaches. J Econ Entomol **56**:119–120.

Zlatkis A, Lichtenstein A, Tishbee A (1973) Concentration and analysis of trace volatile organics in gases and biological fluids with a new solid adsorbent. Chromatographia **6**:67–70.

Chapter 11

Techniques for Purifying, Analyzing, and Identifying Pheromones

R. R. Heath[1] *and J. H. Tumlinson*[2]

I. Introduction

"The ultimate practical goal of pheromone research on insect pests is to place the communication system on a molecular basis and to use the knowledge to detect, survey, trap, or disrupt the population" (Silverstein, 1982). Achievement of this goal requires the combined efforts of entomologists working in behavior and ecology and analytical and synthetic organic chemists. Analytical chemistry is the indispensable link between field ecology, the observation of pheromone induced behavior, and the development of a bioassay on one hand and field application of the synthesized pheromone on the other. Thus the success of a pheromone project often depends on the proper selection of analytical procedures for isolation and identification of the biologically active compound(s) and for evaluation of the purity and structural congruity of the synthesized pheromone.

Selection of procedures to be used for the purification, analysis, and identification of a biologically active compound usually depends on the method used in obtaining the initial sample of active material. For example, the type of solvent (polar or nonpolar) used to extract the source of material may dictate several subsequent steps in the purification process, since columns and procedures for separating polar and nonpolar compounds differ significantly. Most of the insect pheromones identified to date have been extracted from whole insects, specific parts of insects like pheromone glands or abdominal tips, or the

[1,2]Insect Attractants, Behavior, and Basic Biology Research Laboratory, Agricultural Research Service, USDA, Gainesville, Florida.

frass (including fecal material) produced by insect feeding, with solvents ranging in polarity from hexane to dichloromethane or chloroform (Tumlinson et al., 1969; Sower et al., 1973; Klun et al., 1980; Heath et al., 1983). However, it is also possible to collect volatiles emitted by calling females by drawing air over the females and then through a trap (Byrne et al., 1975; Hill et al., 1975; Cross et al., 1976; Ma et al., 1980; Tumlinson et al., 1982). Obviously, the more crude samples obtained by extraction of frass or whole insects require more purification to obtain the biologically active compounds in the required purity for further analysis. On the other hand, it is often possible to analyze volatiles collected from calling females directly by capillary gas chromatography and mass spectrometry (Golub and Weatherston, Chapter 10, this volume).

Most insects produce submicrogram amounts of pheromones that are often mixtures including permutations of geometry, functionality, and molecular size. Thus, except for procedures used for the initial purification of crude samples, techniques that provide the highest possible resolution and sensitivity are required for the separation and identification of these compounds. Many of the techniques described in this chapter were not developed for microanalytical chemistry, and are used by many investigators working with larger amounts of material. In many cases certain modifications or refinements are required to make these methods useful for pheromone analyses.

Our purpose here is to provide a guide to the analytical chemical techniques that can be employed effectively for the purification, analysis, and identification of pheromones. A complete review of all approaches or detailed descriptions of the methods used in pheromone isolation and identification is beyond the scope of this chapter. Readers are referred to Young et al. (1975), Brand et al (1979), and to references under each section in the bibliography for other procedures and greater detail about specific procedures.

II. Purification of Biologically Active Material

The chromatographic procedures used to purify pheromones can be divided into two categories which are based on the nature of the mobile phase, i.e., liquid chromatography (LC) and gas chromatography (GC). Each form of chromatography provides choices as to the nature of the separation, loading capacity, and amount of resolution available (Tables 1,2). Liquid chromatography offers various modes of separation and a higher loading capacity than gas chromatography. The detectors used with LC systems do not have the sensitivity of those available in GC systems. However, this is not of great concern in the initial purification process since bioassays are normally used to determine which fractions are biologically active. Gas chromatography provides higher resolution and sensitivity than LC but generally the column performance degrades when overloaded with crude material. Since a crude sample consists mostly of extraneous material and the active material is present in the parts-per-billion range, LC provides the required sample capacity for enrichment of the initial extract.

Table 1. Examples of liquid chromatographic columns used to purify and analyze insect pheromones

Predominant mode of separation	Mobile phase polarity range	Preparative separations[1]			Analytical separations[2]		
		Column size o.d. X length, cm (material)	Loading capacity (mg)	Efficiency N	Column size o.d. X length, cm (material)	Loading capacity (μg)	Efficiency effective plates
Gel permeation	Hexane tetrahydro-furan	1.27 X 90 (Divinylbenzene polymer, poly-styrene gel)	1–500	2000	Generally not used for pheromone analysis		
Adsorption	Hexane to isopropanol	2.54 X 25	4–1000	8000	1.27 X 25	800	12,000
		1.27 X 25 (2 to 10-μm silica)	1–200	8000	0.64 X 25 (5-μm silica)	200	12,000
Reverse phase	Acetonitrile to H_2O	1.27 X 25 (10-μm silica bonded with hydrocarbons 2, 8, and 18 carbons)	0.1–0.4	8000	0.64 X 25 (5-μm silica bonded with hydrocarbons 2, 8, and 18 carbons)	100	12,000
Ag^+ complexation	Toluene hexane/ether	2.54 X 25	2–500	8000	1.27 X 25	800	10,000
		1.27 X 25 (2 to 10-μm silica + Ag^+)	0.5–200	8000	0.64 X 25 (5-μm silica + Ag^+)	200	10,000

[1] Loading capacity–The lower and upper ranges are the mass of material which results in no loss in column efficiency and the mass of material which results in complete deterioration of a separation, respectively. Upper ranges are typical values and depend on initial separation obtained.

[2] Loading capacity–Typical maximum load resulting in a 10% decrease in column efficiency.

Table 2. Examples of gas chromatographic stationary phases used to purify and analyze insect pheromones

Category	Example of phases	Predominant factor(s) that affect elution (examples of use)
Low polarity	SE-30, OV-101 squalane	Boiling point of the compound (General purpose–initial purification of material by boiling points)
Intermediate polarity	Carbowax 20M, FFAP	Boiling point and the polarity of the compound (General purpose–initial purification of material based on polarity)
High polarity	OV-275, SP-2340 Silar 10CP	Polarity and the boiling point of the compound. Capable of discerning small differences in polarity (Separation of geometrical and positional isomers)
Liquid crystal	Cholesteryl cinnamate and cholesteryl *para* substituted cinnamate derivatives, diethyl 4,4′-azoxydi-cinnamate 4-(*p*-Methoxycinnamyloxy)-4′-methoxyazobenzene	This type of phase shows potential for separating compounds according to very small differences in boiling point caused by the steric nature of the compound (Separation of geometrical and positional isomers)

2.1. Liquid Chromatography

The choice of which liquid chromatography column to use for the initial purification of crude material is seldom unequivocal. Table 1 lists several types of columns available for LC that separate by different modes. Crude extracts may be purified *via* separation according to molecular size (gel permeation-G.P.C.), preferential adsorption on silica gel (adsorption chromatography), preferential partitioning on a bonded phase (reverse-phase chromatography), or specific complexation of functional groups like double bonds with Ag^+ ions ($AgNO_3$ columns). For a particularly crude extract that is likely to contain inactive material of high-molecular-weight GPC may be used for the first fractionation, followed by adsorption or reverse phase chromatography of the active GPC fraction(s). The $AgNO_3$ columns are normally more useful than other types of columns for the subsequent resolution of olefinic compounds in relatively pure fractions.

The amount of material that can be loaded on an LC column (loading capacity) is dependent on the resolution and separation desired. In difficult analytical separations that require optimum column efficiency, the loading capacity limit is considered to be that amount of solute that results in a 10% decrease in column efficiency (Scott and Kucera, 1974). In preparative separations, the loading capacity limit is considered to be the amount of material that results in a significant change in the capacity ratio (k) (Majors, 1972). In the initial purification of crude material it is more effective to use gross overloading on several columns that effect separations by different modes than to be concerned with column efficiencies. The loading capacities of several analytical and preparative columns are compared in Table 1. It should be noted that a mode of separation is not absolute. Reverse-phase columns may contain active sites available for solute adsorption. Similarly the pore size of silica adsorbents may effect some separation based on molecular size due to permeation of the solute into the pores. In many cases when several modes of separation are occurring in a liquid chromatographic column, the loading capacity may be severely limited. Additionally, the solubility of the compound in the mobile phase may severely limit loading capacity.

A. Gel Permeation Chromatography

A GPC column separates compounds by molecular size. The larger molecules permeate fewer pores of the gel and thus are eluted first, while smaller molecules, which can enter more of the openings, follow a more circuitous route and elute later. A useful substrate for pheromone purification is one with a working range of about 100 to 2000 molecular weight (MW) (e.g., Styragel 60A, 37-75 μm). Compounds above 2000 MW are excluded from the pores and elute with the column void volume. It should be noted that separations on GPC columns also are influenced by the solvent used as a mobile phase. With a nonpolar solvent like hexane, adsorption on the gel also occurs, while polar solvents like tetrahydrofuran may form complexes with certain types of molecules, increasing their effective molecular size.

While commercially prepared GPC columns are available, it is relatively easy to pack a glass column that will provide the needed resolution to perform the initial purification of a crude extract. For example, a glass column, 1.27 cm (inside diameter) is packed to a height of about 90 cm with a hexane slurry of the gel. The column is eluted with hexane at a flow rate of 5 ml/min and a column inlet pressure of about 40 psi. A 0.5-ml sample of a concentrated crude extract can be loaded on this column and the entire separation requires less than 1 hr. Fractions can be collected either by time or by volume and several samples can be chromatographed on the column before column performance degrades. Even after the column performance has deteriorated because of crude material not eluted with hexane, it can often be regenerated by reversing the flow and eluting with two or three column volumes of tetrahydrofuran. The components of a standard mixture of (*Z*)-9-tricosene, (*Z*)-9-tetradecen-1-ol acetate, and citronellol will be separated on this column and can be used to determine approximate elution volumes of candidate pheromones.

B. Adsorption Chromatography

The most commonly used support in adsorption chromatography is silica gel, although alumina, Florisil, and other solid materials may be used also. Separations on these columns are effected by the different adsorptivities on the solid support and solubilities in the mobile phase of the compounds in the sample. More polar solvents are required to displace strongly bound compounds from the surface of the support. For example, hydrocarbons can be readily eluted from a silica gel column with hexane, but ether or other more polar solvents are required to elute strongly adsorbed compounds like alcohols or lactones.

The components of crude samples can often be effectively fractionated into broad groups, based on functionality, on gravity flow columns. These columns are slurry packed with 60 to 100-mesh (100 to 250-μm) silica gel and eluted sequentially with batches of solvents of increasing polarity. Typically such a column is eluted first with hexane, then with 95:5 hexane:ether, 90:10 hexane: ether, 50:50 hexane:ether, and finally with pure ether. Gravity flow columns are inexpensive to prepare and the packing material may be discarded after one use. Various sizes of columns may be prepared and used to suit the size of the samples to be fractionated.

While gravity flow LC columns are useful for crude separations, much higher resolution can be obtained with columns packed under high pressure. High-performance LC columns can be prepared by the methods described in Appendix A and the references listed under liquid chromatography. However, the equipment required to pack columns is expensive and the high pressure required is hazardous. Also, attention to detail and considerable care are required to prepare high-quality columns. Thus purchase of one of the many excellent commercially prepared columns now available is recommended unless a number of different sizes and types of columns is required.

High-performance adsorption LC columns of two general types are useful for the purification of pheromones. If large quantities of relatively crude material require purification, a column 2.5 cm (outside diameter) × 25 cm long, packed with 2- to 10-μm silica is very useful. The silica used to pack this column is inexpensive and the efficiency and loading capacity are relatively high (Table 1). When column performance is degraded by build-up of strongly adsorbed crude material, it can usually be regenerated by stripping the column with methanol or another high-polarity solvent.

Although most of the crude material can be removed from a sample by GPC or adsorption LC on one of the aforementioned columns, the sample may still consist of a very complex mixture of compounds. At this point it may be advantageous to fractionate it further by chromatography on a very high-resolution silica LC column (Table 1). The choice of column at this point will probably depend on the size of the sample, since overloading a high-resolution column usually negates any advantage that might have been achieved by using it.

C. Reverse-Phase Chromatography

In a reverse-phase column an organic compound is chemically bonded to the silica support. The most widely used reverse-phase column has octadecane bonded to the silica, but a range of bonded phases, with different chain lengths or functionalities, is available from commercial sources. A polar solvent is used for the mobile phase (typically methanol:H_2O, 90:10), and the compounds partition between the bonded phase and the polar solvent. Gravity flow columns containing supports with particles larger than 200 μm may be used but high pressure columns with particles smaller than 20 μm are generally more useful for pheromone purifications. Columns may be packed by the same methods as adsorption columns. Columns packed with a wide range of bonded phases are available from several commercial sources.

Reverse-phase chromatography is generally used when the compound of interest contains one or more polar functional groups. Although many analytical separations of pheromones which contain oxygenated functionality (acetates, aldehydes, alcohols, and lactones) are possible with reverse-phase columns, their preparative separations are severely limited in loading capacity because of limited solute solubility in the mobile phase.

D. Complexation Liquid Chromatography

Olefinic groups are nearly ubiquitous in insect pheromones. Thus, complexation of the olefinic pi electrons with Ag^+ ions provides a powerful tool for separating and purifying many pheromones. There are two methods of separating compounds with Ag^+ ion complexation chromatography. In one the Ag^+ ions reside on the surface of the solid phase and in the other they are contained in the mobile phase. In both cases columns may be packed in the same manner described for adsorption chromatography. The first type of column is most conveniently prepared by packing (or purchasing) a silica adsorption column which is then coated *in situ* by passing a solution of $AgNO_3$ dissolved in acetonitrile through it (Heath and Sonnet, 1980). Solvents, like benzene and toluene, that complex with the Ag^+ ions and thus compete with the solutes for the ions are typically used. These solvents may be mixed with hexane in varying ratios to form mobile phases with varying eluting powers. Hexane–ether systems also perform well and no deterioration of column performance is noted. However, more polar solvents will leach the Ag^+ from the column and must be avoided.

In the second type of complexation LC, Ag^+ ions are dissolved in the mobile phase in concentrations ranging from 15 μM to 150 mM (Phelan and Miller, 1980). A mobile phase of 20 to 40% methanol in H_2O (containing the Ag^+ ions) is used to elute the sample from a reverse-phase column (bonded with octadecane).

The argentation chromatography method chosen for a separation will depend on the nature of the sample. Polar materials cannot be purified using Ag^+–silica columns. The Ag^+–reverse-phase columns allow a larger choice of inexpensive solvents but the mobile phase contains Ag^+ ions that must be removed before proceeding with further analyses of the fractions. Small amounts of Ag^+ ions also are contained in the mobile phase with Ag^+-silica columns but are easily removed. Both types of columns may be prepared with varying degrees of loading capacity and column efficiency.

2.2. Gas–Liquid Chromatography

Once the pheromone has been enriched by one or more of the LC methods discussed previously, or when a relatively pure material is obtained by careful extraction of a gland or collection of airborne volatiles, further purification by micropreparative GLC is possible. Flame ionization detectors can detect as little as 1 ng of a compound eluting from a packed GLC column, but the material is destroyed in the detector flame. Therefore, the effluent from the column is split, sending 2 to 5% to the detector and the remainder to an efficient collector like the one designed by Brownlee and Silverstein (1968). The samples are collected in 30-cm glass melting point capillary tubes (A. H. Thomas Co., Philadelphia, Pa). A temperature gradient is maintained over the length of the collector tube by a heating block at the end next to the GLC and a cooling block at the other end. The section closest to the GLC is maintained about 10°C above the maximum column temperature used for the run, and the point furthest from the GLC is maintained at dry-ice/methanol or liquid nitrogen temperature. The collection system is efficient with recovery of material usually greater than 90%. The collected material can be sealed in the capillary tube for storage or may be rinsed out with a minimal amount of solvent for subsequent chemical or biological analysis.

Since GLC on packed columns is a well established technique that has been used extensively for the separation and analysis of natural products, a wide variety of columns is available from commercial sources. Also very efficient columns can be packed in the laboratory to meet the requirements for a particular analysis. Leibrand and Dunham (1973) discuss several factors critical for obtaining efficient analyses and separations by packed column GLC. The variables that affect column efficiency and separation include the particle size, surface deactivation, and type of solid support and the percentage loading of the stationary phase. For micropreparative GLC, a 5% loading of the stationary phase on the support is generally useful. Lighter loadings will increase column efficiency but support deactivation becomes more critical and sample capacity is reduced. Higher percentage loadings can be used effectively when a large sample capacity is required. The use of glass columns is desirable to reduce the possibility of degradation of labile compounds.

Over 500 liquid phases are commercially available for GLC. The choice of which phase to use in the isolation of an unknown pheromone can be reduced to four types which may be categorized as low polarity, intermediate polarity, high polarity, and liquid crystal phases (Table 2). Like the separation employed in liquid chromatography, the choice of starting phase is seldom unequivocal. While the order in which phases are used is not critical, a general scheme of purification by GLC employs columns in the order mentioned.

Although the sample may have been purified considerably by LC, further purification on at least three different GLC phases may be required to ensure high purity before spectroscopic analysis is undertaken. If the active component or components are hydrocarbon in nature, the use of a polar column will result in a large degree of peak broadening. In this case a liquid crystal phase may provide a suitable alternative. Liquid crystal phases are discussed further in the capillary GLC section of this chapter, and in references listed under gas chromatography.

III. Chromatographic Analysis of Pheromones

3.1. Establishment of Purity

Prior to the identification of an insect sex pheromone it is crucial to determine the purity of the material. Additionally, after the pheromone is identified and synthesized, the purity of the synthetic product must be established. Failure to ensure that material is pure will result in thwarted identification efforts and, in extreme examples, incorrect identification of the active material. Bioassays of synthetic material whose geometric or enantiomeric composition is not known may result in reduced or no behavioral response. Although there are no guaranteed methods for ensuring that isolated material is 100% pure, criteria for analysis of a compound can be established that will maximize the knowledge of the compound's purity. The effective use of chromatography for the analysis of pheromones and the determination of their purity requires an understanding of the chromatographic process. Appendix B discusses the factors that affect the separation of compounds in chromatography and defines the terms used to measure these factors.

The separation scheme chosen for the isolation and purification of a pheromone will generally determine the degree of purification that is possible. Additionally, by deductive reasoning, the degree of purity of a compound can be determined approximately from knowledge of the resolving and separating powers of the chromatographic processes used in the purification. Thus, knowledge of the purity of material obtained by fractionation on a gravity flow LC column (500 plates) followed by subsequent separation and analysis on two packed GLC columns (3000 plates each) is severely limited. Alternatively, if a substance has been purified on several HPLC columns (>10,000 plates), then on three different packed GLC columns (3000 plates each), and finally analyzed on two or three capillary GLC columns (>150,000 plates each) its degree of purity and considerable information about its structure and functionality should be known with a high degree of confidence. This knowledge will greatly facilitate subsequent spectroscopic and microchemical analyses to discern the more subtle aspects of the compound's structure and define the complete structure.

3.2. Capillary Gas Chromatography

Capillary GLC is probably the most useful method available for pheromone analysis because it provides both the high resolution needed to separate complex mixtures and the sensitivity to detect minute amounts of impurities. Thus it not only provides a way to analyze the purity of compounds separated by other methods and of synthesized compounds but also allows samples obtained from one or a few insects to be analyzed. In the latter case, samples that have been collected in a way that excludes large amounts of high-molecular-weight compounds may be analyzed without previous purification. This procedure may eliminate losses incurred in other purification methods and provide more accurate analyses.

Several of the factors that affect the resolution of compounds on capillary GLC columns are listed and briefly explained in Appendix B. More in-depth discussion of these and related topics may be found in the references listed under gas chromatography. The choice of the column and carrier gas to use depends on the separation required. If maximum resolution is needed the column inside diameter should not exceed 0.25 mm and the length should be greater than 30 m. However, larger inside diameter columns may be desirable when greater loading capacity is needed. Although more expensive, the flexible fused silica columns are easier to connect to the GLC and are more inert with fewer active sites.

Generally, helium is used as the carrier gas because it provides reasonable elution times and column efficiency.

The choice of liquid phase is dictated by the type separation required, but a set of columns including one of each of the types listed in Table 2 will provide sufficient capability to analyze nearly any pheromone sample. The separations of Δ7- and Δ9-tetradecen-1-ol acetates on four capillary columns coated with the different types of stationary phases (Fig. 1) illustrate some of the separation characteristics of these columns. The low-polarity OV-1 and the medium-polarity Carbowax 20M are commercially available (Hewlett-Packard) fused silica capillary columns. The usefulness of these highly efficient (N effective) fused silica columns is limited because the separation factor, α, on Carbowax 20M and OV-1 is insufficient to resolve most positional and geometrical isomers found in the pheromone blends of lepidopteran insects. The high-polarity phases containing large amounts of cyano functionality such as SP-2340 and Silar 10C

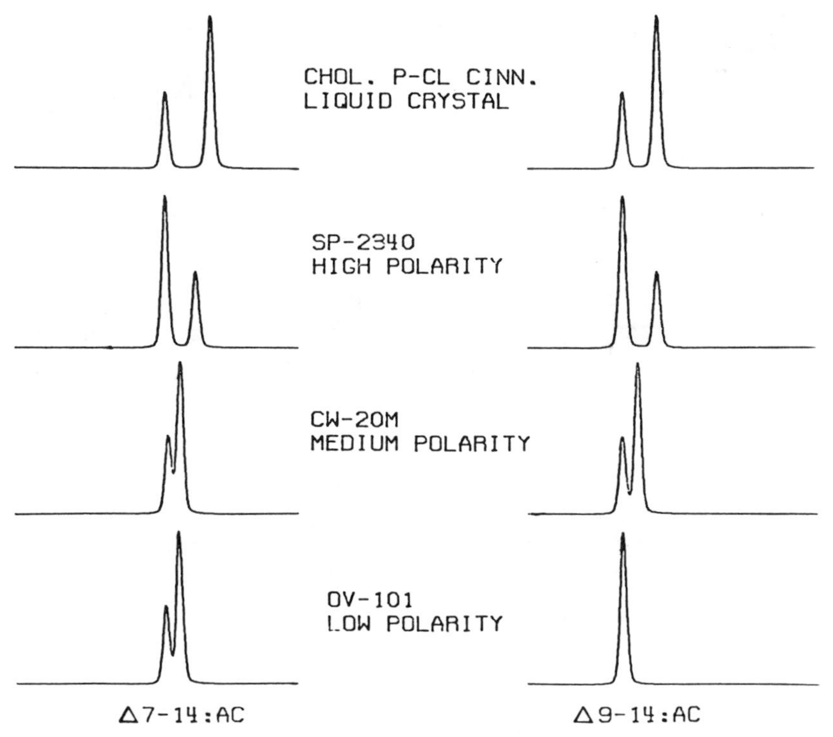

Figure 1. Separation of tetradecenyl acetates on four different types of stationary phases on capillary columns. The OV-1 and Carbowax 20M (CW-20M) are coated on 50 m fused silica and the SP-2340 and cholesteryl *p*-chlorocinnamate on 25-m glass columns. The ratio of *Z:E* is 1:2, respectively, for both (*Z*)- and (*E*)-7-tetradecen-1-ol acetate (Δ7-14:OAc) and (*Z*)- and (*E*)-9-tetradecen-1-ol acetate (Δ9-14:OAc).

provide good resolution of the many positional and geometrical isomers of monounsaturated compounds found in lepidopteran insects. Superior resolution of the geometrical isomers of mono- and diunsaturated compounds is obtained with liquid crystal phases, although the separation of positional isomers is compromised as the double bond position approaches the middle of the hydrocarbon chain in some cases. The comparison of the separations of the analogous series of tetradecen-1-ol acetates on liquid crystal and cyano phase capillary columns is shown in Fig. 2. The *E*-isomers elute prior to *Z*-isomers from a cyano phase (SP-2340). However, on the liquid crystal column, the *Z*-isomers elute first when the olefinic bond is near the middle of the chain. As the double bond is moved toward the hydrocarbon end of the chain, (*Z*)- and (*E*)-11-tetradecen-1-ol acetate (*Z*- and *E*11-14:OAc) coelute and then the elution order reverses for *Z*- and

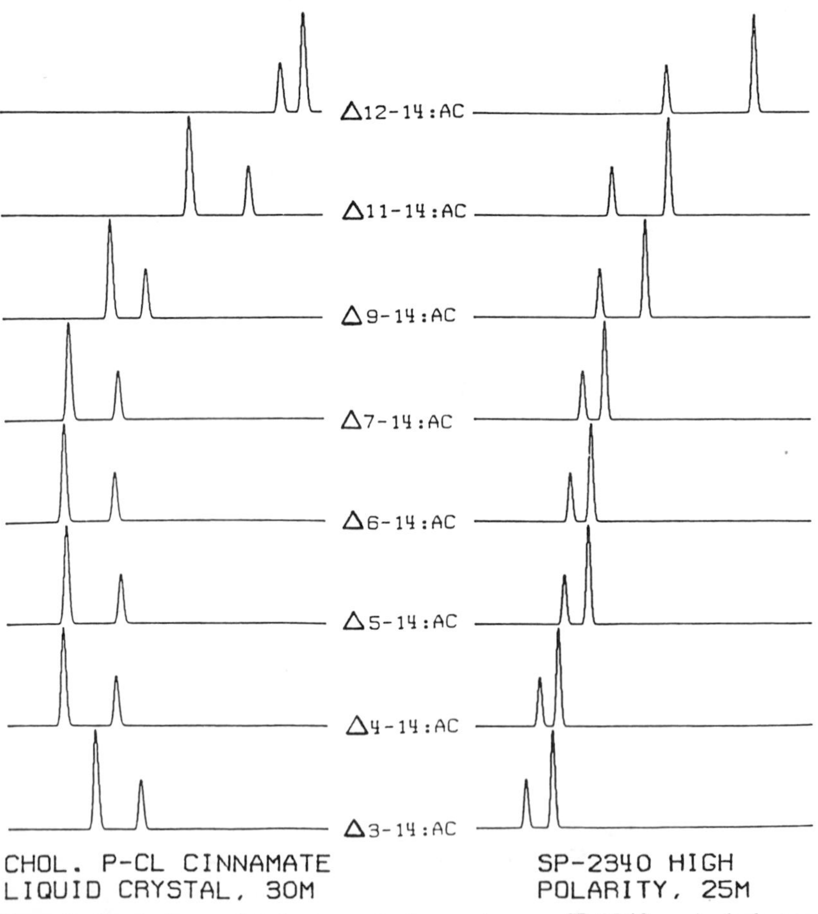

Figure 2. Separation of tetradecen-1-ol acetates on SP-2340 and cholesteryl *p*-chlorocinnamate stationary phases in glass capillary columns. The ratio of *Z*:*E* is 2:1, respectively, in each set of isomers.

*E*12-14:OAc. The use of both the liquid crystal columns and the cyano columns offers a powerful analytical procedure for the identification of compounds like those found in lepidopteran blends. As an example, Fig. 3 illustrates the analysis on the two columns of a complex mixture of positional and geometrical isomers of 16-carbon aldehydes, acetates, and alcohols, many of which are components of the pheromones of *Heliothis* species.

As noted earlier, the elution order of *Z*- and *E*-isomers is opposite on the liquid crystal and cyano phases. Aldehydes elute first on both phases. The

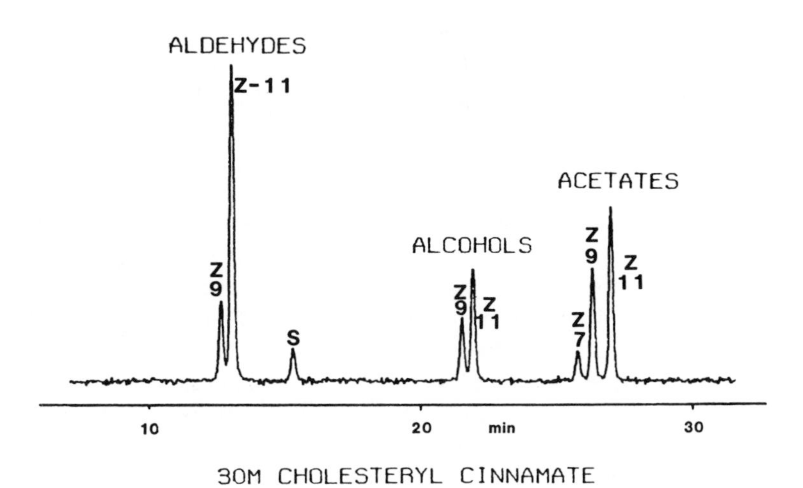

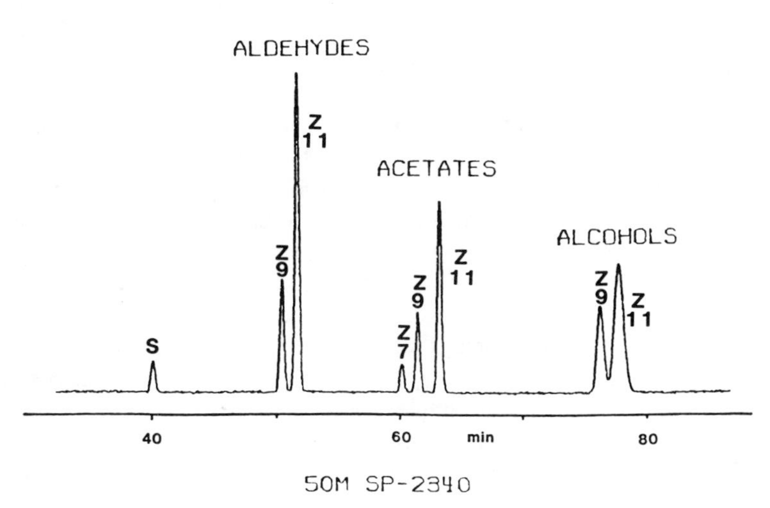

Figure 3. Separation of blends of 16-carbon compounds on a liquid crystal and a cyanosilicon capillary column.

alcohols are retained more than the acetates on the high-polarity cyano phase, but the elution order of alcohols and acetates is reversed on the liquid crystal phase. While neither phase separated all of the components of the synthetic mixture, the combination of separations obtained on both columns allows components of a pheromone blend to be identified on the basis of their retention indices, subject, of course, to confirmation by mass spectrometric analysis.

Special considerations are necessary when liquid crystal columns are used. Since they are ordered phases, they must be operated near their mesophase transition temperature. Increasing the column temperature above the mesophase transition temperature results in decreased separations. However, if the column is heated to the transition temperature and then carefully cooled (supercooled) 10 to 20° below that temperature, improved separations may be obtained. The characteristics and capabilities of liquid crystal columns are discussed in more detail by Heath et al. (1981) and Tumlinson et al. (1982) and in other references listed under gas chromatography.

IV. Spectroscopic and Microchemical Analyses

The chemical structure of the biologically active compound(s) is usually determined by spectroscopic and chemical degradative analysis of the highly purified material. Although useful information can sometimes be obtained by analysis of mixtures, the results of such analyses often can be misleading or confusing. Thus, with the availability of the chromatographic methods discussed in previous sections, attempts to analyze impure materials usually are not worthwhile. An exception to this rule is the direct analysis by capillary GC–mass spectroscopy of pheromone blends from gland extracts or volatiles collected from single insects or small number of insects. Even then, given the high resolution of capillary GC columns, the mass spectra usually are obtained on pure compounds.

In developing an analytical scheme for probing a compound's structure, one should consider the sensitivity of each method and the loss of material that will be incurred in each analysis. Usually only a limited amount of sample is available and every purification and transfer process will further reduce this amount. Although mass spectroscopy is a process which destroys the compound being analyzed, only a few nanograms are required. Infrared (IR) spectroscopy requires a 10-fold increase (ca. 100–400 ng) in sample size. Recovery of the sample is possible although the losses are usually large with volatile samples and polar compounds on KBr pellets. Proton nuclear magnetic resonance (NMR) requires an additional 10-fold increase (1–5 μg) in material over that required for IR, but complete recovery of the sample is possible.

In addition to spectroscopy there are several chemical techniques which provide information about the structure of a molecule. Most of these techniques employ mass spectroscopy to analyze the reaction product and therefore sample size requirements are governed by the sensitivity of the mass spectrometer (1–40 ng).

The order in which spectra are obtained generally follows the amount required for a given spectroscopic method, i.e., mass spectra, infrared spectra, and proton nuclear magnetic resonance spectra, but if less than 2 μg of a compound is available, the order should probably be reversed for IR and NMR. Microchemical techniques are employed when the spectroscopic data are insufficient to define completely the structure and suggest that a proposed structure can be further defined by a microdegradative process.

Interpretation of spectroscopic data is a subject that is too lengthy for discussion in this chapter. For an introduction, the reader is referred to the book by Silverstein et al. (1981) that covers UV, IR, NMR, and mass spectroscopy, and gives enough information under each topic to aid in the identification of many compounds. Other references listed under spectroscopic methods provide more information on spectroscopic identification of organic compounds.

4.1. Mass Spectroscopy

The first spectrum obtained of a sample is nearly always the mass spectrum. Little material is required and the information derived helps maximize the information obtained with other spectroscopic methods. While several methods can be used to ionize and fragment the compound under investigation, electron impact (EI) and chemical ionization (CI) techniques are most frequently used in pheromone identification. The advent of chemical ionization with various reagent gases has improved this technique considerably (Hunt and Ryan, 1971, 1972a,b). The reagent gas that is particularly suitable for a class of compounds and the type of information desired can be selected (Table 3). Generally

Table 3. Characteristics of reagent gases used most frequently for chemical ionization mass spectroscopy of pheromones[1]

Reactant gas	Major reactants (decreasing order)	Type of fragmentation
Methane	CH_5+, C_2H_5+, C_3H_5+,	Weak M+1, large characteristic fragmentation ions (M+29; M+41 indication of M.W.)
iso-Butane	C_4H_5+	Large M+1, limited number of fragmentation ions
Argon-water	$Ar+$, Ar_2+, H_2O	EI like but with large M+1
Deuterium	D_3+, $(D_2O)_2D^+$, $(D_2O)_3D+$	Determination of active hydrogens, differentiation of amines
Ammonia	NH_4+, $(NH_3)_2H+$, $(NH_3)_3H^+$	Determination of molecular weight of polyfunctional organic compounds

[1] Additional reagent gases that may be used include nitrous oxide, oxygen, hydrogen, helium, tetramethylsilane, ethylene oxide, and vinyl methyl ether.

methane is the first reagent gas used to analyze a pheromone. If the ions corresponding to the molecular weight are absent or weak, a less reactive reagent like iso-butane can be used to provide ions diagnostic for the compound's molecular weight. Alternatively, the EI spectrum of a compound usually provides more information about the structure because of increased fragmentation.

The identification of the pheromone of the southern corn rootworm, *Diabrotica undecimpunctata howardi* Barber (Guss et al., 1983) provides an example of the use of both CI and EI mass spectroscopy in determining the structure of a compound. The methane ionization mass spectrum (Fig. 4) established that the molecular weight of the pheromone was 212 with diagnostic peaks at *m/e* 211 (M−1), 213 (M+1), 241 (M+29), and 253 (M+41). The fact that the M+1 peak was the base peak, and that there was almost no fragmentation suggested a fairly stable parent molecule with an oxygen function. The fragmentation pattern obtained using electron impact mass spectrometry (Fig. 5) provided additional information about the compound's structure. The base peak at *m/e* 43, the strong peak at *m/e* 58, and the weak *m/e* 57, plus the moderately strong peak at *m/e* 71 strongly suggest a methyl ketone with no substitution on the carbons α or β to the carbonyl (Budzikiewicz et al., 1967).

A comparison of the EI mass spectra of authentic 2-tetradecanone with the natural pheromone at this point indicated that there were seversal similarities in

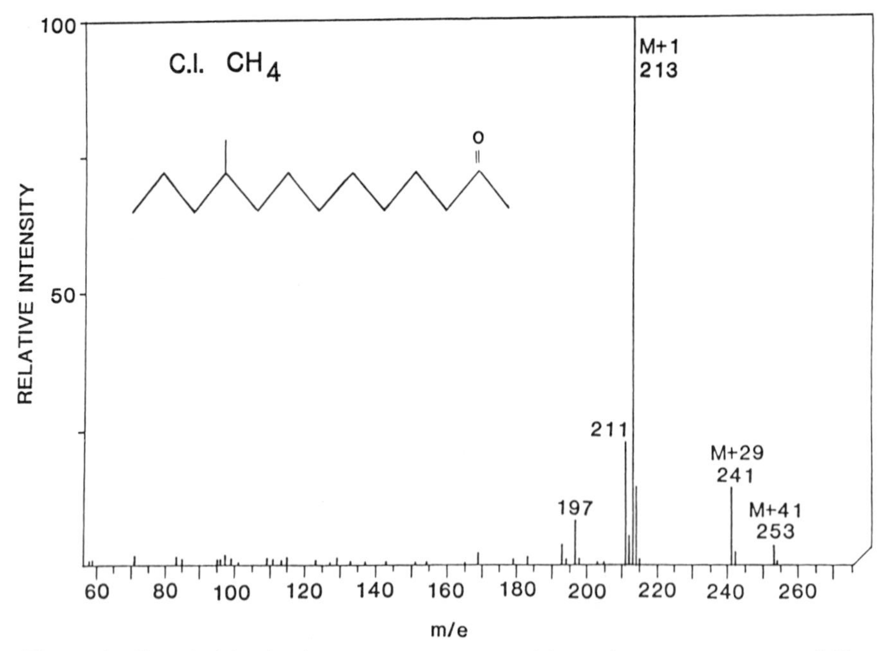

Figure 4. Chemical ionization mass spectrum, with methane reagent gas, of 10-methyl-2-tridecanone, the pheromone of *Diabrotica undecimpunctata howardi* Barber.

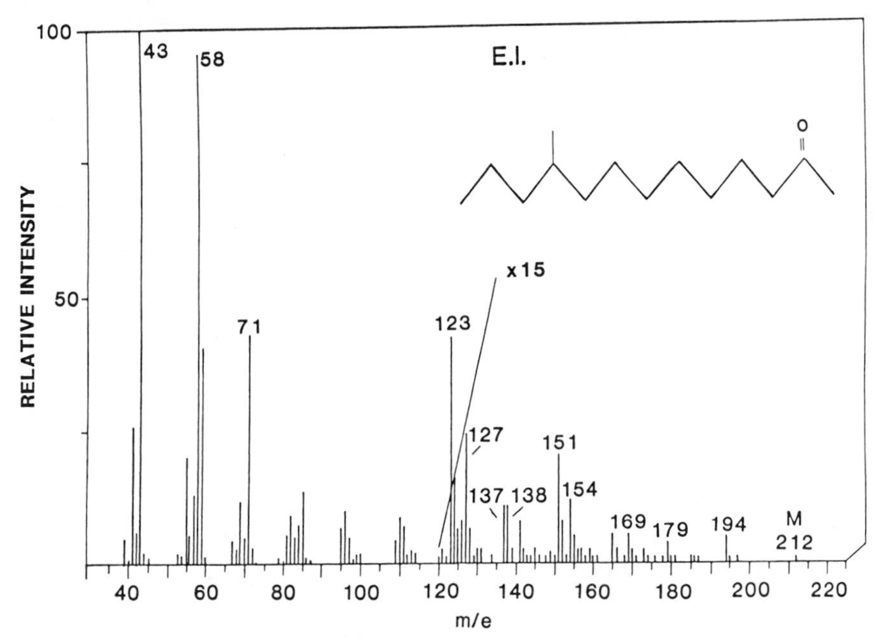

Figure 5. Electron impact mass spectrum of 10-methyl-2-tridecanone. The heights of the peaks beyond *m/e* 120 have been expanded 15-fold relative to the remainder of the spectrum.

the low mass region of the spectra but distinct differences in the high mass region. Additionally, the pheromone eluted prior to 2-tetradecanone on capillary columns suggesting the possibility of a branched-chain 14-carbon methyl ketone. Hydrogenolysis (described in the microchemical technique section) of the pheromone in the gas chromatograph injector leading to the EI source produced a product with the mass spectrum shown in Fig. 6. The peaks at *m/e* 155, 154, 71, and 70 establish the structure of the hydrogenolysis product as 4-methyl tridecane and the pheromone of the southern corn rootworm was therefore identified as 10-methyl-2-tridecanone

4.2. Infrared Spectroscopy

The absorption of infrared radiation in the region between 4000 and 600 cm^{-1} is of particular value in characterizing the structure of a molecule. While matching all the absorption bands in the IR spectrum of an unknown with those of an authentic sample is an accepted method for confirming the structure of a compound, pheromones are often new compounds and thus no authentic sample exists. Fortunately certain groups of atoms absorb at or near the same frequency regardless of the structure of the rest of the molecule. Inspection for these characteristic bands in a spectrum and reference to charts of characteristic

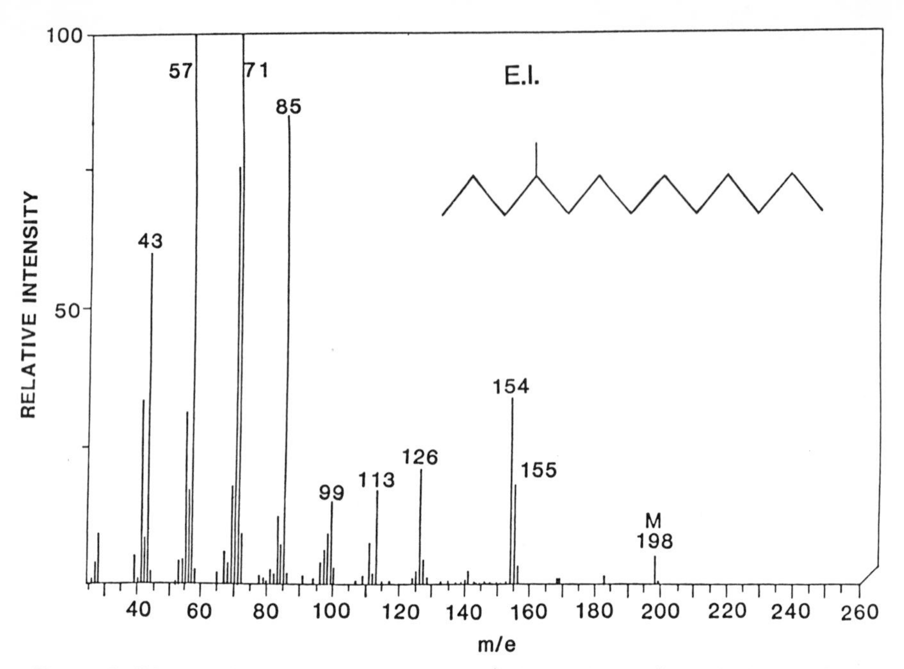

Figure 6. Electron impact mass spectrum of the product of the hydrogenolysis of 10-methyl-2-tridecanone. The hydrogenolysis was carried out in the injector of the gas chromatograph that served the EI source. About 6 cm of a 3.2-mm (inside diameter) stainless-steel tube was filled with neutral Pd catalyst and placed in the injector port ahead of the OV-1 column. The catalyst was maintained at 260°C and H_2 was used as the carrier gas.

group frequencies permit the investigator to obtain useful information regarding the presence or absence of functional groups and other moieties in the molecule. This information is used in conjunction with other spectroscopic, chromatographic, and chemical data to determine the molecular structure.

Recent developments in Fourier transform (FT) infrared spectrometers have drastically reduced the sample size needed to obtain spectra. The advantage of these systems is that a large number of scans which cover the range of frequencies (4000 to 600 cm^{-1}) simultaneously can be time averaged. Repetitive scanning at a rapid rate (less than a second) enables a large signal-to-noise improvement in a relatively short period of time. The signal-to-noise ratio is proportional to the square root of the number of scans obtained. Such systems are based on the Michelson interferometer and utilize a computer for conversion of the time-domain data to frequency-domain. Additional information is contained in the reference section under infrared spectroscopy.

For IR analysis a sample may be deposited on a micro-KBr disc (1 mm diameter) or dissolved in CCl_4 or CS_2 and contained in a microcavity cell. Either the disc or the cell must be held in a beam condenser which focuses the infrared beam on the small area containing the sample. The KBr disc eliminates solvent absorption bands that interfere with certain areas of the sample spectrum. However, it is difficult to obtain good spectra on volatile compounds because they evaporate from the disc too rapidly. Solvent bands in solution spectra can be reduced considerably by ratioing with a solvent blank, but with very dilute samples it is very difficult to eliminate the solvent bands completely. Thus the minimum amount of solvent should be used when samples must be analyzed in solution. The strength of the infrared absorption by the various bonds of the molecule determines the lower limit of sensitivity of IR spectroscopy. The spectum of (*E,Z*)-3,13-octadecadien-1-ol acetate (Fig. 7) was obtained from 100 ng deposited on a micro-KBr disc. In this spectrum the weak band at 970 cm^{-1} resulting from absorption by the *trans* olefinic bond was readily apparent.

An improved method of obtaining IR spectra on small samples involves coupling a gas chromatograph to a light pipe that is installed in a Fourier transform IR spectrometer. The compounds eluting from the GLC flow into the heated light pipe and are rapidly scanned. Since the compounds are in the vapor phase, the spectral bands exhibited are narrower and more intense. With this system good spectra may be obtained with as little as 20 ng of sample and the

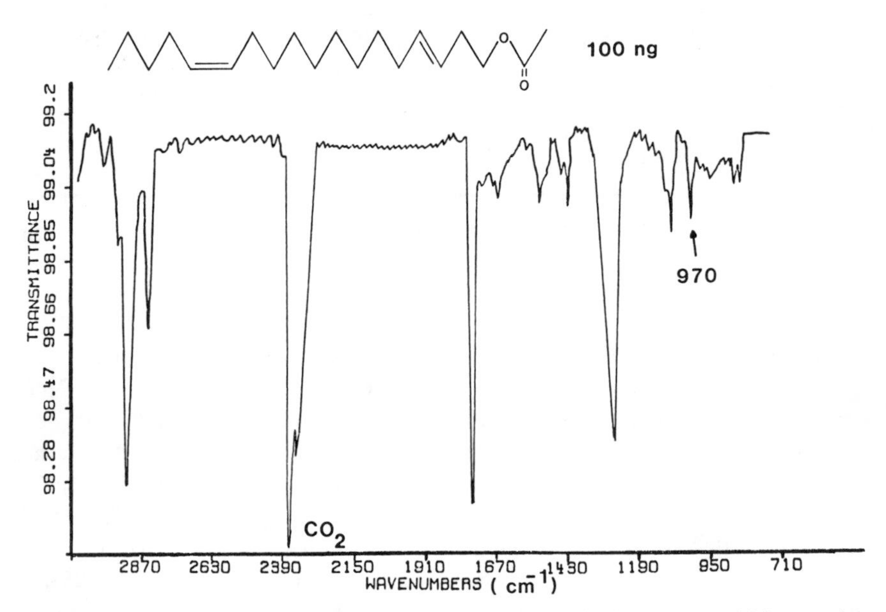

Figure 7. IR spectrum of (*E,Z*)-3,13-octadecadien-1-ol acetate, 100 ng, deposited on a micro-KBr disc (1 mm diam). One-thousand scans of the sample were ratioed against 1000 scans of KBr for background. Total time of the measurement was 741 sec. Resolution was 4 cm^{-1}. The strong peak with a shoulder at 2300 is caused by residual CO_2 in the spectrometer.

sample emerging from the light pipe may be collected the same way samples are collected from micropreparative GLC.

4.3. Nuclear Magnetic Resonance Spectroscopy

Proton NMR spectroscopy is the most definitive but least sensitive method readily available for determining the structure of pheromones. Carbon NMR provides even more structural information but unfortunately due to low natural abundance its lack of sensitivity (ca. 200-μg minimum sample size) precludes its use for the analysis of most pheromones isolated from insects. NMR is basically another form of absorption spectroscopy and the absorption frequencies are characteristic of the electron density around the atoms and thus are influenced by functional groups and other moieties in the molecule. The difference in the absorption position of a proton from the absorption position of a reference proton is called the chemical shift. Chemical shifts are usually given in ppm relative to tetramethylsilane (TMS). In addition to the chemical shift of a resonance frequency, the degree of splitting of the signal caused by spin–spin coupling and the coupling constant, J, give information characteristic of certain features of a molecule and are useful for structure elucidation. Again, charts of chemical shifts and coupling constants are available in the reference texts and more information can be derived from the literature. The reader is referred to the references listed under NMR for more detailed and in-depth discussion of NMR spectroscopy.

The introduction of superconducting magnets operating at 300 MHz and higher, and recent advances in electronics have improved both the resolution and sensitivity of NMR spectrometers significantly. With a 300 MHz Fourier transform NMR (FTNMR) spectrometer it is now possible to obtain excellent proton NMR spectra on 10-μg samples routinely and the analysis takes less than an hour. By scanning a sample overnight, spectra can be obtained on samples as small as 0.5 μg. Sensitivity depends on the degree of proton coupling in the molecule and at this level special consideration of solvent purity is required.

The spectrum of a component of the queen recognition pheromone of the red imported fire ant, *Solenopsis invicta* Buren (Rocca et al., 1983), obtained with about 0.9 μg of sample in a Nicolet 300 MHz spectrometer illustrates the application of this method. The sample was collected from the GLC in a capillary tube with the collector previously described. The end of the capillary was drawn to a fine point and the sample was transferred with minimal loss to a 70-μl external capillary extension tube (Wilmad, Buena, N.J.) with about 25–30 μl of benzene-d_6. Because a strong lock signal is required for long runs and a minimal amount of solvent is desirable in order to minimize solvent impurities, benzene-d_6 is used. The spectra in Fig. 8 illustrate the sensitivity and resolution of a 300 MHz spectrometer. The plot of the spectrum between δ 0.3 and 7.4 ppm shows a significant resonance at 0.4 ppm that represents the residual H_2O

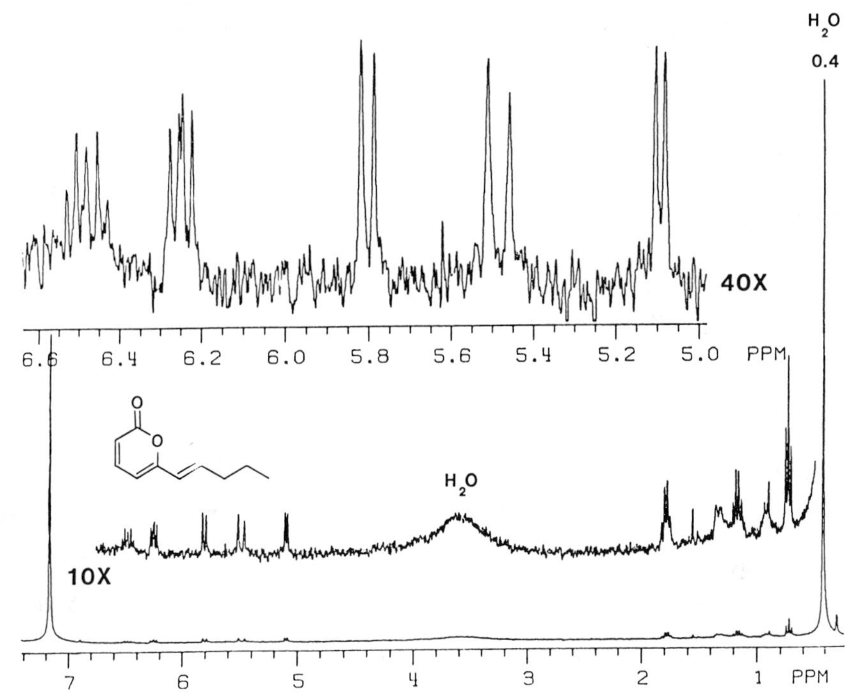

Figure 8. Proton NMR spectrum of about 0.9 μg of (*E*)-6-(1-pentenyl)-2H-pyran-2-one, a component of the queen recognition pheromone of the red imported fire ant. The sample was dissolved in about 30 μl of benzene-d_6 in a 70 μl external cavity cell. Spectrometer frequency was 300.0 MHz and a sweep width of 1500 Hz was used. A total of 16,000 transients were obtained over a 12-hr period.

in the solvent. The resonance at 7.28 is residual undeuterated benzene (0.02%) and serves as the reference for chemical shift assignments. The broad peak at δ 3.6 ppm in the 10X amplification plot results from H_2O contained in the probe. The 10X and 40X amplification plots of the spectrum illustrate that more than adequate sensitivity (signal to noise) was available to enable the structure of the queen pheromone to be deduced.

Since benzene is an anisotropic solvent the resonance signals may be shifted markedly from those obtained with the sample dissolved in CCl_4 or $CDCl_3$. This can have considerable advantages. However, since most of the spectra recorded in the literature were run with the samples dissolved in CCl_4, direct comparison with spectra of known compounds is often difficult. Fig. 9 shows spectra of the same compound in $CDCl_3$ and benzene-d_6. The information obtained from the benzene-d_6 spectrum is obviously greater than that obtained with the sample in $CDCl_3$.

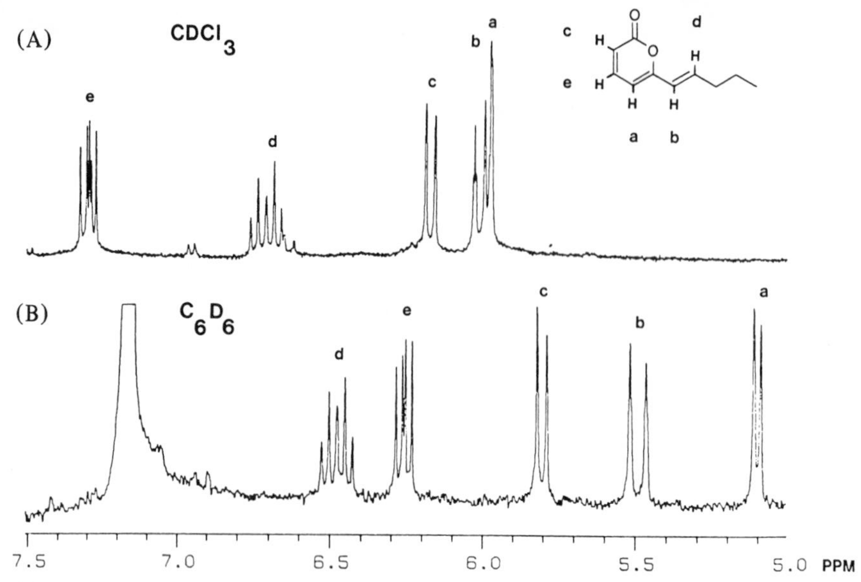

Figure 9. Comparison of the spectra of synthetic (*E*)-6-(1-pentenyl)-2H-pyran-2-one (2.0 μg) in $CDCl_3$ (A) and benzene-d_6 (B). Spectrometer frequency was 300 MHz and a sweep width of ±1500 Hz was used.

4.4. Microchemical Techniques

Beroza (1975) discusses several chemical reactions that are useful in determining the structure of small amounts of natural products.

Two microdegradative techniques that we find most useful in determining the number and position of olefinic bonds are hydrogenation or hydrogenolysis over neutral palladium catalyst and ozonolysis. These techniques were developed on a microscale by Beroza and co-workers and the experimental techniques are described elsewhere (Beroza and Sarmiento, 1966; Beroza and Bierl, 1966, 1967). The advantages of these procedures are that they can be performed with less than a microgram of material, and the products are available for analysis in a few minutes at most.

Hydrogenation is carried out in the inlet of a gas chromatograph at a temperature determined by the use of standards so as not to effect hydrogenolysis of the compound. An injector temperature less than 200°C usually results in adequate hydrogenation of the compound. Columns may be packed or capillary and ideally are coupled directly to a mass spectrometer. The mass spectrum of the product is recorded as it is eluted from the GLC column. Under the proper conditions, this procedure only hydrogenates olefinic bonds; thus a vast amount of information is gained very quickly. For example, the mass spectrum of an unknown compound indicated that it was a straight-chain molecule with a molecular weight of 210. Thus, it could either be a 14-carbon alcohol with two olefinic bonds or a 14-carbon aldehyde with one. Insufficient material was available for an infrared spectrum. Reduction of the compound in the inlet of the gas chromatograph and subsequent mass spectroscopy showed that the product had a molecular weight of 212. Thus, the unknown was a monounsaturated 14-carbon aldehyde.

Hydrogenolysis is conducted in the same manner as hydrogenation, except that longer catalyst beds and higher temperatures are required to reduce the molecule all the way to its hydrocarbon skeleton. It is particularly useful when branching is detected in a molecule but cannot be precisely located by EI mass spectroscopy of the compound. This was illustrated by the example discussed in the mass spectroscopy section (Fig. 6).

Microozonolysis of a pheromone allows the position of an olefinic bond in the molecule to be precisely established and requires less than 50 ng of material. The sample is dissolved in a few microliters of CS_2 or hexane (at least 99 mole% pure) and cooled to about −70°C in a dry ice bath. A fine stream of ozone is bubbled through the sample until the sample is saturated. The ozonide is then reduced with triphenyl phosphine, and the products are analyzed directly by GLC–mass spectroscopy.

V. Future Developments in Analytical Methods

Analytical chemistry is a dynamic science that is rapidly developing new technology and modifications of existing technology to solve new problems as they arise. It is not our intent to infer that classical analytical methods like thin-layer chromatography, ultraviolet spectroscopy, and derivatization have no utility in today's pheromone research. However, new methods, some of which are already in the process of development, will greatly expand the capability of the analytical chemist to separate and analyze the complex mixtures of semiochemicals that insects and other organisms produce in minute quantities.

Examination of current research efforts in analytical chemistry leads to the prediction that significant improvements in separation science and the sensitivity of detection will occur in the near future. Some of these new methods like gas chromatography coupled with tuneable ion detectors and microbore and capillary LC columns coupled to mass spectrometers or tuneable ion detectors will be used routinely in pheromone research. The position and geometry of the ubiquitous olefinic bonds in pheromones will be routinely determined by tandem capillary columns in GLC. Many enantiomers and stereoisomers may be separated chromatographically and their configurations determined without resorting to the chiral derivatives or chiral NMR shift reagents used currently. With the extremely high-field NMR spectrometers and advances in electronics and computers, NMR resolution and sensitivity will be increased even further and ^{13}C NMR may become feasible for analysis of many pheromones.

Hopefully, as analytical techniques improve, we will be able to identify and elucidate some of the more obscure factors that influence chemical communication in insects and other organisms. Knowledge of some of the more subtle nuances of semiochemical blends may help explain and accurately define behavioral responses and thus ultimately contribute to our knowledge of the chemical ecology of these organisms. This, in turn, should allow more biorational and environmentally acceptable methods of pest management to be devised.

VIII. References

Beroza M (1975) Microanalytical methodology relating to the identification of insect sex pheromones and related behavior-control chemicals. J Chromatogr Sci **13**:314–321.

Brand JM, Young JC, Silverstein RM (1979) Insect pheromones; A critical review of recent advances in their chemistry, biology, and application. In: Progress in the Chemistry of Organic Natural Products; Vol 37. Herz W, Grisebach H, Kirby GW (eds), Springer-Verlag, Wien-New York.

Brownlee RG, Silverstein RM (1968) A micro preparative gas chromatograph and a modified carbon skeleton determinator. Anal Chem **40**:2077–2079.

Budzikiewicz H, Djerassi C, Williams DH (1967) Mass Spectrometry of Organic Compounds. Holden-Day, San Francisco, California.

Byrne KJ, Gore WF, Pearce GT, Silverstein RM (1975) Porapak Q collection of airborne organic compounds serving as models for insect pheromones. J Chem Ecol **1**:1–7.

Cross JH, Byler RC, Cassidy RF, Jr, Silverstein RM, Greenblatt, RE, Burkholder WE, Levinson AR, Levinson HZ (1976) Porapak-Q collection of pheromone components and isolation of (*Z*)- and (*E*)-14-methyl-8-hexadecenal, sex pheromone components, from females of four species of *Trogoderma* (Coleoptera:Dermestidae). J Chem Ecol **2**:457–468.

Ettre LS (1965) Open Tubular Columns in Gas Chromatography. Plenum, New York.

Guss PL, Tumlinson JH, Sonnet PE, McLaughlin JR (1983) Identification of female produced sex pheromone from the Southern corn rootworm, *Diabrotica undecimpunctata howardi* Barber. J Chem Ecol **9**:1363–1375.

Heath RR, Sonnet PE (1980) Technique for *in situ* coating of Ag^+ onto silica gel in HPLC columns for the separation of geometrical isomers. J Liquid Chromatogr **3**:1129–1135.

Heath RR, Jordan JR, Sonnet PE (1981) Effect of film thickness and supercooling on the performance of cholesteryl cinnamate liquid crystal capillary columns. High Resol Chromatogr Chromatogr Comm **4**:328–332.

Heath RR, Tumlinson JH, Leppla NC, McLaughlin JR, Dueben B, Dundulis EA, Guy RH (1983) Identification of a sex pheromone produced by female velvetbean caterpillar moths. J Chem Ecol **9**:645–656.

Hill AS, Cardé RT, Kido H, Roelofs WL (1975) Sex pheromone of the orange tortrix moth, *Argyrotaenia citrana* (Lepidoptera:Tortricidae). J Chem Ecol **1**:215–224.

Hunt DF, Ryan JF, III (1971) Chemical ionization mass spectrometry studies. I. Identification of alcohols. Tetrahedron Lett 4535–4538.

Hunt DF, Ryan JF, III (1972a) Argon-water mixtures as reagents for chemical ionization mass spectrometry. Anal Chem **44**:1306–1309.

Hunt DF, Ryan JF, III (1972b) Chemical ionization mass spectrometry studies. Nitric oxide as a reagent gas. J Chem Soc Chem Commun 620–621.

Klun JA, Bierl-Leonhardt BA, Plimmer JR, Sparks AN, Primiani M, Chapman OL, Lepone G, Lee GH (1980) Sex pheromone chemistry of the female tobacco budworm moth, *Heliothis virescens.* J Chem Ecol **6**:177–183.

Leibrand RJ, Dunham LL (1973) Preparing high efficiency packed GC columns. Research/Development, September. 32–38.

Ma M, Hummel HE, Burkholder WE (1980) Estimation of single formiline carpet beetle (*Anthrenus flavipes* LeConte) sex pheromone release by dose-response curve and chromatographic analysis of pentafluorobenzyl derivative of (*Z*)-3-decanoic acid. J Chem Ecol **6**:597–607.

Majors RE (1972) High performance liquid chromatography on small particle silica gel. Anal Chem **44**:1722–1728.

Phelan PL, Miller JR (1981) Separation of isomeric insect pheromonal compounds using reversed-phase HPLC with $AgNO_3$ in the mobile phase. J Chromatogr Sci **19**:13–17.

Rocca JR, Tumlinson JH, Glancey BM, Lofgren C (1983) The queen recognition pheromone of *Solenopsis invicta,* preparation of (*E*)-6-(1-pentenyl)-2H-pyran-2-one. Tetrahedron Lett **24**:1889–1892.

Silverstein RM, Bassler GC, Morrill TC (1981) Spectrometric Identification of Organic Compounds. 4th ed. Wiley, New York.

Silverstein RM (1982) Chemical communications in insects: Background and application. Pure Appl Chem **54**:2480–2481.

Scott RPW, Kucera P (1974) The examination of some commercially available silica gel adsorbents for use in liquid solid chromatography. J Chem Sci **12**: 473–485.

Scott RPW (1960) Gas chromatography, p. 144. Proceedings 3rd Int. Symposium. Butterworths, London.

Sower LL, Coffelt JA, Vick KW (1973) Sex pheromone: A simple method of obtaining relatively pure material from females of five species of moths. J Econ Entomol **66**:1220–1222.

Tumlinson JH, Hardee DD, Gueldner RC, Thompson AC, Hedin PA, Minyard JP (1969) Sex pheromone produced by male boll weevil: Isolation, identification, and synthesis. Science **166**:1010–1012.

Tumlinson JH, Heath RR, Teal PEA (1982) Analysis of chemical communications systems of Lepidoptera. In: Insect Pheromone Technology: Chemistry and Applications. Leonhardt B, Beroza M (eds), ACS Symposium Series, Am. Chem. Soc., Washington, D.C., pp 1–25.

Young JC, Silverstein RM (1975) Biological and chemical methodology in the study of insect communication. In: Methods in Olfactory Research. Moulton DG, Turk A, Johnston JW (eds), Academic Press, New York, pp 75–161.

IX. Subject-Area References

9.1. Purification of Biologically Active Material

Dean JA (1963) Chemical Separation Methods. Reinhold, New York.

Overton KH (1963) Isolation, purification, and preliminary observations. In: Technique of Organic Chemistry. Weissberger A (ed), Interscience, New York.

Pasto DJ, Johnson CR (1969) Physical methods of separation, purification and characterization. In: Organic Structure Determination. Prentice-Hall, New Jersey.

Tumlinson JH, Heath RR (1976) Structure elucidation of insect pheromones by microanalytical methods. J Chem Ecol **2**:87–99.

Young JC, Silverstein RM (1975) Biological and chemical methodology in the study of insect communication. In: Methods of Olfactory Research. Moulton DG, Turk A, Johnston JW (eds), Academic Press, New York, pp 75–161.

9.2. Liquid Chromatography

Adams MA, Nakanishi K (1979) Selected uses of HPLC for the separation of natural products. J Liquid Chromatogr **2**:1097–1136.

Anonymous (1979) Preparative Liquid Chromatography. Waters Associates, Framingham, Massachusetts.

Bakalyan S, Yuen J, Henry D (1977) How to Pack Liquid Chromatography Columns. Spectra Physics Corp, San Jose, California.

Done JN, Knox IH, Loheac J (1982) Application of High-Speed Liquid Chromatography. Wiley-Interscience, New York.

Hamilton RJ, Sewell PA (1982) Introduction to High Performance Liquid Chromatography. Chapman and Hall/Methuen Inc., New York.

Kaiser RE, Oelrich E (1981) Optimization in HPLC. Dr. Alfred Huethig Publishers, New York.

Kirkland II (1971) Report for analytical chemists. Anal Chem **43**:37A–48A.

Laub RJ (1974) Packings for HPLC. Research/Development **14**:24–28.

Schram SB (1980) The LDC Basic Book on Liquid Chromatography. Milton Roy Company, St. Petersburg, Florida.

Snyder LR, Kirkland JJ (1974) Introduction to Modern Liquid Chromatography. Wiley, New York.

Snyder LR (1972) A rapid approach to selecting the best experimental conditions for high-speed liquid column chromatography. Part II. Estimating column length, operating pressure and separation time for some required sample resolution. J Chromatogr Sci **10**:369–379.

Yost RW, Ettre LS, Conlon RD (1980) Practical Liquid Chromatography: An Introduction. Perkin-Elmer, Norwalk, Connecticut.

9.3. Gel Permeation

Anonymous (1983) Chromatography, electrophoresis, immunochemistry and HPLC. Bio-Rad, New York.

Anonymous (1978) Aquapou™ sep columns: Rapid aqueous and non-aqueous size exclusion chromatography. Brownlee Labs, Santa Clara, California.

Anonymous (1973) Liquid Chromatography: Packing Materials and Packed Columns. Waters Associates, Framingham, Massachusetts.

Kirkland JJ, Antle PE (1977) High performance size-exclusion chromatography of small molecules with columns of porous silica microspheres. J Chromatogr Sci **15**:137–147.

Krishen A, Tucker RG (1977) Gel permeation chromatography of low molecular weight materials with high efficiency columns. Anal Chem **49**:898–902.

9.4. Adsorption on Silica Gel

Majors RE (1972) High performance liquid chromatography on small particle silica gel. Anal Chem **44**:1722–1726.

Rabel FM (1975) Microparticle packings for liquid chromatography. Use and care. Am Lab **15**:53–63.

Saunders DL (1974) Solvent selection in liquid chromatography. Anal Chem **46**: 470–473.

Scott RPW, Kucera P (1974) The examination of some commercially available silica gel adsorbents for use in liquid solid chromatography. J Chromatogr Sci **12**:473–485.

9.5. Reverse-Phase Chromatography

Colin H, Guiochon G (1978) Comparison of some packings for reversed-phase high-performance liquid-solid chromatography. II. Some theoretical considerations. J Chromatogr **158**:183–205.

Harrison KH, Millar VI, Yates TL (1982) Comprehensive guide to reverse phase materials for HPLC. Vydac, Hesperia, California.

Schmit RA, Henry RA, Williams RC, Dieckman JF (1971) Application of high speed reversed-phase liquid chromatography. J Chromatogr Sci **9**:645–651.

9.6. Complexation Chromatography

Evershed RP, Morgan ED, Thompson LD (1982) Preparative scale separation of alkene geometric isomers by liquid chromatography. J Chromatogr **237**: 350–354.

Heath RR, Sonnet PE (1980) Technique for *in situ* coating of Ag^{+} onto silica gel in HPLC columns for the separation of geometrical isomers. J Liquid Chromatogr **3**:1129–1135.

Heath RR, Tumlinson JH, Doolittle RE, Proveaux AT (1975) Silver nitrate–high pressure liquid chromatography of geometrical isomers. J Chromatogr Sci **13**:380–382.

Heath RR, Tumlinson JH, Doolittle RE (1977) Analytical and preparative separation of geometrical isomers by high efficiency silver nitrate liquid chromatography. J Chromatogr Sci **15**:10–13.

Houx NWH, Jongen WMF (1974) Purification and analysis of synthetic insect sex attractants by liquid chromatography on a silver loaded resin. J Chromatogr Sci **96**:25–32.

Janak J, Jagaric Z, Dressler M (1970) Use of olefin-Ag^{+} complexes for chromato-

graphic separations of higher olefins in liquid-solid system. J Chromatogr **53**:525–530.
Morris LJ, Wharry DM, Hammond EW (1967) Chromatographic behavior of isomeric long-chain aliphatic compounds. II. Argentation thin-layer chromatography of isomeric octadecenoates. J Chromatogr **31**:69–76.
Phelan PL, Miller JR (1981) Separation of isomeric insect pheromonal compounds using reversed-phase HPLC with $AgNO_3$ in the mobile phase. J Chromatogr Sci **19**:13–17.
Schomburg G, Zegarski K (1975) Separation of olefinic compounds by reversed-phase liquid chromatography with a mobile phase containing π-complexing metal salts like silver nitrate. J Chromatogr **114**:174–178.
Tscherne RJ, Capitano G (1977) High-pressure liquid chromatographic separation of pharmaceutical compounds using a mobile phase containing silver nitrate. J Chromatogr **136**:337–341.
Vonach B, Schomburg G (1978) High-performance liquid chromatography with Ag^+ complexation in the mobile phase. J Chromatogr **149**:417–430.
Warthen JD, Jr (1976) Liquid chromatographic purification of geometric isomers on reverse-phase and silver-loaded macroporous cation exchange columns. J Chromatogr Sci **14**:513–515.

9.7. Gas–Liquid Chromatography

Anonymous (1968) Preparation of coated packings. Use of the HI-EFF® fluidizer. Applied Science Laboratories, State College, Pennsylvania.
Freeman RR (1979) High Resolution Gas Chromatography. Hewlett–Packard, Avondale, Pennsylvania.
Heath RR, Burnsed GE, Tumlinson JH, Doolittle RE (1980) Separation of a series of positional and geometrical isomers of olefinic aliphatic primary alcohols and acetates by capillary gas chromatography. J Chromatogr **189**: 199–208.
Heath RR, Doolittle RE (1983) Derivatives of cholesterol cinnamate: A comparison of the separations of geometrical isomers when used as gas chromatographic stationary phases. High Resol Chromatogr Chromatogr Comm **6**:16–19.
Heath RR, Jordan JR, Sonnet PE (1981) Effect of film thickness and supercooling on the performance of cholesteryl cinnamate liquid crystal capillary columns. HRC CC **4**:328–332.
Heath RR, Jordan JR, Sonnet PE, Tumlinson JH (1979) Potential for the separation of insect pheromones by gas chromatography on columns coated with cholesteryl cinnamate. J High Resol Chromatogr Chromatogr Comm **2**: 712–714.
Jennings W (1978) Gas Chromatography with Glass Capillary Columns. Academic Press, New York.
Kruppa RF, Henly RS, Smead DL (1967) Improved gas chromatography packing with fluidized drying. Anal Chem **39**:851–853.
Lester R, Hall DR (1980) 4-(*p*-Methoxycinnamyloxy)-4′-methoxyazobenzene: A nematic liquid crystal for the gas-liquid chromatographic analysis of the stereochemistry of lepidopterous sex pheromones and related unsaturated fatty alcohols and derivatives. J Chromatogr **190**:35–41.

Lester R (1978) Smectic liquid crystal for the gas-liquid chromatographic separation of lepidopterous sex pheromones and related isomeric olefins. J Chromatogr **156**:55–62.

Littlewood AB (1962) Gas chromatography. Academic Press, New York.

Mon TR (1971) Preparation of large bore open tubular columns for GC. Res Devel **12**:14–17.

Said SD (1981) Theory and Mathematics of Chromatography. Hüthig, Heidelberg, West Germany.

Scott RPW (1970) Determination of the optimum conditions to effect a separation by gas chromatography. In: Advances in Chromatography. Giddings JC, Keller RA (eds), Dekker, New York, Vol 9.

Tumlinson JH, Heath RR, Teal PEA (1982) Analysis of chemical communications systems of Lepidoptera. ACS Symp. Series No. 190 Insect pheromone technology. Chemistry and Applications. Amer Chem Soc., Washington, DC, pp 1–25.

Zielinski WL, Jr (1980) Difficult isomer separations. Res Dev **22**:177–182.

9.8. Spectroscopic Methods

Dyer JR (1965) Applications of Absorption Spectroscopy of Organic Compounds. Prentice-Hall, Englewood Cliffs, New Jersey.

Pasto DJ, Johnson CR (1969) Organic Structure Determination. Prentice-Hall, Englewood Cliffs, New Jersey.

Silverstein RM, Bassler GC, Morrill TC (1981) Spectrometric Identification of Organic Compounds. 4th ed. Wiley, New York, p 444.

9.9. Mass Spectroscopy

Beynon JH, Saunders RA, Williams AE (1968) The Mass Spectra of Organic Molecules. American Elsevier, New York.

Biemann K (1962) Mass spectrometry: Organic chemical applications. McGraw-Hill, New York.

Budzikiewicz H, Buske E (1980) Studies in chemical ionization mass spectroscopy III, CI spectra of olefins. Tetrahedron **36**:255–266.

Budzikiewicz H, Djerassi C, Williams DH (1967) Mass Spectrometry of Organic Compounds. Holden-Day, San Francisco, California.

Foltz RL (1975) Structural analysis via chemical ionization mass spectrometry. Chemtech **14**:39–44.

Hunt DF (1976) Selective reagents for chemical ionization mass spectrometry. Finnigan Spectra **6**:1–8.

Leonhardt BA, De Vilbiss ED, Klun JA (1983) GC-MS indication of double-bond position in monounsaturated primary acetates and alcohols without derivatization. Org Mass Spec **18**:9–11.

Milne GWA (1971) Mass Spectrometry: Techniques and Applications. Wiley-Interscience, New York.

Tumlinson JH, Heath RR, Doolittle RE (1974) Application of chemical ionization mass spectrometry of epoxides for the determination of olefin position in aliphatic chains. Anal Chem **46**:1309–1311.

9.10. Infrared Spectroscopy

Conley RT (1972) Intrared Spectroscopy. 2nd ed. Allyn and Bacon, Boston.
Griffiths PR (1975) Chemical Infrared Fourier Transform Spectroscopy. Wiley-Interscience, New York.
Griffiths PR (1976) On-line measurement of the infrared spectra of gas chromatographic eluents. EPA Report 170, EPA-600/4-76-061, 1-21.
King ST (1973) Application of infrared Fourier transform spectroscopy to the analysis of micro samples. J Agric Food Chem **21**:526–530.
Krishman K, Brown RH, Hill SL, Simonoff SC, Olson ML, Kuehi D (1981) Recent developments in GC/FTIR spectroscopy. Am Lab **22**:122–126.
Nakanishi K, Solomon PH (1977) Infrared absorption spectroscopy-practica. 2nd ed. Holden-Day, San Francisco, California.
Rossiter V (1982) Capillary GC/FTIR. Am Lab **23**:71–80.
Shafer KH, Bjorseth A, Tabon J, Jacobsen RJ (1980) Advancing the chromatography of GC/FT-IR to WCOT capillary columns. High Resol Chromatogr Chromatogr Comm **3**:87–88.

9.11. Nuclear Magnetic Resonance

Abraham RJ, Loftus P (1978) Proton and carbon-13 NMR spectroscopy. Heyden, London.
Becker ED (1969) High resolution NMR. Academic Press, New York.
Bovey FA (1969) NMR Spectroscopy. Academic Press, New York.
Levy GC, Lichter RL, Nelson GL (1980) Carbon-13 Nuclear Magnetic Resonance for Organic Chemists. 2nd ed. Wiley, New York.
Levy GC, Craile DJ (1981) Recent developments in nuclear magnetic resonance spectroscopy. Science **214**:291–299.
Shoolery JN (1972) A Basic Guide to NMR. Varian Assoc., Palo Alto, California.
Stothers JB (1972) Carbon-13 NMR Spectroscopy. Academic Press, New York.

9.12. Microtechniques

Beroza M (1970) Determination of the chemical structure of organic compounds at the microgram level by gas chromatography. Accounts Chem Res **3**:33–49.
Beroza M (1975) Microanalytical methodology relating to the identification of insect sex pheromones and related behavior-control chemicals. J Chromatogr Sci **13**:314–321.
Beroza M, Bierl B (1966) Apparatus for ozonolysis of microgram to milligram amounts of compound. Anal Chem **38**:1976–1977.
Beroza M, Bierl B (1967) Rapid determination of olefin position in organic compounds in microgram range by ozonolysis and gas chromatography. Anal Chem **39**:1131–1135.
Beroza M, Coad RA (1966) Reaction gas chromatography. J Gas Chromatogr **3**:199–216.
Beroza M, Sarmiento R (1966) Apparatus for reaction chromatography. Instantaneous hydrogenation of unsaturated esters, alcohols, ethers, ketones, and

other compound types and determination of their separation factors. Anal Chem **38**:1042–1047.
Huwyler S (1973) Ultramikromethoden. Katalytische Hydrierung. Experientia **29**:1310–1311.
Huwyler S (1976) Ultramicromethods. VI. Lithium aluminumhydride reduction of aldehydes and ketones. J Microchem **21**:135–139.
Huwyler S (1977) Ultramicromethods. VII. Vacuum distillation. Improvement of a previously described apparatus. J Microchem 22:236–237.
Huwyler S (1979) Ultramicromethods. VII. Ozonolysis. J Microchem **24**:468–474.
Tumlinson JH, Heath RR (1976) Structure elucidation of insect pheromones by microanalytical methods. J Chem Ecol **2**:87–99.

VI. Appendix A

6.1. Procedure for Packing High-Performance Liquid Chromatography Columns

High-pressure slurry packing procedures are designed to provide a homogeneous bed of material in which the separation of different size particles is minimized. This method of packing requires a high-pressure, high-volume pump capable of delivering in excess of 200 ml of solvent/min and operating at 10,000 to 15,000 psi. Extreme care should be taken to ensure that all fittings are correctly and securely connected and that safety procedures are followed at all times.

The packing system (see Fig. 10) consists of a solvent reservoir for the pump, an air-driven pump, pressure gauge, a valve to shut off flow to the column, and a check valve to prevent backflow in the system. The reservoir for the packing slurry contains a removable cap for the introduction of the slurry. A precolumn (10 cm in length) couples the reservoir to the column. Although a commercial column terminator with a flat surface may be used to terminate the column, the cone shape terminator previously described (Heath et al., 1978) results in improved column performance when packing 1.27- and 2.54-cm (outside diameter) columns. This terminator requires no special machining (Swagelok) and can be used on 0.635 cm outside diameter columns. An air-operated on/off valve (Swagelok) located at the end of the column permits the system to be pressurized to the desired pressure and when opened rapidly forces the packing material against the frit located in the column terminator.

Prior to packing a column the internal surfaces of the 0.635- and 1.27-cm outside diameter columns must be polished to minimize band spreading encountered in columns having a noninfinite diameter. The carbon tetrachloride used to prepare the packing slurry should be dried by passing it through activated silica. Silica gel used for packing material should be dried for 24 hr at 150°C at reduced pressure, and the reverse-phase material for 24 hr at 50°C.

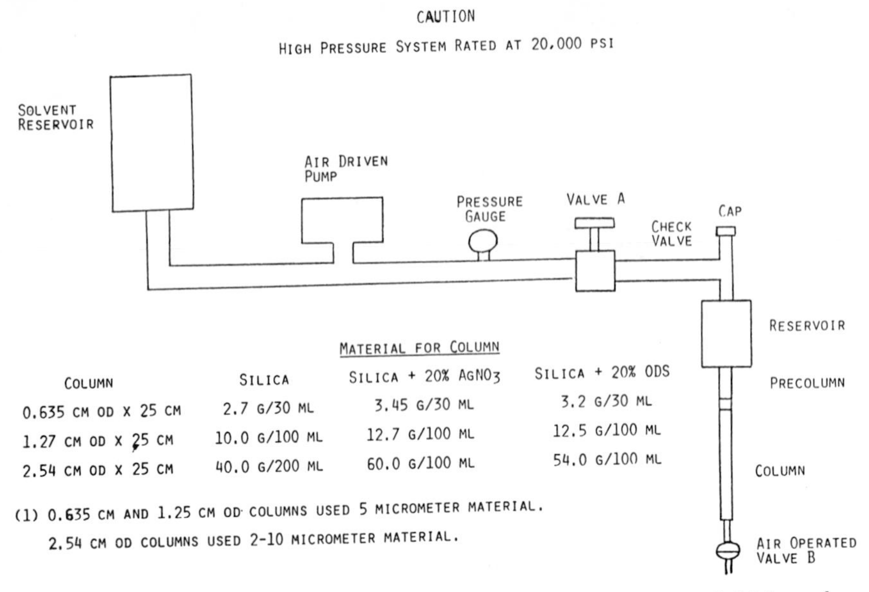

Figure 10. Packing apparatus and list of material required to pack high-performance liquid chromatography columns.

The packing procedure is carried out in a hood with a protective barrier and has been previously described (Heath et al., 1977). The procedure consists of the following steps.

1. Assemble packing system consisting of pump, reservoir, precolumn, and column.
2. Slurry packing material (see Fig. 10) in dry CCl_4 and sonicate for ca. 5 min; slurry and degas with a vacuum.
3. Turn air-operated on–off valve located at the bottom of the column off and introduce the slurry via the top T located on the reservoir. Add an additional 50 ml of CCl_4 and fill remainder with hexane. Place cap on snugly but allow for a small amount of leakage; pressurize pump to 1000 psi and slowly open valve between pump and reservoir and expel any air by using the T cap.
4. Tighten cap and open valve completely. Put safety shield in place. Close hood and pressurize system to 9000 psi.
5. Open air-operated on–off valve and expel the CCl_4 plus an additional 300 ml of hexane.
6. Reduce pump pressure, close valve connecting pump to reservoir, and allow several minutes for the column to depressurize.
7. Remove column from precolumn and on–off valve and install injector and detector connectors.

VII. Appendix B

7.1. Factors that Affect the Separation of Compounds in Chromatography

The degree of separation obtained for two compounds is a function of two parameters. Although many factors affect these two parameters the separation obtained by the chromatography system is the result of the separation factor, α, and the column efficiency.

To calculate α, it is necessary to determine the amount of time each compound spends interacting with the stationary phase or support, excluding the time spent in the mobile phase (gas or liquid). This is defined as the adjusted retention t'_r:

$$t'_r = t_r - t_m,$$

where t_r = measured retention of the compound and t_m = measured retention of an unretained compound, referred to as the void volume. The separation factor of the compound is then determined as

$$\alpha = \frac{t'_r \text{ second eluting compound} = t'_{rB}}{t'_r \text{ first eluting compound} = t'_{rA}}$$

Alpha is always greater than unity and is a function of the stationary phase or adsorbent.

Calculation of the columns' effective efficiency (termed effective theoretical plates = N) requires the measurement of the adjusted retention of the compound and the peak width. The efficiency of a column can be calculated two ways.

$$N = 16 \times \left(\frac{t'_r \text{ of the compound}}{W_b \text{ width at the base}} \right)^2$$

or

$$N = 5.54 \times \left(\frac{t'_r \text{ of the compound}}{W_{1/2} \text{ width at peak } 1/2 \text{ height}} \right)^2$$

W must be measured in the same units as t'_r, i.e., time or distance on chart paper. N is a function of the efficiency with which the column performs the separation and a function of column length.

Another important parameter in chromatography is the resolution, R_s, obtained for two compounds on a column. The resolution of two peaks may be calculated from the retentions and the peak widths at the base (using triangulation):

$$R_s = \frac{2(t_{rB} - t_{rA})}{W_{\text{base A}} + W_{\text{base B}}}$$

An additional factor affecting resolution (which is taken into account when physically measuring the parameters required in the resolution formula) is the ratio of the compounds adjusted retention, t'_r, compared to the retention of an unretained compound. The retention ratio, k, is calculated as

$$k = \frac{t'_r}{t_m}$$

k also is termed the partition ratio. Knowing α, N, and k is of particular importance in the chromatographic separations of compounds that do not have a high R_s value. Rewriting the R_s formula which now includes the partition ratio for compounds eluting close to each other ($W_{\text{base A}} = W_{\text{base B}}$) results in

$$R_s = \frac{N^{1/2}}{4} \times \frac{\alpha - 1}{\alpha} \times \frac{k}{k+1}$$

Based on the formula, increasing k beyond a value of 5 gives little improvement in R_s, while it increases the time required for the analysis. Fig. 11 illustrates the effect of varying α and N for a given separation.

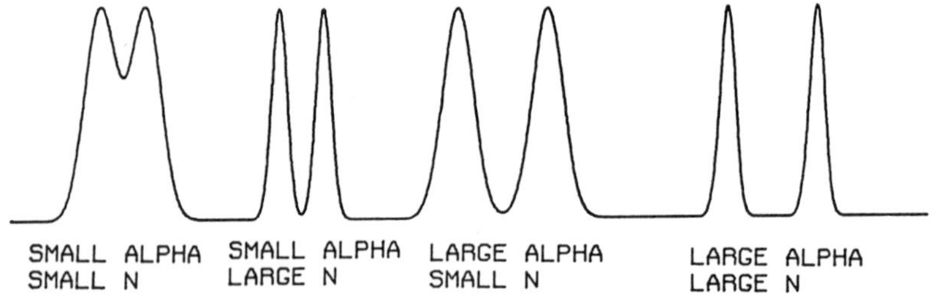

Figure 11. The effect of varying α and N on the separation of two hypothetical peaks.

The need for resolution of compounds has been discussed in the text of this chapter. Criteria of purity are based on the ability to resolve one component from another and are often complicated by the occurrence of one of the compounds in a minor amount. The effect this has on the required resolution is illustrated in Fig. 12.

The greatest change in the resolution of components is obtained by using a column that provides the largest α. There are several other factors that influence the performance of the chromatography column and should be considered when either purchasing or packing columns.

With LC columns, the total effective plates obtained will depend on the length of the column, the particle size with which it was packed, and the flow rate used in the analysis (Table 4). Use of a smaller particle size requires a higher pumping pressure. Similarly, while use of a low flow rate increases the column efficiency, it also increases the analysis time.

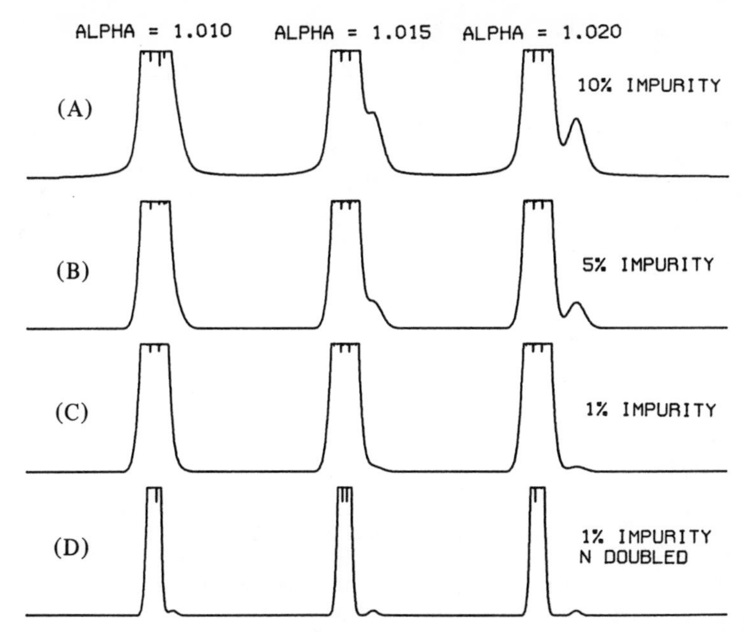

Figure 12. A comparison of the separation obtained for a minor component (impurity) when it occurs at three different percentages using an α of 1.01, 1.015 and 1.02. D illustrates the effect on the separation of a 1% impurity when the column length is increased by a factor of 4.

For capillary GLC columns, in addition to the column length, the total number of effective plates will depend on the column internal diameter and the carrier gas used. A comparison of the column efficiencies obtained with various diameter columns and carrier gases is shown in Table 5. The effect of reducing

Table 4. Comparison of particle size, column efficiency and column pressure, for 0.46-cm inside diameter X 25-cm HPLC silica columns[1]

Particle size (μm)	N_{eff} per meter ($k = 2.0$)	Pressure in psi required to pump at 1 ml/min [2]	
		Pentane	H_2O
3	100,000	2300	7000
5	40,000	1250	2800
10	4,000	600	1250
20	850	125	850

[1]Irregular silica–spherical silica will result in increased efficiency, but with reduced loading capacity.

[2]Increasing column diameter by 2 greatly reduces pressure required to deliver 1 ml/min.

Table 5. Effect of column diameter and carrier gas on the efficiency of capillary columns

Effect of column radius[1]		Effect of carrier gas[2]			
Diameter (mm)	N max/m [3]	Phase	Gas	Linear flow (cm/sec)	Neff/m [3]
0.1	7700	Low polarity	N_2	9.2	3500
0.2	3850		He	25.0	2000
0.25	3125		H_2	40.0	1470
0.50	1560				
0.75	1040	High polarity	N_2	9.5	1790
			He	16.0	1560
			H_2	34.0	1250

[1] Based on formulae derived by Ettre (1965).

[2] Scott (1960).

[3] Based on the authors' findings, medium-polarity and liquid crystal phases have an optimum flow that appears to be between the low-polarity and high-polarity phases. It should be noted that the theoretical column efficiencies listed in this table assume 100% coating efficiency. While low polar phases usually obtain 100% coating efficiency, the medium- and high-polarity coating efficiencies are typically 90%.

the column diameter is similar to that of reducing the particle size in liquid chromatography. It should be noted that column pressure increases with smaller inside diameter as a solute loading capacity decreases. Choosing a carrier gas that will give higher column efficiency will result in a longer analysis time.

Acknowledgments. Mention of a commercial or proprietary product does not constitute an endorsement by the USDA.

Chapter 12

The Significance of Chirality: Methods for Determining Absolute Configuration and Optical Purity of Pheromones and Related Compounds

Kenji Mori[1]

I. Introduction

Since the beginning of pheromone science, the importance of geometrical isomerism in controlling the biological activity of an olefinic pheromone has well been recognized due to the brilliant work of Hecker and Butenandt described in Chapter 1. The importance of optical isomerism or chirality in pheromone perception by insects, however, remained obscure until the mid-1970s. Although Kafka's (1973) pioneering work demonstrated the behavioral discrimination of the two enantiomers of 4-methylhexanoic acid by the honeybee, a great deal of effort among synthetic chemists was necessary before the complicated stereo-

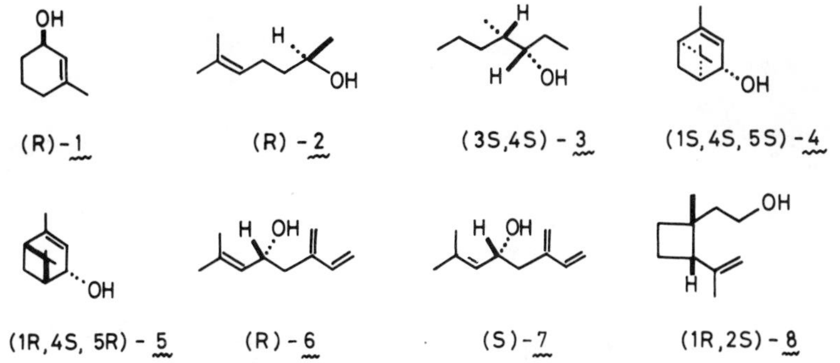

(R)-1 (R)-2 (3S,4S)-3 (1S,4S,5S)-4

(1R,4S,5R)-5 (R)-6 (S)-7 (1R,2S)-8

Figure 1. Pheromone alcohols.

[1] Department of Agricultural Chemistry, The University of Tokyo, Bunkyo-ku, Tokyo, 113, Japan.

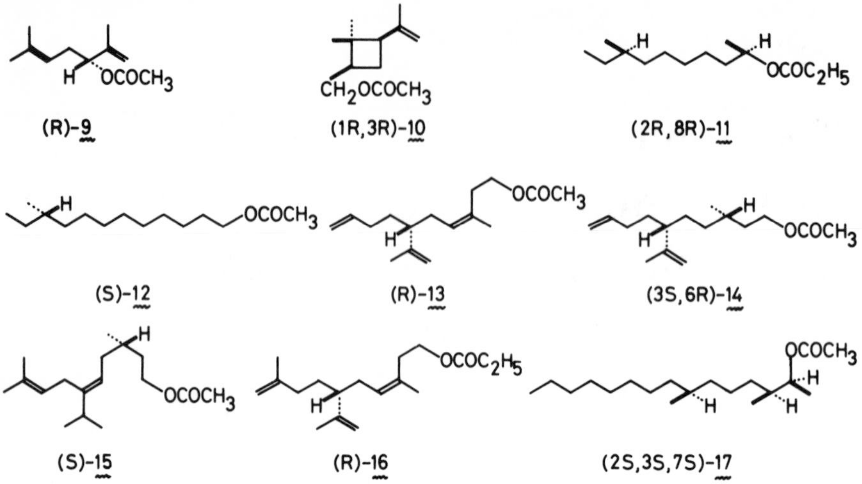

(R)-9 (1R,3R)-10 (2R,8R)-11

(S)-12 (R)-13 (3S,6R)-14

(S)-15 (R)-16 (2S,3S,7S)-17

Figure 2. Pheromone acetates and propionates.

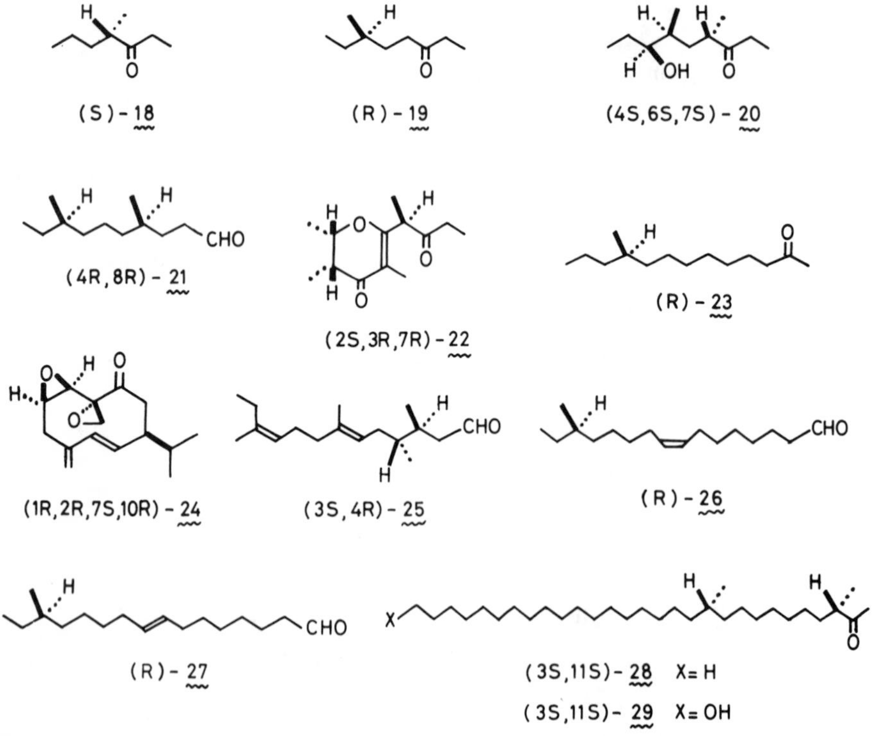

(S)-18 (R)-19 (4S,6S,7S)-20

(4R,8R)-21 (2S,3R,7R)-22 (R)-23

(1R,2R,7S,10R)-24 (3S,4R)-25 (R)-26

(R)-27 (3S,11S)-28 X= H

(3S,11S)-29 X= OH

Figure 3. Pheromone aldehydes and ketones.

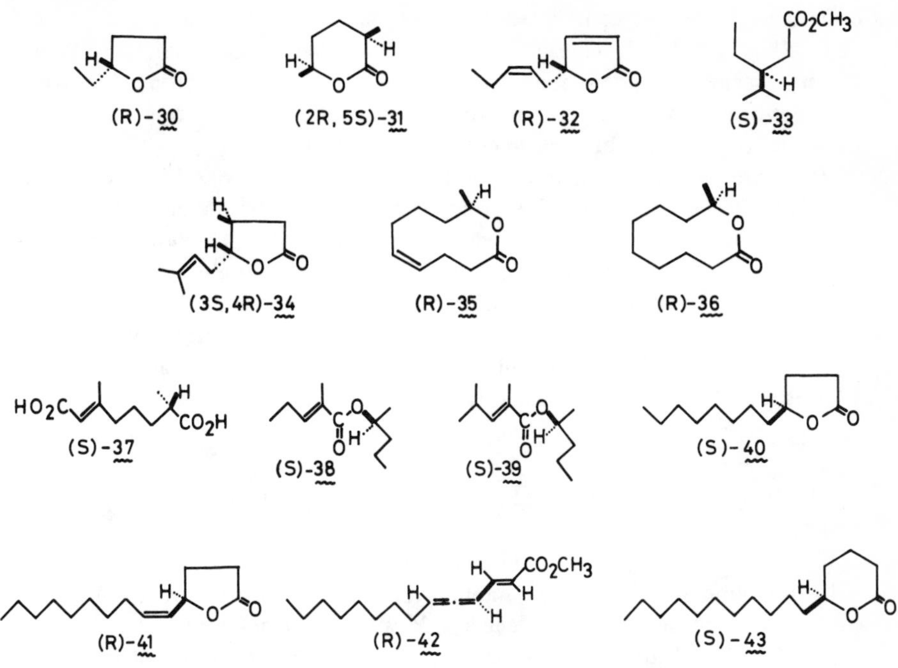

Figure 4. Pheromone acids, esters, and lactones.

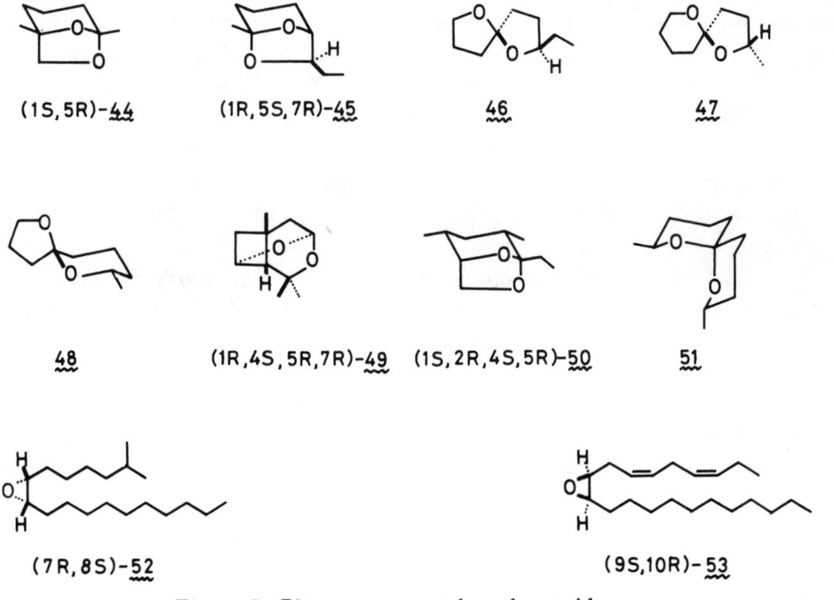

Figure 5. Pheromone acetals and epoxides.

chemistry–pheromone activity relationship became clear. Only after the synthesis of a highly optically pure pheromone is it possible to know something about the stereochemistry–pheromone activity relationship through proper bioassay. The significance of chirality in pheromone perception is now being recognized by proper combination of chiral synthesis and bioassay.

There are two reasons that account for the rapid growth of our knowledge in this area. One is the remarkable progress in synthetic methodology for chiral synthesis applicable to the pheromone field (Mori, 1981). The other is the advent of various new techniques in determining the absolute configuration and optical purity of chiral compounds. This chapter deals with these new techniques. The structures of pheromones discussed in this chapter are shown in Figs. 1 (alcohols), 2 (acetates and propionates), 3 (aldehydes and ketones), 4 (acids, esters, and lactones), and 5 (acetals and epoxides).

II. Methods for Determining Absolute Configuration of Pheromones

Without doubt the most reliable method for the determination of the absolute configuration of an organic molecule is X-ray crystallography (review: Parthasarathy, 1977). Unfortunately the oily nature of most of the pheromones precludes the use of the X-ray method. Another difficulty encountered in pheromone work is that the available amount of the natural product is so limited as to exclude the degradative works leading to a compound of known absolute configuration. These two were the reasons why nothing was known about the absolute configuration of a chiral pheromone until 1973 when Mori (1973, 1974a) synthesized the antipode of the dermestid beetle pheromone artifact **55** (Fig. 6) from (*S*)-(–)-2-methyl-1-butanol and established the absolute configuration of the pheromone artifact to be (*R*). Since then, a synthesis starting from a compound of known absolute configuration has become the standard method for the determination of the absolute configuration of a chiral pheromone if its chiroptical properties such as $[\alpha]_D$ value or ORD/CD data are known.

In Tables 1–5, fifteen pheromones are listed whose absolute configuration was determined by comparing their chiroptical data with those of the synthetic enantiomers. A review of the ORD/CD method is available (Legrand and Rougier, 1977).

Even in the absence of such chiroptical data, the absolute configuration of a pheromone may be clarified by combining a synthesis of the pure enantiomers with their bioassay results. Of course the bioactive enantiomer is the natural pheromone. This is the most popular approach. The absolute configuration of more than 30 pheromones was clarified by this method. However, in the case of a pheromone whose antipode is also bioactive, the bioassay gives no information on the stereochemistry of the natural pheromone. In fact, the final clue to establish the absolute configuration of the German cockroach pheromones **28** and **29**

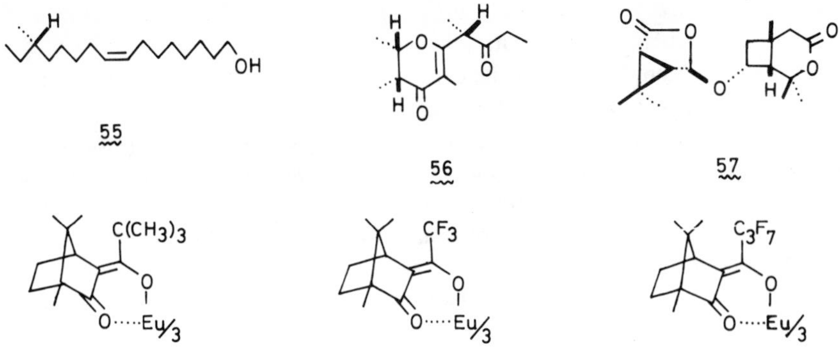

55 56 57

58 59 Eu(facam)$_3$ = Eu(tfc)$_3$ 60 Eu(hfbc)$_3$ = Eu(hfc)$_3$

Figure 6. Structural formulas–1.

was the classical mixture melting point determination (Mori et al., 1981a), since all of the four possible stereoisomers are equally bioactive.

Very recently two groups employed X-ray crystallographic analysis in pheromone studies. Hoffmann et al. (1981) assigned (2*S*, 3*R*, 7*R*)-stereochemistry to the drugstore beetle pheromone (stegobinone) **22** by an X-ray analysis of 7-epistegobinone **56**, which was synthesized from (3*S*)-3-hydroxy-2-methylbutyric ester. Mori et al. (1982b) established the absolute configuration of (+)-lineatin (the pheromone of *Trypodendron lineatum*) to be (1*R*, 4*S*, 5*R*, 7*R*) by an X-ray analysis of an optically active synthetic intermediate **57**. Since the left part of the molecule **57** was derived from (+)-*trans*-chrysanthemic acid, the X-ray study established the stereochemistry of the right-half of **57**. Until now no single natural pheromone was analyzed by the X-ray method.

Chromatographic and nuclear magnetic resonance (NMR) spectroscopic methods have recently been used for determining the absolute configuration of pheromones. These techniques will be discussed in sections IV and V.

III. Problems in Determining Optical Purity of Pheromones

In order to know the stereochemistry–pheromone activity relationship, pheromone enantiomers of high optical purity are required to minimize the cross-contamination with the other enantiomer which may give incorrect bioassay results. It is therefore important to have reliable methods for determining optical purity of pheromones.

The original definition of optical purity as found in a textbook is based on classical polarimetric methods as shown below.

$$\text{Percentage optical purity} = \frac{[\alpha] \text{ mixture*}}{[\alpha] \text{ one enantiomer}} \times 100.$$

*–Specific rotatory power $[\alpha] = \alpha/(c \times l)$, where α is the measured angle of rotation of the plane of polarization in degrees, l is the length of the cell (in dm) containing the sample, and c is the concentration of the sample (in g/ml), which is the density in the case of a pure liquid (cf. Mori et al., 1982a).

In those fortunate cases in which the specific rotation of an optically pure enantiomer is known and the value is large enough, this definition causes no trouble as an experimentally measurable quantity. But it is often difficult to know whether the $[\alpha]_D$ value described in an old literature reference is correct or not. Moreover, in the case of compounds with small $[\alpha]_D$ values, trace amounts of impurities with large $[\alpha]_D$ values can cause errors in polarimetrically determined optical purities with misleading consequences. Indeed the $[\alpha]_D$ value for (+)-disparlure **52**, the gypsy moth pheromone, is too small to be measured correctly (+0.48°-+0.8°).

An alternative definition of optical purity is more suitable for our purpose.

$$\text{Percentage optical purity} = \text{percentage enantiomeric purity} = \frac{M_+ - M_-}{M_+ + M_-} \times 100.$$

where M_+ is the mole fraction of the dextrorotatory enantiomer (here the predominant one), and M_- the mole fraction of the levorotatory one. If there is any method by which one can measure M_+ and M_- directly, this definition gives a more exact result than the polarimetric one. Methods based on NMR spectroscopy and gas or liquid chromatography belong to this category by which one can directly estimate M_+ and M_-. Application of these two methods in the pheromone field will be detailed below.

There are some reviews on the determination of optical purity. Raban and Mislow (1967) reviewed the principle of modern methods for determining optical purity. Reviews are also available in connection with optical resolution (Wilen et al., 1977; Jacques et al., 1981; Newman, 1981). A comprehensive treatise has recently been published (Morrison, 1983) and those who want to study more deeply should read it.

IV. Nuclear Magnetic Resonance Spectroscopic Methods for Determining Optical Purity of Pheromones and Related Compounds

The two enantiomers of an optically active compound show identical NMR spectra if measurements are carried out under achiral conditions such as in an achiral solvent (CCl_4, $CDCl_3$, etc.). In a chiral environment, however, two enantiomers are distinguishable. Symmetry criteria in NMR spectroscopy were discussed by Mislow and Raban (1967). Gaudemer (1977) reviewed various aspects of the determination of optical purity and absolute configuration by NMR spectroscopy.

Three NMR methods are available for determining optical purity: (i) use of chiral shift reagents, (ii) conversion of enantiomers into diastereomers prior to the spectroscopic study, and (iii) use of chiral solvents and chiral solvating reagents.

4.1. Use of Chiral Shift Reagents

Chiral lanthanide shift reagents were first introduced by Whitesides and Lewis (1971). Their chiral shift reagent was tris[3-(*t*-butylhydroxymethylene)-*d*-camphorato] europium (III) **58**. This was modified by Goering et al. (1971, 1974) and by Fraser et al. (1971) to tris[3-(trifluoromethylhydroxymethylene)-*d*-camphorato] europium (III) **59** [Eu(facam)$_3$ or Eu(tfc)$_3$] and tris[3-(heptafluoropropylhydroxymethylene)-*d*-camphorato] europium (III) **60** [Eu(hfbc)$_3$ or Eu(hfc)$_3$]. These two fluorinated reagents are now widely used for the determination of optical purity.

The method consists of the measurement of the NMR spectrum of a mixture of enantiomers in the presence of a chiral shift reagent in an achiral solvent. Complexation of the enantiomers takes place which induces contact and/or pseudocontact shifts, $\Delta\delta$, of different magnitudes for enantiotopic protons of the complexing enantiomers. This causes separation of signals due to enantiotopic protons of the enantiomers. Integration of the peak area of the two signals allows one to know M_+ and M_- directly. This enables one to calculate the optical purity of the enantiomeric mixture. A large $\Delta\delta$ value is desirable for the purpose of correct integration. The magnitude of $\Delta\delta$ depends on the properties of the substrate, the shift reagent, and the solvent. Eu(hfc)$_3$ in CCl_4 is believed to be the best combination for the differentiation of various enantiomeric amines, alcohols, sulfoxides, ketones, esters, and ethers. The use of chiral shift reagents is reviewed by Sullivan (1978).

A ^{1}H-NMR study of our synthetic (±)-, (+)-, and (−)-lineatin **49** is shown in Fig. 7. In the absence of a chiral shift reagent, three CH_3 signals due to (±)-**49** show no splitting and a doublet due to C-1 proton as well as a triplet due to C-5 proton are observable (Fig. 7A). When Eu(hfc)$_3$ was added to (±)-**49**, the C-7 CH_3 signal as well as a CH_3 signal at C-3 splitted into a pair of two singlets and the signals due to C-1 and C-5 protons of two enantiomers were completely separated (Fig. 7B). Separation of the signals like this makes it possible to determine the optical purity of a lineatin sample. In Fig. 7C and 7D, ^{1}H-NMR spectra of (+)-**49** and (−)-**49** in the presence of Eu(hfc)$_3$ are shown in which no splitting of the signal is observable confirming the high optical purity of our optically active lineatin **49**.

Several application of this technique for the estimation of the optical purity of the naturally occurring insect pheromones were reported by Silverstein's group. Their work revealed the optical purity of *trans*-verbenol **5** form *Dendroctonus frontalis* to be 20%, seudenol **1** from *Dendroctonus pseudotsugae* to be 0% and (−)-ipsdienol **6** from *Ips pini* (Idaho) to be 100% (Plummer et al., 1976). Determinations were done on 50∼500 μg of pheromone alcohols. According to

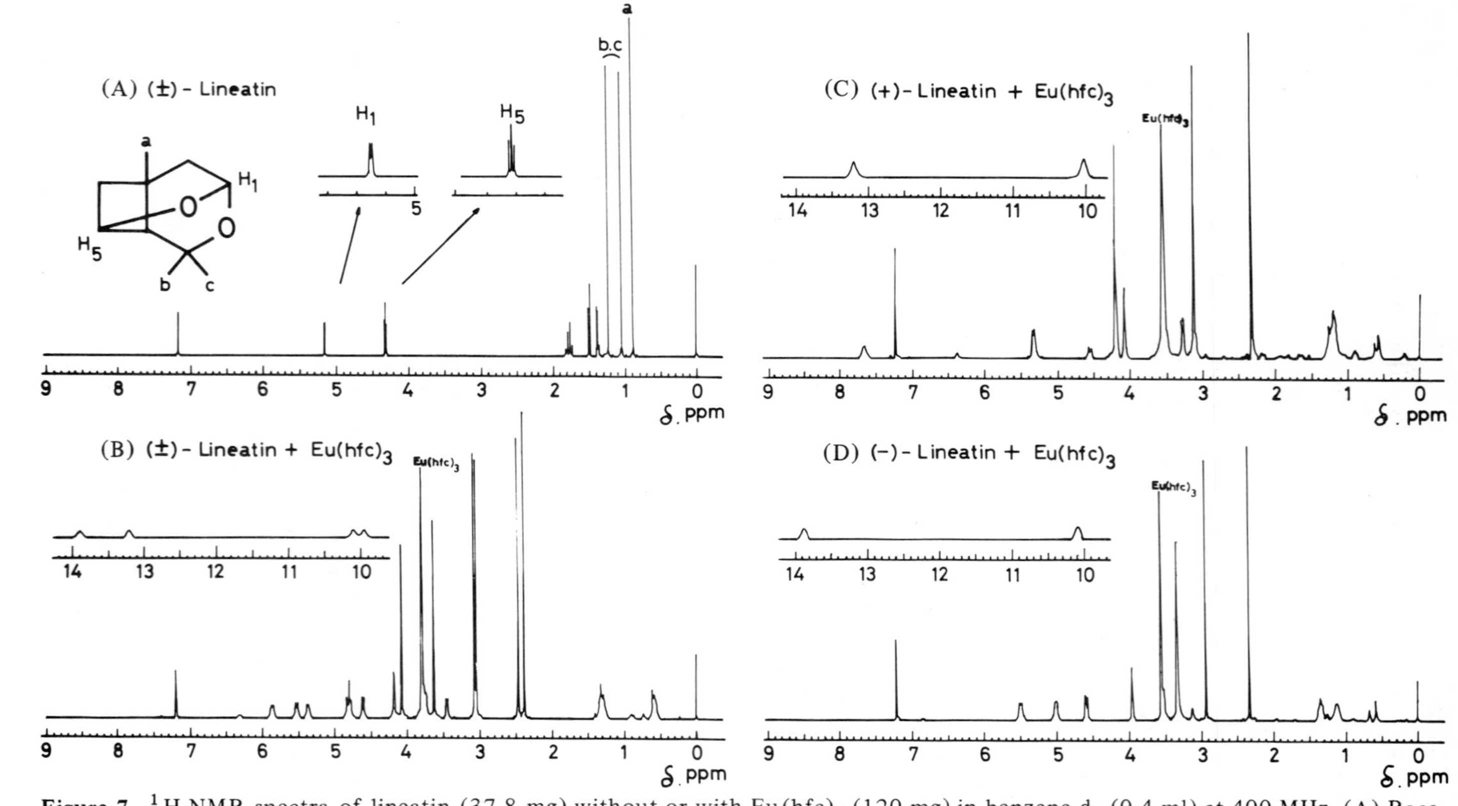

Figure 7. ^{1}H-NMR spectra of lineatin (37.8 mg) without or with $Eu(hfc)_3$ (120 mg) in benzene-d_6 (0.4 ml) at 400 MHz. (A) Racemate without $Eu(hfc)_3$, (B) racemate with $Eu(hfc)_3$, (C) (+)-enantiomer with $Eu(hfc)_3$, (D) (−)-enantiomer with $Eu(hfc)_3$.

them, chiral shift reagents have the following advantages: (i) relative ease of experimental procedure and interpretation of the spectra, (ii) larger enantiomeric shift differences, (iii) smaller amounts of sample, and (iv) quantitative recovery of the sample by GLC. Disadvantages of the procedure are that (i) the sample must be a strong enough Lewis base to coordinate with the europium chelate, and that (ii) the experimentation to determine the best shift reagent/substrate ratio is tedious.

The optical purity of several bicyclic acetal pheromones was also determined by this method (Stewart et al., 1977). *exo*-Brevicomin **45** isolated from female beetles of *Dendroctonus brevicomis* was shown to be the (+)-enantiomer by the measurement of ^{1}H-NMR spectrum in the presence of $Eu(hfc)_3$. The optical purity of (−)-frontalin **44** isolated from *Dendroctonus frontalis* females was revealed to be 70% with only 5 μg of the natural pheromone.

The optical purity of sulcatol **2** produced by *Gnathotrichus retusus* was estimated by this method to be ⩾99% optically pure (*S*)-(+)-enantiomer employing 100 μg of the natural pheromone (Borden et al., 1980). In these NMR studies with $Eu(hfc)_3$, peak assignments were made by spiking the natural material with (±)-sulcatol **2**, since peak positions cannot be reliably reproduced with very small amounts of material. A spectrum of (±)-**2** is shown in Fig. 8A. Fig. 8B shows the collapse to a singlet of the $\mathbf{CHCH_3}$ doublet on decoupling the **CHOH** proton. An undecoupled spectrum of (±)-**2** and $Eu(hfc)_3$ in $CDCl_3$ is shown in Fig. 8C. A ratio of $Eu(hfc)_3$ to substrate of ca. 2.5 produced the optimum separation of peaks in the spectrum. Two doublets at δ 5.1∿5.4 are due to the separation of the $\mathbf{CHCH_3}$ doublets. Decoupling the **CHOH** proton collapsed these doublets to singlets (Figs. 8D-1).

Under the same conditions, 100 μg of synthetic (*S*)-(+)-sulcatol **2** spiked with 20 μg of (±)-**2** produced two peaks with an area ratio of ca. 11:1 (Fig. 8D-2). The spectrum obtained from 100 μg of natural sulcatol from *G. retusus* (Fig. 8D-3) did not display the two singlets characteristic of decoupled racemic sulcatol. When the natural pheromone from *G. retusus* was spiked with 22 μg of (±)-**2** (Fig. 8D-4), the spectrum was almost identical with that obtained from racemic-spiked (*S*)-(+)-**2**. In these NMR measurements detection of less than about 1% of the antipode is precluded by the limit of sensitivity of the spectrometer. *G. retusus* therefore produces at least 99% optically pure (*S*)-(+)-sulcatol. As to the determination of absolute configuration by NMR spectroscopy, a synthetic reference sample with known absolute configuration is always necessary as shown in this example.

Use of chiral shift reagents in ^{13}C-NMR spectroscopy of pheromones was first attempted by Pearce et al. (1976) to determine the optical purities of the synthetic α-multistriatin enantiomers **50**. The noise-decoupled ^{13}C-NMR spectrum of (±)-α-multistriatin **50** with $Eu(hfc)_3$ in deuteriobenzene displayed observable enantiomeric separations for 8 out of 10 carbons, at the molar ratio of $Eu(hfc)_3$/(±)-**50** = 0.22∿0.29, with the largest differences of 0.7∿1.9 ppm observed for the C-1 resonances. The complexity caused by spin-spin coupling often encoun-

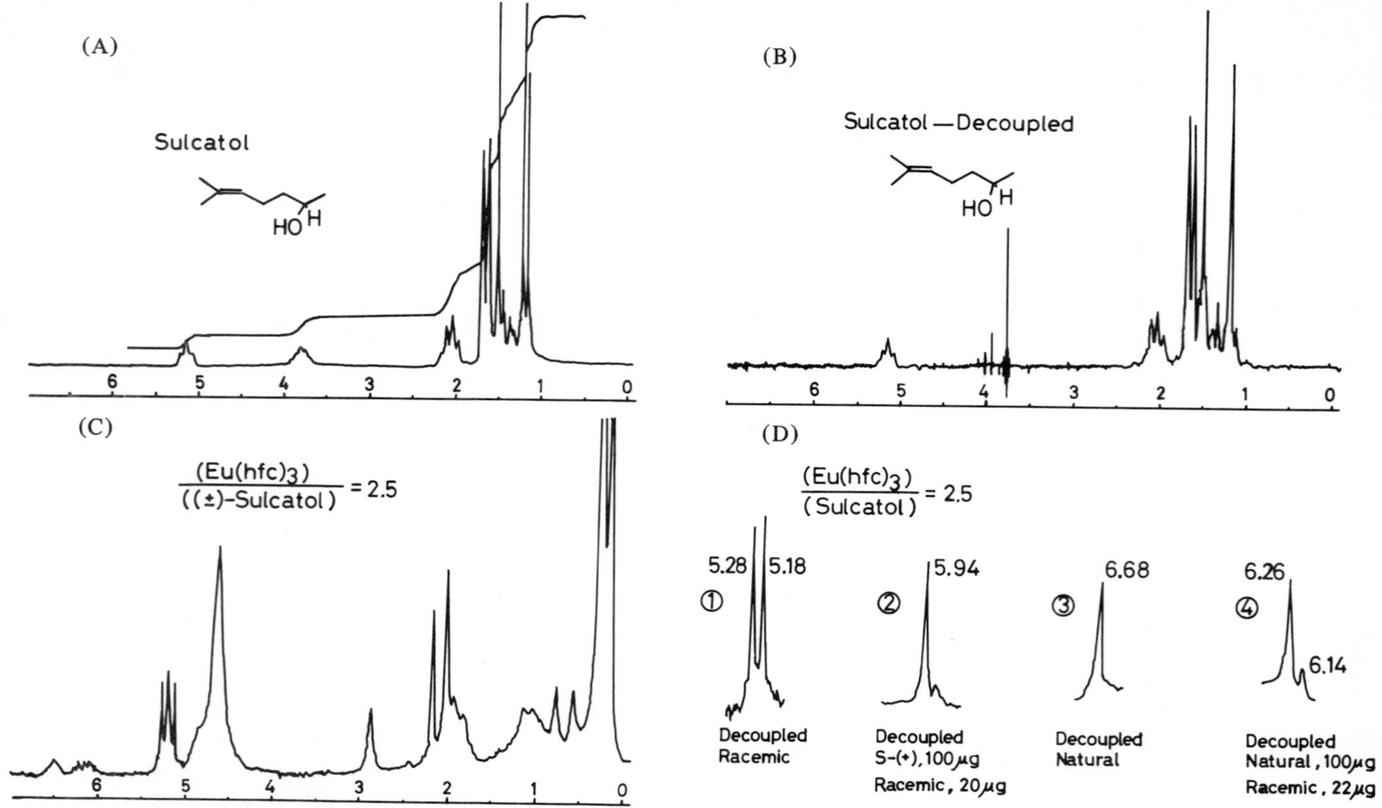

Figure 8. ^{1}H-NMR spectra of sulcatol in chloroform-d at 100 MHz. (A) Racemate, (B) racemate decoupled at C-2, (C) racemate (0.004 *M*), and Eu(hfc)$_3$ (0.010 *M*), [shift reagent]/[substrate] = 2.5, (D) portions of the decoupled NMR spectra of (') racemic, (2) (*S*)-(+) with racemic, (3) natural from *G. retusus*, (4) natural from *G. retusus* with racemic (Borden et al., 1980). Reproduced with the permission of Plenum Publishing Corporation and Professor R. M. Silverstein.

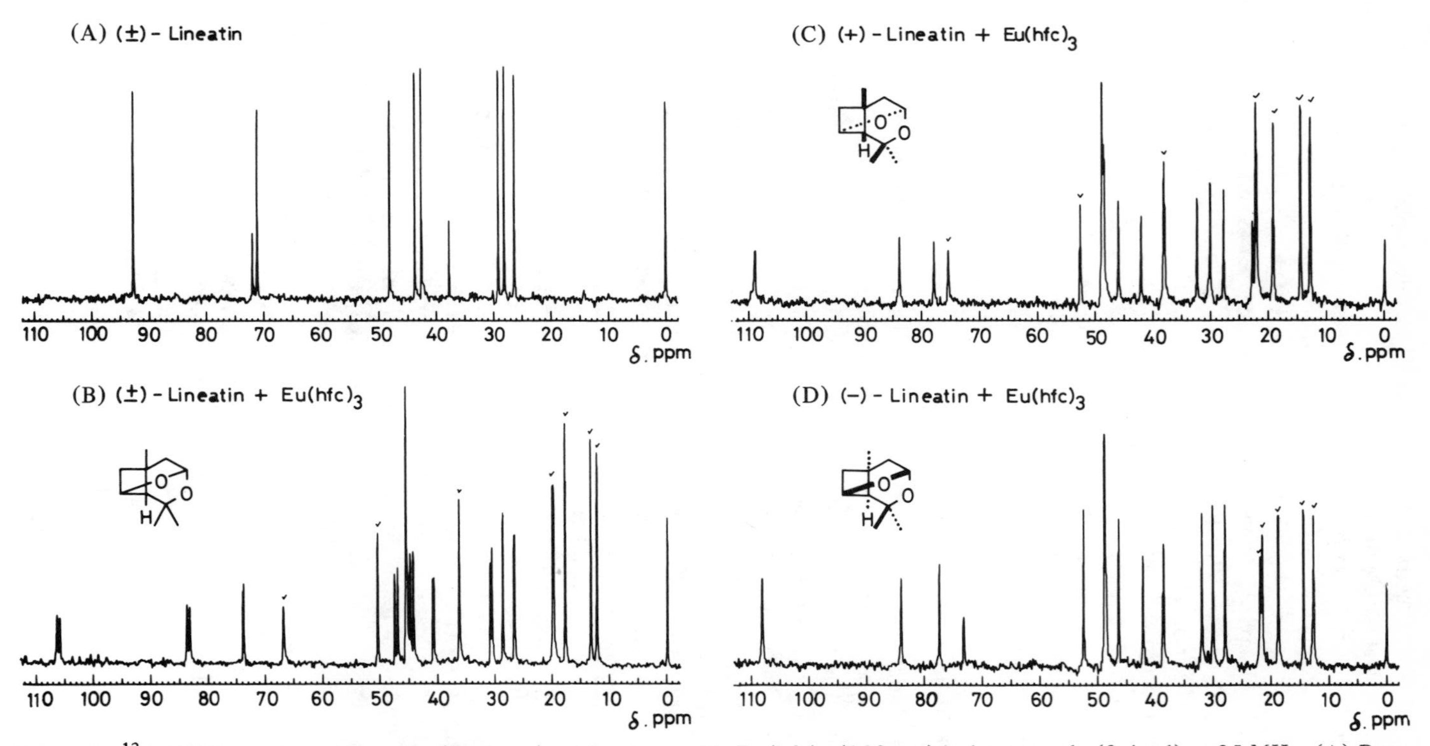

Figure 9. ^{13}C-NMR spectra of lineatin (37.8 mg) without or with Eu(hfc)$_3$ (120 mg) in benzene-d$_6$ (0.4 ml) at 25 MHz. (A) Racemate without Eu(hfc)$_3$, (B) racemate with Eu(hfc)$_3$, (C) (+)-enantiomer with Eu(hfc)$_3$, (D) (−)-enantiomer with Eu(hfc)$_3$.

tered in ^{1}H-NMR spectra is absent in the ^{1}H spin-decoupled ^{13}C-NMR spectrum due to the singlet nature of the ^{13}C-signals.

A ^{13}C-NMR spectrum of (±)-lineatin **49** in deuteriobenzene is shown in Fig. 9A. In the presence of Eu(hfc)$_3$, 6 out of 10 carbons of (±)-**49** exhibited enantiomeric separations (Fig. 9B). Signals due to C-1 and C-5 showed separations most easily observable. In the cases of (+)-**49** and (−)-**49**, no separation was observable in the presence of Eu(hfc)$_3$ (Figs. 9C,D). This supported the high optical purity of our lineatin enantiomers (Mori et al., 1982b).

For the determination of optical purity of chiral lactones Jacovac and Jones (1979) reported a method in which the use of a chiral shift reagent was incorporated. They treated lactones with an excess of methyllithium and the enantiomeric ratio of the resulting dimethyldiol was estimated by measuring its ^{1}H-NMR spectrum in the presence of Eu(tfc)$_3$.

4.2. Use of the Mosher's Ester

By using chiral derivatizing agents, a pair of enantiomers can be converted into a pair of diastereomers which can be analyzed by NMR spectroscopy or chromatography. The most widely used chiral derivatizing agent is Mosher's acid, α-methoxy-α-trifluoromethylphenylacetic acid **61** (Fig. 10), which is abbreviated as MTPA (Dale et al., 1969; Dale and Mosher, 1973). This acid affords diastereomeric derivatives from alcohols or amines (see **62** + **63** → **64** + **65**) that often show sufficient NMR chemical shift differences for diastereotopic nuclei to allow determination of the ratio of diastereomeric products. Generally speaking, this diastereomeric ratio will be equal to the original enantiomeric ratio of the alcohol or amine provided that (i) MTPA is optically pure, (ii) the derivatives are configurationally stable, and (iii) no kinetic resolution or fractionation has taken place during the preparation of the diastereomers.

Many natural and synthetic pheromones were analyzed by the MTPA-NMR, MTPA-gas–liquid chromatography (GLC), or MTPA-high-performance liquid chromatography (HPLC) methods. For example, in order to determine the optical purity of sulcatol 2 isolated from *Gnathotrichus sulcatus,* Plummer et al. (1976) studied in detail the ^{1}H-NMR spectra of the MTPA esters derived from synthetic and natural sulcatol 2. The ^{1}H-NMR spectrum of the (+)-MTPA ester of (±)-2 is shown in Fig. 11A. The enantiomeric composition was reflected only on the signal due to —CH(OR)**CH$_3$**. As shown in Fig. 11B, decoupling by irradiation of —**CH**(OR)CH$_3$ collapsed the —CH(OR)**CH$_3$** triplet to a pair of singlets at δ 1.26 and 1.33, showing that the original triplet was actually two overlapping doublets, each representing an enantiomer. In the ^{1}H-NMR spectrum (Fig. 11C) of the (+)-MTPA ester of (+)-sulcatol 2, the —CH(OR)**CH$_3$** signal was a doublet, which collapsed to a singlet at δ 1.33 when —**CH**(OR)CH$_3$ was irradiated. The (+)-MTPA ester of the natural sulcatol (120 μg) exhibited a spectrum in which —CH(OR)**CH$_3$** appeared as an unsymmetrical triplet (Fig. 11D). After decoupling, this signal collapsed to two singlets of unequal intensity, the singlet at δ 1.33 and the smaller singlet at δ 1.26. The optical purity of the natural sample **2** was thus 30% (*S*)-(+). In the same manner, (−)-4-methyl-3-heptanol **3** isolated from the smaller European elm bark beetle, *Scolytus multistriatus,* was shown to be optically pure upon ^{1}H-NMR and ^{19}F-NMR measurements (Plummer et al., 1976).

The NMR chemical shift differences for diastereotopic nuclei in an MTPA ester becomes more remarkable if an achiral lanthanide shift reagent is added to the sample solution. This technique was first used by Marumo in the pheromone field to determine the optical purity of the synthetic disparlure **52** (Iwaki et al.,

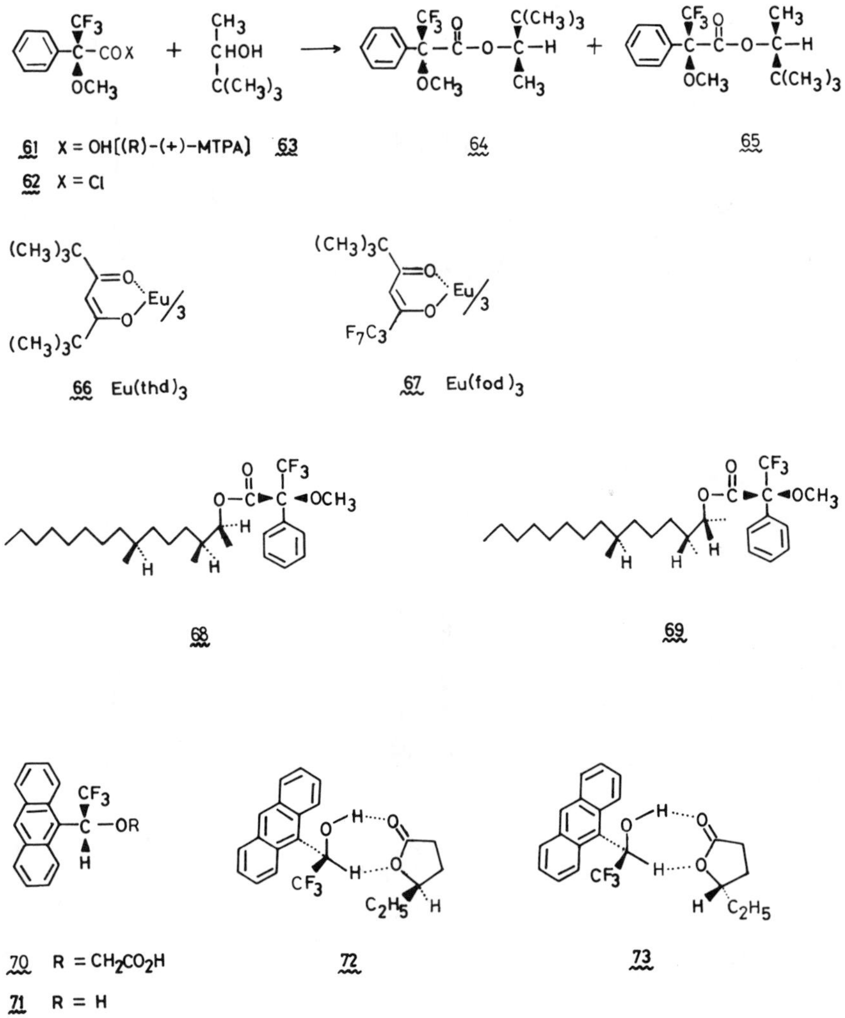

61 X = OH[(R)-(+)-MTPA] 63 64 65

62 X = Cl

66 Eu(thd)$_3$ 67 Eu(fod)$_3$

68 69

70 R = CH_2CO_2H 72 73

71 R = H

Figure 10. Structural formulas–2.

1974) and widely used later. Shift reagents used for this purpose are tris(2,2,6,6-tetramethyl-3,5-heptanedionato) europium (III) **66** [$Eu(DPM)_3$ or $Eu(thd)_3$] and the so-called Sievers's reagent, tris(6,6,7,7,8,8,8-heptafluoro-2,2-dimethyl-3,5-octanedionato) europium (III), **67** [$Eu(fod)_3$]. A concise review of Kime and Sievers (1977) is a practical guide for employing shift reagents.

The MTPA-shift reagent method was used for the determination of the opti-

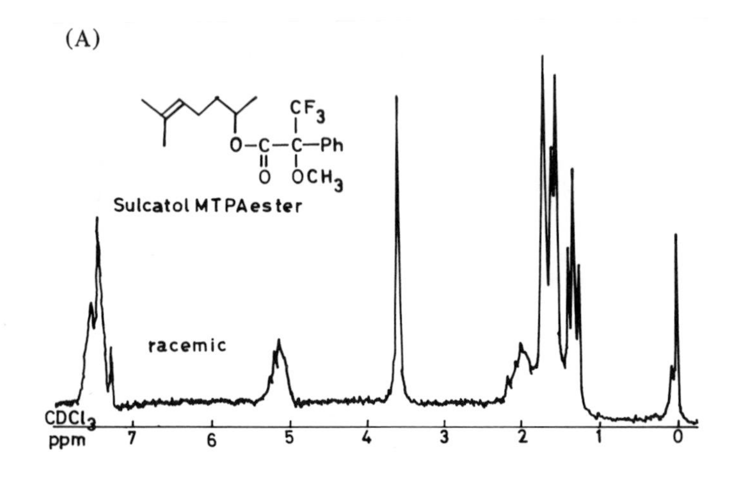

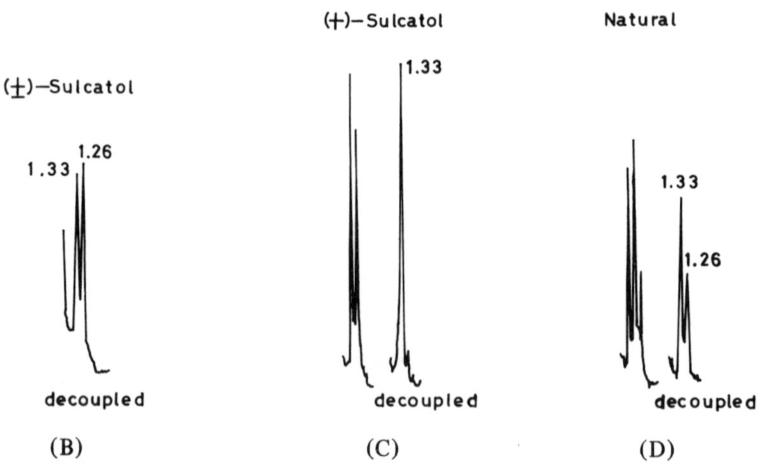

Figure 11. ^{1}H-NMR spectra of the (*R*)-(+)-MTPA ester of sulcatol samples at 100 MHz. (A) Racemate, (B) the C-1-CH_3 signal from the decoupled spectrum of the racemate, (C) the C-1-CH_3 signal from the decoupled spectrum of (+)-sulcatol, (D) the C-1-CH_3 signal from the decoupled spectrum of the natural sulcatol from *Gnathotrichus sulcatus* (Plummer et al., 1976). Reproduced with the permission of Plenum Publishing Corporation and Professor R. M. Silverstein.

cal purity of the synthetic pine sawfly pheromone **17** (Mori and Tamada, 1979). Fig. 12A shows the low-field portion of the ^{1}H-NMR spectrum of a mixture of **68** and **69** in the presence of Eu(fod)$_3$ in CCl_4. Signals due to OCH_3 of **68** and **69** are observed as completely separated two singlets. A part of the signals due to aromatic protons is also seen as a pair of two broad signals. However, only a singlet due to the OCH_3 was exhibited in the spectrum of **68** or **69** even in the presence of Eu(fod)$_3$ (Figs. 12B, C). This confirmed the high optical purity of the synthetic stereoisomers of the pine sawfly pheromone. In Tables 1, 2, and 5

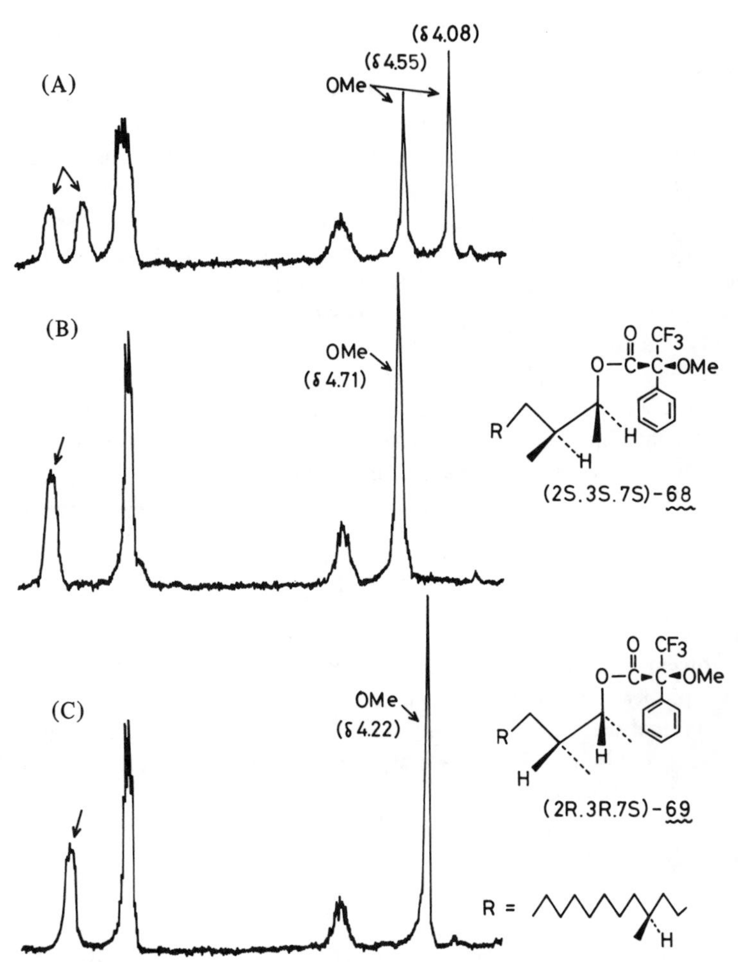

Figure 12. ^{1}H-NMR spectra of the (*S*)-(−)-MTPA ester of 3,7-dimethyl-2-pentadecanol at 100 MHz. (A) (2*S*, 3*S*, 7*S*)-isomer (50 mg), (2*R*, 3*R*, 7*S*)-isomer (50 mg), and Eu(fod)$_3$ (20 mg) in CCl_4 (0.4 ml), (B) (2*S*, 3*S*, 7*S*)-isomer (50 mg) and Eu(fod)$_3$ (20 mg) in CCl_4 (0.4 ml). (C) (2*R*, 3*R*, 7*S*)-isomer (50 mg) and Eu(fod)$_3$ (20 mg) in CCl_4 (0.4 ml).

are listed 10 examples of the use of the MTPA-NMR method for determining optical purity.

Determination of absolute configuration by MTPA-NMR method deserves a short comment, since it is applicable to the pheromone problems. Dale and Mosher (1973) have offered a conformational model for correlating NMR behavior with the absolute configurations of the diastereomers like **64** and **65**. The conformational model, however, represents a weighted average of the effect of all conformations contributing to the diastereotopic nonequivalence and not the exclusive populaton of a single conformation. The use of lanthanide shift reagents reduces the conformational mobility of the diastereomers like **64** and **65** and the resultant chelate-like complexes also make more certain the assignment of absolute configurations to the derivatized alcohols from NMR data. Yamaguchi et al. (1976) found that the ^{1}H-NMR signal from the OCH_3 group of (*R*,*R*)-diastereomer like **64** shifted further downfield with a specified molar ratio of Eu(fod)$_3$ than that of (*R*,*S*)-diastereomer like **65**. The absolute configuration and optical purity of primary carbinols with a chiral center at the C-2 position (Yasuhara and Yamaguchi, 1977) and hydroxycarboxylic esters (Yasuhara and Yamaguchi, 1980) were determined by the present MTPA-Eu(fod)$_3$ method. Indeed, as shown in Fig. 12, the OCH_3 signal of the (*S*,*S*)-diastereomer **68** shifted further downfield than that of the (*S*,*R*)-diastereomer **69**.

Finally it should be added that Pirkle's new chiral derivatizing reagent, (*R*)-[1-(9-anthryl)-2,2,2-trifluoroethoxy]acetic acid **70**, seems to be more useful than MTPA in assigning absolute configuration to alcohols because of the greater diamagnetic anisotropy of the anthryl group and more fixed solution conformation of the esters derived from alcohols and **70** (Pirkle and Simmons, 1981).

4.3. Use of Chiral Solvents and Chiral Solvating Reagents

Enantiomers must be in different average environments when in an optically active solvent. In asymmetrical environments, enantiomers need not possess identical properties toward symmetrical agents. Such differences in behavior toward symmetrical agents may be detected by a sensitive environmental probe like NMR spectroscopy. This was first postulated by Raban and Mislow (1965) and found experimentally by Pirkle (1966). Since then determination of optical purity by NMR spectroscopy using a chiral solvent or a chiral solvating reagent to render the spectra of the enantiomers nonequivalent has mainly been studied by Pirkle (Gaudemer, 1977), although only very few applications in the pheromone field are recorded.

Pirkle's chiral solvating reagent, (*R*)-(−)-2,2,2-trifluoro-1-(9-anthryl)ethanol **71**, has proved to be a useful tool for the determination of optical purity of γ-lactones (Pirkle et al., 1977). Ravid et al. (1978) employed this reagent for the determination the optical purity of their synthetic (*S*)-(−)-γ-caprolactone **30**, a pheromone component of a dermestid beetle, *Trogoderma granarium*. As shown in the solvation models **72** and **73**, the C_2H_5 substituents of the lactone enantiomers respond differently to the shielding effect of the anthryl substituent of **71**. (±)-γ-Caprolactone **30** (0.0035 mmole) and **71** (0.0362 mmole) were dissolved in CCl_4 (130 μl) and C_6D_6 (20 μl). The CH_2CH_3 signal appeared in the NMR spectrum as two overlapping triplets (0.65 and 0.66 ppm), reflecting the racemic composition of the lactone. The NMR spectrum of (*S*)-(−)-**30** (0.0022 mmole) with **71** (0.029 mmole) shows only one triplet of CH_2CH_3, indicating the high optical purity of (*S*)-(−)-**30**.

Recently Webster et al. (1982) reported an absolute procedure based on ^{13}C-NMR spectroscopy for monitoring resolution of carboxylic acids. The diastereomeric quinine salts of nine acids were investigated by ^{13}C-NMR spectroscopy, and in each case the diastereomers could be distinguished. This technique may be useful in pheromone synthesis.

In summary, NMR spectroscopy is a good analytical tool in determining optical purity. A limitation, however, is the difficulty encountered in discerning a signal from noise when the measurements must be done at the ultra-micro scale.

V. Chromatographic Methods for Determining Optical Purity of Pheromones and Related Compounds

Under achiral condition two enantiomers of an optically active compound cannot be separated by chromatographic methods such as gas–liquid chromatography or high-performance liquid chromatography. Separation is possible, however, by converting the enantiomers into diastereomers prior to the chromatographic study on an achiral stationary phase or by employing a chiral stationary phase. Application of these methods in pheromone studies will be described below. For a more general and comprehensive treatise see J. D. Morrison's book (Morrison, 1983), in which V. Schurig and W. Pirkle reviewed the techniques in a detailed manner. Blaschke's short review (1980) is also instructive.

5.1. Use of Gas–Liquid Chromatography on Achiral Stationary Phases

An enantiomeric mixture of a chiral alcohol can readily be converted into a diastereomeric mixture by introducing another asymmetric center by acylation with an optically pure acid. The resultant diastereomeric mixture can be separated by GLC on a conventional achiral stationary phase. Application of this technique on the naturally occurring pheromones was reported by Francke and Kruse (1979; Kruse et al., 1979; Francke and Kruse, 1980). They acylated chiral alcohols with *N*-trifluoroacetyl (TFA)-*L*-alanine and found the resulting esters to be readily separable on glass capillaries with OV-17, SE-30, or Emulphor. A typical procedure for acylation is to keep a mixture of an alcohol (1 mg), a 10%

solution of *N*-TFA-*L*-alanine in methylene chloride (25 μl) and a 10% solution of dicyclohexylcarbodiimide in methylene chloride (10 μl) for 12 hr at room temperature. The enantiomeric composition of an enantiomeric mixture of seudenol **1**, sulcatol **2**, 4-methyl-3-heptanol **3**, *cis*-verbenol **4**, *trans*-verbenol **5**, ipsdienol **6**, and ipsenol **7** was determined by this method. Francke and Kruse (1980) analyzed a raw pentane extract of males of *Ips amitinus* by this method and found that the insect uses 95% optically pure (*R*)-(−)-ipsdienol **6** as an aggregation pheromone.

A more popular acylating agent is (+)- or (−)-α-methoxy-α-trifluoromethylphenylacetyl chloride (MTPA-Cl) **62**. The MTPA esters were analyzed by capillary GLC to estimate the optical purity of either natural or synthetic pheromones. The pheromone of the Comstock mealybug was shown to be optically pure by this method (Bierl-Leonhardt et al., 1980). In Tables 1–5, five examples of the MTPA ester–GLC method are listed.

Saucy et al. (1977) proposed GLC analysis of ortho esters as a convenient and general method for determining the optical purities of chiral δ-lactones. The derivatization-step is the conversion of the lactones to their ortho esters with (−)-(2*R*, 3*R*)-2,3-butanediol and *p*-toluenesulfonic acid in benzene.

5.2. Use of Gas–Liquid Chromatography (GLC) on Chiral Stationary Phases

Quite recently it has become possible to separate enantiomers directly by taking advantage of chiral recognition exhibited by some well-designed chiral stationary phases.

To date Schurig's "complexation GLC" is the most successful method in this area (Schurig, 1980; Schurig and Bürkle, 1982; Morrison, 1983). "Complexation GLC" is the technique that utilizes the rapid and the reversible coordination equilibrium between a substrate and the solution of a metal coordination compound in a nonvolatile liquid. Quantitative separation of the enantiomeric pairs of chalcogran **46** was achieved on an optically active chelate, bis[6-heptafluorobutyryl-(*R*)-pulegonato] nickel (II) **74** (Fig. 13) (0.12 *M* in squalane) using a 100 m × 0.5 mm nickel capillary column (Koppenhoefer et al., 1980). Spiroacetal pheromones such as methyl-1,6-dioxaspiro [4.5] decanes **47** and **48** were analyzed on bis[3-heptafluorobutyryl-*d*-camphorato] manganese (II) **75** (Hintzer et al., 1981). GLC analysis of spiroacetals on **74** or **75** is also recorded (Weber et al., 1980).

Schurig's method was successfully applied to resolve lineatin enantiomers **49** on bis[3-heptafluorobutyryl-*d*-camphorato] copper (II) (Weber and Schurig, 1981). Fig. 14 shows the analysis of our synthetic lineatin enantiomers **49** (Mori et al., 1982b) by complexation GLC (Schurig and Weber, 1982). The optical purity of lineatin enantiomers was estimated to be >99 (±0.5) % for (1*R*, 4*S*, 5*R*, 7*R*)-(+)-**49** and 98.4 (±0.5) % for (1*S*, 4*R*, 5*S*, 7*S*)-(−)-**49**. By this method Schurig et al. (1982) proved that all three *Trypodendron* spp. (*T. lineatum, T. domesticum,* and *T. signatum*) enantioselectively produce (+)-lineatin **49** with an optical purity of 99 ± 0.5%. Enantiomers of sulcatol **2**, *threo*-4-methyl-3-heptanol **3**, frontalin **44** and *exo*-brevicomin **45** were separable by complexation GLC (Schurig and Weber, 1982; Schurig et al., 1983).

Other chiral stationary phases were also developed by Ôi et al. (1982a). *O*-Lauroyl-(*S*)-mandelic acid (*R*)-1-(1-naphthyl)ethylamide was especially useful in separating an enantiomeric mixture of ethyl chrysanthemate or some

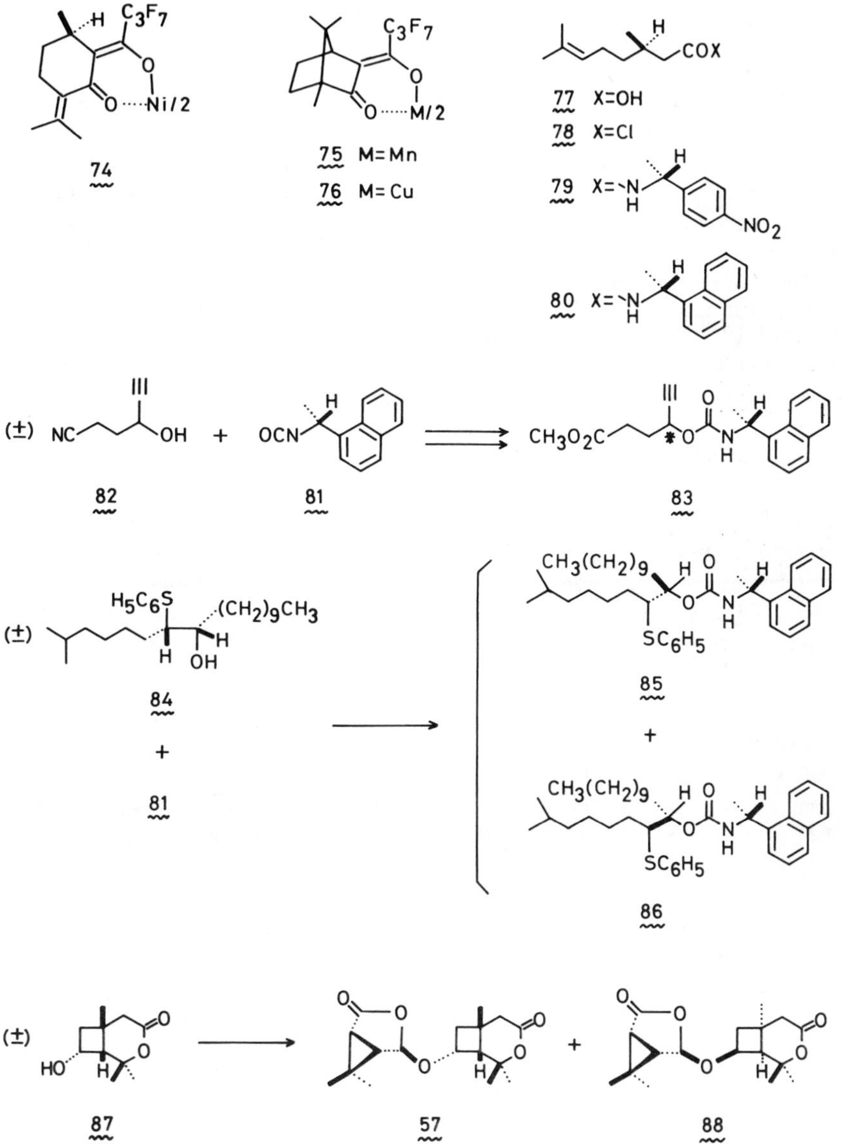

Figure 13. Structural formulas–3.

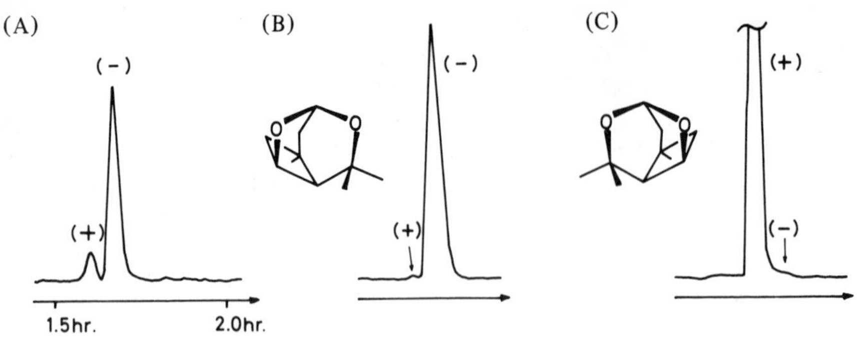

Figure 14. Determination of optical purity of lineatin enantiomers by complexation GLC. (A) (—)-Lineatin was coinjected with (±)-lineatin on 18 m × 0.3 mm soft glass capillary column coated with a solution of 0.027 *M* bis[3-heptafluorobutanoyl-(1*R*)-camphorato] copper (II) in OV 101; oven temperature: 35°C, 0.35 bar N_2, split: 1/50. (B) (—)-Lineatin was injected on the same column under the same condition. Optical purity: 98.4 (±0.5) %. (C) (+)-Lineatin was injected on 62 m × 0.3 mm Pyrex glass capillary column coated with a solution of 0.05 *M* bis[3-heptafluorobutanoyl-(1*R*)-camphorato] copper (II) in OV 101; oven temperature 45°C, 0.4 bar N_2, split: 1/50. Optical purity 99 (±0.5) %.

carboxylic acid amides (Ôi et al., 1982a). Better separation of alkyl carboxylate enantiomers was achieved by using *O*-(1*R*, 3*R*)-*trans*-chrysanthemoyl-(*S*)-mandelic acid (*R*)-1-(1-naphthyl)ethylamide (Ôi et al., 1982b). An illustration of the successful use of the latter chiral stationary phase is the determination of the absolute configuration of phoracantholides I **36** and J **35**, the major components of the metasternal gland secretion of the eucalypt longicorn, *Phoracantha synonyma* (Moore and Brown, 1976).

The synthesis of (±)-, (*R*)-, and (*S*)-phoracantholides I and J was carried out by us (Kitahara et al., 1983). A minute amount of the natural secretion was available from Dr. Moore, which was insufficient for the chiroptical measurements. The natural products were therefore compared with the synthetic samples by GLC on *O*-(1*R*, 3*R*)-*trans*-chrysanthemoyl-(*S*)-mandelic acid (*R*)-1-(1-naphthyl)ethylamide (Ôi, 1982). As shown in Fig. 15A, (±)-phoracantholide I **36** was resolvable. (*S*)-**36** was eluted later as revealed by the coinjection experiment (Fig. 15B). (±)-Phoracantholide J **35** was also resolvable (Fig. 15C) and the later eluted one was (*S*)-**35** (Fig. 15D). Analysis of a mixture of the natural phoracantholides I and J indicated that they are enantiomerically pure (Fig. 15E). The compound with a shorter retention time was phoracantholide I **36** as shown by the coinjection of (±)-**36** with the natural products (Fig. 15F). By the coinjection of (±)-**35** and (±)-**36** with the natural products, the compound with a longer retention time was proved to be phoracantholide J **35** (Fig. 15G). The (*R*)-configuration was assigned to the natural products by proving the identity of synthetic (*R*)-**35** and (*R*)-**36** with natural phoracantholides J and I upon coinjection (Fig. 15H).

The latest development in this area is the microscale resolution of chiral alcohols on glass capillary columns coated with XE-60-(*S*)-valine-(*S*)-α-phenylethylamide (Benecke and König, 1982; König et al., 1982). Samples of 0.5 mg or less of racemic alcohols were treated with isopropyl isocyanate to give stable isopropyl urethanes. By this derivatization, the polarity of alcohols and their enantioselective intermolecular interaction with the chiral stationary phase was sufficiently enhanced resulting in enantiomer separation with moderate retention times (<1 hr on a 40-m Pyrex glass capillary column at 120°). The following pheromone alcohols were successfully analyzed: sulcatol **2**, *threo*-4-methyl-3-heptanol **3**, *trans*-verbenol **5**, and ipsdienol **6**.

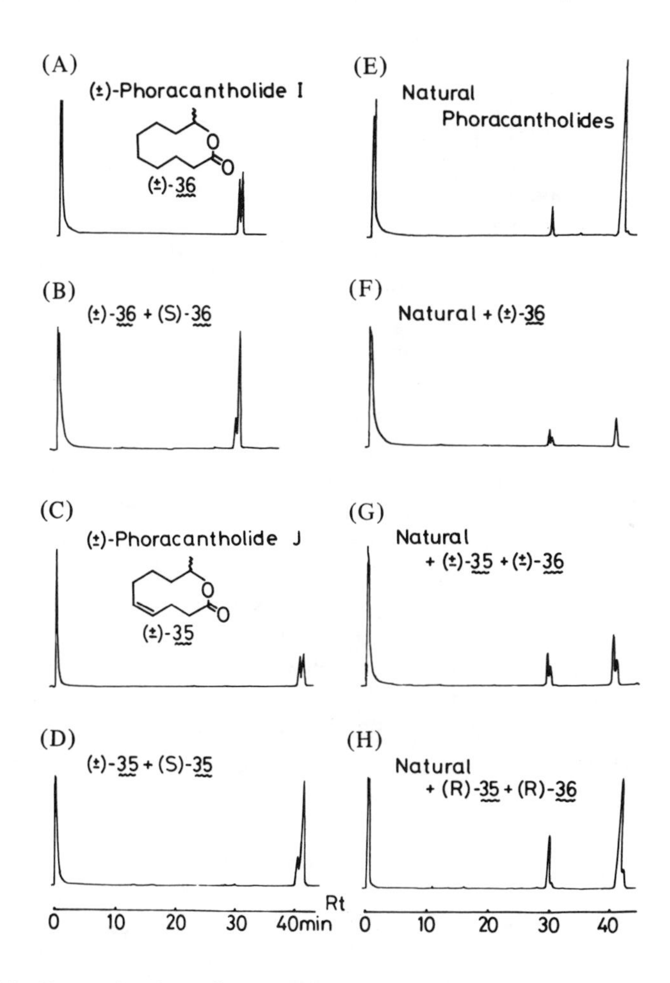

Figure 15. Determination of absolute configuration of phoracantholide I and J by GLC on a chiral stationary phase. The sample was analyzed on 40 m × 0.25 mm glass capillary column coated with *O*-(1*R*, 3*R*)-*trans*-chrysanthemoyl-(*S*)-mandelic acid (*R*)-1-(1-naphthyl)ethylamide; oven temperature: 95°C; carrier gas, He, 0.47 ml/min; split: 1/170. For details see the text.

5.3. Use of High-Performance Liquid Chromatography on Achiral Stationary Phases

An enantiomeric mixture of a chiral alcohol or a chiral acid can be converted into a diastereomeric mixture by introducing another asymmetric center by treatment with an optically pure derivatizing reagent. The resultant diastereomeric mixture can be separated by HPLC on a conventional achiral stationary phase. This technique was used for the determination of optical purity of some intermediates in pheromone syntheses as well as for preparative-scale separation of diastereomers.

(*R*)-(+)-Citronellic acid **77** is a versatile starting material for the syntheses of vitamins and pheromones. Valentine et al. (1976) analytically separated the diastereomeric amides **79** (+ its diastereomer) obtained by the reaction of (*R*)-(+)-1-(*p*-nitrophenyl)ethylamine with citronelloyl chloride **78** (see also Scott et al., 1976). An improved procedure of this method was to use commercially available (+)-1-(1-naphthyl)ethylamine (Bergot et al., 1978). (*R*)-(+)-Citronellic acid **77** was thus converted to an amide **80**, which was chromatographed on a Zorbax SIL microparticulate silica column with ultraviolet detection at 254 nm. The enantiomeric composition is then derived from the diastereomeric ratio. Very low loading ($\leqslant 1$ μg) was required because of a high molar extinction coefficient due to the naphthalene ring. Analysis of several chiral monoterpene acids and related compounds gave no conclusive correlation between elution order and absolute configuration (Bergot et al., 1978; Mori et al., 1982a). Other chiral acids and lactones were also analyzed by HPLC as their 1-(1-naphthyl)ethylamides (Roelofs et al., 1978; Heath et al., 1980; Mori and Ueda, 1982).

Some of Pirkle's extensive works on HPLC were concerned with the preparative-scale separation of intermediates in pheromone synthesis. For the separation of alcohol enantiomers, a racemic alcohol was converted to a diastereomeric mixture of carbamates by treatment with (*R*)-1-(1-naphthyl)ethyl isocyanate **81**. For example, in their synthesis of the sex pheromone of the dried bean beetle **42**, Pirkle and Boeder (1978) converted an alcohol **82** into a carbamate by treatment with **81**. It was then treated with hydrogen chloride in methanol-ether followed by water to give **83**. HPLC separation of **83** (11.7 g) yielded 5.8–5.9 g (quantitative) of two pure diastereomers of **83**.

Another example of Pirkle's separation of carbamate diastereomers was reported in connection with their synthesis of disparlure enantiomers (Pirkle and Rinaldi, 1979). A racemic hydroxy sulfide **84** was converted to a mixture of **85** and **86**, which was separated by HPLC. Silanolysis of **85** or **86** with trichlorosilane yielded both enantiomers of **84**. The optical purity of **84** was determined by the measurement of its ^{13}C-NMR spectrum in the presence of Eu(hfc)$_3$. HPLC analysis of the corresponding MTPA ester of a chiral alcohol was also useful in determining optical purity (Anderson et al., 1980).

Helmchen et al. (1979a,b) converted enantiomeric carboxylic acids and lactones into the diastereomeric amides by treatment with (*R*)-(−)-phenylglycinol. The diastereomeric amides were separable by HPLC even on a preparative scale.

This method was used in the synthesis of the tsetse fly pheromone (Ade et al., 1980).

In our synthesis of (1*R*, 4*S*, 5*R*, 7*R*)-(+)-lineatin **49**, a racemic intermediate **87** was converted to a diastereomeric mixture of **57** and **88**. This mixture was separable by chromatography (Mori et al., 1982b).

5.4. Use of High-Performance Liquid Chromatography on Chiral Stationary Phases

The direct HPLC separation of enantiomers on a column packed with a chiral stationary phase is a highly desirable technique from both analytical and preparative standpoints. If a chiral stationary phase is to show a greater affinity for one solute enantiomer than the other, the preferentially bound enantiomer must undergo a minimum of three simultaneous interactions with the chiral stationary phase, with at least one of the interactions being dependent upon the stereochemistry of that solute enantiomer.

Using this chiral recognition rationale, Pirkle et al. devised chiral fluoroalcoholic bonded stationary phases **89** and **90** (Fig. 16) (Pirkle and House, 1979; Pirkle et al., 1980). These proved capable of resolving a large number of racemates including sulfoxides, lactones, and 3,5-dinitrophenyl or 3,5-dinitrobenzoyl derivatives of alcohols, amines, amino acids, hydroxy acids, and mercaptans.

Pirkle et al. developed another chiral stationary phase **91**, which was suitable for the resolution of arylalkylcarbinols and bi-β-naphthols (Pirkle et al., 1981;

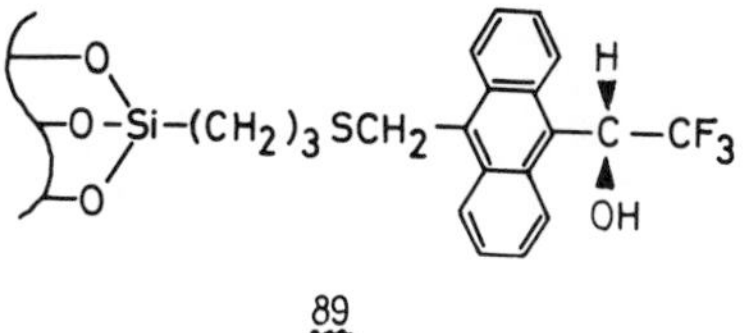

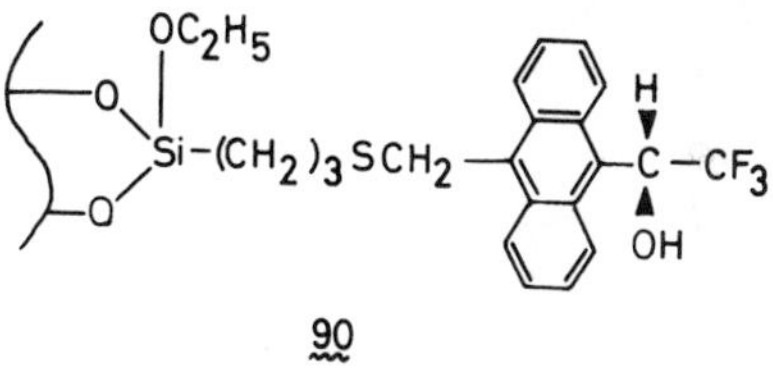

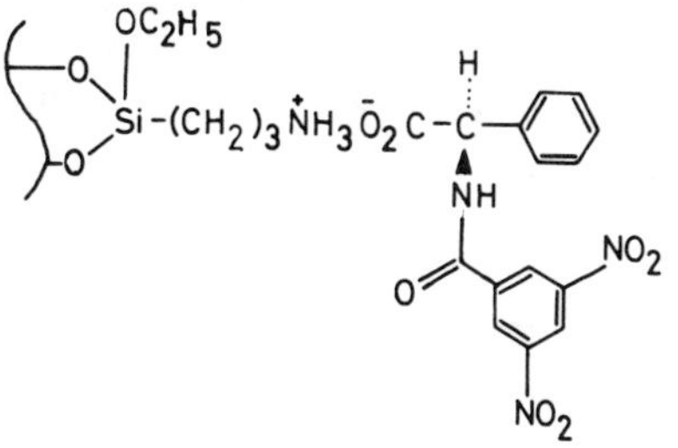

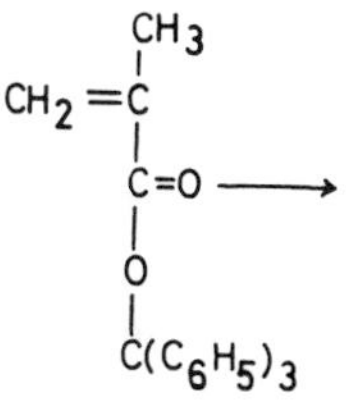

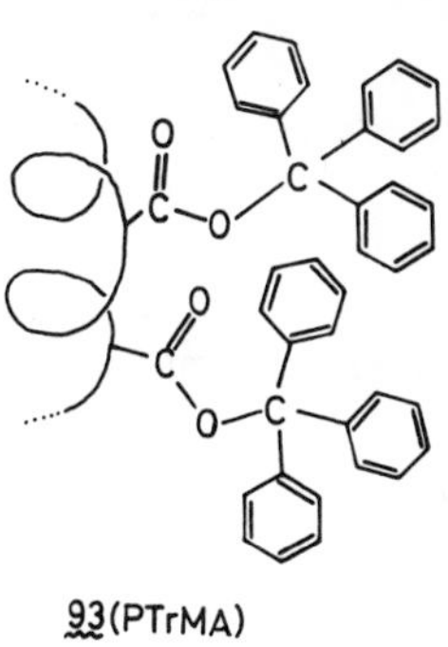

Figure 16. Structural formulas—4.

Pirkle and Finn, 1981; Pirkle and Schreiner, 1981). By employing the chiral stationary phase **91**, Pirkle determined the optical purity of disparlure, the gypsy moth pheromone, as follows (Pirkle, 1982). Disparlure **52** was treated with 9-anthryl thiol in pyridine. The resulting regioisomeric hydroxy sulfides were treated with *m*-chloroperbenzoic acid to give a mixture of hydroxy sulfoxides. This was chromatographed upon 5 μm spherisorb silica. The major hydroxy sulfoxide diastereomer (actually a mixture of regioisomers) is eluted before the minor diastereomer. A portion of the chromatographed major diastereomer was then chromatographed on a chiral stationary phase **91**. (±)-Disparlure **52** gave four equal area peaks (two regioisomers, each of which is racemic) at this stage. Optically pure disparlure gave two peaks. Partially resolved pheromone gave two unequally sized pairs of peaks. The optical purity of disparlure **52** could thus be determined. As little as 0.2% of the minor enantiomer could be detected. Preparative resolution of racemates by HPLC on **91** was also reported by Pirkle and Finn (1982).

Very recently Okamoto et al. (1981a,b) found that optically active poly(triphenylmethyl methacrylate) [(+)-PTrMA, **93**] is an efficient chiral packing material for optical resolution by HPLC. The chiral stationary phase **93** was obtainable by polymerization of triphenylmethyl methacrylate (TrMA, **92**) with *n*-butyllithium-(−)-sparteine. PTrMA has the chirality caused only by helicity and is capable of resolving various racemic aromatic compounds. It has not yet been employed in pheromone studies. Recently amino acid esters were resolved by HPLC on a new chiral stationary phase employing a hexahelicene derivative (Kim et al., 1982). Two monographs appeared (Eliel and Otsuka, 1982; Leonhardt and Beroza, 1982), in which chromatographic separation of enantiomers is briefly discussed.

VI. Conclusion—The Significance of Chirality in Pheromone Perception

The existing data on the determination of absolute configuration and optical purity of pheromones are summarized in Tables 1 (alcohols), 2 (acetates and propionates), 3 (aldehydes and ketones), 4 (acids, esters, and lactones), and 5 (acetals and epoxides).

The relationship between chirality and pheromone activity of each pheromone is also indicated in the tables. The capital letter A, B, C, D, or E after the name and origin of the pheromone indicates the category to which that individual pheromone belongs.

In the case of those pheromones in group A, only one enantiomer is biologically active and no inhibitory action was observed with the inactive antipode. The majority of chiral pheromones belongs to this group.

Similarly, in the case of those in group B, only one enantiomer is biologically active. The inactive antipode, however, inhibits the action of the correct enantiomer. Especially in the case of the pheromone of the Japanese beetle (*R*)-**41**, its racemate lacks biological activity due to the inhibition caused by the wrong enantiomer (Tumlinson et al., 1977; Doolittle et al., 1980).

In the case of pheromones in group C, insects do not discriminate between stereoisomers. It is indeed surprising that the unnatural (1*S*, 2*R*)-grandisol **8** is biologically active (Mori et al., 1978b).

Ipsdienol **6** is the only pheromone which belongs to group D. Different species of *Ips* use different enantiomers of **6** and the chirality of the pheromone is quite important in establishing and maintaining a particular *Ips* species (Birch et al., 1980).

Sulcatol (group E) is the only pheromone both of whose enantiomers are required for pheromone activity (Borden et al., 1976).

In summary, the relationship between chirality and pheromone activity is rather complicated. In some cases only one enantiomer is biologically active, but in others both enantiomers are fully active. The situation changes from insect to insect. The precise meaning of this diversity requires more study about the nature of pheromone perception.

Recent developments in the methods for determining absolute configuration and optical purity of pheromones have been extraordinarily rapid. In the near future, determination of the enantiomeric composition of a natural pheromone will be routinely carried out even with a limited amount of sample. This will certainly give us many new data to know more about the significance of chirality in pheromone perception.

The literature survey for this chapter was made up to the end of April 1982.

VII. Appendix: Selected Additional References

Two isomeric γ-lactones were isolated from the male-produced pheromone blend of Mexican fruit fly (*Anastrepha ludens*) and characterized by X-ray analyses (Stokes et al., 1983). This is the first example of the application of the X-ray method to structure elucidation of natural pheromones, although only their relative stereochemistries were assigned.

Stokes JB, Uebel EC, Warthen Jr JD, Jacobson M, Flippen-Anderson JL, Gilardi R, Spishakoff LM, Wilzer KR (1983) Isolation and identification of novel lactones from male Mexican fruit flies. J Agric Food Chem **31**:1162–1167.

Blight et al. (1979) studied 4-methyl-3-heptanol **3** produced by *Scolytus scolytus.*

Blight MM, Wadhams LJ, Wenham MJ (1979) The stereoisomeric composition of the 4-methyl-3-heptanol produced by *Scolytus scolytus* and the preparation and biological activity of the four synthetic diastereoisomers. Insect Biochem **9**:525–533.

Table 1. Determination of absolute configuration and optical purity of pheromones (alcohols)

Molecular formula (Structure)	Name and origin of the pheromone	Stereochemistry-bioactivity relationship	Absolute configuration of the natural pheromone	Method for determining the absolute configuration of the natural pheromone	Method for determining the optical purity of the synthetic pheromone or its starting material
$C_7H_{12}O$ (1)	Seudenol *Dendroctonus pseudotsugae*	C (McKnight, 1978)	Both (*R*) and (*S*)[1] 1:1	Synthesis (Mori et al., 1978c) and NMR (shift reagent) (Plummer et al., 1976)	NMR (MTPA-shift reagent) (Mori et al., 1978c)
$C_8H_{16}O$ (2)	Sulcatol *Gnathotrichus sulcatus*	E (Borden et al., 1976)	Both (*R*) and (*S*)[1] 35:65	Synthesis (Mori, 1975b) and NMR (MTPA) (Plummer et al., 1976)	NMR (MTPA) (Mori, 1975b; Schuler and Slessor, 1977)
$C_8H_{18}O$ (3)	(−)-*threo*-4-Methyl-3-heptanol *Scolytus multistriatus*	A (Lanier et al., 1977)	(3*S*,4*S*)[1]	Synthesis and $[\alpha]_D$ (Mori, 1977)	NMR (MTPA) (Mori and Iwasawa, 1980)

$C_{10}H_{16}O$ (4)	*cis*-Verbenol *Ips paraconfusus*	B (Vité et al., 1976b)	(1*S*,4*S*,5*S*)	Synthesis and $[\alpha]_D$ (Mori et al., 1976)	NMR (MTPA) (Mori et al., 1976)
$C_{10}H_{16}O$ (5)	*trans*-Verbenol *Dendroctonus frontalis*	–	Both (1*R*,4*S*,5*R*)[1] and (1*S*,4*R*,5*S*) 60:40	NMR (shift reagent) (Plummer et al., 1976)	NMR (MTPA) (Mori, 1976a)
$C_{10}H_{16}O$ (6)	Ipsdienol *Ips pini* (Idaho) *Ips paraconfusus*	D (Vité et al., 1978; Birch et al., 1980)	(*R*) *Ips pini* (Idaho)[1] (*S*) *Ips paraconfusus*	Synthesis and $[\alpha]_D$ (Mori, 1976c; Mori et al., 1979c)	$[\alpha]_D$ of the starting material (Ohloff and Giersch, 1977)
$C_{10}H_{18}O$ (7)	Ipsenol *Ips paraconfusus*, etc.	A (Vité et al., 1976a)	(*S*)	Synthesis and $[\alpha]_D$ (Mori, 1976b; Mori et al., 1979c)	$[\alpha]_D$ of the starting material (Mori, 1976b; Mori et al., 1979c)
$C_{10}H_{18}O$ (8)	Grandisol *Anthonomus grandis*	C (Mori et al., 1978b)	(1*R*,2*S*)	Synthesis and $[\alpha]_D$ (Hobbs and Magnus, 1976; Mori, 1978)	$[\alpha]_D$ of the starting material (Hobbs and Magnus, 1976) NMR (shift reagent) (Mori, 1978)

[1]The optical purity of the natural pheromone was determined by either NMR (shift reagent) or NMR (MTPA) method (Plummer et al., 1976).

Table 2. Determination of absolute configuration and optical purity of pheromones (acetates and propionates)

Molecular formula (Structure)	Name and origin of the pheromone	Stereochemistry-bioactivity relationship	Absolute configuration of the natural pheromone	Method for determining the absolute configuration of the natural pheromone	Method for determining the optical purity of the synthetic pheromone or its starting material
$C_{11}H_{18}O_2$ (**9**)	Acetate of 2,6-dimethyl-1,5-heptadien-3-ol *Pseudococcus comstockii*	A (Mori and Ueda, 1981b)	(*R*)[1]	Synthesis and $[\alpha]_D$ (Mori and Ueda, 1981b)	GLC (MTPA) (Mori and Ueda, 1981b)
$C_{12}H_{20}O_2$ (**10**)	Acetate of 2,2-dimethyl-3-isopropenylcyclobutylmethanol *Planococcus citri*	A (Bierl-Leonhardt et al., 1981)	(1*R*,3*R*)	Synthesis and $[\alpha]$ (Bierl-Leonhardt et al., 1981)	Comparison of $[\alpha]_{3130}$ (Bierl-Leonhardt et al., 1981)
$C_{14}H_{28}O_2$ (**11**)	Propionate of 8-methyl-2-decanol *Diabrotica virgifera virgifera*	Both (2*R*,8*R*) and (2*S*,8*R*)-isomers are bioactive (Tumlinson, 1982)	Unknown	Synthesis and bioassay (Tumlinson, 1982)	
$C_{15}H_{30}O_2$ (**12**)	Acetate of 10-methyl-1-dodecanol *Adoxophyes* sp.	C (Tamaki et al., 1980)	Unknown	Synthesis and bioassay (Suguro and Mori, 1979b; Sonnet and and Heath, 1982)	HPLC (naphthylethylamide) (Mori et al., 1981a) HPLC (prolinolamide) (Sonnet and Heath, 1982)

$C_{16}H_{26}O_2$ **(13)**	Acetate of (*Z*)-6-isopropenyl-3-methyl-3,9-decadien-1-ol *Aonidiella auranti*	A (Roelofs et al., 1978)	(*R*)	Synthesis and bioassay (Roelofs et al., 1978)	HPLC (naphthylethylamide) (Roelofs et al., 1978)
$C_{16}H_{28}O_2$ **(14)**	Acetate of 6-isopropenyl-3-methyl-9-decen-1-ol *Aonidiella auranti*	A (Anderson et al., 1980)	(3*S*,6*R*)	Synthesis and bioassay (Anderson et al., 1980)	HPLC (MTPA) (Anderson et al., (1980)
$C_{17}H_{30}O_2$ **(15)**	Acetate of (*E*)-6-isopropyl-3,9-dimethyl-5,8-decadien-1-ol *Aonidiella citrina*	A (Roelofs et al., 1982)	(*S*)	Synthesis and bioassay (Mori and Kuwahara, 1982)	HPLC (naphthylethylamide) (Mori et al., 1981a)
$C_{18}H_{30}O_2$ **(16)**	Propionate of (*Z*)-6-isopropenyl-3,9-dimethyl-3,9-decadien-1-ol *Pseudaulacaspis pentagona*	A (Heath et al., 1980)	(*R*)	Synthesis and bioassay (Heath et al., 1979, 1980)	HPLC (naphthylethylamide) (Heath et al., 1980)
$C_{19}H_{38}O_2$ **(17)**	Acetate of 3,7-dimethyl-2-pentadecanol *Neodiprion lecontei* Propionate (*Diprion similis*)	A (Kraemer et al., 1981) C[2]	(2*S*,3*S*,7*S*)	Synthesis and bioassay (Mori and Tamada, 1979; Byström et al., 1981)	NMR (MTPA, shift reagent) (Mori and Tamada, 1979) NMR (MTPA) (Byström et al., 1981)

[1] The natural pheromone was shown to be optically pure by GLC (MTPA) (Bierl-Leonhardt et al., 1980, 1982).

[2] Both (2*R*,3*R*,7*R*)- and (2*S*,3*S*,7*S*)-isomers were active (Longhurst et al., 1980).

Table 3. Determination of absolute configuration and optical purity of pheromones (aldehydes and ketones)

Molecular formula (Structure)	Name and origin of the pheromone	Stereochemistry-bioactivity relationship	Absolute configuration of the natural pheromone	Method for determining the absolute configuration of the natural pheromone	Method for determining the optical purity of the synthetic pheromone or its starting material
$C_8H_{16}O$ (**18**)	4-Methyl-3-heptanone *Atta texana*	A (Riley et al., 1974)	(*S*)	Synthesis and bioassay (Riley et al., 1974)	NMR (SAMP–hydrazone–Shift reagent) (Enders and Eichenauer, 1979)
$C_9H_{18}O$ (**19**)	6-Methyl-3-octanone *Crematogaster* sp.	–	–	–	$[\alpha]_D$ of the starting material (Rossi and Salvadori, 1979)
$C_{11}H_{22}O_2$ (**20**)	Serricornin *Lesioderma serricorne*	A (Mori et al., 1982c)	(4*S*,6*S*,7*S*)	Synthesis, $[\alpha]_D$ and bioassay (Mori et al., 1982c)	GLC (MTPA) (Mori et al., 1982c)
$C_{12}H_{24}O$ (**21**)	4,8-Dimethyldecanal *Tribolium castaneum* *Tribolium confusum*		(4*R*,8*R*)	Synthesis and bioassay (Mori et al., 1982c)	HPLC (naphthylethylamide (Mori et al., 1981a)
$C_{13}H_{20}O_3$ (**22**)	Stegobinone *Stegobium paniceum*	A (Hoffmann et al., 1981)	(2*S*,3*R*,7*R*)	Synthesis, X-ray, CD, and bioassay (Hoffmann et al., 1981)	CD (Hoffmann et al., 1981)

$C_{14}H_{28}O$ (**23**)	10-Methyl-2-tridecanone *Diabrotica undecimpunctata howardi*	A (Tumlinson, 1982)	(*R*)	Synthesis and bioassay (Tumlinson, 1982)	
$C_{15}H_{20}O_3$ (**24**)	Periplanone-B *Periplanata americana*	A (Adams et al., 1979)	(1*R*,2*R*,7*S*,10*R*)	Synthesis and bioassay (Still, 1979; Adams et al., 1979)	HPLC (MTPA) (Adams et al., 1979)
$C_{17}H_{30}O$ (**25**)	Faranal *Monomorium pharaonis*	A (Kobayashi et al., 1980)	(3*S*,4*R*)	Synthesis and bioassay (Kobayashi et al., 1980; Mori and Ueda, 1981a, 1982)	HPLC (naphthylethylamide) (Mori and Ueda, 1981a, 1982)
$C_{17}H_{32}O$ (**26,27**)	Trogodermal *Trogoderma granarium*	A (Levinson and Mori, 1980; Silverstein et al., 1980; Levinson et al., 1981)	(*R*)	Synthesis and bioassay (Mori et al., 1978a, 1982a; Suguro and Mori, 1979a)	HPLC (naphthylethylamide) (Mori et al., 1982a)
$C_{31}H_{62}O$ (**28**)	3,11-Dimethyl-2-nonacosanone *Blattella germanica*	C (Mori et al., 1981a)	(3*S*,11*S*)	Synthesis, $[\alpha]_D$ and mixed mp (Mori et al., 1981a)	HPLC (naphthylethylamide) (Mori et al., 1981a)
$C_{31}H_{62}O_2$ (**29**)	29-Hydroxy-3,11-dimethyl-2-nonacosanone *Blattella germanica*	C (Mori et al., 1981a)	(3*S*,11*S*)	Synthesis, $[\alpha]_D$ and mixed mp (Mori et al., 1981a)	HPLC (naphthylethylamide) Mori et al., 1981a)

Table 4. Determination of absolute configuration and optical purity of pheromones (acids, esters, and lactones)

Molecular formula (Structure)	Name and origin of the pheromone	Stereochemistry-bioactivity relationship	Absolute configuration of the natural pheromone	Method for determining the absolute configuration of the natural pheromone	Method for determining the optical purity of the synthetic pheromone or its starting material
$C_6H_{10}O_2$ (**30**)	4-Hexanolide *Trogoderma glabrum*	B (Ravid et al., 1978)	(*R*)	Synthesis and bioassay (Ravid et al., 1978)	NMR (chiral solvating agent) (Ravid et al., 1978)
$C_7H_{12}O_2$ (**31**)	2-Methyl-5-hexanolide *Xilocopa hirutissima*	—	—	—	HPLC (naphthylethylcarbamate) (Pirkle and Adams, 1979)
$C_9H_{12}O_2$ (**32**)	(2*Z*,6*Z*)-2,6-Nonadien-4-olide *Aphomia gularis*	—	(*R*)	Synthesis and ORD (Miyashita and Mori, 1981)	GLC (MTPA) (Miyashita and Mori, 1981)
$C_9H_{18}O_2$ (**33**)	Methyl 3-isopropylpentanoate *Formica rufa*, *Formica polyctena*	—	—	—	HPLC (naphthylethylamide) (Tanida and Mori, 1981)
$C_{10}H_{16}O_2$ (**34**)	3,7-Dimethyl-6-octen-4-olide *Eldana saccharina*	—	(3*S*,4*R*)	Synthesis and CD (Vigneron et al., 1982; Uematsu et al., 1983)	HPLC (naphthylethylamide) (Mori et al., 1981a)

$C_{10}H_{16}O_2$ (**35**)	Phoracantholide J *Phoracantha synonyma*	–	(*R*)[1]	Synthesis and GLC (Kitahara et al., 1983)	GLC (chiral phase) (Kitahara et al., 1983)
$C_{10}H_{18}O_2$ (**36**)	Phoracantholide I *Phoracantha synonyma*	–	(*R*)[1]	Synthesis and GLC (Kitahara et al., 1983)	GLC (chiral phase) (Kitahara et al., 1983)
$C_{10}H_{16}O_4$ (**37**)	Callosobruchusic acid *Callosobruchus chinensis*	C (Mori et al., 1983a)	–	–	GLC (prolinolamide) (Mori et al., 1983a)
$C_{11}H_{20}O_2$ (**38**)	Dominicalure 1 *Rhyzopertha dominica*	C (Williams et al., 1981)	(*S*)[2]	Synthesis, $[\alpha]_D$, and NMR (Williams et al., 1981)	$[\alpha]_D$ and NMR (shift reagent) (Williams et al., 1981)
$C_{12}H_{22}O_2$ (**39**)	Dominicalure 2 *Rhyzopertha dominica*	C (Williams et al., 1981)	(*S*)[2]	Synthesis, $[\alpha]_D$, and NMR (Williams et al., 1981)	$[\alpha]_D$ and NMR (shift reagent) (Williams et al., 1981)
$C_{12}H_{22}O_2$ (**40**)	4-Dodecanolide *Bledius mandibularis*	–	–	–	HPLC (naphthylethylcarbamate) (Pirkle and Adams, 1979) NMR (shift reagent) (Solladié and Matloubi-Moghadam, 1982)
$C_{14}H_{24}O_2$ (**41**)	(*Z*)-5-Tetradecen-4-olide *Popillia japonica*	B (Tumlinson et al., 1977)	(*R*)	Synthesis and bioassay (Tumlinson et al., 1977; Doolittle et al., 1980)	HPLC (naphthylethylcarbamate) (Pirkle and Adams, 1979) $[\alpha]_D$ (Baker and Rao, 1982)

Table 4. (continued)

Molecular formula (Structure)	Name and origin of the pheromone	Stereochemistry-bioactivity relationship	Absolute configuration of the natural pheromone	Method for determining the absolute configuration of the natural pheromone	Method for determining the optical purity of the synthetic pheromone or its starting material
$C_{15}H_{24}O_2$ (**42**)	Methyl (*E*)-2,4,5-tetradecatrienoate *Acanthoscelides obtectus*	–	(*R*)	Synthesis and $[\alpha]_D$ (Pirkle and Boeder, 1978; Mori et al., 1981b)	HPLC (naphthylethylcarbamate) (Pirkle and Boeder, 1978) GLC (MTPA), NMR (MTPA) (Mori et al., (1981b)
$C_{16}H_{30}O_2$ (**43**)	5-Hexadecanolide *Vespa orientalis*	–	–	–	HLPC (naphthylethylcarbamate) (Pirkle and Adams, 1979) NMR (shift reagent) (Solladié and Matloubi-Moghadam, 1982)

[1] The natural product was optically pure by GLC (chiral phase) (Kitahara et al., 1983).

[2] The natural pheromone was optically pure by NMR (shift reagent) (Williams et al., 1981).

Table 5. Determination of absolute configuration and optical purity of pheromones (acetals and epoxides)

Molecular formula (Structure)	Name and origin of the pheromone	Stereochemistry-bioactivity relationship	Absolute configuration of the natural pheromone	Method for determining the absolute configuration of the natural pheromone	Method for determining the optical purity of the synthetic pheromone or its starting material
$C_8H_{14}O_2$ (**44**)	Frontalin *Dendroctonus frontalis*	A (Wood et al., 1976)	Both (1*S*,5*R*) and (1*R*,5*S*)[1] 85:15	Synthesis, NMR and bioassay (Mori, 1975a; Stewart et al., 1977)	NMR (shift reagent) (Mori, 1975a; Stewart et al., 1977)
$C_9H_{16}O_2$ (**45**)	*exo*-Brevicomin *Dendroctonus brevicomis*	A (Wood et al., 1976)	(1*R*,5*S*,7*R*)[1]	Synthesis, NMR, and bioassay (Mori, 1974b; Stewart et al., 1977)	NMR (shift reagent) (Stewart et al., 1977)
$C_9H_{16}O_2$ (**46**)	Chalcogran *Pityogenes chalcographus*	–	–	–	NMR (MTPA–shift reagent) (Mori et al., 1979a)
$C_9H_{16}O_2$ (**47**)	2-Methyl-1,6-dioxa-spiro[4.5]decane *Paravespula vulgaris*	–	–	–	GLC (chiral phase) (Hintzer et al., 1981)
$C_9H_{16}O_2$ (**48**)	7-Methyl-1,6-dioxa-spiro[4.5]decane *Paravespula vulgaris*	–	–	–	GLC (chiral phase) (Hintzer et al., 1981)

Table 5. (continued)

Molecular formula (Structure)	Name and origin of the pheromone	Stereochemistry-bioactivity relationship	Absolute configuration of the natural pheromone	Method for determining the absolute configuration of the natural pheromone	Method for determining the optical purity of the synthetic pheromone or its starting material
$C_{10}H_{16}O_2$ (**49**)	Lineatin *Trypodendron lineatum*	A (Slessor et al., 1980)	(1*R*,4*S*,5*R*,7*R*)	Synthesis, X-ray, and bioassay (Slessor et al., 1980; Mori et al., 1982b)	HPLC (naphthylethyl-carbamate) (Slessor et al., 1980) NMR (shift reagent), GLC (chiral phase) (Mori et al., 1982b)
$C_{10}H_{18}O_2$ (**50**)	α-Multistriatin *Scolytus multistriatus*	A (Elliott et al., 1979)	(1*S*,2*R*,4*S*,5*R*)	Synthesis, $[\alpha]_D$, and bioassay (Mori, 1976d; Pearce et al., 1976; Elliott et al., 1979)	NMR (shift reagent) (Pearce et al., 1976)

$C_{11}H_{20}O_2$ **(51)**	2,8-Dimethyl-1,7-dioxaspiro[5.5]-undecane *Andrena wilkella*	–	–	–	GLC (MTPA) (Mori and Tanida, 1981)
$C_{19}H_{38}O$ **(52)**	Disparlure *Porthetria dispar*	B (Iwaki et al., 1974; Vité et al., 1976b)	(7*R*,8*S*)	Synthesis and bioassay (Iwaki et al., 1974)	NMR (MTPA–shift reagent) (Iwaki et al., 1974; Mori et al., 1979b)
$C_{21}H_{38}O$ **(53)**	(3*Z*,6*Z*)-*cis*-9,10-Epoxy-3,6-heneicosadiene *Estigmene acrea*	–	(9*S*,10*R*)	Synthesis, CD, and EAG (Mori and Ebata, 1981)	–

[1] The optical purity of the natural pheromone was determined by NMR (shift reagent) (Stewart et al., 1977).

A pheromone from the head of the ant *Tetramorium impurum* was shown to be (3*R*,4*S*)-4-methyl-3-hexanol by GLC (derivatization) analysis (Pasteels JM et al., 1981).

Pasteels JM, Verhaeghe JC, Ottinger R, Braekman JC, Daloze D (1981) Absolute configuration of (3*R*,4*S*)-4-methyl-3-hexanol, a pheromone from the head of the ant *Tetramorium impurum* Foerster. Insect Biochem **11**:675–678.

Acknowledgments. I am grateful to Professors W. Francke, W. H. Pirkle, F. V. Schurig, and R. M. Silverstein, and Dr. N. Ôi, for kindly informing me of their unpublished results. I acknowledge with thanks the permission of Plenum Publishing Corporation and Professor R. M. Silverstein to reproduce Figs. 8 and 11. I thank my junior colleagues for preparing the figures.

VIII. References

Adams MA, Nakanishi K, Still WC, Arnold DEV, Clardy J, Persoons CJ (1979) Sex pheromone of the American cockroach: Absolute configuration of periplanone-B. J Am Chem Soc **101**:2495–2498.

Ade E, Helmchen G, Heiligenmann G (1980) Synthesis of the stereoisomers of 17,21-dimethylheptatriacontane–Sex recognition pheromone of the tsetse fly. Tetrahedron Lett **21**:1137–1140.

Anderson RJ, Adams KG, Chinn HR, Henrick CA (1980) Synthesis of the optical isomers of 3-methyl-6-isopropyl-9-decen-1-yl acetate, a component of the California red scale pheromone. J Org Chem **45**:2229–2236.

Baker R, Rao VB (1982) Synthesis of optically pure (*R,S*)-5-dec-1-enyloxacyclopentan-2-one, the sex pheromone of the Japanese beetle. JCS Perkin **I**:69–71.

Benecke I, König WA (1982) Isocyanate as universal reagents for the formation of derivatives for gas chromatographic enantiomer separation. Angew Chem Int Ed Engl **21**:709.

Bergot BJ, Anderson RJ, Schooley DA, Henrick CA (1978) Liquid chromatographic analysis of enantiomeric purity of several terpenoid acids as their 1-(1-naphthyl)ethylamide derivatives. J Chromatogr **155**:97–105.

Bierl-Leonhardt BA, Moreno DS, Schwarz M, Forgerlund J, Plimmer JR (1981) Isolation, identification and synthesis of the sex pheromone of the citrus mealybug, *Planococcus citri* (RISSO). Tetrahedron Lett **22**:389–392.

Bierl-Leonhardt BA, Moreno DS, Schwarz M, Forster HS, Plimmer JR, DeVilbiss ED (1980) Identification of the pheromone of the comstock mealybug. Life Sci **27**:399–402.

Bierl-Leonhardt BA, Moreno DS, Schwarz M, Forster HS, Plimmer JR, DeVilbiss ED (1982) Isolation, identification, synthesis, and bioassay of the pheromone of the Comstock mealybug and some analogs. J Chem Ecol **8**:689–699.

Birch MC, Light DM, Wood DL, Browne LE, Silverstein RM, Bergot BJ, Ohloff G, West JR, Young JC (1980) Pheromonal attraction and allomonal interruption of *Ips pini* in California by the two enantiomers of ipsdienol. J Chem Ecol **6**:703–717.

Blaschke G (1980) Chromatographic resolution of racemates. Angew Chem Int Ed Engl **19**:13–24.

Borden JH, Chong L, McLean JA, Slessor KN, Mori K (1976) *Gnathotrichus sulcatus*: Synergistic response to enantiomers of the aggregation pheromone sulcatol. Science **192**:894–896.

Borden JH, Handley JR, McLean JA, Silverstein RM, Chong L, Slessor KN, Johnston BD, Schuler HR (1980) Enantiomer-based specificity in pheromone communication by two sympatric *Gnathotrichus* species. J Chem Ecol **6**:445–456.

Byström S, Hogberg H-E, Norin T (1981) Chiral synthesis of (2*S*, 3*S*, 7*S*)-3,7-dimethylpentadecan-2-yl acetate and propionate, potential sex pheromone components of the pine saw-fly *Neodiprion sertifer.* Tetrahedron **37**:2249–2254.

Dale JA, Dull DL, Mosher HS (1969) α-Methoxy-α-trifluoromethylphenyl acetic acid, a versatile reagent for the determination of enantiomeric composition of alcohols and amines. J Org Chem **34**:2543–2549.

Dale JA, Mosher HS (1973) Nuclear magnetic resonance enantiomer reagents. Configurational correlations *via* nuclear magnetic resonance chemical shifts of diastereomeric mandelate, O-methyl-mandelate and α-methoxy-α-trifluoromethylphenylacetate (MTPA) esters. J Am Chem Soc **95**:512–519.

Doolittle RE, Tumlinson JH, Proveaux AT, Heath RR (1980) Synthesis of the sex pheromone of the Japanese beetle. J Chem Ecol **6**:473–485.

Eliel EL, Otsuka S (eds) (1982) Asymmetric reactions and processes in chemistry, ACS Symposium Series 185. American Chemical Society, Washington, D.C., pp 300.

Elliott WJ, Hromnak G, Fried J, Lanier GN (1979) Synthesis of multistriatin enantiomers and their action on *Scolytus multistriatus.* J Chem Ecol **5**: 279–287.

Enders D, Eichenauer H (1979) Asymmetric synthesis of ant alarm pheromones—α-Alkylation of acyclic ketones with almost complete asymmetric induction. Angew Chem Int Ed Engl **18**:397–399.

Francke W, Kruse K (1979) Gas chromatographic determination of enantiomeric ratios in bark beetle pheromone alcohols. EUCHEM Conference on Insect Chemistry (Öland, Sweden), Abstracts.

Fraser RR, Petit MA, Saunders JK (1971) Determination of enantiomeric purity by an optically active nuclear magnetic resonance shift reagent of wide applicability. Chem Commun 1450–1451.

Gaudemer A (1977) Determination of optical purity and absolute configuration by nuclear magnetic resonance. In: Stereochemistry. Kagan HB (ed), Georg Thieme, Stuttgart. Vol 1, pp 117–136.

Goering HL, Eikenberry JN, Koermer GS (1971) Tris[3-(trifluoromethylhydroxymethylene)-*d*-camphorato] europium (III). A chiral shift reagent for direct determination of enantiomeric compositions. J Am Chem Soc **93**:5913–5914.

Goering HL, Eikenberry JN, Koerner GS, Lattimer CJ (1974) Direct determination of enantiomeric compositions with optically active nuclear magnetic resonance lanthanide shift reagents. J Am Chem Soc **96**:1493–1501.

Heath RR, Doolittle RE, Sonnet PE, Tumlinson JH (1980) Sex pheromone of the white peach scale: Highly stereoselective synthesis of the stereoisomers of pentagonol propionate. J Org Chem **45**:2910–2912.

Heath RR, McLaughlin JR, Tumlinson JH, Ashley TR, Doolittle RE (1979) Identification of the white peach scale sex pheromone, an illustration of micro techniques. J Chem Ecol **5**:941–953.

Helmchen G, Nill G, Flockerzi D, Schühler W, Youssef MSK (1979a) Extreme liquid chromatographic separation effects in the case of diastereomeric amides containing polar substituents. Angew Chem Int Ed Engl **18**:62–63.

Helmchen G, Nill G, Flockerzi D, Youssef MSK (1979b) Preparative directed resolution of enantiomeric carboxylic acids and lactones *via* liquid chromatography and neighboring-group assisted hydrolysis of diastereomeric amides. Angew Chem Int Ed Engl **18**:63–65.

Hintzer K, Weber R, Schurig V (1981) Synthesis of optically active 2*S*- and 7*S*-methyl-1,6-dioxaspiro [4.5] decane, the pheromone components of *Paravespula vulgaris* L. from *S*-ethyl lactate. Tetrahedron Lett **22**:55–58.

Hobbs PD, Magnus PD (1976) Studies on terpenes 4. Synthesis of optically active grandisol, the boll weevil pheromone. J Am Chem Soc **98**:4594–4600.

Hoffmann RW, Ladner W, Steinbach K, Massa W, Schmidt R, Snatzke G (1981) Absolute Konfiguration von Stegobinon. Chem Ber **114**:2786–2801.

Iwaki S, Marumo S, Saito T, Yamada M, Katagiri K (1974) Synthesis and activity of optically active disparlure. J Am Chem Soc **96**:7842–7844.

Jacques J, Collet A, Wilen SH (1981) Enantiomers, Racemates and Resolutions. Wiley, New York, pp 447.

Jakovac IJ, Jones JB (1979) Determination of enantiomeric purity of chiral lactones. A general method using nuclear magnetic resonance. J Org Chem **44**:2165–2168.

Kafka WA, Ohloff G, Schneider D, Vareschi E (1973) Olfactory discrimination of two enantiomers of 4-methylhexanoic acid by the migratory locust and the honeybee. J Comp Physiol **87**:277–284.

Kim YH, Balan A, Tishbee A, Gil-Av E (1982) Chiral differentiation by the P-(+)-hexahelicene-7,7′-dicarboxylic acid disodium salt. Resolution of N-2,4-dinitrophenyl-α-aminoacid esters by high performance liquid chromatography. JCS Chem Commun 1336–1337.

Kime KA, Sievers RE (1977) A practical guide to uses of lanthanide NMR shift reagents. Aldrichim Acta **10**:54–62.

Kitahara T, Koseki K, Mori K (1983) Synthesis and absolute configuration of phoracantholide I and J, the secretion of *Phoracantha synonyma*. Agric Biol Chem **47**:389–393.

Kobayashi M, Koyama T, Ogura K, Seto S, Ritter FJ, Brüggemann-Rotgans IEM (1980) Bioorganic synthesis and absolute configuration of faranal. J Am Chem Soc **102**:6602–6604.

König WA, Francke W, Benecke I (1982) Gas chromatographic enantiomer separation of chiral alcohols. J Chromatogr **239**:227–231.

Koppenhoefer B, Hintzer K, Weber R, Schurig V (1980) Quantitative separation of the enantiomeric pairs of the pheromone 2-ethyl-1,6-dioxaspiro [4.4] nonane by complexation chromatography on an optically active metal complex. Angew Chem Int Ed Engl **19**:471–472.

Kraemer ME, Coppel HC, Matsumura F, Wilkinson RC, Kikukawa T (1981) Field and EAG responses of the red-headed pine sawfly, *Neodiprion lecontei* (FITCH), to optical isomers of sawfly sex pheromones. J Chem Ecol **7**: 1063–1072.

Kruse K, Francke W, König WA (1979) Gas chromatographic separation of chiral alcohols, amino alcohols and amines. J Chromatogr **170**:423–429.

Lanier GN, Gore WE, Pearce GT, Peacock JW, Silverstein RM (1977) Response of the European elm bark beetle, *Scolytus multistriatus*, to isomers and components of its pheromone. J Chem Ecol **3**:1–8.

Legrand M, Rougier MJ (1977) Application of the optical activity to stereochemical determinations. In: Stereochemistry. Kagan HB (ed), Georg Thieme, Stuttgart. Vol 2, pp 33–183.

Leonhardt BA, Beroza M (1982) Insect pheromone technology: Chemistry and applications, ACS Symposium Series 190. American Chemical Society, Washington, D.C., pp 260.

Levinson HZ, Levinson AR, Mori K (1981) Olfactory behavior and receptor potentials of two khapra beetle strains induced by enantiomers of trogodermal. Naturwiss **67**:480–481.

Levinson HZ, Mori K (1980) Pheromone activity of chiral isomers of trogodermal for male khapra beetle. Naturwiss **67**:148.

Longhurst C, Baker R, Mori K (1980) Response of the sawfly *Diprion similis* to chiral sex pheromones. Experientia **36**:946–947.

McKnight RC (1978) Unpublished results cited in Professor J. P. Vité's personal communication to KM dated June 15.

Mislow K, Raban M (1967) Stereoisomeric relationships of groups in molecules. In: Topics in Stereochemistry. Allinger NL, Eliel EL (eds), Interscience, New York, Vol 1, pp 1–38.

Miyashita Y, Mori K (1981) Synthesis of both the enantiomers of (2*Z*, 6*Z*)-2,6-nonadien-4-olide, the possible male-secreted sex pheromone from a pyralid moth, *Aphomia gularis* SELLER. Agric Biol Chem **45**:2521–2526.

Moore BP, Brown WV (1976) The chemistry of metasternal gland secretion of the eucalypt longicorn *Phoracantha synonyma*. Aust J Chem **29**:1365–1374.

Mori K (1973) Absolute configurations of (—)-14-methyl-*cis*-8-hexadecen-1-ol and methyl (—)-14-methyl-*cis*-8-hexadecenoate, the sex attractant of female dermestid beetle. Tetrahedron Lett 3869–3872.

Mori K (1974a) Absolute configurations of (—)-14-methylhexadec-8-*cis*-en-1-ol and methyl (—)-14-methylhexadec-8-*cis*-enoate, the sex pheromone of female dermestid beetle. Tetrahedron **30**:3817–3820.

Mori K (1974b) Synthesis of *exo*-brevicomin, the pheromone of western pine beetle, to obtain optically active forms of known absolute configuration. Tetrahedron **30**:4223–4227.

Mori K (1975a) Synthesis of optically active forms of frontalin, the pheromone of *Dendroctonus* bark beetles. Tetrahedron **31**:1381–1384.

Mori K (1975b) Synthesis of optically active forms of sulcatol, the aggregation pheromone in the Scolytid beetle *Gnathotrichus sulcatus.* Tetrahedron **31**: 3011–3012.

Mori K (1976a) Synthesis of optically pure (+)-*trans*-verbenol and its antipode, the pheromone of *Dendroctonus* bark beetles. Agric Biol Chem **40**:415–418.

Mori K (1976b) Synthesis of optically active forms of ipsenol, the pheromone of *Ips* bark beetles. Tetrahedron **32**:1101–1106.

Mori K (1976c) Absolute configuration of (+)-ipsdienol, the pheromone of *Ips paraconfusus* Lanier, as determined by the synthesis of its (*R*)-(−)-isomer. Tetrahedron Lett 1609–1612.

Mori K (1976d) Synthesis of (1*S*, 2*R*, 4*S*, 6*R*)-(−)-α-multistriatin, the pheromone in the smaller European elm bark beetle, *Scolytus multistriatus.* Tetrahedron **32**:1979–1981.

Mori K (1977) Absolute configuration of (−)-4-methylheptan-3-ol, a pheromone of the smaller European elm bark beetle as determined by the synthesis of its (3*R*, 4*R*)-(+)- and (3*S*, 4*R*)-(+)-isomers. Tetrahedron **33**: 289–94.

Mori K (1978) Synthesis of the both enantiomers of grandisol, the boll weevil pheromone. Tetrahedron **34**:915–920.

Mori K (1981) The synthesis of insect pheromones. In: The Total Synthesis of Natural Products. ApSimon J (ed), Wiley, New York, Vol 4, pp 1–183.

Mori K, Ebata T (1981) Synthesis of optically active pheromones with an epoxy ring, (+)-disparlure and the saltmarsh caterpillar moth pheromone (*Z,Z*)-3,6-cis-9,10-epoxyheneicosadiene. Tetrahedron Lett **22**:4281–4282.

Mori K, Ebata T (1983) Synthesis of enantiomers of epoxy diene pheromones of *Estigmene acrea* and *Hyphantria cunea.* Tetrahedron, In press.

Mori K, Iwasawa H (1980) Preparation of the both enantiomers of *threo*-2-amino-3-methylhexanoic acid by enzymatic resolution and their conversion to optically active forms of *threo*-4-methylheptan-3-ol, a pheromone component of the smaller European elm bark beetle. Tetrahedron **36**: 2209–2213.

Mori K, Ito T, Honda H, Yamamoto I (1983a) Synthesis and biological activity of optically active forms of (*E*)-3,7-dimethyl-2-octene-1,8-dioic acid (callosobruchusic acid), a component of the copulation release pheromone (erectin) of the azuki bean weevil. Tetrahedron **39**:2303–2306.

Mori K, Kuwahara S (1982) Synthesis of optically active forms of (*E*)-6-isopropyl-3,9-dimethyl-5,8-decadienyl acetate, the pheromone of the yellow scale. Tetrahedron **38**:521–525.

Mori K, Kuwahara S, Levinson HZ, Levinson AR (1982a) Synthesis and biological activity of both (*E*)- and (*Z*)-isomers of optically pure (*S*)-14-methyl-8-hexadecenal (trogodermal), the antipodes of the pheromone of the khapra beetle. Tetrahedron **38**:2291–2297.

Mori K, Kuwahara S, Ueda H (1983b) Synthesis of all of the four possible stereoisomers of the pheromone of flour beetles *Tribolium custaneum* and *Tribolium confusum.* Tetrahedron **39**:2439–2444.

Mori K, Masuda S, Suguro T (1981a) Stereocontrolled synthesis of all of the possible stereoisomers of 3,11-dimethylnonacosan-2-one and 29-hydroxy-3,11-dimethylnonacosan-2-one, the female sex pheromone of the German cockroach. Tetrahedron **37**:1329–1340.

Mori K, Mizumachi N, Matsui M (1976) Synthesis of optically pure (1*S*, 4*S*, 5*S*)-2-pinen-4-ol (*cis*-verbenol) and its antipode, the pheromone of *Ips* bark beetles. Agric Biol Chem **40**:1611–1615.

Mori K, Nomi H, Chuman T, Kohno M, Kato K, Noguchi M (1982c) Synthesis and absolute stereochemistry of serricornin [(4*S*, 6*S*, 7*S*)-4,6-dimethyl-7-hydroxy-3-nonanone], the sex pheromone of the cigarette beetle. Tetrahedron **39**:3705–3711.

Mori K, Nukada T, Ebata T (1981b) Synthesis of optically active forms of methyl (*E*)-2,4,5-tetradecatrienoate, the pheromone of the male dried bean beetle. Tetrahedron **37**:1343–1347.

Mori K, Sasaki M, Tamada S, Suguro T, Masuda S (1979a) Synthesis of optically active 2-ethyl-1,6-dioxaspiro [4.4] nonane (chalcogran), the principal aggregation pheromone of *Pityogenes chalcographus* L. Tetrahedron **35**:1601–1605.

Mori K, Suguro T, Uchida M (1978a) Synthesis of optically active forms of (*Z*)-14-methylhexadec-8-enal, the pheromone of female dermestid beetle. Tetrahedron **34**:3119–3123.

Mori K, Takigawa T, Matsui M (1979b) Stereoselective synthesis of the both enantiomers of disparlure, the pheromone of the gypsy moth. Tetrahedron **35**:833–837.

Mori K, Takigawa T, Matsuo T (1979c) Synthesis of optically active forms of ipsdienol and ipsenol, the pheromone components of *Ips* bark beetles. Tetrahedron **35**:933–940.

Mori K, Tamada S (1979) Stereocontrolled synthesis of all of the four possible stereoisomers of *erythro*-3,7-dimethylpentadec-2-yl acetate and propionate, the sex pheromone of the pine sawflies. Tetrahedron **35**:1279–1284.

Mori K, Tamada S, Hedin PA (1978b) (−)-Grandisol, the antipode of the boll weevil pheromone, is biologically active. Naturwiss **65**:653.

Mori K, Tamada S, Uchida M, Mizumachi N, Tachibana Y, Matsui M (1978c) Synthesis of optically active forms of seudenol, the pheromone of Douglas fir beetle. Tetrahedron **34**:1901–1905.

Mori K, Tanida K (1981) Synthesis of three stereoisomeric forms of 2,8-dimethyl-1,7-dioxaspiro [5.5] undecane, the main component of the cephalic secretion of *Andrena wilkella*. Tetrahedron **37**:3221–3225.

Mori K, Ueda H (1981a) Synthesis of optically active forms of faranal, the trail pheromone of Pharaoh's ant. Tetrahedron Lett **22**:461–464.

Mori K, Ueda H (1981b) Synthesis of the optically active forms of 2,6-dimethyl-1,5-heptadien-3-ol acetate, the pheromone of the Comstock mealybug. Tetrahedron **37**:2581–2583.

Mori K, Ueda H (1982) Synthesis of optically active forms of faranal, the trail pheromone of Pharaoh's ant. Tetrahedron **38**:1227–1233.

Mori K, Uematsu T, Minobe M, Yanagi K (1982b) Synthesis and absolute configuration of lineatin, the pheromone of *Trypodendron lineatum*. Tetrahedron Lett **23**:1921–1924.

Morrison JD (ed) (1983) Ways To Obtain Chiral Compounds and Determine Their Enantiomeric Composition. In: Asymmetric synthesis, a multivolume treatise. Academic Press, New York, Vol 1, Part 1.

Newman P (1981) Optical Resolution of Acids by Chromatographic Method: Section 5, Methods for Determining Optical Purity. In: Optical resolution procedure for chemical compounds. Optical Resolution Information Center, Manhattan College, Riverdale, New York, Vol 2, Part 2, Section 3.

Ohloff G, Giersch W (1977) Access to optically active ipsdienol from verbenone. Helv Chim Acta **60**:1496–1500.

Ôi N (1982) Unpublished results.

Ôi N, Kitahara H, Doi T (1982b) Direct Separation of Optical Isomers of Chrysanthemic Esters by GLC with a Chiral Stationary Phase. Abstracts of papers, 43rd symposium on analytical chemistry, Yamagata, Japan.

Ôi N, Kitahara H, Inda Y, Doi T (1982a) Some N-acyl derivatives of 1-(α-naphthyl)ethylamine as stationary phases for the separation of optical isomers in gas chromatography. J Chromatogr **237**:297–302.

Okamoto Y, Honda S, Okamoto I, Yuki H, Murata S, Noyori R, Takaya H (1981b) Novel packing material for optical resolution: (+)-Poly (triphenylmethyl methacrylate) coated on macroporous silica gel. J Am Chem Soc **103**:6971-6973.

Okamoto Y, Okamoto I, Yuki H (1981a) Chromatographic resolution of enantiomers having aromatic group by optically active poly(triphenylmethyl methacrylate). Chem Lett 835–838.

Parthasarathy R (1977) The determination of relative and absolute configurations of organic molecules by X-ray diffraction methods. In: Stereochemistry. Kagan HB (ed), Georg Thieme, Stuttgart, Vol 1, pp 181–234.

Pearce GT, Gore WE, Silverstein RM (1976) Synthesis and absolute configuration of multistriatin. J Org Chem **41**:2797–2803.

Pirkle WH (1966) The nonequivalence of physical properties of enantiomers in optically active solvents. Differences in nuclear magnetic resonance spectra I. J Am Chem Soc **88**:1837.

Pirkle WH (1982) Personal communication to KM dated May 21, 1982.

Pirkle WH, Adams PE (1979) Broad-spectrum synthesis of enantiomerically pure lactones. 1. Synthesis of sex pheromones of the carpenter bee, rove beetle, Japanese beetle, black-tailed deer and oriental hornet. J Org Chem **44**: 2169–2175.

Pirkle WH, Boeder CW (1978) Synthesis and absolute configuration of (−)-methyl (*E*)-2,4,5-tetradecatrienoate, the sex attractant of the male dried bean weevil. J Org Chem **43**:2091–2093.

Pirkle WH, Finn JM (1981) Chiral high-pressure liquid chromatographic stationary phases. 3. General resolution of arylalkylcarbinols. J Org Chem **46**: 2935–2938.

Pirkle WH, Finn JM (1982) Preparative resolution of racemates on a chiral liquid chromatography column. J Org Chem **47**:4037–4040.

Pirkle WH, Finn JM, Schreiner JL, Hamper BC (1981) A widely useful chiral stationary phase for the high-performance liquid chromatography separation of enantiomers. J Am Chem Soc **103**:3964–3966.

Pirkle WH, House DW (1979) Chiral high-pressure liquid chromatographic stationary phases. 1. Separation of the enantiomers of sulfoxides, amines, amino acids, alcohols, hydroxy acids, lactones, and mercaptans. J Org Chem **44**:1957–1960.

Pirkle WH, House DW, Finn JM (1980) Broad-spectrum resolution of optical isomers using chiral high-performance liquid chromatographic bonded phases. J Chromatogr **192**:143–158.

Pirkle WH, Rinaldi PL (1979) Synthesis and enantiomeric purity determination of the optically active epoxide disparlure, sex pheromone of the gypsy moth. J Org Chem **44**:1025–1028.

Pirkle WH, Schreiner JL (1981) Chiral high-pressure liquid chromatographic stationary phases. 4. Separation of the enantiomers of bi-β-naphthols and analogues. J Org Chem **46**:4988–4991.

Pirkle WH, Sikkenga DL, Pavlin MS (1977) Nuclear magnetic resonance determination of enantiomeric composition and absolute configuration of γ-lactones using chiral 2,2,2-trifluoro-1-(9-anthryl)ethanol. J Org Chem **42**: 384–387.

Pirkle WH, Simmons KA (1981) Nuclear magnetic resonance determination of enantiomeric composition and absolute configuration of amines, alcohols, and thiols with α-[1-(9-anthryl)-2-2-2-trifluoroethoxy]acetic acid as a chiral derivatizing agent. J Org Chem **46**:3239–3246.

Plummer EL, Stewart TE, Byrne K, Pearce GT, Silverstein RM (1976) Determination of the enantiomeric composition of several insect pheromone alcohols. J Chem Ecol **2**:307–331.

Raban M, Mislow K (1965) The determination of optical purity by Nuclear Magnetic Resonance Spectroscopy. Tetrahedron Lett 4249–4253.

Raban M, Mislow K (1967) Modern methods for the determination of optical purity. In: Topics in Stereochemistry. Allinger NL, Eliel EL (eds), Interscience, New York, Vol 2, pp 199–230.

Ravid U, Silverstein RM, Smith LR (1978) Synthesis of the enantiomer of 4-substituted γ-lactones with known absolute configuration. Tetrahedron **34**:1449–1452.

Riley RG, Silverstein RM, Moser JC (1974) Biological responses of *Atta texana* to its alarm pheromone and the enantiomer of the pheromone. Science **183**:760–762.

Roelofs W, Gieselmann M, Cardé A, Tashiro H, Moreno DS, Henrick CA, Anderson RJ (1978) Identification of the California red scale sex pheromone. J Chem Ecol **4**:211–224.

Roelofs WL, Gieselmann MJ, Mori K, Moreno DS (1982) Sex pheromone chirality comparison between sibling species–California red scale and yellow scale. Naturwiss **69**:348.

Rossi R, Salvadori PA (1979) Synthesis of both enantiomers of 6-methyl-3-octanone, a component of the alarm pheromone of ants in the genus *Crematogaster.* Synthesis 209–210.

Saucy G, Borer R, Trullinger DP, Jones JB, Lok KP (1977) Gas chromatographic analysis of ortho esters as a convenient new general method for determining the enantiomeric purities of chiral δ-lactones. J Org Chem **42**:3206–3208.

Schuler HR, Slessor KN (1977) Synthesis of enantiomers of sulcatol. Can J Chem **55**:3280–3287.

Schurig V (1980) Resolution of enantiomers and isotopic compositions by selective complexation gas chromatography on metal complexes. Chromatographia **13**:263–270.

Schurig V, Bürkle W (1982) Extending the scope of enantiomer resolution by complexation gas chromatography. J Am Chem Soc **104**:7573–7580.

Schurig V, Weber R (1982) Unpublished results.

Schurig V, Weber R, Klimetzek D, Kohnle U, Mori K (1982) Enantiomeric composition of 'lineatin' in three sympatric Ambrosia beetles. Naturwiss **69**: 602–603.

Schurig V, Weber R, Nicholson GJ, Oehlschlager AC, Pierce Jr H, Pierce AM, Borden JH, Ryker LC (1983) Enantiomer composition of natural *exo*- and *endo*-brevicomin by complexation gas chromatography/selected ion mass spectrometry. Naturwiss **70**:92.

Scott CG, Petrin MJ, McCorkle T (1976) The liquid chromatographic separation of some acyclic terpenoid acid enantiomers via diastereomer derivatization. J Chromatogr **125**:157–161.

Silverstein RM, Cassidy RF, Burkholder WL, Shapas TJ, Levinson HZ, Levinson AR, Mori K (1980) Perception by *Trogoderma* species of chirality and methyl branching at a site far removed from a functional group in a pheromone component. J Chem Ecol **6**:911–917.

Slessor KN, Oehlschlager AC, Johnston BD, Pierce Jr HD, Grewal SK, Wickremesinghe KG (1980) Lineatin: Regioselective synthesis and resolution leading to the chiral pheromone of *Trypodendron lineatum*. J Org Chem **45**:2290–2297.

Solladié G, Matloubi-Moghadam F (1982) Asymmetric synthesis of five- and six-membered lactones from chiral sulfoxides: Application to the asymmetric synthesis of insect pheromones, (*R*)-(+)-δ-n-hexadecanolactone and (*R*)-(+)-γ-n-dodecanolactone. J Org Chem **47**:91–94.

Sonnet PE, Heath RR (1982) Synthesis of (±)-10-methyl-1-dodecanol acetate, the chiral component of the smaller tea tortrix moth (*Adoxophyes* sp), with an option for asymmetric induction. J Chem Ecol **8**:41–53.

Stewart TE, Plummer EL, McCandless LL, West JR, Silverstein RM (1977) Determination of enantiomer composition of several bicyclic ketal insect pheromone components. J Chem Ecol **3**:27–43.

Still WC (1979) (±)-Periplanone-B. Total synthesis and structure of the sex excitant pheromone of the American cockroach. J Am Chem Soc **101**: 2493–2495.

Suguro T, Mori K (1979a) Synthesis of optically active forms of (*E*)-14-methyl-8-hexadecenal (trogodermal). Agric Biol Chem **43**:409–410.

Suguro T, Mori K (1979b) Synthesis of optically active forms of 10-methyl-dodecyl acetate, a minor component of the pheromone complex of the smaller tea tortrix moth. Agric Biol Chem **43**:869–870.

Sullivan GR (1978) Chiral lanthanide shift reagent. In: Topics in Stereochemistry. Allinger NL, Eliel EL (eds), Interscience, New York, Vol 10, pp 287–329.

Tamaki Y, Noguchi H, Sugie H, Kariya A, Arai S, Ohba M, Terada T, Suguro T, Mori K (1980) Four-component synthetic sex pheromone of the smaller tea tortrix moth: Field evaluation of its potency as an attractant for male moth. Jap J Appl Ent Zool **24**:221–228.

Tanida K, Mori K (1981) Synthesis of both enantiomers of methyl 3-isopropyl-pentanoate, a volatile substance isolated from two ant species, *Formica rufa* L. and *Formica polyctena* Forst. J Chem Soc Jpn (Nippon Kagaku Kaishi) 635–638.

Tumlinson JH (1982) The chemical basis for communication between the sexes in *Heliothis virescens* and other insect species. In: Les médiateurs chimiques agissant sur le comportement des insectes (les Colloques de l'INRA, No. 7), Institut National de la Recherche Agronomique, Paris, pp 193-201.

Tumlinson JH, Klein MG, Doolittle RE, Ladd TL, Proveaux AT (1977) Identification of the female Japanese beetle sex pheromone: Inhibition of male response by an enantiomer. Science **197**:789-792.

Uematsu T, Umemura T, Mori K (1983) Synthesis of both the enantiomers of eldanolide, the wing gland pheromone of the male African sugar-cane borer. Agric Biol Chem **47**:597-601.

Valentine Jr D, Chan KK, Scott CG, Johnson KK, Toth K, Saucy G (1976) Direct determinations of *R/S* enantiomer ratios of citronellic acid and related substances by nuclear magnetic resonance spectroscopy and high pressure liquid chromatography. J Org Chem **41**:62-65.

Vigneron JP, Méric R, Larchevêque M, Debal A, Kunesch G, Zagatti P, Gallois M (1982) Absolute configuration of eldanolide, the wing gland pheromone of the male African sugar cane borer, *Eldana saccharina* (Wlk.). Syntheses of its (+) and (−) enantiomers. Tetrahedron Lett **23**:5051-5054.

Vité JP, Hedden R, Mori K (1976a) *Ips grandicollis*: Field response to the optically pure pheromone. Naturwiss **63**:43.

Vité JP, Klimetzek D, Loskant G, Hedden R, Mori K (1976b) Chirality of insect pheromones: Response interruption by inactive antipodes. Naturwiss **63**: 582-583.

Vité JP, Ohloff G, Billings RF (1978) Pheromonal chirality and integrity of aggregation response in southern species of the bark beetle *Ips* sp. Nature (London) **272**:817-818.

Weber R, Hintzer K, Schurig V (1980) Enantiomer resolution of spiroketals. Complexation gas chromatography on an optically active metal complex. Naturwiss **67**:453-455.

Weber R, Schurig V (1981) Analytical enantiomer resolution of lineatin by complexation gas chromatography. Naturwiss **68**:330-331.

Webster FZ, Zeng X-N, Silverstein RM (1982) Following the course of resolution of carboxylic acids by ^{13}C NMR spectrometry of amine salts. J Org Chem **47**:5225-5226.

Whitesides GM, Lewis DW (1971) The determination of enantiomeric purity using chiral lanthanide shift reagents. J Am Chem Soc **93**:5914-5916.

Wilen SH, Collet A, Jacques J (1977) Strategies in optical resolutions. Tetrahedron **33**:2725-2736.

Williams HJ, Silverstein RM, Burkholder WE, Khorramshanic A (1981) Dominicalure 1 and 2: Components of aggregation pheromone from male lesser grain borer *Rhyzopertha dominica* F. J Chem Ecol **7**:759-780.

Wood DL, Browne LE, Ewing B, Lindahl K, Bedard WD, Tilden PE, Mori K, Pitman GB, Hughes PR (1976) Western pine beetle: Specificity among enantiomers of male and female components of an attractant pheromone. Science **192**:896-898.

Yamaguchi S, Yasuhara F, Kabuto K (1976) Use of shift reagent with diastereomeric MTPA esters for determination of configuration and enantiomeric purity of secondary carbinols in ^{1}H NMR spectroscopy. Tetrahedron **32**: 1363-1367.

Yasuhara F, Yamaguchi S (1977) Use of shift reagent with MTPA derivatives in

^{1}H NMR spectroscopy III. Determination of absolute configuration and enantiomeric purity of primary carbinols with chiral center at the C-2 position. Tetrahedron Lett 4085–4088.
Yasuhara F, Yamaguchi S (1980) Determination of absolute configuration and enantiomeric purity of 2- and 3-hydroxycarboxylic acid esters. Tetrahedron Lett **21**:2827–2830.

Chapter 13

Tabulations of Selected Methods of Syntheses That Are Frequently Employed for Insect Sex Pheromones, Emphasizing the Literature of 1977-1982

Philip E. Sonnet[1]

I. Introduction

The molecular structures of insect sex pheromones are quite diverse. They vary from the very simple straight-chain alcohols, acetates, and aldehydes that are predominant in the repertory of lepidopteran sexual communication to the more elaborate isoprenoids employed by scale insects. Synthesis of the simpler pheromones has been concerned primarily with chain elongation and the establishment of the required geometry about one or more double bonds. The more varied fare of other insect pheromones has focused considerable attention on their three-dimensional nature.

One might surmise that several areas of organic natural product chemistry, including insect chemistry, have induced an avalanche of exciting new synthetic chemistry to meet the challenges imposed by some of these structures. While our repeated syntheses of the simplest of pheromone structures may make biologists wonder if chemists enjoy driving tacks with sledges, a number of very imaginative synthetic routes and new procedures have been developed in recent years. Actually, the general strategies and precise methods employed in synthesizing insect pheromones constitute a collection almost as broad as the entire scope of organic synthesis.

Timely reviews of the subject of insect pheromone synthesis written from different perspectives have appeared (Table 1). An examination of these reviews will produce some insight into the strategies that were, and still are, being em-

[1]Research Chemist, Agricultural Research Service, USDA, Gainesville, Florida.

Table 1. Reviews of insect pheromone synthesis

Emphasis	References
Achiral pheromones	Rossi (1977)
Application of newer methods	Katzenellenbogen (1977)
Lepidopteran pheromones	Henrick (1977)
Chiral pheromones	Rossi (1978a)
Chemistry, biology, application	Brand et al. (1979)
Applications of Wittig reaction	Bestmann and Vostrowsky (1979)
Uses of glutamic acid	Smith and Williams (1979)
General	Mori (1981)
	Henrick et al. (1982)

ployed in synthesizing pheromones. This chapter, therefore, is written with several additional purposes in mind. First, it is hoped that this will serve as a guide to the recent literature of insect sex pheromones, i.e., those years since 1977–1978 when the first reviews of pheromone synthesis appeared. The chapter is particularly intended for the biologist to serve as a basis for discussion with chemists as well as for the chemist seeking to collaborate with biologists. It is envisioned that the reader intending to perform synthetic experiments has some familiarity with general organic laboratory procedures, is familiar with material balance in chemical reactions, and is aware of the hazards (toxicity, flammability) of the many chemicals commonly employed as solvents and reagents. Two experimental protocols are given which, hopefully, shall serve as guides to conducting experiments in general.

The bulk of this chapter has been devoted to tabulating the synthetic methods employed so that the laboratory scientists may quickly find references that exemplify the desired transformation. Generally these reactions provide yields greater than 70%, though one should bear in mind that in most cases the scientists had not made an effort to optimize each yield. Rather, the reactions employed were chosen by them for their known broad utility. Yields of these reactions will vary to some degree with the skill of the scientist, but largely will depend on the molecular structures in the particular instance for which the reaction is being employed.

It should be noted that many other synthetic methods exist in the body of chemical literature that could be fruitfully applied to the cause of pheromone synthesis. Although most of the tabulated citations provide experimental detail sufficient for the trained and educated chemist, others do not. They are included, however, for the sake of greater completeness in referencing. In addition they often lead *via* their own citations to the details sought.

Most importantly, synthetic chemistry is not a compilation of "how to –," but rather each transformation is based on an understanding of the mechanism of the reaction, its scope, and its limitations. Hence, the additional citations for a given reaction will indicate to some extent the variability of experimental detail for that reaction and its compatability with other functional groups that may be present. Some general references are also supplied in the Appendix.

II. Alkylation Reactions

The usual first step in proposing a synthetic sequence to obtain a target structure is the establishment of the required carbon skeleton. Of necessity, this means making, or breaking, carbon–carbon bonds. A great deal of chemical literature, therefore, concerns itself directly with alkylation reactions, and syntheses of sex pheromone structures likewise partake of that broad spectrum of reactions. They have been segregated in the following manner. Table 2 lists acetylene alkylations (the products of which generally retain the alkyne triple bond); Table 3 lists Wittig condensations (the products would be alkenes); Table 4 gives alkylations of moderately stable organic anions; Tables 5 and 6 show the reactions of organomagnesium and organolithium reagents, respectively. Additional examples of carbon–carbon bond forming reactions are provided in Table 7. Although it is fallacious to assert that a general procedure may be given for the conduct of an organic chemical reaction, certain considerations, e.g., exposure to air, freedom from moisture, means of bringing reagents into contact are common considerations to all reactions.

The following procedures are generally applicable to reactions involving moisture/air sensitive reagents, and are therefore usually employed in alkylation reactions. Organic anions such as acetylides are prepared in an inert atmosphere, and the following statements have general application in synthesis involving organometallics.

For most purposes oven-drying glassware for 30 min at 130°C is adequate. The assembly is placed under nitrogen (argon is preferred if dealing with metallic lithium) that is flowing through a tube of indicator-colored drying agent (Drierite, for example) and, after passage over the reaction glassware, through a bubbler of paraffin oil. Solvents for such reactions are dried according to standard methods (Appendix) and are generally injected through a rubber stopple into the vessel. If a reagent is to be added dropwise to the vessel, it can be similarly injected into a pressure-equalized dropping funnel to which is then added (by injection) a further amount of solvent. The reactions are most conveniently stirred magnetically, and the reaction temperatures can be controlled by cooling baths whose composition determines bath temperature (Gordon and Ford, 1972).

Perhaps the most frequently employed base to generate an organic anion is commercially available butyllithium (BuLi). Sold in hexane solution with a labeled molarity, it can be stored for long periods at 0°C. When in use, the bottle is capped with a stopple and kept in a hood. Samples are withdrawn by syringing an equal volume of nitrogen into the bottle to replace the volume of solution withdrawn. If the opened bottle remains unused for a period of time, it may be useful to determine solution molarity again by titration (Bergbreiter and Pendergrass, 1981). Although BuLi is highly reactive toward moisture, it is relatively safe to handle.

Other bases, such as secondary BuLi, tertiary BuLi, and ethereal methyl-

Table 2. Alkylations of acetylenes

Reaction type	References
$RC{\equiv}CM + R'X \rightarrow RC{\equiv}CR'$	Ideses et al. (1982), Mori et al. (1978a,b, 1981a), Place et al. (1978), Rossi and Carpita (1977), Rossi et al. (1979, 1980, 1981), Sato et al. (1979), Sonnet and Heath (1980), Uchida et al. (1977)
$RC{\equiv}CM + R'CH{=}CHCH_2X$ $\rightarrow$ 1,4-enyne	Rossi and Carpita (1977)
$RC{\equiv}CH + R'CH{=}CHCH_2X$ $\rightarrow$ 1,4-enyne (Pd^0,Cu^I)	Rossi et al. (1982)
$RC{\equiv}CM$ + epoxide $\rightarrow$ β-hydroxyalkyne	Pirkle and Adams (1979)
$RC{\equiv}CM$ + aldehyde $\rightarrow$ α-hydroxyalkyne	Hall et al. (1980), Kunesch et al. (1981), Pirkle and Adams (1979), Tsuboi et al. (1982)
$RC{\equiv}CM$ + ketone $\rightarrow$ α-hydroxyalkyne	Mori (1978)
$RC{\equiv}CM$ + acid halide $\rightarrow$ α-alkynone	Baker and Rao (1982), Midland and Nguyen (1981)
$RC{\equiv}CM + CO_2 \rightarrow RC{\equiv}CCO_2M$	Kunesch et al. (1981)
$MC{\equiv}CCO_2M$ + epoxide $\rightarrow HO(CH_2)_2C{\equiv}CCO_2H$	Pirkle and Adams (1978)
$RC{\equiv}CM + CH_2{=}CHBr(Cu^I)$ $\rightarrow$ conj. enyne	Rossi et al. (1981)
Dialkyl dialkynyl borate $\rightarrow$ conj. diyne	Sinclair and Brown (1976)
$RC{\equiv}CH + R'C{\equiv}CBr(Cu^I)$ $\rightarrow$ conj. diyne	Sonnet and Heath (1980)
Propargyl derivative + R_2CuLi $\rightarrow$ alkylated allene	Pirkle and Boeder (1978)
Internal alkyne $\rightarrow$ terminal alkyne	Negishi and Abramovitch (1977)
Trialkyl alkynylborate $\rightarrow RC{\equiv}CR'$	Bestmann and Li (1981)

Table 2. (continued)

Reaction type	References
Dialkyl vinyl alkynyl borate → (*Z*)-conj. enyne	Negishi and Abramovitch (1977)
$RC{\equiv}CH + R'MgBr(Cu^{I}) \rightarrow RR'C{=}CH_2$	Anderson et al. (1979, 1981)
$HC{\equiv}CH + RMgBr(Cu^{I})$, then $RC{\equiv}CH \rightarrow$ conj. enyne	Normant et al. (1975)
$RC{\equiv}CH + R'MgBr(Cu^{I})$, then $CO_2 \rightarrow RR'C{=}CHCO_2H$ (R' *cis* to CO_2H)	Anderson et al. (1979)
$RC{\equiv}CCO_2CH_3 + R'Li(Cu^{I}) \rightarrow RR'C{=}CHCO_2CH_3$ (R *cis* to CO_2CH_3)	Uchida et al. (1981)
$HC{\equiv}CH + R_2CuLi$, then vinyloxirane → conj. (*E,Z*)-2,4-dienol	Alexakis et al. (1978)

Table 3. Wittig condensations

Reaction type	References
$Ph_3P{=}CHR + R'CHO \rightarrow R\ CH{=}CHR'$	Anderson and Henrick (1975), Bartlett et al. (1982), Bestmann et al. (1977–1979), Bierl-Leonhardt et al. (1980, 1982), Doolittle et al. (1980), Horiike et al. (1980), Mori et al. (1981a), Novak et al. (1979), Redlich et al. (1981)
$Ph_3P{=}CH(CH_2)_xOH + RCHO \rightarrow HO(CH_2)_xCH{=}CHR$	Horiike et al. (1982)
$Ph_3P{=}CHR$ + lactol $\rightarrow HO(CH_2)_xCH{=}CHR$	Bestmann and Li (1981), Mori et al. (1979)
$Ph_3P{=}CHR + R'O_2C(CH_2)_xCHO \rightarrow R'O_2C(CH_2)_xCH{=}CHR$	Bestmann et al. (1979d,e), Hammoud and Descoins (1978)
$Ph_3P{=}CHR$ + epoxyaldehyde → vinylepoxide	Rossiter et al. (1981)

Table 3. (continued)

Reaction type	References
$2Ph_3P{=}CHR$ + dialdehyde → diene	Hoshino and Mori (1980)
$Ph_3P{=}CHR$ + αβ-unsat. aldehyde → conj. diene	Bestmann et al. (1979b), Hall et al. (1980), Ideses et al. (1982), Wollenberg and Peries (1979)
$Ph_3P{=}CRR' + R''CHO \rightarrow RR'C{=}CHR''$	Masuda et al. (1981), Mori and Kuwahara (1982), Heath et al. (1980) (directed), Roelofs et al. (1978) (undirected)
$Ph_3P{=}CHCH{=}CHR + R'CHO$ → conj. diene	McDonough et al. (1982)
$Ph_3P{=}CHCHO + R'CHO$ → αβ-unsat. aldehyde	Bestmann et al. (1979b), Canevet et al. (1980), McDonough et al. (1982)
$Ph_3P{=}CHR + R'R''CO \rightarrow RCH{=}CHR'R''$	Heath et al. (1980), Magnusson (1977), Mori (1978), Uchida et al. (1981)
$(EtO)_2OP{=}CRCO_2Et + R'CHO \rightarrow R'CH{=}CRCO_2Et$	Anderson and Henrick (1979)
Cyclic ylid + RCHO → intermediate → 1,5-dienes	Muchowski and Venuti (1981)
A cyclic enediol diacetate + $Ph_3P{=}CH_2$ → a methylene cyclohexene	Hanessian et al. (1981)

Table 4. Alkylations of carbonyl-, or other-, stabilized anions and related reactions

Reagents	References
Ketone, alkyl halide	Chuman et al. (1981), Kocienski and Ansell (1977), Mori and Nomi (1981)
Ketone (anion), aldehyde	Kozhich et al. (1982)
Ketone (anion), ester	Sakakibara and Mori (1979)
Ester, allylic halide	Alexakis et al. (1978)

Table 4. (continued)

Reagents	References
Ester (anion), ester	Baker et al. (1980)
β-Ketoester, alkyl halide	Naoshima et al. (1980a,b)
Cyclic β-ketoester, alkyl halide	Bacardit and Moreno-Manas (1980)
Malonate ester, alkyl halide	Ade et al. (1980), Jewett et al. (1978), Matsumura et al. (1979), Mori et al. (1981a)
Malonate ester, epoxide	Byström et al. (1981)
αβ-Unsat. ketone (anion), αβ-unsat. aldehyde	Still (1979)
αβ-Unsat. ester, alkyl halide	Still and Mitra (1978)
Acid dianion, alkyl halide	Sonnet and Heath (1982a), Sonnet (1982)
Acid dianion, αβ-unsat. aldehyde	Trost and Fortunak (1980)
β-Diketone dianion, ester	Hoffmann and Ladner (1979)
β-Diketone dianion, anhydride	Mori et al. (1981c)
α-Acetyl-δ-valerolactone dianion, epoxide	Hintzer et al. (1981)
ββ-Triketone trianion, aldehyde	Sakakibara and Mori (1979)
Enamine, αβ-unsat. ketone	Chuman et al. (1979)
Phosphonate ester, alkyl halide	Canevet et al. (1980)
2-Picoline, alkyl halide	Dressaire and Langlois (1980)
Sulfonium salt (ylid), ketone	Still (1979)
Phenylalkyl sulfide, alkyl halide	Babler and Haack (1982), Pirkle and Rinaldi (1979)
Pyridylalkyl sulfide, allylic halide	Mori and Ueda (1981)
Dithiane, alkylhalide	Mori et al. (1982)

Table 5. Alkylations with organomagnesium compounds

Reagents	References
RMgX, sulfinate ester	Farnum et al. (1977)
RMgX, allylic halide	Rossi and Carpita (1977)
RMgX, carbon dioxide	Rossi et al. (1979)
RMgX, oxetane	Rossi et al. (1980)
RMgX, aldehyde	Babler and Invergo (1979), Chuman et al. (1979), Kocienski and Ansell (1977), Labovitz et al. (1975), Mori and Kuwahara (1982), Pirkle and Boeder (1978), Pirkle and Adams (1979), Place et al. (1978), Suguro et al. (1981)
$CH_2{=}CHMgX$, aldehyde	Masuda et al. (1981)
RMgX, $\alpha\beta$-unsat. aldehyde	Hammoud and Descoins (1978), Place et al. (1978), Still and Mitra (1978)
$(CH_3)_3SiCH_2MgCl$, ketone($\rightarrow RR'C{=}CH_2$)	Still (1979)
RMgX, $\alpha\beta$-unsat. ketone	Mori (1978)
RMgX, ester ($\rightarrow$ tertiary alcohol)	Suguro et al. (1981)
RMgX, orthoester ($\rightarrow$ acetal)	Bartlett et al. (1982)
RMgX, β-ketoester ($\rightarrow$ alkylated δ-lactone)	Rossi and Marasco (1980)
RMgX, nitrile	Tamada et al. (1978)
The following reactions were mediated by copper as cuprates	
RMgX, alkyl halide or tosylate	Descoins et al. (1977), Guss et al. (1983), Mandai et al. (1979), Mori et al. (1978b, 1981a), Rossi (1978b), Shani (1979), Sonnet and Heath (1982a,b), Suguro and Mori (1979b)

Table 5. (continued)

Reagents	References
RMgX, allylic halide	Kondo and Murahashi (1979), Mori et al. (1978a)
RMgX, allylic phenylether	Zakharkin and Petrushkina (1982)
RMgX, 3-acyloxy-4-en-1-yne (→ conj. enyne)	Cassani et al. (1979)
RMgX, quaternary ammonium salt	Descodts et al. (1979)
RMgX, (*E,E*)-2,4-diene acetate (→ conj. *E,E*-diene)	Samain et al. (1978)
RMgX, acid halide (→ ketone)	Kondo and Murahashi (1979)
RMgX, $\alpha\beta$-unsat. ketone (1,4-addition)	Guss et al. (1982)
The following reaction was conducted in the presence of Pd^0	
RMgX, vinyl halide	Kondo and Murahashi (1979)

lithium will *ignite* as they bleed from the syringe. On–off adapters should be employed when handling organometallics in syringes in any case. It is amusing to note that such minor pyrophorics can drive sidewalk superintendents from the laboratory, but one should approach the use of these materials with great caution and appreciate their potential hazards. Additionally, since many organic reagents are liquids, one generally labels them with their molarity so that one can measure them out by syringe volume rather than by weighing.

2.1. Alkylation of a 1-Alkyne with 1-Bromo-3-Chloropropane

The 1-alkyne (25 mmole) is injected into a reaction vessel that is under nitrogen. Then dry tetrahydrofuran (THF) (15 ml) is injected, and the vessel is cooled in dry-ice acetone (−78°C). The BuLi (25 mmole) is then injected dropwise which immediately produces the acetylide anion. The 1-bromo-3-chloropropane (25 mmole), and dry hexamethylphosphoric triamide (HMPT) (10 ml) are injected dropwise over about 5 min, and the resulting solution is stirred at −78°C for 1 hr. The bath is removed and, after another 0.5 hr is transferred to a separatory funnel with hexane for the usual workup of an organic reaction–in this case washing with water is sufficient. After drying ($MgSO_4$) and stripping the solvent,

Table 6. Alkylations with organolithium compounds

Reagents	References
RLi, ester	Magnusson (1977)
RLi, lactone (→ ketoalcohol)	Byström et al. (1981)
RLi, acid (→ ketone)	Rossi and Marasco (1980)
CH_2=CHLi, alkyl halide	Roelofs et al. (1978)
CH_2=CHLi, ketone	Still (1979)
Allylic Li, alkyl halide	Anderson et al. (1981)
β-Ethoxyvinyl Li, ketone	Wollenberg and Peries (1979)
The following reactions were mediated by copper as cuprates	
RLi, alkyl halide or tosylate	Mori and Tamada (1979), Mori et al. (1979), Mori et al. (1978a), Novak et al. (1979), Ravid et al. (1978)
RLi, quaternary ammonium salt	Dressaire and Langlois (1980)
RLi, allylic acetate	Anderson and Henrick (1979), Still (1979)
RLi, epoxide	Anderson et al. (1980), Chuman et al. (1981), Kunesch et al. (1981), Mori et al. (1981c), Mori and Nomi (1981)
RLi, Δ2-butenolide	Kunesch et al. (1981)
CH_2=CHLi, alkyl halide	Anderson and Henrick (1979)

the crude product is generally distilled to produce a 1-chloro-4-alkyne suitable for further transformation. A specific example is the conversion of 1-heptyne to 1-chloro-4-decyne in 81% yield, bp 46–52°C (0.02 mm).

Reaction of acetylides with 1-bromoalkanes can be conducted using an ice bath. The lower temperature just described for the dihalide was to maximize selectivity for bromide displacement. Condensations of acetylide anions with carbonyl substituents are also exothermic and should be externally cooled. Reactions with terminal epoxides also proceed well and are generally given reaction

Table 7. Additional examples of carbon–carbon bond forming reactions

Type of reaction	References
R_2Cd, acid halide (→ ketone)	Pirkle and Adams (1979)
Alkenyl borepane (→ [*Z*]-7-alken-1-ols)	Basavaiah and Brown (1982)
Dialkylboranes, alkynes (→ [*Z*]-7-alkenes) *Z*-Allylboronate, aldehyde (→ erythro homoallyl alcohols)	Brown and Basavaiah (1982), Schlosser and Fujita (1982)
Kolbe electrolysis	Jewett et al. (1978), Klunenberg and Schäfer (1978)
1-Alkene, $Al(CH_3)_3$, $TiCl_4$ (→ *E*-2-alkenes)	Schlosser and Fujita (1982)
Electrocylic rearrangements	
Claisen rearrangement	Samain and Descoins (1979)
Claisen ortho ester rearrangement	Anderson et al. (1979), Hammoud and Descoins (1978), Kunesch et al. (1981), Labovitz et al. (1975), Suguro et al. (1981), Tsuboi et al. (1982)
Oxycope rearrangement	Place et al. (1978), Still (1979)
Allyloxymethide → homoallyloxide	Masuda et al. (1981), Mori and Kuwahara (1982), Still and Mitra (1978)

times of several hours at room temperature. Ethylene oxide, incidentally, is conveniently added to a reaction from a precooled syringe.

The 1-chloro-4-alkyne so prepared may be converted to a phosphorane (see the Wittig Condensation in the next section) and allowed to react with an aldehyde to produce 1,5-enynes. Reduction of the triple to a double bond would provide 1,5-dienes such as the pheromone structures of the Angoumois grain moth and pink bollworm moth. Alternatively a magnesio derivative of the 1-chloro-3-alkyne could be treated with a nitrile that upon reduction of the triple bond would yield 1,5-enones such as the peach fruit moth sex pheromone components.

2.2. Wittig Condensation: Preparation of (Z)-6-Octadecene

Probably no single chemical reaction has received as much study or found as much application as that which bears the name of its discoverer, Georg Wittig, namely, the condensation of an ylid, which can be derived from a phosphonium salt by deprotonation, with an aldehyde to give a 1,2-disubstituted olefin. Vedejs has provided considerable insight into the nature of this reaction and has given leading references (Vedejs et al., 1981). An older review (Maercker, 1965) provides much useful tabular data and some procedures both for the condensation itself and for the preparation of the phosphonium salts employed for the ylids.

The triphenylphosphonium salt of 1-bromododecane can be prepared by heating one equivalent each of triphenylphosphine and the halide in acetonitrile (reagent grade) under reflux. A good ratio of solvent to phosphine is 2 ml/g although this is quite flexible. The solvent is removed with a flash evaporator and dry ether is added to induce crystallization. If a phosphonium salt is slow to crystallize, the ether is decanted thereby removing residual acetonitrile. Addition of fresh ether and agitation usually succeeds. In the case of a salt which remains an oil, one can dry the residue (vacuum, P_2O_5) and then make up a standard solution in, for example, dry THF kept under nitrogen and stoppeled. Knowledge of the salt's weight and that of the THF coupled with a density determination of the solution (weight of a filled 1 ml syringe minus the syringe weight after expulsion of solution is generally adequate) allows a calculation of solution molarity. The dodecyl salt is crystalline, however, and is processed by suction filtration. The solid is washed with ether, and then dried (vacuum, P_2O_5).

The condensation itself is carried out in an inert atmosphere as described for the acetylene alkylation. An equivalent of the phosphonium salt is placed into the vessel and dry THF ($\sim$3-5 ml/g) is injected. The mixture is cooled in an ice bath, and BuLi (one equiv.) is slowly injected. Since the ylid is brightly colored, the persistence of the orange color indicates not only its formation but also the absence of moisture. Therefore one normally injects BuLi until the color persists and then injects the amount required by reaction stoichiometry. The resulting mixture is stirred for 15 min and then is cooled in dry ice-acetone. Addition of HMPT (2 equiv. to tie up the lithium) may be done before or after cooling. Finally, hexanal (freshly distilled, one equiv.) is injected dropwise to keep the temperature as low as possible. The mixture is stirred allowing the bath to come to room temperature. It is then partitioned in a separatory funnel between hexane and water. The crude product contains unwanted triphenylphosphine oxide and can be purified in this case by passage with hexane through a small column of silica gel ($\sim$10 g/g olefin). The olefin is obtained in >90% yield and contains 3% *trans* as judged by gas chromatography (GC) (capillary column coated with cholesterol *para*-chlorocinnamate) (Heath et al., 1979).

This procedure can be employed for many pheromone structures that possess isolated *cis*-olefin units by suitable change of reagents and protection of base-sensitive functional groups. Protective groups for this purpose are listed in Table 11 of this chapter.

III. Other Reactions Frequently Employed for Sex Pheromone Syntheses

In addition to establishing a desired carbon skeleton, adjustments of functional groups must be accomplished. Reductions are tabulated (Table 8) showing conversions of alkynes to alkenes or alkanes, carbonyl to alkane, etc. Because the carbonyl-to-carbinol conversion was so common in pheromone literature it was not listed.

A number of insect sex pheromone structures contain a conjugated diene unit. Although these structures show up as alkylation products occasionally in Tables 1-9, a separate listing of additional useful references to this specific topic was made in Table 9. Table 10 deals with oxidation reactions including several carbon–carbon bond breakers. A synthetic effort is often geared to building the desired structure, but sometimes benefits by perceiving synthesis as a directed degradation, and a desired intermediate is formed by fragmenting a larger, or a cyclic, molecule. Oxidations of alcohols to aldehydes and ketones were again so common that they are not listed. In fact, a number of other simple reactions are arbitrarily omitted to conserve space, e.g., esterifications, hydride reductions of esters, use of tetrahydropyranyl ethers as protecting groups.

Table 11 lists the other protecting groups employed in recent pheromone syntheses. A discussion of these and other functional group blocks (preparation, stability to reaction conditions, removal) is provided in reviews (McOmie, 1973, 1979) and a book (Green, 1981). The tabulations of chemical reactions is rounded out with some additional transformations in Table 12.

Because many natural products have asymmetric centers, the relationship of stereostructure (three-dimensional configuration) to biology has generated intense interest. Synthesis of candidate structures has therefore concerned itself with the several methods of generating centers of asymmetry and obtaining a high configurational bias in the product. Table 13 indicates both the techniques and the (commercially available) chiral reagents employed for sex pheromone synthesis in recent years. A listing of companies from which these and other reagents may be purchased is provided in the Appendix. It should be noted that the chemical marketplace for chiral materials is in need of update, both on the matter of naming configurations (standard chemical nomenclature for configuration is not in uniform use) and, very importantly, on the subject of configurational purity. For a thorough discussion of the latter, crucial topic, I refer you to Professor Mori's Chapter 12 in this book.

Table 8. Reductions

Reactant → product	References
Alkyne → (*Z*)-alkene	Baker and Rao (1982), Coke and Richon (1976), Hall et al. (1980), Horiike et al. (1982), Midland and Tramontano (1980), Midland and Nguyen (1981), Mori et al. (1978a), Nishizawa et al. (1981), Pirkle and Adams (1979), Rossi (1978b), Rossi et al. (1979, 1980), Sato et al. (1979), Uchida et al. (1977, 1979)
Alkyne → (*E*)-alkene	Hall et al. (1980), Rossi et al. (1980), Suguro and Mori (1979)
Alkyne → alkane	Nishizawa et al. (1981)
Alkene → alkane	Guss et al. (1982, 1983), Jewett et al. (1978), Magnusson (1977), Hoshino and Mori (1980), Kozhich et al. (1982), Matsumura et al. (1979), Mori et al. (1978b, 1979, 1981a), Schlosser and Fujita (1982)
Alkenylepoxide → alkylepoxide	Rossiter et al. (1981)
Conj. (*E*)-enyne → conj. (*E,Z*)-diene	Labovitz et al. (1975), Negishi and Abramovitch (1977)
Conj. diyne → conj. (*Z,Z*)-diene	Sonnet and Heath (1980)
1-en-3-yne → (*Z*)-1,3-Diene	Rossi et al. (1981)
RI → RH	Hoshino and Mori (1980)
RCH_2OH → RCH_3	Mori (1978), Redlich et al. (1981), Sonnet and Heath (1982a,b), Sonnet (1982)
Aldehyde → alkane	Mori (1978)
Ketone → alkane	Byström et al. (1981), Mori et al. (1978a), Naoshima et al. (1980a,b), Redlich et al. (1981)
Sulfoxide → alkane	Solladie and Matloubi-Moghadam (1982)

Table 8. (continued)

Reactant → product	References
Acid halide → aldehyde	Doolittle et al. (1980)
Nitrile → aldehyde	Kocieski and Ansell (1977)
Lactone → lactol	Mori et al. (1979)
3-Iodoepoxide → allylic alcohol	Mori and Ueda (1981)
Allylic phosphonate → (*E*)-alkene	Bestmann et al. (1979), Canevet et al. (1980)

Table 9. Further examples for conjugated dienes

Reagent(s)	References
Conj. diyne → conj. (*Z,Z*)-diene	Zweifel and Polston (1970)
Alkenyl copper, alkenyl halide (→ *E,Z*;*E,E*)	Jabri et al. (1981)
Alkenyl magnesium, alkenyl halide (→ *E,Z*;*E,E*)	Dang and Linstrumelle (1978)
Alkenyl aluminum, alkenyl halide (→ *E,Z*;*E,E*)	Baba and Negishi (1976)
Alkenyl zirconium, alkenyl halide (→ *E,Z*;*E,E*)	Okukado et al. (1978)
Alkyl alkenyl α-chloroalkenyl borane (→ *E,E*)	Negishi and Yoshida (1973)
Dialkyl alkynyl alkenyl borate (→ [*E*]-enyne)	Negishi et al. (1973)
Elimination of unsat. quaternary ammonium salt (→ *E,Z*)	Dressaire and Langlois (1980)
α-Allene ester, Al_2O_3 → 2E,4Z-ester	Tsuboi et al. (1982)
Allylic sulfoxide → [allylic sulfenate] → conj. diene	Babler and Haack (1982)

Table 10. Some useful oxidation reactions

Reactant → product (oxidizing agent)	References
α,β-Unsat. ketone → epoxyketone (H_2O_2, NaOH)	Roelofs et al. (1978)
Allylic alcohol → α,β-unsat. aldehyde (MnO_2)	Hall et al. (1980)
Ester → αβ-unsat. ester (several steps)	Mori et al. (1981b), Pirkle and Boeder (1978)
Ketone → ester (peracid)	Magnusson (1977)
1,2-Diol → aldehydes ($Pb[OAc]_4$)	Roelofs et al. (1978)
1,2-Diol → aldehydes (HIO_4)	Rossi et al. (1979)
Epoxide → aldehydes (HIO_4)	Hoshino and Mori (1980), Mori (1978), Mori et al. (1978a, 1981a), Mori and Tamada (1979), Sato et al. (1979)
Alkene → aldehydes/ketones (O_3)	Bestmann et al. (1979a–e), Heath et al. (1980), Uchida et al. (1977)
Alkene → acetals (O_3, ROH)	Mori and Kuwahara (1982)
Alkene → aldehydes/ketones (OsO_4, $NaIO_4$)	Wollenberg and Peries (1979)
Alkene → cleaved alcohols (O_3, LAH)	Mori and Tamada (1979)
Alkene → allylic alcohol (PhSeOH, t-BuO_2H)	Mori and Kuwahara (1982)
Cyclic enol ether → aldehyde ester (O_3)	Bestmann et al. (1979b,d,e)
1,4-Cyclohexadiene → (*Z*)-3-hexen-1,6-diol	Uchida et al. (1979)
Phenyl → carboxyl (O_3, H_2O_2)	Mori et al. (1981c)
Sulfide → sulfoxide (RCO_3H)	Babler and Haack (1982)

Table 11. Protecting groups employed in recent pheromone literature

Derivative (group protected)	References
Methyl ether (OH)	Mori et al. (1979), Sonnet and Heath (1982a,b)
Benzyl ether (OH)	Chuman et al. (1981), Hanessian et al. (1981), Mori and Kuwahara (1982), Mori and Nomi (1981)
t-Butyl ether (OH)	Hanessian et al. (1981)
Ethoxyethyl ether (OH)	Hammoud and Descoins (1978), Still (1979), Mori et al. (1982)
MEM-Ether (OH)	Anderson and Henrick (1979), Anderson et al. (1980), Mori and Kuwahara (1982)
Trimethylsilyl ether (OH)	Anderson et al. (1979), Hoffmann and Ladner (1979)
Dimethyl-*t*-butylsilyl ether (OH)	Chuman et al. (1981), Mori and Nomi (1981), Still (1979)
Diphenyl-*t*-butylsilyl ether (OH)	Hanessian et al. (1981)
"Acetonide" (1,2-diol)	Chuman et al. (1981), Mori and Nomi (1981)
Dimethyl acetal (aldehyde)	Heath et al. (1980), Pirkle and Boeder (1978)
Diethyl acetal (aldehyde)	Roelofs et al. (1978)
Ethylene ketal (ketone)	Mori (1978)
Dithiane (aldehyde, ketone)	Mori et al. (1982), Redlich et al. (1981)
3-Trimethylstannyltrimethylsilyl enol ether ($\alpha\beta$-unsat. ketone)	Still (1979)
Trimethylsilyl (RC≡CH)	Hammoud and Descoins (1978)

Table 12. Miscellaneous useful transformations

Transformation	References
$RCH{=}CH_2 \rightarrow RCH_2CH_2OH$	Masuda et al. (1981), Rossi et al. (1979), Sonnet (1982), Sonnet and Heath (1982a,b), Zakharkin et al. (1982)
$RCH{=}CH_2 \rightarrow RCHOHCH_3$	Mori et al. (1981a), Sonnet (1982)
$RCH{=}CH_2 \rightarrow RC{\equiv}CH$	Mori et al. (1978a)
$RCH_2CH_2OH \rightarrow RCHOHCH_3$	Mori et al. (1978b)
$RCH_2CH_2OH \rightarrow RCH{=}CH_2$	Still (1979)
$RCH{=}CH_2 \rightarrow RCH_2CH_2Br$	Rossi and Carpita (1977)
$RCH_2CH_2Br \rightarrow RCH{=}CH_2$	Masuda et al. (1981), Mori et al. (1981a)
$RC{\equiv}CH \rightarrow (Z){-}RCH{=}CHI$	Ravid et al. (1978)
$RCH(OAc)CH(CO_2H)R' \rightarrow (E){-}RCH{=}CHR'$	Trost and Fortunak (1980)
Epoxide $\rightarrow RCH{=}CHR'$	Sonnet (1980b)
Carbamate $\rightarrow$ ROH	Pirkle and Adams (1979)
Amide $\rightarrow RCO_2H$	Sonnet (1982)
Cyclopropyl alkyl carbinol → homoallylic halide (Julia reaction)	Descoins et al. (1977), Guss et al. (1982, 1983)
ROH → RBr, RI	Bartlett et al. (1982), Byström et al. (1981), Chuman et al. (1981), Mori et al. (1978a, 1981a), Mori and Tamada (1979), Mori and Nomi (1981), Mori and Ueda (1981), Roelofs et al. (1978), Rossi et al. (1981), Shani (1979), Still and Mitra (1978), Uchida et al. (1977)

Table 12. (continued)

Transformation	References
ROH → RCN	Mori and Tamada (1979), Mori et al. (1981b), Solladie and Matloubi-Moghadam (1982), Tamada et al. (1978)
1,2-Diol → epoxide	Chuman et al. (1981), Mori et al. (1979), Mori and Nomi (1981)
β-Hydroxysulfide → epoxide	Farnum et al. (1977), Pirkle and Rinaldi (1979)
Epoxide → allylic alcohol	Bierl-Leonhardt et al. (1980, 1982), Mori et al. (1978a,b), Uchida et al. (1981)
Allylic alcohol → allylic nitrile (transposition)	Hoshino and Mori (1980)
$RBr \rightarrow RSC_6H_5$	Babler and Haack (1982)
Allylic sulfide → allylic alcohol (transposition)	Mori and Ueda (1981)
$RCO_2H \rightarrow RBr$, RI (Hunsdiecker)	Ade et al. (1980), Masuda et al. (1981), Mori et al. (1981a)
$RC{\equiv}CCH_2OTHP \rightarrow RCH{=}CHCHO(E)$	Ideses et al. (1982)
Ketoxime → nitrile	Kocienski and Ansell (1977)
Olefin inversion	Descodts et al. (1979), Sonnet (1980a), Suguro et al. (1981)

Table 13. Methods for obtaining configurational bias for chiral sex pheromones

Method	Chiral reagents	References
1. Purchased chiral reagent for synthesis	(a) Citronellol (isopulegol pulegone)	Masuda et al. (1981), Mori et al. (1982, 1978b), Mori and Kuwahara (1982), Mori and Tamada (1979), Suguro and Mori (1979a,b)
	(b) Carbohydrates	Hanessian et al. (1981), Mori (1982), Redlich et al. (1981)
	(c) Ethyl lactate	Hintzer et al. (1981)
	(d) Glutamic acid	Doolittle et al. (1980)
	(e) Isoleucine	Sonnet and Heath (1982a,b)
	(f) Limonene	Heath et al. (1980)
	(g) 2-Methyl-1-butanol	Sonnet and Heath (1982a,b)
	(h) β-Pinene	Hobbs and Magnus (1976)
	(i) Tartaric acid	Mori and Tamada (1979), Mori et al. (1981c)
	(j) Verbenone	Bierl-Leonhardt et al. (1981)
2. Fractional crystallization of diastereomers	(a) Brucine, cinchonine quinine and/or quinidine	Matsumura et al. (1979), Mori (1978), Mori and Tamada (1978)
	(b) Menthol	Farnum et al. (1977), Nishizawa et al. (1981)
	(c) α-Naphthylethylamine	Mori et al. (1981b), Pirkle and Rinaldi (1979)
	(d) α-Phenylethylamine	Baker and Rao (1982), Sonnet (1982), Sato et al. (1979)
3. HPLC preparative separation of diastereomers	(a) Methoxytrifluoromethyl-phenylacetic acid	Anderson et al. (1980), Pirkle and Adams (1978, 1979)
	(b) α-Naphthylethylamine	Pirkle and Boeder (1978)
	(c) α-Phenylethylamine	Sonnet and Heath (1982a,b)
	(d) Phenylglycinol	Ade et al. (1980)
	(e) Tartaric acid	Coke and Richon (1976)

4. GLC preparative separation of enantiomers	(a) Chiral manganese phase (camphor)	Hintzer et al. (1981)
5. Asymmetric induction; alkylation of: Chiral boronate	(a) Camphor	Hoffmann and Ladner (1979)
Chiral anion	(a) 2-Amino-1-phenyl-1,3-propanediol	Byström et al. (1981), Rossi and Marasco (1980)
	(b) Bornanediol	Ade et al. (1980)
	(c) Menthol	Farnum et al. (1977), Solladie and Matloubi-Moghadam (1980)
	(d) Prolinol	Sonnet and Heath (1980)
6. Catalytic hydrogenation	(a) Tartaric acid	Matsumura et al. (1979)
7. Claisen ortho ester rearrangement	(a) α-Naphthylethylamine	Mori et al. (1981b)
8. Epoxidation	(a) Diethyl tartrate	Rossiter et al. (1981), Mori and Ueda (1981)
9. Hydride reduction	(a) Menthol	Nishizawa et al. (1981)
	(b) α-Pinene	Midland and Tramontano (1980), Midland and Nguyen (1981), Baker and Rao (1982)
10. Spontaneous resolution of a racemate	–	Adams et al. (1979)

IV. Appendix 1

4.1. Some Useful Books and Leading References to General Laboratory Practice and Synthetic Chemistry

Apsimon and Sequin (1979) Asymmetric Synthesis.
Bergbreiter and Pendergrass (1981) Titration of Organometallics
Brandsma (1971) Acetylene Chemistry
Brown (1975) Organoborane Chemistry
Burfield, Klee and Smithers (1977) Drying of solvents
Cahn, Ingold and Prelog (1966) Stereonomenclature
Fieser and Fieser (1967) Reagents for Organic Synthesis (first issue of a series)
Gordon and Ford (1972) Compilations of Data Useful for Lab Practice
Harrison and Harrison (1971) Synthetic Methods Tabulated (first issue of a series)
Jones, Sih and Perlman (1976) Biological Systems in Organic Chemistry
McMurray and Miller (1971) An Annual Survey of New Synthetic Methods
Posner (1980) Organocopper Chemistry
Prelog and Helmchen (1982) Stereonomenclature
Riddick and Bunger (1970) Organic Solvents
Roberts et al. (1968) Introduction to Organic Lab Practice
Szabo and Lee (1980) Chiral Reagents

V. Appendix 2

5.1. Some Commercial Sources of Chiral Chemicals

Aldrich Chemical Co.
940 West Saint Paul Avenue
Milwaukee, WI 53233

Eastern Chemical Co.
230 Marcus Boulevard
Hauppauge, NY 11787

Givaudan Corporation
100 Delawanna Avenue
Clifton, NJ 07014

Glidden-Durkee, Division of SCM Corporation
P.O. Box 389
Jacksonville, FL 33201

Sigma Chemical Co.
P.O. Box 14508
St. Louis, MO 63178

Tridom Chemical Inc.
255 Oser Avenue
Hauppauge, NY 11787

VI. References

Adams MA, Nakanishi K, Still WC, Arnold EV, Clardy J, Persoons CJ (1979) (±)-Periplanone-B. Total synthesis and structure of the sex excitant pheromone of the American cockroach. J Am Chem Soc **101**:2495–2498.
Ade A, Helmchen G, Heilegenmann G (1980) Syntheses of the stereoisomers of

17,21-dimethylheptatriacontane–Sex recognition pheromone of the tsetse fly. Tetrahedron Lett **21**:1137–1140.

Alexakis A, Cahiez G, Normant JF (1978) Highly stereoselective synthesis of the insect sex pheromone of *Phthorimaea operculella* and of propylure. Tetrahedron Lett **19**:2027–2030.

Anderson RJ, Adams KG, Chinn HR, Henrick CA (1980) Synthesis of the optical isomers of 3-methyl-6-isopropenyl-9-decen-1-yl acetate, a component of the California red scale pheromone. J Org Chem **45**:2229–2236.

Anderson RJ, Chinn HR, Gill K, Henrick CA (1979) Syntheses of 7-methyl-3-methylene-7-octen-1-yl propanoate and (*Z*)-3,7-dimethyl-2,7-octadien-1-yl propanoate, components of the sex pheromone of the San Jose scale. J Chem Ecol **5**:919–927.

Anderson RJ, Gieselmann MJ, Chinn HR, Adams KG, Henrick CA, Rice RE, Roelofs WL (1981) Synthesis and identification of a third component of the San Jose scale sex pheromone. J Chem Ecol **7**:695–706.

Anderson RJ, Henrick CA (1975) Stereochemical control in Wittig olefin synthesis. Preparation of the pink bollworm sex pheromone mixture, gossyplure. J Am Chem Soc **97**:4327–4334.

Anderson RJ, Henrick CA (1979) Synthesis of (*E*)-3,9-dimethyl-6-isopropyl-5,8-decadien-1-yl acetate, the sex pheromone of the yellow scale. J Chem Ecol **5**:773–779.

Apsimon JW, Sequin RP (1979) Tetrahedron report number 69. Recent advances in asymmetric synthesis. Tetrahedron **35**:2797–2842.

Baba S, Negishi E (1976) A novel stereospecific alkenyl-alkenyl cross-coupling by a palladium–or Nickel-catalyzed reaction of alkenylalanes with alkenyl halides. J Am Chem Soc **98**:6729–6731.

Babler JH, Haack RA (1982) A facile stereoselective route to the sex pheromone of the codling moth via thermolysis of an allylic sulfoxide. J Org Chem **47**: 4801–4803.

Babler JH, Invergo BJ (1979) A convenient stereoselective route to the sex pheromone of the red bollworm moth via an allylic sulfenate to sulfoxide rearrangement. J Org Chem **44**:3723–3724.

Bacardit R, Moreno-Manas M (1980) Hydrogenations of triacetic acid lactone. A new synthesis of the carpenter bee (*Xylocopa hirsutissima*) sex pheromone. Tetrahedron Lett **21**:551–554.

Baker R, Herbert R, Howse PE, Jones OT (1980) Identification and synthesis of the major sex pheromone of the olive fly (*Dacus oleae*). J Chem Soc Chem Comm 52–53.

Baker R, Rao VB (1982) Synthesis of optically pure (*R,Z*)-5-dec-1-enyloxacyclopentan-2-one, the sex pheromone of the Japanese beetle. J Chem Soc Perkin **I**:69–71.

Bartlett PA (1980) Tetrahedron report number 70. Stereocontrol in the synthesis of acyclic systems: Applications to natural products synthesis. Tetrahedron **36**:3–72.

Bartlett RJ, Jones RL, Kulman HM (1982) Hydrocarbon components of the yellow-headed spruce sawfly sex pheromone: A series of (*Z,Z*)-9,19-dienes. J Chem Ecol **8**:95–114.

Basavaiah D, Brown HC (1982) Pheromone synthesis via organoboranes: A stereospecific synthesis of (*Z*)-7-alken-1-ols. J Org Chem **47**:1792.

Bergreiter DE, Pendergrass E (1981) Analysis of organomagnesium and organolithium reagents using N-phenyl-1-naphthylamine. J Org Chem **46**:219–220 (and references cited).

Bestmann HJ, Li K (1981) Pheromone XXVII. Eine stereospezifische Synthese der Komponenten der Sexualpheromone von *Antheraea polyphemus*, (*E*)-6, (*Z*)-11-hexadecadienylacetat und (*E*)-6, (*Z*)-11-hexadecadienal. Tetrahedron Lett **22**:4941–4944.

Bestmann HJ, Koschatzky KH, Platz H, Brosche T, Kantardjiew I, Rheinwald M, Knauf W (1978) (*Z*)-5-Decenyl acetate, a sex attractant for the male turnip moth *Agrotis segetum* (Lepidoptera). Angew Chem Int Ed Engl **17**:768.

Bestmann HJ, Koschatzky KH, Vostrowsky O (1979d) Notiz zur Synthese der Sexuallockstoffe (*Z*)-7-dodecenylacetat und (*Z*)-7-tetradecenylacetat. Chem Ber **112**:1923–1925.

Bestmann HJ, Süss J, Vostrowsky O (1979a) Pheromone XXI. Methylenunterbrechung konjugierter Doppelbindugen: Eine stereoselektive Synthese von (*E*)-n, (*E*)-(n+3)-alkadienen. Tetrahedron Lett **20**:245–246.

Bestmann HJ, Süss J, Vostrowsky O (1979b) Pheromone XXVI. Synthese der Sexuallockstoffe (*E*)-7, (*Z*)-9-dodecadienylacetat, (*E*)-9,11-dodecadienylacetat und (*Z*)-9, (*E*)-11-tetradecadienylacetat. Tetrahedron Lett **20**:2467–2470.

Bestmann HJ, Vostrowsky O (1979) Synthesis of pheromones by stereoselective carbonyl olefination: A unitized construction principle. Chem Phys Lipids **24**:335–389.

Bestmann HJ, Vostrowsky O, Paulus H, Billmann W, Stransky W (1977) Eine Aufbaumethode für konjugierte (*E*),(*Z*)-diene. Synthese des Bombykols, seiner Derivate und Homologen. Tetrahedron Lett **18**:121–124.

Bestmann HJ, Vostrowsky O, Platz H, Brosche T, Koschatzky KH, Knauf W (1979e) Pheromone XXIII. (*Z*)-7-Tetradecenylacetat, ein Sexuallockstoff für Männchen von *Amathes C-nigrum* (Noctuidae, Lepidoptera). Tetrahedron Lett **20**:497–500.

Bestmann HJ, Wax R, Vostrowsky O (1979c) Stereoselective Synthese von (*Z*)-13-octadecenal einer Komponente des Pheromones des Reisstengel-Bohrers *Chilo suppressalis* (Lepidoptera). Chem Ber **112**:3740–3742.

Bierl-Leonhardt BA, Moreno DS, Schwarz M, Fargerlund J, Plimmer JR (1981) Isolation, identification and synthesis of the sex pheromone of the citrus mealybug, *Planococcus citri* (Risso). Tetrahedron Lett **22**:389–392.

Bierl-Leonhardt BA, Moreno DS, Schwarz M, Forster HS, Plimmer JR, DeVilbiss ED (1980) Identification of the pheromone of the Comstock mealybug. Life Sci **27**:399–402.

Bierl-Leonhardt BA, Moreno DS, Schwarz M, Forster HS, Plimmer JR, DeVilbiss ED (1982) Isolation, identification, synthesis and bioassay of the pheromone of the Comstock mealybug and some analogs. J Chem Ecol **8**:689–699.

Brand JM, Young J Chr, Silverstein RM (1979) Insect pheromones: A critical review of recent advances in their chemistry, biology and application. In: Progress in the Chemistry of Organic Natural Products vol 37. Herz W, Grisebach H, Kirby GW (eds), Springer, New York.

Brandsma L (1971) Preparative Acetylenic Chemistry. Elsevier, New York.

Brown HC (1971) Organic Synthesis Via Boranes. Wiley, New York.

Brown HC, Basaviah D (1982) A general and stereospecific synthesis of *cis* alkenes via stepwise hydroboration: A simple synthesis of muscalure, the sex pheromone of the housefly (*Musca domestica*). J Org Chem **47**:3806–3807.

Burfield DR, Klee KH, Smithers RH (1977) Desiccant efficiency in solvent drying. A reappraisal by application of a novel method for solvent water assay. J Org Chem **42**:3060–3065.

Byström S, Högberg H-E, Norin T (1981) Chiral synthesis of (2S,3S,7S)-3,7-dimethylpentadecan-2-yl acetate and propionate, potential sex pheromone components of the pine saw fly *Neodiprion sertifer* (Geoff). Tetrahedron **37**:2249–2254.

Cahn RS, Ingold C, Prelog V (1966) Specification of molecular chirality. Angew Chem Int Ed Engl **5**:385–415.

Canevet C, Röder T, Vostrowsky O, Bestmann HJ (1980) Pheromone XXVIII. Stereoselektive Synthese (*E*)-olefinischer Schmetterlingspheromone. Chem Ber **113**:1115–1120.

Cassani G, Massardo P, Piccardi P (1979) New simple synthesis of internal conjugated (*Z*)-enynes. Tetrahedron Lett **20**:633–634.

Chuman T, Kato K, Noguchi M (1979) Synthesis of (±)-serricornin, 4,6-dimethyl-7-hydroxy-nonan-3-one, a sex pheromone of cigarette beetle (*Lasioderma serricorne* F.). Agric Biol Chem **43**:2005.

Chuman T, Kohno M, Kato K, Noguchi M, Nomi H, Mori K (1981) Stereoselective synthesis of *erythro*-serricornin (4R, 6R, 7S)- and (4S, 6R, 7S)-4,6-dimethyl-7-hydroxynonan-3-one, stereoisomers of the sex pheromone of cigarette beetle. Agric Biol Chem **45**:2019–2023.

Coke JL, Richon AB (1976) Synthesis of optically active δ-*n*-hexadecalactone, the proposed pheromone from *Vespa orientalis*. J Org Chem **41**:3516–3517.

Dang HP, Linstrumelle G (1978) An efficient stereospecific synthesis of olefins by the palladium-catalyzed reaction of Grignard reagents with alkenyl iodides. Tetrahedron Lett **19**:191–194.

Descodts G, Dressaire G, Langlois Y (1979) Pyridines as precursors of conjugated diene pheromones. Synthesis 510–513.

Descoins C, Samain D, Labanne-Cassou B, Gallois M (1977) Synthèse stéréosélectives des acétoxy-1-dodécadienes 7E,9E et 7E,9Z, attractif sexuel pour le mâle de L'eudemis de la vigne: *Lobesia* (*polychrosis*) *botrana* Den et Schiff, lepidoptere Tortricidae. Bull Soc Chim Fr (9–10) **II**:941–946.

Dressaire G, Langlois Y (1980) Pyridines as precursors of conjugated diene pheromones (II). Stereoselective synthesis of (*7E,9Z*)-dodecadien-1-yl acetate, sex pheromone of *Lobesia botrana*. Tetrahedron Lett **21**:67–70.

Doolittle RE, Tumlinson JH, Proveaux AT, Heath RR (1980) Synthesis of the sex pheromone of the Japanese beetle. J Chem Ecol **6**:473–485.

Farnum DG, Veysoglu T, Cardé RT (1977) A stereospecific synthesis of (+)-disparlure, sex attractant of the gypsy moth. Tetrahedron Lett **18**:4009–4012.

Fieser LF, Fieser M (1967) Reagents for organic synthesis (Vol 1). Wiley, New York (and subsequent volumes).

Gordon AJ, Ford RA (1972) The chemist's companion. Wiley, New York.

Green TH (1981) Protective groups in organic synthesis. Wiley, New York.

Guss PL, Tumlinson JH, Sonnet PE, McLaughlin JR (1983a) Identification of a female produced sex pheromone from the southern corn rootworm, *Diabrotica undecimpunctata howardi* Barber. J Chem Ecol **9**:1363–1375.

Guss PL, Tumlinson JH, Sonnet PE, Proveaux AT (1982) Isolation and identification of a sex pheromone of the western corn rootworm. J Chem Ecol **8**:845–853.

Hall DR, Beevor PS, Lester R, Nesbitt BF (1980) (*E,E*)-10,12-Hexadecadienal: A component of the female sex pheromone of the spiny bollworm, *Earias insulana* (Boisd.) (Lepidoptera: Noctuidae). Experientia **36**:152–154.

Hammoud A, Descoins CE (1978) Stereoselective synthesis of 1-acetoxy-7*Z*, 11*E*-hexadecadiene or angoulure, the sex pheromone of the Angoumois grain moth, *Sitotroga cerealella* Oliv. Bull Soc Chim Fr (5–6). **II**:299–303.

Hanessian S, Demailly G, Chapleur Y, Leger S (1981) Facile access to chiral 5-hydroxy-2-methylhexanoic acid lactones (pheromones of the carpenter bee). J Chem Soc Chem Comm 1120–1126.

Harrison IT, Harrison S (1971) Compendium of organic synthetic methods, I. Wiley, New York (and subsequent volumes).

Heath RR, Doolittle RE, Sonnet PE, Tumlinson JH (1980) Sex pheromone of the white peach scale: Highly stereoselective synthesis of the stereoisomers of pentagonol propionate. J Org Chem **45**(14):2910–2912.

Heath RR, Jordan JR, Sonnet PE, Tumlinson JH (1979) Potential for the separation of insect pheromones by gas chromatography on columns coated with cholesteryl cinnamate, a liquid-crystal phase. HRC & CC **2**:712–714.

Henrick CA (1977) Tetrahedron report number 34. The synthesis of insect sex pheromones. Tetrahedron **33**:1845–1889.

Henrick CA, Carney RL, Anderson RJ (1982) Some aspects of the synthesis of insect sex pheromones. In: Insect Pheromone Technology: Chemistry and Applications. Bierl-Leonhardt BA, Beroza M (eds), ACS Symposium Series 190, American Chemical Society, Washington, D.C., pp 27–60.

Hintzer K, Weber R, Schurig V (1981) Synthesis of optically active 2S- and 7S-methyl-1,6-dioxa-spiro [4-5] decane, the pheromone components of *Paravespula vulgaris* (L.), from S-ethyl lactate. Tetrahedron Lett **22**:55–58.

Hobbs PD, Magnus PD (1976) Studies on terpenes. 4. Synthesis of optically active grandisol, the boll weevil pheromone. J Am Chem Soc **98**:4594–4600.

Hoffmann RW, Ladner W (1979) On the absolute stereochemistry of C-2 and C-3 in stegobinone. Tetrahedron Lett **20**:4653–4656.

Horiike M, Hirano C, Tamaki Y (1982) An easy synthesis of the sex pheromones 10-methyldodecyl acetate and its homologs. Agric Biol Chem **46**: 1927–1929.

Horiike M, Tanouchi M, Hirano C (1980) A convenient method for synthesizing (*Z*)-alkenols and their acetates. Agric Biol Chem **44**:257–261.

Hoshino C, Mori K (1980) Synthesis of a diastereomeric mixture of 15,19,23-trimethylheptatriacontane, the most active component of the sex pheromone of the female tsetse fly, *Glossina morsitans morsitans.* Agric Biol Chem **44**:3007–3009.

Ideses R, Klug JT, Shani A, Gothilf S, Gurevitz E (1982) Sex pheromone of the

European grapevine moth, *Lobesia botrana* Schiff (Lepidoptera; Tortricidae): Synthesis and effect of isomeric purity on biological activity. J Chem Ecol **8**:195–200.

Izumi Y, Tai A (1977) Stereodifferentiating reactions. The nature of asymmetric reactions. Academic Press, New York.

Jabri N, Alexakis A, Normant JF (1981) Vinyl copper derivatives XIII. Synthesis of conjugated dienes of very high stereoisomeric purity. Tetrahedron Lett **22**:959–962.

Jewett D, Matsumura F, Coppel HC (1978) Preparation and use of sex attractants for four species of pine sawflies. J Chem Ecol **4**:277–287.

Jones JB, Sih CJ, Perlman D (1976) Application of biological systems to organic chemistry. Wiley, New York.

Kagan HB, Fiaud JC (1978) New approaches in asymmetric synthesis. In: Topics in Stereochemistry. Eliel EL, Allingen NL (eds), Wiley, New York, Vol 10, pp 175–286.

Katsuki T, Sharpless KB (1980) The first practical method for asymmetric epoxidation. J Am Chem Soc **102**:5975–5976.

Katzenellenbogen JA (1977) Insect pheromone synthesis: New methodology. Science **194**:139–148.

Klunenberg H, Schäfer HJ (1978) Synthesis of disparlure by Kolbe electrolysis. Angew Chem Int Ed Engl **17**:47–48.

Kocieski PJ, Ansell JM (1977) A synthesis of 3,7-dimethylpentadec-2-yl acetate, the sex pheromone of the pine sawfly *Neodiprion lecontei.* J Org Chem **42**: 1102–1103.

Kondo K, Murahashi S (1979) Selective transformation of organoboranes to Grignard reagents by using pentane-1,5-di(magnesium bromide). Syntheses of the pheromones of southern armyworm moth and Douglas fir tussock moth. Tetrahedron Lett **20**:1237–1240.

Kozhich OA, Sepal GM, Torpoy EV (1982) Synthesis of 2-ethyl-1,6-dioxaspiro[4.4]nonane, main component of the pheromone of the chalcograph beetle, *Pityogenes chalcographus* (L.). Izv Akad Nauk SSSR, Ser Khim 325–328.

Kunesch G, Zagatti P, Lallemand JY, Debal A, Vigneron JP (1981) Structure and synthesis of the wing gland pheromone of the male African sugar-cane borer: *Eldana saccharina* (Wlk.) (Lepidoptera, Pyralidae). Tetrahedron Lett **22**:5271–5274.

Labovitz JN, Henrick CA, Corbin VL (1975) Synthesis of (7*E*,9*Z*)-7,9-dodecadienyl acetate, a sex pheromone of *Lobesia botrana.* Tetrahedron Lett **16**: 411–414.

Maercker A (1965) The Wittig reaction. In: Organic Reactions. Cope AC (editor-in-chief), Wiley, New York, Vol 14, pp 270–490.

Magnusson G (1977) Pheromone synthesis. Preparation of *erythro*-3,7-dimethylpentadecan-2-ol, the alcohol from the pine sawfly (Hymenoptera: Diprionidae). Tetrahedron Lett **18**:2713–2716.

Mandai T, Yasuda H, Kaito M, Tsuji J, Yamaoka R, Fukami H (1979) Synthesis of 12-acetoxy-1,3-dodecadiene, an insect sex pheromone of the red bollworm moth, from a butadiene telomer. Tetrahedron **35**:309–311.

Masuda S, Kuwahara S, Suguro T, Mori K (1981) Stereoselective synthesis of

(*S,E*)-6-isopropyl-3,9-dimethyl-5,8-decadienyl acetate, the (*S*)-enantiomer of the yellow scale pheromone. Agric Biol Chem **45**:2515–2520.

Matsumura F, Tai A, Coppell HC, Imaida M (1979) Chiral specificity of the sex pheromone of the red-headed pine sawfly, *Neodyprion lecontei*. J Chem Ecol 5:237–249.

McDonough LM, Hoffmann MP, Bierl-Leonhardt BA, Smithhisler CL, Bailey JB, Davis HG (1982) Sex pheromone of the avocado pest, *Amorbia cuneana* (Walsingham). J Chem Ecol 8:255–266.

McMurry J, Miller RB (1971) Annual reports in organic synthesis–1970. Academic Press, New York (and subsequent issues).

McComie JFW (1973) Protective Groups in Organic Chemistry. Plenum, New York.

McComie JFW (1979) Recent developments with protective groups. Chem Ind (15 Sept) 603–609

Midland MM, Nguyen NH (1981) Asymmetric synthesis of γ-lactones. A facile synthesis of the sex pheromone of the Japanese beetle. J Org Chem **46**: 4107–4108.

Midland MM, Tramontano A (1980) The synthesis of naturally occurring 4-alkyl- and 4-alkenyl-γ-lactones using the asymmetric reducing agent *B*-3-pinanyl-9-borabicyclo(3•3•1)nonane. Tetrahedron Lett **21**:3549–3552.

Mori K (1978) Synthesis of the both enantiomers of grandisol, the boll weevil pheromone. Tetrahedron **34**:915–920.

Mori K (1981) The synthesis of insect pheromones. In: Total Synthesis of Natural Products. Apsimon JW (ed), Wiley, New York, Vol 4, pp 1–183.

Mori K, Chuman T, Kohno M, Kato K, Noguchi M (1982) Absolute stereochemistry of serricornin, the sex pheromone of cigarette beetle, as determined by the synthesis of its (4S,6R,7R)-isomer. Tetrahedron Lett **23**: 667–670.

Mori K, Ebata T, Sabakibara M (1981c) Pheromone synthesis XXXVIII. Synthesis of (2S,3R,7RS)-stegobinone [2,3-dihydro-2,3,5-trimethyl-6-(1-methyl-2-oxobutyl)-4H-pyran-4-one] and its (2R,3S,7RS) isomer. The pheromone of the drugstore beetle. Tetrahedron **37**:709–713.

Mori K, Kuwahara S (1982) Synthesis of optically active forms of (*E*)-6-isopropyl-3,9-dimethyl-5,8-decadienyl acetate, the pheromone of the yellow scale. Tetrahedron **38**:521–525.

Mori K, Masuda S, Matsui M (1978b) A new synthesis of a stereoisomeric mixture of 3,7-dimethylpentadec-2-yl acetate, the sex pheromone of the pine sawflies. Agric Biol Chem **42**:1015–1018.

Mori K, Masuda S, Suguro T (1981a) Stereocontrolled synthesis of all of the possible stereoisomers of 3,11-dimethylnonacosan-2-one and 29-hydroxy-3,11-dimethylnonacosan-2-one. Tetrahedron **37**:1329–1340.

Mori K, Nakuda T, Ebata T (1981b) Synthesis of optically active forms of methyl (*E*)-2,4,5-tetradecatrienoate, the pheromone of the male dried bean beetle. Tetrahedron **37**:1343–1347.

Mori K, Nomi H (1981) Determination of the absolute configuration at C-6 and C-7 of serricornin (4,6-dimethyl-7-hydroxy-3-nonanone), the sex pheromone of the cigarette beetle. Tetrahedron Lett **22**:1127–1130.

Mori K, Suguro T, Uchida M (1978a) Synthesis of optically active forms of

(*Z*)-14-methylhexadec-8-enal, the pheromone of female dermestid beetle. Tetrahedron **34**:3119–3123.

Mori K, Takigawa T, Matsui M (1979) Stereoselective synthesis of optically active disparlure, the pheromone of the gypsy moth (*Porthetria dispar* L.). Tetrahedron **35**:833–837.

Mori K, Tamada S (1978) (—)-Grandisol, the antipode of the boll weevil pheromone, is biologically active. Naturwiss **65**:653–654.

Mori K, Tamada S (1979) Stereocontrolled synthesis of all of the four possible stereoisomers of *erythro*-3,7-dimethylpentadec-2-yl acetate and propionate, the sex pheromone of the pine sawflies. Tetrahedron **35**:1279–1284.

Mori K, Ueda H (1981) Synthesis of the optically active forms of 2,6-dimethyl-1,5-heptadien-3-ol acetate, the pheromone of the Comstock mealybug. Tetrahedron **37**:2581–2583.

Muchowski JM, Venuti MC (1981) Cyclic Phosphonium ylids. A short synthesis of gossyplure. J Org Chem **46**:459–461.

Naoshima Y, Yamamoto T, Wakabayashi S, Hayashi S (1980b) A facile synthesis of 2-methylheptadecane, sex pheromone of tiger moths (Arctiidae). Agric Biol Chem **44**:2231–2232.

Naoshima Y, Yamamoto T, Wakabayashi S, Hayashi S (1980b) A facile synthesis of 2-methylheptadecane, sex pheromone of tiger moths (Arctiidae). Agric Biol Chem **44**:2231–2232.

Negishi E, Abramovitch A (1977) A highly efficient chemo-, regio-, and stereoselective synthesis of (7*E*,9*Z*)-dodecadien-1-yl acetate, a sex pheromone of *Lobesia botrana,* via a functionalized organoborate. Tetrahedron Lett **18**: 411–414.

Negishi E, Lew G, Yoshida T (1973) Stereoselective synthesis of conjugated *trans*-enynes readily convertible into conjugated *cis,trans*-dienes and its application to the synthesis of the pheromone bombykol. J Chem Soc Chem Comm 874–875.

Negishi E, Yoshida T (1973) A highly stereoselective and general synthesis of conjugated *trans,trans*-dienes and *trans*-alkylketones via hydroboration. J Chem Soc Chem Comm 606–607.

Nishizawa M, Yamada M, Noyori R (1981) Highly enantioselective reduction of alkynyl ketones by a binaphthol-modified aluminum hydride reagent. Asymmetric synthesis of some insect pheromones. Tetrahedron Lett **22**:247–250.

Normant JF, Commercon A, Villieras J (1975) Synthese d'enynes et de dienes conjugues a l'aide d'organocuivre vinyliques. Application a la synthese du bombykol. Tetrahedron Lett **16**:1465–1468.

Novak L, Toth M, Balla J, Szantay C (1979) Sex pheromone of the cabbage armyworm, *Mamestra brassicae*; isolation, identification, and stereocontrolled synthesis. Acta Chim Acad Scient Hung (Tomus) **102**:135–140.

Okukado N, Van Horn DE, Klima WL, Negishi E (1978) A highly stereo-, regio-, and chemoselective synthesis of conjugated dienes by the palladium-catalyzed reaction of (*E*)-1-alkenylzirconium derivatives with alkenyl halides. Tetrahedron Lett **19**:1027–1030.

Pirkle WH, Adams PE (1978) Synthesis of the carpenter bee pheromone. Chiral 2-methyl-5-hydroxyhexanoic acid lactone. J Org Chem **43**:378–379.

Pirkle WH, Adams PE (1979) Broad-spectrum synthesis of enantiomerically pure lactones. 1. Synthesis of sex pheromones of the carpenter bee, rove beetle, Japanese beetle, black-tailed deer, and Oriental hornet. J Org Chem **44**: 2169–2175.

Pirkle WH, Boeder CW (1978) Synthesis and absolute configuration of (–)-methyl (*E*)-2,4,5-tetradecadienoate, the sex attractant of the male dried bean weevil. J Org Chem **43**:2091–2093.

Pirkle WH, Rinaldi PL (1979) Synthesis and enantiomeric purity determination of the optically active epoxide disparlure, sex pheromone of the gypsy moth. J Org Chem **44**:1025–1028.

Place P, Roumestant M-L, Gore J (1978) New synthesis of 3,7-dimethyl-pentadec-2-yl acetate, sex pheromone of the pine sawfly *Neodiprion lecontei.* J Org Chem **43**:1001.

Posner GH (1980) An Introduction to Synthesis Using Organocopper Reagents. Wiley, New York.

Prelog V, Helmchen G (1982) Basic principles of the CIP system and proposals for a revision. Angew Chem Int Ed Engl **21**:567–583.

Ravid U, Silverstein RM, Smith LR (1978) Synthesis of the enantiomers of 4-substituted γ-lactones with known absolute configuration. Tetrahedron **34**: 1449–1452.

Redlich H, Xiang-Jun J, Paulsen H, Francke W (1981) Darstellung von (*S*)-2,5-dimethyl-2-isopropyl-2,3-dihydrofuran, einem der beiden Enantiomeren des Sexuallockstoffs des Werftkäfers *Hylecoetus dermestoides* L. Tetrahedron Lett **22**:5043–5046.

Riddick JA, Bunger WB (1970) Organic solvents. In: Techniques of Chemistry. Weissberger A (ed), Wiley, New York, Vol II.

Roberts RM, Gilbert JC, Rodewald LB, Wingrove AS (1969) An introduction to modern experimental organic chemistry. Holt, Rinehart & Winston, New York.

Roelofs W, Gieselmann M, Cardé A, Toshiro H, Moreno DS, Henrick CA, Anderson RJ (1978) Identification of the California red scale pheromone. J Chem Ecol **4**:211–224.

Rossi R (1977) Insect pheromones. I. Synthesis of achiral components of insect pheromones. Synthesis 817–836.

Rossi R (1978a) Insect pheromones. II. Synthesis of chiral components of insect pheromones. Synthesis 413–434.

Rossi R (1978b) Stereoselective synthesis of (*Z*)-9-dodecen-1-yl acetate, the sex pheromone of some tortricoid moths. Chim Ind (Milan) **60**:652–653.

Rossi R, Carpita A (1977) Insect pheromones. Synthesis of chiral sex pheromone components of several species of Trogoderma (Coleoptera: Dermestidae). Tetrahedron **33**:2447–2450.

Rossi R, Carpita A, Guadenzi ML (1981) Highly stereoselective synthesis of (*Z*)-9,11-dodecadien-1-yl acetate: A sex pheromone component of *Diparopsis castanea* Hmps. Synthesis 359–361.

Rossi R, Carpita A, Guadenzi L, Quirici MG (1980) Insect sex pheromones. Stereoselective synthesis of several (*Z*)- and (*E*)-alken-1-ols, their acetates, and of (9Z,12E)-9,12-tetradecadien-1-yl acetate. Gazz Chim Ital **110**:237–246.

Rossi R, Carpita A, Quirici MG, Gaudenzi ML (1982) Insect sex pheromones:

Palladium-catalyzed synthesis of aliphatic 1,3-enynes by reaction of 1-alkynes with alkenyl halides under phase transfer conditions. Tetrahedron **38**:631–637.

Rossi R, Marasco M (1980) Insect pheromones by asymmetric synthesis. La Chim L'Ind **62**:314–316.

Rossi R, Salvadori PA, Carpita A, Niccoli A (1979) Synthesis of the (R)(−) enantiomers of the pheromone components of several species of Trogoderma (Coleoptera: Dermestidae). Tetrahedron **35**:2039–2042.

Rossiter BE, Katsuki T, Sharpless KB (1981) Asymmetric epoxidation provides shortest routes to four chiral epoxy alcohols which are key intermediates in syntheses of methymycin, erythromycin, leukotriene C-1, and disparlure. J Am Chem Soc **103**:464–465.

Sakakibara M, Mori K (1979) Synthesis of a stereoisomeric mixture of 2,3-dihydro-2,3,5-trimethyl-6-(1-methyl-2-oxobutyl)-4H-pyran-4-one, the pheromone of the drugstore beetle. Tetrahedron Lett 2401–2402.

Samain D, Descoins C (1979) Etude de la stéréosélectivité de la transposition de Claisèn appliquée aux énynols secondaires. Synthesés stéréoselective du bombykol: Hexadécadiéne-10E,12Z-ol-1 et de ses dérivés: Bombykal et acétate de bombykol. Bull Soc Chim Fr (1–2) **II**:71–76.

Samain D, Descoins C, Commercon A (1978) A short stereoselective synthesis of 8E,10E-dodecadien-1-ol, the sex pheromone of the codling moth, *Laspeyresia pomonella,* L. Synthesis 388–389.

Sato K, Nakayama T, Mori K (1979) Pheromone synthesis. Part XXXII. New synthesis of the both enantiomers of (*Z*)-5-(1-decenyl) oxacyclopentan-2-one, the pheromone of the Japanese beetle. Agric Biol Chem **43**:1571–1575.

Schlosser M, Fujita K (1982) Eine stereokontrollierte Syntheses des Pheromons 4-methyl-3-heptanol: Neue und selektive CC-verknüpfende Aufbaureaktionen. Angew Chem (Suppl) 646–653.

Shani A (1979) An efficient synthesis of muscalure from jojoba oil or oleyl alcohol. J Chem Ecol **5**:557–654.

Sinclair JA, Brown HC (1976) Synthesis of unsymmetrical conjugated diynes via the reaction of lithium dialkynyldialkylborates with iodine. J Org Chem **41**: 1078–1079.

Smith LR, Williams HJ (1979) Glutamic acid in pheromone synthesis: A useful chiral synthon. J Chem Ed **56**:696–698.

Solladie G, Matloubi-Moghadam F (1982) Asymmetric synthesis of five- and six-membered lactones from chiral sulfoxides: Application to the asymmetric synthesis of insect pheromones, (*R*)-(+)-δ-n-hexadecalactone and (*R*)-(+)-γ-n-dodecanolactone. J Org Chem **47**:91–94.

Sonnet PE (1974) A practical synthesis of the sex pheromone of the pink bollworm. J Org Chem **39**:3793–3794.

Sonnet PE (1979) Synthesis of the male stable fly polyene (*Z,Z*)-1,7,13-pentacosatriene and its geometrical isomers. J Chem Ecol **5**:415–422.

Sonnet PE (1980a) Olefin inversion. 3. Preparations and reductions of *vic*-halohydrin trifluoroacetates. J Org Chem **45**:154–157.

Sonnet PE (1980b) A convenient stereospecific reduction of epoxides and iodohydrins to olefins. Synthesis 828–829.

Sonnet PE (1982) Syntheses of the stereoisomers of the sex pheromones of the southern corn rootworm and lesser tea tortrix moth. J Org Chem **47**:3793–3796.

Sonnet PE, Heath RR (1980) Stereospecific synthesis of (*Z*,*Z*)-11,13-hexadecadienal, a female pheromone of the navel orangeworm, *Amyelois transitella* (Lepidoptera: Pyralidae). J Chem Ecol **6**:221–228.

Sonnet PE, Heath RR (1982a) Synthesis of (±)-10-methyl-1-dodecanol acetate, the chiral component of the smaller tea tortrix moth (Adoxophyes spp.) with an option for asymmetric induction. J Chem Ecol **8**:41–53.

Sonnet PE, Heath RR (1982b) Chiral insect sex pheromones: Some aspects of synthesis and analysis. In: Insect Pheromone Technology: Chemistry and Applications. Bierl-Leonhardt BA, Beroza M (eds), ACS Symposium Series 190. American Chemical Society, Washington, D.C.

Still WC (1979) (±)-Periplanone-B. Total synthesis and structure of the sex excitant pheromone of the American cockroach. J Am Chem Soc **101**: 2493–2495.

Still WC, Mitra A (1978) Highly stereoselective synthesis of (*Z*)-trisubstituted olefins via (2,3)-sigmatropic rearrangement. Preference for a pseudoaxially substituted transition state. J Am Chem Soc **100**:1927–1928.

Suguro T, Mori K (1979a) Synthesis of optically active forms of (*E*)-14-methyl-8-hexadecenal (trogodermal). Agric Biol Chem **43**:409–410.

Suguro T, Mori K (1979b) Synthesis of optically active forms of 10-methyldodecyl acetate, a minor component of the pheromone complex of the smaller tea tortrix moth. Agric Biol Chem **43**:869–870.

Suguro T, Roelofs WL, Mori K (1981) Stereoselective synthesis of (±)-(*E*)-isopropyl-3,9-dimethyl-5,8-decadienyl acetate, the racemate of the yellow scale pheromone, and its (*Z*)-isomer. Agric Biol Chem **45**:2509–2514.

Szabo WA, Lee HT (1980) Chiral starting materials and reagents. Aldrichim Acta **13**:13–18.

Tamada S, Mori K, Matsui M (1978) Pheromone synthesis. Part XIX. Simple synthesis of (*Z*)-7-eicosen-11-one and (*Z*)-7-nonadecen-11-one, the pheromone of the peach fruit moth. Agric Biol Chem **42**:191–192.

Trost BM, Fortunak JM (1980) A new diene synthesis via organopalladium chemistry. J Am Chem Soc **102**:2841–2843.

Tsuboi S, Masuda T, Takeda A (1982) Highly stereocontrolled synthesis of (2*E*,4*Z*)-dienoic esters with alumina catalyst. Its application to total syntheses of flavor components and insect pheromones. J Org Chem **47**:4478–4482.

Uchida M, Mori K, Matsui M (1977) Synthesis of (*Z*,*Z*)-3,13-octadecadienyl acetate and its (*E*,*Z*)-isomer, the attractant for the cherry tree borer. Agric Biol Chem **12**:1067–1070.

Uchida M, Nakagawa K, Mori K (1979) Stereoselective synthesis of (*Z*,*Z*)-3,13-octadecadienyl acetate, the attractant for the smaller clear wing moth. Agric Biol Chem **43**:1919–1922.

Uchida M, Nakagawa K, Negishi T, Asano S, Mori K (1981) Syntheses of 2,6-dimethyl-1,5-heptadien-3-ol acetate, the pheromone of the Comstock mealybug *Pseudococcus comstockii* Kuwana and its analogs. Agric Biol Chem **45**:369–372.

Valentine D Jr, Scott JW (1978) Asymmetric synthesis. Synthesis 329–356.
Vedejs E, Meier GP, Snoble KAJ (1981) Low-temperature characterization of the intermediates in the Wittig reaction. J Am Chem Soc **103**:2823–2831.
Wollenberg RH, Peries R (1979) Efficient syntheses of insect sex pheromones emitted by the boll weevil and the red bollworm moth. Tetrahedron Lett **20**:297–300.
Zakharkin LI, Petrushkina EA (1982) Synthesis of (*E*)-6-nonen-1-ol, a sex pheromone component of Mediterranean fruit fly, *Ceratitis capitata.* Izv Akad Nauk SSSR Ser Khim 1181–1182.
Zweifel G, Polston NL (1970) Selective hydroboration of conjugated diynes with dialkylboranes. A convenient route to conjugated *cis*-enynes, α,β-acetylenic ketones, and *cis,cis*-dienes. J Am Chem Soc **92**:4067–4071.

Chapter 14

Survey of Pheromone Uses in Pest Control

D. G. Campion[1]

I. Introduction

Pheromones are chemicals produced by one organism which influence the behavior of other members of the same species. Chemical communication of this kind is well established among insects. The two classes of pheromones most exploited in pest control situations are the sex pheromones employed by insects during mating and the aggregation pheromones which bring both sexes together for feeding and reproduction. Pheromone usage has developed in three main ways:

1. mating disruption, whereby the pheromone permeates the atmosphere so as to prevent communication among the sexes and hence subsequent mating.
2. mass-trapping, where large numbers of traps are used to reduce population levels.
3. monitoring insect populations with pheromone baited traps.

Such developments, involving identification of chemical structure and synthesis, have occurred mainly during the past 10 years.

This chapter summarizes the practical achievements so far attained and discusses problems that have been encountered.

The literature involved is large and continues to grow at a rapid rate. Detailed information and references will be found in other recent reviews: Birch (1974),

[1]Tropical Development and Research Institute, College House, Wright's Lane, London W8 5SJ, United Kingdom.

Tamaki (1974, 1980), Shorey and McKelvey (1977), Roelofs and Cardé (1977), Baker (1979), Minks (1979), Konodrat'ev Yu A et al. (1978), Ritter (1979), Piccardi (1980), Boness (1980), Brand et al. (1979), Campion and Nesbitt (1981), Hall (1981), Mitchell (1981), Nordlund et al. (1981), Silverstein (1981), Kydonieus and Beroza (1982).

II. Mating Disruption

During the years following the isolation of the sex attractant of the silkworm moth *Bombyx mori* (L.) by Butenandt et al. in 1959, over 670 more pheromones have been identified (Klassen et al., 1982). Only one of these so far has been used as a mating disruptant for the control of an insect pest on a relatively large commercial scale, namely, gossyplure which is used to control an important cotton pest, the pink bollworm *Pectinophora gossypiella* (Saunders) (Doane and Brooks, 1981). It may be instructive to consider why things have taken so long, and in doing so to review the many difficulties; biological, chemical, technological, and commercial that have had to be overcome before such a system based on mating disruption could be successfully adopted. In this way it may be possible to assess the likelihood of further future successes for the control of other important pest species by this control technique.

Mating disruption is assumed to be achieved by permeating the area under treatment with a synthetic pheromone so as to reduce mate finding or aggregation; the end result being mating suppression. The mechanisms involved to achieve this effect are still not really known but could consist of one or a combination of any of the following:

1. The constant exposure of the insect to a relatively high level of pheromone leads to adaptation of the antennal receptors and habituation of the central nervous system. Under such circumstances, the responding insect would be unable to respond to any normal level of the stimulus.
2. A sufficiently high background level of the applied pheromone masks the natural pheromone plume and therefore trail following is impossible.
3. The synthetic pheromone is applied in a relatively large number of discrete sources so that insects flying within the treatment area can be diverted from the naturally occurring plumes.

Knowledge as to which disruption mechanisms are operating is important, since otherwise the design of appropriate formulations remains just guesswork. Formulations can be designed to create in effect a uniform background to cope with mechanisms 1 and 2 or provide point-source, so as to create trail following as postulated for mechanism 3. In practice both types of formulation have been developed, and at the present time it is still not certain which is the most appropriate. In fact the situation may well differ for individual insect species.

The first pheromone which was isolated, bombykol, was thought to be the only component produced by the insect that was required to elicit the necessary response. The search for similar single-component systems was therefore undertaken. However, it was generally found that although such single components might cause insect excitation in the laboratory, when exposed in the field in traps, no long-range attraction could be demonstrated. Silverstein et al. (1966) found that the pheromone produced by the male bark beetle, *Ips paraconfusus* Lanier, which attracts both male and female beetles was a mixture of three compounds. Similarly Moorhouse et al. (1969) showed that male moths of the red bollworm *Diparopsis castanea* (Hmps.) responded in the laboratory to five substances isolated from the female and that three fractions separated by gas chromatography (GC) produced an electroantennogram (EAG) response from male antennae. Subsequently during the 1970s, most moth pheromones discovered were found to be multicomponent, including even *Bombyx* (Kaissling et al., 1978).

Since optimum attraction could only occur by producing the correct ratios of the various components of the pheromone blend, it was speculated that mating disruption might also be achieved by permeating the air with only one component. In this instance, it is assumed that disruption is achieved by drastically altering the pheromone blend from the natural ratio. If this is the case, then mechanisms 1 and 2 proposed above could be involved, since habituation and masking of the natural pheromone volume would occur.

Mating disruption may also be achieved by the use of parapheromones which are synthetic chemicals that also attract the target species and antipheromones which interfere in some way with the perception by the insect of the natural pheromone (Gaston et al., 1972). Again mechanisms 1 and 2 may be involved, but mechanism 3 is unlikely unless the parapheromone is more attractive than the naturally occurring one; a situation so far not found in practice.

The pheromone of *Pectinophora* was identified as a 1:1 mixture of (*ZZ*)- and (*ZE*)-7,11-hexadecadienyl acetate (Hummel et al., 1973; Bierl et al., 1974). Earlier a parapheromone hexalure, (*Z*)-7-hexadecenyl acetate (Green et al., 1969) was evaluated in the field. To demonstrate mating disruption the lure was set out by hand to provide numerous point sources using dispensing systems, originally developed for use in traps. The procedures developed for field evaluation were subsequently also used for gossyplure which was found to be much more biologically active (McLaughlin et al., 1972; Shorey et al., 1976; Gaston et al., 1977).

Permeation with pheromone at adequate concentration levels may well stop mating within the treatment areas. The question then arises whether insects from surrounding areas may migrate into the treatment area and if gravid females will nullify the effects of treatment. This possibility exposes a lack of knowledge concerning both insect behavior and insect movement of most important insect pest species. It is not generally known, for example, what distances adult insects regularly fly or whether mating occurs before or after flight; it is not even known in some cases whether mating occurs within the crop. Realistic assessment of mating disruption must depend on the acquisition of such information. It may be argued that insect species with a restricted range of food plants may be generally more localized in their movements than polyphagous ones. On this basis *Pectinophora* is a good candidate since the larvae are virtually restricted to cotton as their host plant. The information available on the flight potential of *Pectinophora* moths is contradictory, however. Glick and Noble (1961) carried out pioneering studies on the widespread aerial movement of *Pectinophora* and showed that they could be found at altitudes of 610 m and widely dispersed. It was not stated whether the dispersing female moths caught were mated or unmated. Flint and Merkle (1981), however, contend that, within cotton growing areas at least, the movements of *Pectinophora* are generally localized.

Whatever the scale of insect movement is considered likely it has generally been agreed that it is important to treat as large an area as possible or to treat "semi-isolated" areas, so as to reduce the possibility of immigration. This emphasis has generally resulted in mating disruption trials consisting of large unreplicated treatment areas and therefore statistically acceptable comparisons with control areas have often not been possible. This is one reason why many trials using mating disruption have produced equivocal results.

The need for large-scale trials to reduce the possibility of gravid female entry has stimulated the search for formulations which would provide a sustained uniform release of the pheromone over a long period that could also be easily broadcast over wide areas, preferably by aircraft. In the United States, the Controlled Release Division of Albany International (formerly Conrel) has since 1974 produced a formulation consisting of hollow plastic fibers about 1 cm long with an inside diameter of 0.2 mm. The pheromone is held within the fiber by capillary action and is released by evaporation from the open end (Golub and Weatherston, Chapter 10, this volume). To make the fibers stick to the foliage they are mixed with a special glue and sprayed using specially designed applicators attached to aircraft (Funkhouser, 1979).

An alternative formulation was produced by another American company, the Hercon Division of Health-Chem Corporation. It consists of small flakes of plastic laminate consisting of a porous layer impregnated with pheromone, sandwiched between two layers of a plastic chosen so that the pheromone diffuses through it at an appropriate rate. The system also requires a strong adhesive to ensure adherence to the foliage and specially designed applicators (Quisumbing and Kydonieus, 1982).

Both types of formulation have been successfully used in mating disruption trials for the control of *Pectinophora* (Brooks et al., 1979; Henneberry et al., 1981). These discrete pheromone sources are applied at rates of between 800 to 40,000/hectare and this assumes that trail following or mechanism 3 discussed above is the major disruptive mechanism. A third type of formulation used for *Pectinophora* control is by microencapsulating the pheromone by coacervation or interfacial polymerization techniques. The capsule shells have consisted of gelatin, polyurea, polyamide, or polyurea crosslinked with polyamide. Release rates from the microcapsules can be controlled by changing the permeability of the polymer shell. The sizes of the microcapsules have ranged from 2 to 400 μm. The advantage of this type of formulation is that it can be sprayed with conventional applicators and generally requires no special adhesives to ensure retention on the foliage.

Earlier studies with microencapsulated formulations were generally unsatisfactory (Campion et al., 1978). This was subsequently shown to be due, at least in some cases, to the degradation by sunlight of both the capsule wall and the contained pheromone. More recently stabilized microencapsulated formulations have been developed in the United Kingdom (Campion et al., 1981a,b; Hall et al., 1982) and when aerially applied have given satisfactory control of *Pectinophora* in Egypt (Campion and Nesbitt, 1982a; Critchley et al., 1983). This implies that mechanisms 1 and 2 are involved in mating disruption since such a formulation provides a uniform background or fog of pheromone throughout the crop.

Further trials were conducted in Egypt during 1982. Two 50 ha blocks of cotton were treated with the microencapsulated pheromone formulation applied by helicopter. Comparisons of effectiveness in 50 ha blocks of cotton were also made with the laminate flake and hollow fiber formulations applied by hand. For each pheromone formulation, paired comparisons were made with similar areas of cotton treated with conventional insecticides. In all instances the pheromone treatments were as good as insecticides for the control of *Pectinophora* and highly satisfactory yields of cotton were obtained (Campion and co-workers, unpublished data).

Another important factor contributing to the commercial development of the mating disruption technique for *Pectinophora* control is the relatively large potential market in which such a system can be sold. It is well established that a high percentage of the total pesticide usage on crops is devoted to cotton pest control. Within the cottom complex, *Pectinophora* holds an important position with a worldwide distribution.

The large potential market for the control of *Pectinophora* by mating disruption has in turn led to improved large-scale synthesis of the pheromone and a consequent dramatic reduction in price of what was formerly an exotic and hence very expensive commodity. Chemical companies involved in the production of *Pectinophora* pheromone on both small and large scales include Albany International, Bend Research Inc., the Zoecon Corporation, of the United States, Borregard Industries Ltd. in Norway, the Montedison Group of Italy, the Shin-Etsu Chemical Company, Takeda Chemical Company of Japan, the Wolfson Unit of Chemical Ecology, Southampton University, United Kingdom, TNO in the Netherlands, and the Chemada Chemical Company in Israel.

Several procedures have been developed for determining the efficacy of the formulated pheromone in the field, in terms of the level and persistence of its disruptive action. The procedure most often adopted is to compare the reduction of trapped insects in pheromone-baited traps within the treated area with the catches in similar traps positioned in control areas. The absence or relative absence of catch as the result of treatment is assumed to indicate a high level of mating disruption. The significance and relevance of this procedure have been much debated. For *Pectinophora* at least, a significant relationship has been established between the absence of catch of male moths in traps sited in treatment areas and mating suppression of tethered virgin females. This latter technique has now been widely adopted and consists of exposing virgin female moths overnight in pheromone-treated and control areas. The presence of a spermatophore is taken as evidence of successful mating in assessing its frequency (see McVeigh et al., 1983, for recent review of literature).

Further work with *Pectinophora* has shown that the level of catch in traps located in pheromone-treated areas, at pheromone concentrations known to cause a high reduction of mating of tethered females, can be raised by increasing the emission level of the pheromone from the source (Doane and Brooks, 1981). This observation may also shed some light on the disruption mechanism involved, since it implies that male orientation to a pheromone source can be achieved if the concentration gradient from the source is greater than the background of pheromone created by the treatment. Disruption may therefore be associated with mechanism 2 discussed previously. Nevertheless, mechanism 3 is strongly advocated by Brooks et al. (1979) since there is evidence that male *Pectinophora* attempt mating with individual hollow fibers in the field. It is not at the moment certain whether this occurs so extensively as to be of great significance in control. This concept does, however, have further implications. If indeed trail following does occur extensively, then admixture of the pheromone point sources with insecticides may achieve control both by mating disruption and by annihilation of a significant proportion of the attracted insects (Dean and Lingren, 1982). Evidence from wind-tunnel studies conducted with *Adoxophyes orana* (F.v.R.) (Kennedy et al., 1981) and *Choristoneura fumiferana* (Clemens) (Sanders, 1982) lend some support to the concept of trail following (see also David et al., 1982). Similar studies for *Pectinophora* have not yet been conducted. The degree to which mechanisms 2 and 3 are important may be related to the amount of active ingredient applied and the controversy may be resolved by economic factors (see Table 1).

The economics of using pheromones as an alternative to conventional insecticides can only be determined if there are reliable criteria by which the effect of the target species on crop yields can be assessed. Fortunately these are well established for cotton.

Although *Pectinophora* control by pheromone is slightly more expensive than conventional pesticides, Brooks et al. (1979) showed that both improved yield and quality of crop more than compensated. To this may be added the advantage of avoiding the problems of pest resistance to insecticides and destruction

Table 1. Summary table of pheromone formulations used in mating disruption trials for control of pink bollworm *Pectinophora gossypiella*

Formulation	Number of sources per hectare	Grams per hectare of pheromone applied	Main disruption mechanism
Laminated flakes	800–10,000	3	False trail following
Hollow fibers	12,000–40,000	3	False trail following and trail masking
Microcapsules	Uniform distribution	5–10	Trail masking and habituation

of beneficial insects that may lead to eruption of other pest species (e.g., *Bemisia*) (Reynolds et al., 1982).

The method lends itself best to control by governments or large cooperatives rather than by individual farmers, because to be effective the areas treated must be very large. This offers a considerable market for *Pectinophora* control where cotton is grown under large-scale irrigation. Between 1980 and 1982 the areas treated with first hollow fibers and later with flake formulations as well rose from 50 to 100,000 ha in the United States and South America, with further pilot projects elsewhere.

Another major hurdle that had to be overcome was usage registration. Formerly pheromones were classified as insecticides and therefore registration costs of use approval were very high because of the elaborate toxicity testing procedures necessary. The relatively small market for pheromones compared with insecticides had therefore inhibited commercial interests. More recently the Environmental Protection Agency in the United States has agreed to consider pheromones as biorational pesticides and as such require less rigorous evaluation procedures (Zweig et al., 1982). Since then several others in addition to *Pectinophora* pheromone have been registered for use.

Assuming the growers are convinced and that environmental safety is agreed, the final group to be satisfied is the Crop Protection Industry. The patentability of pheromones or pheromone usage seems controversial. Companies entering the field have, however, successfully patented formulations including hollow fibers by Albany International, flakes by the Health-Chem Corporation, and microcapsules by ICI. It is to be anticipated that in the area of parapheromones or antipheromones synthetic patentable products may be sought (Djerassi et al., 1974).

In summary the important factors leading to the commercial usage of *Pectinophora* pheromone include the following:

1. a basically stable pheromone structure with no complicating secondary pheromone components.
2. an insect which is a key pest with a narrow range of food plants.
3. an insect with a limited migratory capability.

Table 2. Ranking of pest species with regard to suitability for control with pheromones[1]

		Pheromone development					Pest biology					Pest status				
Pest species	Need for substitute for other control techniques	Identification of components	Ease of synthesis	Ease of formulation	Adequate application procedures	Average	Knowledge of general biology	Knowledge of behavior in relation to pheromones	Narrow range of host plants	Low migratory potential	Average	Key pest in ecosystem	Economic importance	Adequate measure of larval density and crop loss	Average	Total of averages
Pectinophora	3	5	5	5	5	5	4	4	5	5	4.5	5	5	5	5	17.5
Eucosoma	5	5	5	5	5	5	4	2	5	5	4.0	2	2	5	3	17.0
Lymantria	5	5	3	4	4	4.3	5	4	5	3	4.3	5	5	2	4	16.6
Platyptilia	4	5	5	5	5	5	3	3	5	5	4.0	4	2	4	3.3	16.3

Choristoneura	4	5	5	3	5	4.5	5	3	5	2	3.8	5	5	2	4	16.3
Spodoptera littoralis	4	5	3	2	5	3.8	4	4	2	3	3.3	5	5	3	4.3	15.4
Lobesia	4	5	5	3	3	4.0	4	3	5	2	3.5	4	3	4	3.7	15.2
Laspeyresia	3	5	5	3	5	4.5	4	3	5	3	3.8	3	4	4	3.7	15.0
Zeiraphera	4	3	3	4	4	3.5	5	2	5	2	3.5	5	5	2	4.0	15.0
Diparopsis spp.	3	5	3	3	5	4	4	4	5	3	4	5	1	5	3.7	14.7
Earias spp.	3	3	3	3	5	3.5	4	2	5	4	3.8	4	4	5	4.3	14.6
Grapholita spp.	3	5	5	5	5	5	4	3	3	3	3.3	3	3	4	3.3	14.6
Ostrinia	3	3	5	3	4	3.8	4	2	5	3	3.5	4	5	4	4.3	14.6
Eupoecilia	4	5	5	3	3	4.0	4	1	5	3	3.2	3	3	3	3.0	14.2
Dendroctonus spp.	4	5	3	3	4	3.8	4	2	5	2	3.3	3	5	1	3	14.1
Chilo spp.	3	5	5	3	4	4.3	4	2	5	3	3.5	2	3	4	3	13.8
Heliothis spp.	4	5	1	1	4	2.8	4	2	1	1	2.0	4	5	4	4.3	13.1
Trichoplusia	3	5	5	5	4	4.8	4	2	2	3	2.8	2	3	2	2.3	12.9
Adoxophyes	3	5	5	3	3	4	3	2	4	3	3.0	2	3	3	2.7	12.7
Synanthedon spp.	3	5	4	5	4	4.5	2	2	4	3	2.8	2	2	2	2	12.3
Spodoptera frugiperda	3	2	4	5	4	3.8	4	3	1	1	2.3	2	4	2	2.7	11.8
Ephestia spp.	1	5	5	2	2	3.5	4	2	4	5	3.8	1	2	3	2.0	10.3

[1]Factors scored on a basis of 1 (= low) to 5 (= high).

4. a basic biological knowledge of the insect is already available.
5. adequate measures are available both for measuring disruption and for estimating larval populations and crop yield.
6. the availability of adequate formulations and appropriate methods of application.
7. the economic importance of the pest insect is great enough to warrant commercial interest.
8. there are convincing reasons for changing from the presently available pest strategy.

In the following section the status of mating disruption is reviewed for the control of other cotton insects and other major groups of insect pests. Table 2 provides a list of candidate pest species ranked in order of their suitability for control by mating disruption on the basis of several relevant critiera. It is realized that this is very much a subjective assessment but serves to provide some sense of perspective to highlight which insect targets seem particularly promising.

2.1. Other Cotton-Pests

Much research has been devoted to establishing the feasibility of the mating disruption approach for control of *Heliothis* spp. The number and importance of up to seven components of the pheromone blends is still not clearly confirmed (Klun et al., 1979; Tumlinson et al., 1982). The functional groups of the major attractant components are aldehydes, and hence more reactive and difficult to stabilize in the field than the acetate functional groups of the *Pectinophora* pheromone. All the *Heliothis* species are polyphagous and the adult moths highly migratory. Perhaps for these reasons attacks of cotton by both Old World and New World species are often unpredictable. Their status as major pests of cotton and also important food crops and the great potential commercial market for a novel pest control agent must be the essential stimuli for this work. The likelihood of success at the present time is therefore not rated very high.

Another cotton pest of similar habitat to *Pectinophora* is the spiny bollworm *Earias insulana* (Boisd.). This is a late-season pest; unlike *Pectinophora* there is no diapause stage and therefore the insect has to survive the winter period on a restricted range of wild host plants. These habits suggest it to be an excellent candidate for mating disruption. Although the pheromone consists of a single component (*E,E*)-10,12-hexadecadienal (Hall et al., 1980), the aldehyde functional group and the conjugated double bond may pose problems of stability as shown in small-scale trials in progress in Egypt (Campion et al., unpublished data). The pheromone of the related insect *Earias vittella* (F.) which is an important pest of cotton in India and Pakistan has not yet been isolated, but work is in progress (BF Nesbitt, R Baker, private communication).

A study on the mating disruption technique for the control of the red bollworm *Diparopsis castanea* (Hmps.), a key pest of cotton in southeastern Africa, was undertaken by Marks et al. (1978, 1981).

There is no stabilized pheromone formulation for *D. castanea* and the commercial interests for a specific control technique are small. However, it is closely related to the Sudan bollworm *Diparopsis watersi* (Roths.), a widespread pest on cotton in west Africa, and the pheromones are likely to be very similar (Beevor et al., 1973), so that a pheromone formulation appropriate for both species may be possible.

The above-mentioned cotton pests all attack the bolls or fruiting parts of the plant, and estimation of larval numbers is therefore fairly straightforward and achieved by acceptable methods of boll sampling. Sampling for leaf-eating caterpillars is more difficult, but an important pest of such habit occurs in the Mediterranean region, the Egyptian cotton leafworm *Spodoptera littoralis* (Boisd.). The pheromone blend of up to four components (Nesbitt et al., 1973) varies from one part of the region to another (Campion et al., 1980), and this will pose formulation problems. Relatively stable pheromone formulations have been developed (Campion et al., 1981b; Hall et al., 1982), but isomerization of the major attractant component occurs in the field (Shani and Klug, 1980). *S. littoralis* is a key pest of cotton and the marketing potential for a mating disruptant is therefore encouraging. The insect is, however, polyphagous and of indeterminant flight range, features which have been suggested to be of disadvantage for the eventual success of the mating disruption technique.

2.2. Forest Pests

Forest pests have been considered suitable targets for control by mating disruption since the host range is invariably narrow and large forest plantations make area wide application possible. A major problem, with a few notable exceptions, has been the absence of suitable methods to assess the significance of the damage caused by the pests in anything other than outbreak situations. Migration may be a problem in some species but little is known about the majority.

The gypsy moth *Lymantria dispar* (L.) is an important forest pest in the United States. The female is wingless and produces a single pheromone (*Z*)-7,8-epoxy-2-methyloctadecane (disparlure) (Bierl et al., 1970). The windborne movements of young larvae on silk threads achieve the spread of the insect. Thus far the possibilities of success for control by mating disruption seem encouraging. Initial trials in the early 1970s were conducted in relatively large areas before either effective dose levels or adequately stabilized formulations were available (Cameron, 1979). This emphasizes the need for careful laboratory and small-scale field evaluation before attempting major control programs. A parapheromone 2-methyl-(*Z*)-7-octadecene was also found to be ineffective, probably for similar reasons. There was certainly evidence for some level of mating disruption as measured by the tethered female technique but the sparse and uneven distribution of the larval population made realistic assessments difficult.

At the time the major trials were conducted, large-scale pheromone synthesis had not been achieved and the high costs prevailing did not encourage belief in an economic level of control. Costs have subsequently declined, although a further problem is the chiral nature of the pheromone (see Mori, Chapter 12, this volume) and if it is found that the chirally pure enantiomer is necessary for disruption, then costs of control could remain prohibitively high. Later studies on the aerial distribution of disparlure when emitted from fiber, flake, or microencapsulated formulations suggested that all three provided concentrations adequate to cause mating disruption for periods of 30 days postapplication (Caro et al., 1981).

Another widespread and important pest of the North American forests is the spruce budworm *Choristoneura fumiferana* (Clem.). The pheromone consists of a blend of (*E*) and (*Z*)-11-tetradecenal (Sanders and Weatherston, 1976), and the highest level of disruption was found when used close to the naturally occurring ratio (Sanders, 1981a). The aldehyde functional groups are relatively unstable and may have posed unsuspected problems in early trials. Treatment of an area of 250 ha with a hollow fiber formulation was reported to be moderately successful. When an area of 10 ha was treated, gravid female immigration was suspected, which for an insect of such considerable migration potential is perhaps not surprising (Sanders, 1981b). Improved stabilized hollow fiber, flake, and microencapsulated formulations are at present being evaluated. Strong environmentalist pressures supply the need for an alternative to conventional pesticide control. If a convincing trial can be mounted, the question of economic cost will have to be carefully considered.

Considerable knowledge exists of the incidence and distribution of the larch budmoth *Zeiraphera diniana* (Guenée) in the subalpine larch forests of Switzerland. Low populations give rise at intervals of approximately 10 years to explosive and damaging outbreaks. The impact of a mating disruptant was assessed by treating a total area of 300 ha with a microencapsulated formulation of the pheromone. The results were equivocal, perhaps due to the relatively uncharacterized nature of the formulation (Baltensweiler and Delucchi, 1979). Small-scale disruption trials for the control of this pest have also been attempted in Czechoslovakia (Vrkoch et al., 1981).

The results obtained from disruption trials for the control of the Western Pine Shoot borer *Eucosoma sonomana* (Kearfott) are much more encouraging. The pheromone is a mixture of (*Z*) and (*E*)-9-dodecenyl acetate which are stable configurations (Sower et al., 1979). The insect which attacks ponderosa pine *Pinus ponderosa* has only one generation each year, and population levels are generally low. The eggs are laid under the scales of young shoots and the larvae burrow straight into the growing point causing stunted growth; insect damage can therefore be easily and accurately assessed. Results of trials in areas ranging from 5 to 600 ha, using pheromone formulated in hollow fibers, flakes, or PVC pellets have all been successful. No alternative control method exists and the problem to be resolved is whether the damage caused by the insect is worth more than the treatment (Overhulser et al., 1980; Daterman, 1982; Sower et al., 1982).

Other lepidopterous forest pests for which mating disruption studies have reached some level of progress include the Douglas-Fir Tussock moth *Orygia pseudotsugata* (McDunnough) (Sower, 1982) and the pine beauty moth *Panolis flammea* (Schiff.) (R Baker, private communication).

The utilization of pheromones by forest Coleoptera differs from that of the Lepidoptera, because the release of an attractant typically results in the location and colonization of the tree where the insects feed, mate, and reproduce. The attractant may elicit a synchronized invasion by several thousand beetles. Where successful aggregation does not occur the insects fail to reproduce. Disruption in this case is therefore directed toward reducing aggregation either by the permeation method using the pheromone or by use of antiaggregation compounds, to a level below that which would kill the tree.

Examples of such work include the permeation of pheromone to reduce aggregation of the southern pine beetle *Dendroctonus frontalis* Zimmerman (Payne, 1981) or the lowered tree attacks by the Douglas-Fir beetle *Dendroctonus pseudotsugae* Hopkins by use of an antiaggregation compound (Furniss et al., 1977, 1981). The difficulties involved in the acquisition of adequate test data are discussed fully by Wood (1979).

Although some preliminary results were reported which seemed encouraging, no attempt to control any Coleopterous forest pest has been conclusively successful. This is in part due both to the familiar problems of inadequate sampling procedures and to the use of uncharacterized formulations.

2.3. Orchard Pests

Orchard pests usually have a restricted host range and in certain areas one or two species occur as key pests. They include the codling moth *Laspeyresia pomonella* (L.), the summer fruit tortrix moth *Adoxophyes orana* (F.v.R.), the Oriental fruit moth *Grapholita molesta* (Busck.), the plum fruit moth *Grapholita funebrana* (Treitshke), the peach tree borer *Synanthedon exitiosa* (Say.), and the clear applewing moth *Synanthedon myopaeformis* (Borkhausen).

Pheromones are available for these species and mating disruption trials have generally involved the hand placement of the pheromone formulated in either plastic fibers, plastic flakes, or rubber tubes. Fruit damage assessment procedures were already well established and adequate methods for assessing mating disruption devised. Using such techniques, some level of success was reported for *Laspeyresia* control in France by Audemard (1979), while complete success was claimed by Charmillot (1980, 1982a) and Mani and Wildbolz (1982) from trials carried out in Switzerland and at a cost equivalent to that of conventional treatment.

Mating disruption trials for the control of *Laspeyresia* in Italy gave poor results in areas of high infestation, and the method was thought to be only useful in well-isolated orchards where population density is low (Maini et al., 1982). Such variability of result could also be explained by gravid female entry into the treatment areas perhaps in relation to differences in locality.

In Switzerland *Laspeyresia* coexists with *Adoxophyes* and further mating disruption trials for the control of the latter using similar techniques were also encouraging, although the efficiency was not good enough to maintain the pest under the tolerance level (Charmillot, 1982b).

Minks et al. (1976) reported results from the Netherlands using a microencapsulated formulation of an antipheromone against *Adoxophyes,* with equivocal results due no doubt in part to limitations in the formulations (see Campion et al., 1981b). Trials using laminated flake formulations were unsuccessful (Vanwetswinkel and Paternotte, 1982).

Successful trials for the control of *C. molesta* and *G. funebrana* by the hand application technique have been reported by Rothschild (1975), Arn et al. (1976), Cardé et al. (1977), and Gentry et al. (1982), although only limited control was achieved by Audemard (1982). A disruption trial for control of *G. funebrana* at high population was unsuccessful (Mani et al., 1978).

Effective mating disruption was also achieved for the grape moth *Lobesia botrana* Schiff, using the pheromone formulated either in rubber dispensers or in hollow fibers. The level of control in plots of between 0.8 and 1.5 ha was sometimes as good as that obtained by conventional insecticides. In other instances poor control occurred and this was again attributed, at least in part, to the immigration of gravid female moths into the treatment areas (Roehrich and Carles, 1982; Vita and Caffarelli, 1982). Unsatisfactory levels of control were also reported in trials using hollow fiber formulations for the control of the European grape berry moth *Eupoecilia ambiguella* (Hübner) (Schruft, 1982).

A variety of slow release mechanisms containing the pheromone of *Synanthedon exitiosa* have been evaluated. The level of control obtained with hand-placed hollow fibers in an orchard of 24 ha compared favorably with an orchard of similar size where conventional insecticidal control methods were used (Yonce and Gentry, 1982). Attempts to control the apple clearwing moth *Synanthedon myopaeformis* using hollow fiber pheromone formulations were unsuccessful (Minks and Voerman, 1982).

2.4. Vegetable Pests

The pheromone of the cabbage looper *Trichoplusia ni* (Hübner) appeared to consist of one component (*Z*)-7-dodecenyl acetate, although more recently a second pheromone component, dodecyl acetate, was reported (Bjostad et al., 1980). Despite the early demonstrations of mating disruption (Shorey et al., 1972), followed by relatively large-scale trials, no further adoption of the technique has occurred. A limitation may be the polyphagous nature of insect, since this makes it necessary to treat a wide range of host plants in the treatment area which may be uneconomic.

Similar considerations may apply to the fall armyworm *Spodoptera frugiperda* (JE Smith), together with the problem of known widespread migration of the adult moths. The pheromone of this species has also not yet been fully identified (McLaughlin et al., 1981), although mating and oviposition were reduced in areas of 12 ha using a hollow fiber formulation of a pheromone analog (Mitchell and McLaughlin, 1982).

The pheromone of the artichoke plume moth *Platyptilia carduidactyla* (Riley), (*Z*)-11-hexadecenal formulated in black celcon hollow fibers applied aerially was effective in controlling this important pest of artichokes in the United States. It is being offered for sale under an Environmental Protection Agency experimental use permit (Haworth et al., 1982). A flake formulation of the pheromone similarly applied is also reported to be successful in controlling the insect (AF Kydonieus, private communication).

2.5. Crop Borers

Host plants of this group include maize, rice, and sugarcane. The European corn borer, *Ostrinia nubilalis* (Hübner) is an important maize pest in both Europe and North America. The pheromone blend, a mixture of (*E*) and (*Z*) isomers of 11-tetradecenyl acetate, varies from one place to another (Klun et al., 1975; Anglade, 1977), necessitating different formulations for different regions. During the course of preliminary trials, it was also discovered that mating mostly occurred outside the host crop in the surrounding long grass (Showers et al., 1976; Buchi et al., 1981). This is a good example of how important biological knowledge is only acquired as the result of this kind of work. Control by mating disruption is therefore only feasible if area-wide applications are made to include such mating habitats.

Mating disruption trials for the attempted control of the striped stem borer *Chilo suppressalis* Walker have been conducted in the Philippines and Japan. There are two pheromone components (*Z*)-11-hexadecenal and (*Z*)-13-octadecenal (Nesbitt et al., 1975). Because of the instability of the aldehyde functional groups, a number of synthetic analogs were tested for disruptant activity (Beevor and Campion, 1979; Kanno et al., 1980; Beevor et al., 1981; Tatsuki and Kanno, 1981). More recently stabilized microencapsulated formulations have been developed (Hall et al., 1982) and areas of up to 30 ha treated (P Beevor, private communication).

Where *Chilo* occurs as one of a complex of rice pests, mating disruption is unlikely to be effective. There are areas in its distribution such as Japan and Korea where *Chilo* is a key pest, but commercial interests would require wider areas of application. The pheromone of the other major rice borer *Scirpophaga*

(*Tryporyza*) *incertulas* (Walker) has not yet been identified. The long-term possibility would be to disrupt simultaneously both species as suggested for other coexistent species by Mitchell (1975), Campion and Nesbitt (1983).

2.6. Stored Product Pests

Mating disruption trials have been conducted for species of *Ephestia* and *Plodia* in warehouse situations by permeation with (*Z,E*)-9,12-tetradecadienyl acetate. The isolated environment within such large stores certainly provides a situation where immigration is virtually excluded. The relatively high population densities that occur within such enclosures do not, however, suggest this approach to be promising (Sower et al., 1975; Barrer, 1976; Haines and Read, 1977).

2.7. Conclusions

The limited successes so far achieved suggest that future prospects for control by mating disruption are good for those insects of key pest status with a limited range of host plants. Examples have been recorded for varied crop situations including cotton, vegetables, orchards, and forests, so that no one particular habitat is likely to be most appropriate.

The apparently low migration potential of the pests in these examples may merely indicate that these early attempts have been limited in size and scope by the relative expense of the pheromones, and this has precluded treatment over wider areas. Increased confidence in the already available techniques should lead to increasing areas of the target pest species being treated, as is already occurring for *Pectinophora* control.

Control of insects species of greater migratory potential may also be considered and in particular those in forest localities would offer a promising goal for the future.

III. Mass-Trapping

3.1. Introduction

Compared with the unknown mechanisms involved in mating disruption, the concept of mass-trapping seems simple enough. A powerful, highly specific insect attractant, if deployed in traps, should catch a sufficiently large number of the target insect species to reduce population increase to economically acceptable levels. The immediate question arises as to what proportion of the wild population needs to be trapped to achieve such a result. For Lepidoptera at least where only males are trapped, it is generally assumed that trapping efficiency should be as high as 80-95% (Knipling and McGuire, 1966).

Early attempts at mass-trapping were conducted before the pheromone blends of many target insect species had been fully characterized and therefore trapping efficiency was low. Designs of traps have been developed, generally on an empirical basis (see reviews by Minks, 1977; Cardé, 1979). Various release mechanisms have been developed to dispense the pheromones at a controlled rate (Campion et al., 1978). Environmental factors such as temperature and windspeed affect both emission rate and shape of the pheromone plume (Lewis and Macaulay, 1976; Murlis and Bettany, 1977; Murlis et al., 1982).

Part of the problem is to quantify the number of traps necessary per unit area to achieve control. This has ranged from 1 to 700 ha^{-1}, the trap density selected either on an empirical basis, or as the result of computer simulation modeling (or, one suspects, from sheer desperation!). The upper limit is, moreover, fixed by the economic factors of trap costs and the subsequent maintenance costs of the trapping network.

Assuming a highly efficient trapping system has been achieved and an adequate trapping regime established, then the problems remain of assessing accurately the effects of the treatment. The problems involved are essentially the same as those discussed for the mating disruption trials. The treatment areas must be either isolated or large so as to reduce the possibility of gravid female entry; the migration potential of the target insect must therefore be realistically assessed. The need for large- or single-area treatment has in turn resulted in many cases of trials consisting of unreplicated blocks, so that statistically acceptable comparisons with control areas have not been possible.

Mass-trapping has been attempted for a wide variety of insect pests of field, orchard, and forest on scales ranging from a few hectares to several thousands of hectares. There are indications of success from some of these trials but at the present time no fully defined, statistically acceptable mass-trapping system has achieved success at an economic cost.

To illustrate the problems involved in the assessment of such trials, the case histories presented in this section are mass-trapping for control of *S. littoralis* and *S. litura* (F.).

Traps at densities of 1 to 2 ha^{-1} in areas of up to 3000 ha resulted in reductions in the number of insecticide applications required to control the *S. littoralis* in Israel by 30–40% (Teich et al., 1979). By 1981 more than 20,000 ha were similarly treated (Shani, 1982). Similar encouraging results were reported from Japan for control of the closely related species *Spodoptera litura* (F.), a major pest of vegetables and sweet potatoes (Nemoto et al., 1980; Tamaki, 1980). However, measurement in both instances was mainly by the indirect method of comparing the frequency of insecticide usage in treatment and control areas.

At the same time mass-trapping trials were also being conducted in Greece and Egypt to determine whether a significant reduction of eggs and larvae of *S. littoralis* could be shown as the result of treatment. To limit the possibility of mated female moths migrating into the treatment area, the first trials were conducted in a semi-isolated area of 150 ha in Crete.

The question of effective trap density was approached experimentally following a preliminary computer simulation exercise suggesting that in the absence of immigration, a trap density of 25 ha^{-1} would be necessary to achieve control (Symmons and Rosenberg, 1978). It was found, however, that trap densities above 9 ha^{-1} produced no increase in the total catch which was merely shared among a larger number of traps (EM McVeigh, unpublished data).

For practical convenience an arbitrary 5 traps ha^{-1} was taken as a maximum density. Trap densities of 1, 3, and 5 traps ha^{-1} were used in successive years in the experimental site and the traps were maintained throughout the year. Despite a progressive reduction of moths in the trapping area with increased trap density, the number of eggs and larvae sampled were not significantly lower than those in two control areas (Campion et al., unpublished data).

The results in Crete had therefore failed to agree with those obtained for *S. littoralis* in Israel and *S. litura* in Japan. It was possible that a limiting factor was the relatively small size of the treatment area in Crete compared with those in Israel and Japan. Although the trials in Crete were in a semi-isolated area, the immigration of mated females may still have nullified the effects of the traps. Larger-scale trials were therefore undertaken in Egypt in a semi-isolated area of 600 ha.

S. littoralis is an important pest of cotton and other crops in Egypt. Control of the insect is often achieved by the hand picking of egg-masses from the young cotton plants. Teams of small children go through the cotton crop once every 3 days to collect the egg masses which are counted and then destroyed (Bishara, 1934; Isa, 1981). An economic appraisal of the mass-trapping method by Gubbins and Campion (1982) suggested that provided it was technically feasible, it would be considerably cheaper than the system of egg-mass collection.

Mass-trapping trials at a density of 3 and 5 traps ha^{-1} failed to reduce significantly the number of egg-masses collected in the treatment areas compared with those collected in areas without traps (Hosny et al., 1979; LJ McVeigh et al., unpublished data). Moreover at the higher trap density it was shown that the mating frequency of tethered female moths was unaffected (EM McVeigh, unpublished data).

More recent mass-trapping experiments for control of *S. litura* in Japan suggest that at least 15 traps ha^{-1} are required to achieve a significant reduction in the mating rate of tethered females (Kobayashi et al., 1981). However, at this trap density considerable trap interactions seem likely (Wall and Perry, 1978). The results of these various mass-trapping trials are summarized in Table 3.

Such differences in the outcome of trials in Greece and Egypt compared with those in Israel could in some circumstances be attributed to the varying role of the beneficial insects. The level of *Spodoptera* attack in Egypt as measured by egg-mass numbers can vary from year to year from a daily average of 10 ha^{-1} in a light infestation year to more than 10,000 ha^{-1} in an outbreak year, with

Table 3. Summary table of mass-trapping trials for attempted control of *S. littoralis* and *S. litura*

Trial location	Trap density employed per hectare	Area under treatment (ha)	Mating reduction of tethered females (%)	Effect of treatment
Crete	1–5	50–150	–	No significant reduction in larval population
	9	10	–	Reduced use of insecticides
Egypt	3–5	400–600	None	No significant reduction in egg mass trapping
Israel	2	2000–20,000	–	Reduced number of insecticide sprays
Japan	2	500–2000	–	Reduced number of insecticide sprays
	15–30	20	50–75	–

the frequency of outbreaks about 1 year in 10. Even where considerable egg-mass numbers do occur they may be heavily attacked by predators and possibly egg-parasites (WR Ingram, private communication). As long as the egg-mass collectors are active, no insecticide sprays are used and so the beneficial insects are preserved. It is therefore possible that successes claimed for mass-trapping are sometimes the result of the beneficial insects since in the presence of the traps, insecticides are less likely to be used.

3.2. Other Cotton Pests

The highly polyphagous feeding habit of the *Spodoptera* larvae is a major problem with the mass-trapping approach, since traps have to be deployed throughout the area under treatment and not confined to a specific crop. Such a problem is not found with the pink bollworm *Pectinophora gossypiella* where cotton is the major host plant.

Several attempts have been made to control this insect in the United States by mass-trapping. Initial trials were conducted before the pheromone was fully characterized and the results were unsuccessful or equivocal (Graham et al., 1966; Guerra et al., 1969). A reduction in the level of larval boll infestation was claimed following mass-trapping with the correct pheromone in an area of 36 ha at a trap density of 5 ha^{-1}, although some late-season insecticide applications were still necessary (Flint et al., 1976). In the absence of adequate controls the infestation levels found during the mass-trapping period were compared with those in the same area for several years prior to treatment.

A similar method of assessment was used by Huber et al. (1979) in a much larger area of 3000 ha using trap densities ranging from 5 to 10 ha^{-1}. A reduction in the level of boll infestation was claimed as the result of the mass-trapping, although the levels of infestation both prior to and during the trials were too low to warrant conventional control practices.

The spiny bollworm *Earias insulana* occurs in countries of the Near East. Like *Pectinophora* it is also virtually confined to cotton. Unlike *Pectinophora* the insect does not diapause and the insect is therefore particularly vulnerable in regions with cold winters. A significant reduction in the level of boll infestation attributable to *Earias* was achieved in Syria in 30 ha of cotton at a trap density of 5 ha^{-1} (Campion et al., 1981a), although an unacceptable level of trap predation occurred.

The boll weevil *Anthonomus grandis* Boheman is an important cotton pest in the New World. The male beetle after feeding on the cotton plant emits a pheromone which causes the aggregation of both sexes during the overwintering period and as a mating attractant for female *Anthonomus* during the summer. Several traps have been developed on a commercial basis for the early-season detection of the insect (Mitchell and Hardee, 1974; Cross et al., 1971). Despite attempts made over a number of years, no clearly defined success for mass-trapping as a control technique has yet been achieved. From results reported by Lloyd et al. (1981) using mark, release, recapture techniques, and computer simulation models, it is suggested that at trap densities of 14 ha^{-1} a high proportion of the female population would be caught. It remains to be seen whether such conclusions can be translated into a commercially viable control strategy.

3.3. Orchard Pests

The control by mass-trapping of the red banded leafroller *Argyrotaenia velutinana* (Walker), an important orchard pest in the United States, was achieved in an area of 8 ha containing low levels of infestation. A trap density of 200 to 300 ha^{-1} was considered necessary but this number was too high to be commercially viable (Trammel et al., 1974). Since this insect forms part of a complex of economically important pest species within the orchards, there would seem to be no future for such a control strategy, even if success with lower trap densities could be achieved.

Similar high trap densities have been used for the control of the citrus flower moth *Prays citri* Mill. in Israel in an area of up to 800 ha. In this instance the insect is a key pest and it was considered that the savings in insecticide usage and the preservation of beneficial insects for the control of other pest species more than compensated for the cost of the traps. Some problems of trap maintenance were, however, encountered (Sternlicht, 1972; Shani, 1982).

Several attempts have been made to control the codling moth *Laspeyresia pomonella* (L.) by mass-trapping in various parts of the world at trap densities ranging from 5 to 44 ha^{-1} (Charmillot and Baggiolini, 1975; Proverbs et al., 1975; MacLellan, 1976; Hagley, 1978; Madsen and Carty, 1979). The general conclusions from these studies were that a reduction in the level of infestation could only be achieved when insect populations were low. The level of control achieved even then was not great enough to be commercially acceptable.

Promising results of mass-trapping for the control of the Oriental fruit moth *Grapholita molesta* (Busk.) in Japan reported by Negishi et al. (1977) were not confirmed in the United States by Wilson and Trammel (1980). Damage caused by a fruit tree roller *Archips podana* (Scopoli) was reduced in a mass-trapping trial in Italy (Boness, 1976). Control of the smaller tea tortrix *Adoxophyes* was only achieved at a massive trap density of 700 ha^{-1} (Negishi et al., 1980; Shimada, 1980). Control of the peach tree borer *Synanthedon exitiosa* (Say.) on the other hand was reported at low trap densities of 2.5 and 5 ha^{-1} (Gentry, 1981).

In an attempt to overcome the limitations set by trapping one of a complex of pest insects, combined mass-trapping for several fruit tortricids including *Grapholita* and *Laspeyresia* was attempted by Wilson and Trammel (1980). At trap densities of 30 to 40 ha^{-1} the attempt was unsuccessful in that the fruit infestations were not reduced to commercially acceptable levels.

3.4. Forest Pests

Results of large-scale mass-trapping trials for the control of the gypsy moth *Lymantria dispar* in the United States have been either equivocal or failed, possibly because of either inadequately characterized pheromone or inadequate release mechanisms (Cameron, 1979). Results of smaller trials conducted in Yugoslavia, Spain, and Romania were also inconclusive due essentially to inadequate controls (Maksimovic et al., 1974; Boness et al., 1977; Dissescu, 1978). The pheromone (*Z*)-7,8-epoxy-2-methyloctadecane was identified by Bierl et al. (1970), but it was later shown that the insect produces only one of two possible enantiomers, the 7 R, 8 S isomer known as (+) disparlure. Traps baited with chirally pure material (Iwaki et al., 1974) caught more moths than those baited with the racemic mixture. An improved synthesis of (+)-disparlure that does not require the separation of the diasterioisomers was reported by Rossiter et al. (1981). For these reasons interest has once more been revived in this method of gypsy moth control (Webb, 1982).

The general lack of success of mass-trapping for the control of lepidopterous pests, except perhaps at very high trap densities, may be because of the need to catch such a high proportion of the male population. The female moths that produce the eggs and hence the subsequent damaging larvae are not attracted. Beetle aggregation pheromones attract both sexes and therefore may have greater promise for use in mass-trapping control programs.

The absence of reliable methods to assess accurately the efficacy of treatment has been a major difficulty (Wood, 1979). Such a limitation has not inhibited ambitious and expensive mass-trapping programs for the attempted control of beetle pests of forests. The largest is that for the control of the spruce bark beetle *Ips typographus* (L.), an important pest in Scandanavian forests. A joint project financed by Norway and Sweden started in 1979 and involved the use of 1 million traps, which during the season trapped 2.9 billion beetles at an estimated cost of 23 million U.S. dollars. The economic argument in favor of this program was the possible loss of up to 59% of the spruce stands in future years (Lie and Bakke, 1981; Bakke, 1982).

An enlarged trapping program in Norway alone the following year resulted in the catch of an estimated 4.5 billion beetles. The campaign has been claimed to be successful since the predicted catastrophic pest outbreak did not occur, while insect numbers have subsequently continued to decline. In the absence of control areas it is not possible to be sure whether this decline is due to the traps or to environmental factors. Some critics have therefore suggested that such large-scale programs without critical methods are premature. On the other hand, the only alternative control strategy is the widespread application of insecticides which has been opposed on the grounds of environmental pollution. No other remotely possible control strategy has been suggested, so the trapping continues.

Earlier mass-trapping programs conducted in the United States for the control of the western pine beetle *Dendroctonus brevicomis* LeConte and the Elm bark beetle *Scolytus multistriatus* (Marsh) were only moderately successful in relatively small areas of low infestation while larger trials produced equivocal results. A limiting factor for both species for the success of this technique is that there is a dispersal flight of several kilometers of the emerging adult beetles (Bedard and Wood, 1981; Lanier, 1981; Peacock et al., 1981; Roelofs, 1981). Attempts to trap a defined population in an artificially contrived isolated area indicated a low rate of recapture (Birch et al., 1982).

A more modest approach is the use of pheromone traps to protect timber and sawed wood in Canadian saw mills from attacks by ambrosia beetles *Gnathotrichus sulcatus* (LeConte), *G. retusus* (LeConte), and *Trypodendron lineatum* (Olivier). These insects bore pin holes into the sap wood of many coniferous tree species. Traps baited with pheromones of these species placed strategically around the perimeter of the yards intercept the beetles before they reach the felled timber. Results obtained so far suggest this to be an acceptable method of maintaining control of this important group of pest insects (Borden et al., 1980; McLean and Borden, 1979).

3.5. Conclusions

The problems encountered in the evaluation of mass-trapping trials are very similar to those described earlier for the work on mating disruption. The additional problem of trap deployment and maintenance poses in many instances a further economic constraint in countries where labor costs are high.

At the present time, successes in the use of mass-trapping for insect control are even less than those achieved by mating disruption. For Lepidoptera only the control of the citrus flower moth *Prays citri* in Israel has been successful and at economic costs, but using a very high trapping density. Mass-trapping programs for a few other lepidopterous species show some promise and will undoubtedly continue. Improvement in trap design and pheromone characterization must be anticipated.

Greater promise in the future may well be with the continued utilization of beetle pheromones where both sexes are attracted to the traps, instead of only males as it occurs in the case of lepidopterous pheromones. A convincing outcome of the mass-trapping campaign for *Ips typographus* in Scandanavia will be required to stimulate further programs against related forest pests.

IV. Monitoring

4.1. Introduction

Pheromone traps can be used to detect both the presence and the density of certain pest species, with the aim of obtaining a more precise control so that insecticides are used only when necessary. Such detection may be either qualitative or quantitative.

Pheromone traps generally catch when pest numbers are very low and so they can be used qualitatively to provide an early warning of pest incidence. They can also be useful to define areas of infestation of a pest, particularly where the overall distribution and life cycle are poorly understood.

Monitoring with pheromone traps becomes more difficult when attempts are made to relate numbers of insects caught in the traps to some economic threshold of damage. In the case of lepidopterous pest species, the traps catch adult moths while the damaging stage is the larva. A time interval of possibly several weeks is therefore involved between the two stages and such a lack of precision is perhaps not surprising.

Nevertheless pheromone-baited traps for an ever increasing number of pest species are being used to provide information on their abundance at critical times of the year when the crop under survey might be at risk.

4.2. The Codling Moth

The efforts that have been made to establish a pheromone monitoring system for the codling moth *Laspeyresia pomonella* (L.) in fruit orchards in various parts of the world provide an instructive case history.

The *Laspeyresia* pheromone was first isolated and synthesized by Roelofs et al. (1971) as (*E*,*E*)-8,10-dodecadien-1-ol. Significant correlations were demonstrated between moth catch numbers in pheromone-baited traps and subsequent levels of fruit infestation in Canada, by Riedl and Croft (1974), who stressed their validity only for the year in which the experiments were conducted and the localities where the observations were made.

Attempts to develop forecasting systems of practical use to farmers were undertaken in various parts of the world. Thus Madsen and Vakenti (1973), Riedl and Crofts (1974) in Canada, Alford et al. (1979) in the United Kingdom, and Tanskii and Bulgak (1981) in the USSR reported a trap catch threshold of 5 moths/trap/week. When this regime was followed, a 50–75% reduction in the number of spray applications needed for adequate control was achieved. Similar systems using threshold values slightly higher or lower than this figure with equally successful results were reported from New Zealand by Charles and Wearing (1979), from Romania by Iacob et al. (1980), from the Soviet Union by Abashidze and Kipiani (1980), Matvieski (1980), and Gontarenko et al. (1981), from Canada by Madsen (1981), and from the United Kingdom by Glen and Brain (1982).

Such encouraging reports suggest that pheromone traps do provide useful information on when action can be taken to ensure adequate protection of the crop. Other workers are, however, less optimistic. Audemard (1979) in France found that forecasts based on moth catches in the pheromone traps often resulted in a greater number of treatments than normally recommended, although it was admitted that adjustments to the critical threshold could be made. A major problem was the attraction of moths from neighboring orchards and without additional traps in these areas it was not considered that forecasting was possible. A similar problem was noted by Madsen (1981) in Canada, who also found that moth catches generally increased in the traps after insecticide application which he considered were caused by enhanced mobility induced by insecticidal irritability. Since the period of protection following the application of insecticide was 3 weeks, the catches in the traps were therefore allowed to exceed the prescribed threshold during this time.

Competition for the available males between the traps and the variable numbers of wild females was considered a major problem by Touzeau (1979). Thus in his view only early flights when overall populations are low may be followed accurately by the pheromone traps.

Differences in threshold values and trapping efficiency are likely to be related to greatly varying differences in trap design which unfortunately is far from being standardized (Paradis et al., 1979; Arsura et al., 1979, 1980; McNally and Barnes, 1980; Riedl, 1980). The choice of trap location within the tree canopy has also been shown to have a significant effect on catch (Riedl et al., 1979). The purity of pheromone and the type of dispensing system used are also important to ensure an adequate pheromone release over prolonged periods (see Campion et al., 1978; Duran, 1980).

It is possible to conclude thus far that there is no difficulty in determining when the first flights occur (qualitative monitoring). To achieve greater precision (quantitative monitoring) the questions arise of the scale of operation, the density of traps used throughout the area to be monitored, and the assurance that immigrating moths from neighboring areas can be dealt with.

Estimates of the required trap density to ensure adequate coverage range from 1 per 10 ha in Switzerland (Charmillot, 1979), 1 per 2.5 ha in New Zealand (Charles and Wearing, 1979), and 1 per ha in South Africa and Canada (Myburgh et al., 1974; Vakenti and Madsen, 1976).

A regional forecasting system presupposes a considerable amount of coordinated effort by responsible extension officials and is far from a reliance on a simple monitoring device. Area-wide forecasting by government extension services in turn poses difficulties, since differences in population density, local ecological characteristics, and microclimate may vary considerably (Cranham, 1978; Solomon, 1978; Charmillot, 1979). Indeed as suggested by Alford et al. (1979), regional forecasts can only be a guide and when related to any particular orchard; the spray date suggested could be wrong by up to 2 weeks or more. Each individual orchard should therefore have its own trap or traps and the information acquired should be used to modify the general district forecast: an ideal compromise.

All such work presupposes *Laspeyresia* to be the key pest; in situations where this is not the case, monitoring one pest species in isolation, as, for example, reported in Sweden by Andersson (1979); the information acquired may not be of value. The successful use of pheromone traps to time insecticidal applications for control of codling moth was, however, reported elsewhere in Sweden (Stenmark, 1978).

4.3. Other Orchard Pests

The initiation of the flights of the summer tortrix moth *Adoxophyes orana* is monitored in the orchards of the Netherlands by the Horticultural Advisory Service throughout an area of 25,000 ha (Minks and Dejong, 1975; Dejong, 1980). The total number of insecticide sprays has thus been reduced from five to seven each season to three or four, and similar results were reported in orchards of 600 ha in Canada. No correlation between moth catch in traps and fruit damage was found in the United Kingdom by Alford et al. (1979), in contrast to the report of Berling et al. (1980) from Germany that peaks of moth capture occurred 2 to 3 weeks earlier than the appearance of the maximum number of larvae.

Populations of the Oriental fruit moth *Grapholita molesta* in peach orchards in Japan were studied by Tanaka and Yabuki (1978) using pheromone traps. Five generations each year were recognized and the threshold of development was found to be 11.1°C with a thermal constant of 387.7 day degrees above this threshold. The use of such a constant enabled subsequent moth flights to be predicted. The distribution of *Grapholita molesta* in Czechoslovakia was reported by Hdry et al. (1979), in Italy by Graziano and Viggiani (1981), and in Brazil by Silveira Nato et al. (1981). Reports in Sweden state that pheromone traps have been useful in following flights of *G. funebrana* and *G. grapholita* (Stenmark, 1978). Monitoring stations for these and related insects have also been established in France, West Germany, Romania, and Switzerland (Audemard et al., 1976; Minks, 1979).

The synthetic pheromone of the olive moth *Prays oleae* Bern. was used in olive groves in France, Greece, and Spain to monitor the flight of the phyllophagous, anthophagous, and carpophagous generations of the moth. It was found that the catches faithfully reflected the emergence curves and reproductive activity (Pravalorio et al., 1981; Polyrakis, private communication; Ramos et al., 1981). Such relationships were not found in similar trials conducted in Sicily, using the pheromone of the citrus flower moth *Prays citri* Mill. and the method was not recommended for timing chemical treatments (Mineo et al., 1980).

Attractants of various kinds have long had use in the control of fruit flies (Cino and Vita, 1980). The status of sex pheromones for this insect group is not so well advanced, but major attractant components of the olive fly *Dacus oleae* and the melon fly *Dacus cucurbitae* Coq. have been identified, synthesized and field tested (Baker et al., 1980, 1982).

4.4. Vegetable Pests

Spodoptera litura F. is an important pest of sweet potato and vegetables in Japan. A forecasting system was developed to relate the yield of the former crop to the number of male moths caught in the pheromone traps. Parameters used included moth numbers, number of egg masses, hatchability, development period from hatch to fourth instar, and survival rate. The injury threshold density was 4.8 fourth instar larvae/m^2 and it was shown that the control threshold was 190 males/trap/day in July and 160 males/trap/day in August (Nakasuji and Kiritani, 1978). There has been no indication that such information has been used as a practical warning system for farmers. The influence of trap location on catch was reported by Hirano (1981).

The pea moth *Cydia nigricana* (S.) is a major pest of peas in the main pea growing areas of the United Kingdom and other parts of Europe. In the past, populations of this insect were difficult to monitor effectively, the best method being the laborious task of searching for eggs on the foliage. Since 1977 in the United Kingdom, pheromone traps have been used to provide an early warning system for the presence of the insect (Lewis and Macaulay, 1976; Macaulay and Lewis, 1977). A commercial monitoring system was marketed by Oecos Ltd. direct to the growers for localized monitoring and the results were coordinated by the Agricultural Development and Advisory Service (ADAS) to provide area-wide spray warnings. This has resulted in a considerable reduction in the number of insecticide applications required to control the pest insect. Pilot monitoring schemes have also been successful in Czechoslovakia and Sweden (Horak et al., 1980; Morner, 1981). The development of a warning system to growers in France of imminent attacks of the leek tineid *Acrolepiopsis assectella* (Zell.) using pheromone traps baited with (*Z*)-11-hexadecenal was reported by Rahn (1980).

4.5. Cotton Pests

The pink bollworm *Pectinophora gossypiella* is an established cotton pest throughout the world. Extensive surveys are conducted in the United States to detect the possible spread of the insect from Arizona and New Mexico into adjacent states (see Kennedy, 1981). All cotton fields (60,000 ha) in Israel have been monitored by pheromone traps for *Pectinophora* since 1975. Insecticide treatments are limited to the beginning of the season in order to control the adult moths. Usually no more than two sprays are applied, whereas formerly farmers sprayed 10-15 times each season (Shani, 1982). *Pectinophora* pheromone trap catches were shown to be closely correlated with boll damage in experiments carried out on sea-island cotton in Barbados. Catches of 8-9 moths in a night represented a 10% level of boll damage 10 days later (Ingram, 1980), while according to Taneja and Jayaswal (1981), cotton in India should be sprayed 24-48 hr after catches of 8 moths/trap/night.

The attractant pheromone of *Pectinophora* has been combined with those of the red bollworm *Diparopsis castanea* (Hmps.) in an attempt to develop a dual monitoring systems (Marks, 1976, 1977) for *Pectinophora* and *Diparopsis* in the Shire Valley cotton growing region of Malawi.

Pheromone-baited traps of the American bollworm *Heliothis armigera* (Hb.) and the spiny bollworm *Earias insulana* have been used as monitoring agents in 5000 ha of cotton in Israel (Shani, 1982). Preliminary trials on the use of these pheromones have been undertaken in Egypt and Syria (Critchley et al., unpublished data; Campion et al., 1981a). The relative density of field populations of *Heliothis virescens* (F.) was estimated from early-season catches in pheromone traps (Hartstack and Witz, 1981).

Significant correlations were found between numbers of moths of the Egyptian cotton leafworm *Spodoptera littoralis* in pheromone traps and numbers of egg-masses on cotton 3 days later in Egypt (Mohnem et al., unpublished data). The presence of biological control mechanisms that occur within the cotton crop of Egypt provides a complicating factor. In some instances few larvae result from these egg-masses and this has been attributed to a high rate of egg predation (WR Ingram, unpublished data). In such circumstances, the traps are of limited value in decision making for insecticide spray application. In countries where insecticides are more widely used to control the cotton pest complex, biological control mechanisms of a type encountered in Egypt may be less important.

4.6. Forest Pests

The gypsy moth *Lymantria dispar* L. was introduced into the United States just over 100 years ago, although only 10% of the available forest has so far been infested. The qualitative distribution of the male moth has been intensively surveyed in recent years. According to Kennedy (1981), during 1979, 95,000 traps were deployed in 38 states with an additional 1500 traps employed along the leading edge of the spreading population.

The traps cannot help in the timing of insecticide applications since the annual flight of the male moth occurs several weeks after the damaging larva stage. The primary aim of the trapping program has therefore been the detection of populations in areas where numbers are generally sparse. According to Silverstein (1981) a recent infestation in San Francisco was successfully controlled as the result of early detection by the pheromone traps. As was mentioned earlier, the pheromone molecule possesses a chiral center and its (+) enantiomer was subsequently found to be 10 times more attractive to male moths than the racemic mixture (Cardé et al., 1977).

The nun moth *Lymantria monacha* (L.) is also attracted to the same synthetic pheromone. The flight activity of this destructive pest of European forests has been studied using pheromone traps in Switzerland and Austria by Maksymov (1978) and Ferenczy and Holzschuh (1976).

Pheromone traps for the detection of the spruce budworm *Choristoneura fumiferana* (Clemens) in Canadian forests have been operational for 20 years. Originally traps were baited with virgin females but subsequently synthetic lures have been utilized. Studies undertaken during this period have indicated that the numbers of moths caught in the traps are paralleled by larval densities (Sanders, 1981a). However, at present most control strategies are directed toward high-density populations numbering 50 million or more ha^{-1} and these can easily be detected from aerial survey by the associated defoliation. The future possibility of control by reduction of incipient outbreaks requires a more sensitive monitor for which purpose the pheromone trap will be of great service. A survey of the Douglas-Fir tussock moth *Orygia pseudotsugata* (McDonnough) using pheromone traps defined the limits of its distribution and its known range was much expanded (Livingstone and Daterman, 1977).

Another survey with pheromone traps spread over several thousands of hectares is being used to study the population of the larch bud moth *Zeiraphera diniana*, a pest of sub-alpine forests of Switzerland and Czechoslovakia (Baltensweiler and Delucchi, 1979; Zhdarek et al., 1981). Monitoring for the pine beauty moth *Panolis flammea* in European forests is also in progress (Baker, 1979; Knauf et al., 1979).

Pheromones have not generally been used to detect bark beetle populations for the timing of control measures, nor have relationships between trapping data and the size and location of beetle populations been reported (Wood, 1979). The difficulties involved in measuring populations are great, but it is perhaps surprising that so little is known in comparison with the knowledge acquired in the pheromone chemistry of the group which includes five species of *Dendroctonus* and five species of *Ips*. Pheromone traps are used as general survey tools; a good example is that undertaken for the smaller European elm bark beetle *Scolytus multistriatus* (Marsham), a principal vector of Dutch elm disease. The traps can be used to check flight activity and for the estimation of the effectiveness of sanitary control measures (Vité et al., 1976; Minks and Van Deventer, 1978).

4.7. Stored Product Pests

Stored food-stuffs are particularly susceptible to insect infestation, and stringent requirements for the absence of insects in such commodities necessitate sensitive insect monitoring tools. An extensive pheromone monitoring program for Dermestid beetles has been in progress in the United States for several years. Detection occurs both in warehouses within the country and the monitoring of incoming cargoes at major U.S. ports (Burkholder, 1981). A similar approach has been made in the USSR (Smetnik, 1978).

Monitoring of storage moth species in warehouses including species of *Ephestia, Plodia,* and *Sitotroga* has also been widely conducted (Levinson and Levinson, 1979; Levinson and Buchelos, 1981; Hoppe and Levinson, 1979; Reichmuth et al., 1978; Cogburn and Vick, 1981; Sifner and Zdarek, 1982).

4.8. Conclusions

From this less than exhaustive survey it may be concluded that where precise population estimates are not required, pheromone-baited traps have been successful in delineating the presence or relative absence of the target pest species in a wide variety of crop situations. Quarantine monitoring as, for example, at warehouses and ports has also been undertaken. The problem of relating trap catch to subsequent economic damage has so far been overcome for only a few insect pest species. This approach requires much painstaking work over prolonged periods, and of the many pheromones presently available, relatively few studies are in progress. It is important that such work is continued and it is certain that further successes will be achieved in the future.

V. General Conclusions

World crop losses from pests, diseases, and weeds are thought to be about 30% of yields (Haskell, 1981). Conventional pesticides, used on a vast scale, are meeting with increasing natural resistance problems, and costs escalate continually. Integrated pest management aims at minimal use of insecticides through improved ecological knowledge of the pests, preservation of their natural enemies, and alternative means of reducing their populations. Pheromones offer opportunities in all three aspects of IPM and consequently receive much research attention. Their specificity is important in conferring environmental acceptability, and as advances are made in formulations with improved slow release, commercial applications will increase.

Monitoring as an aid to decision taking is already widespread, but much remains to be done to relate trap catches to pest populations and damage thresholds. Behavioral responses to pheromone plumes can still throw light on improved trap designs, and the need for careful fundamental research is evident, with every prospect of significant advances to be made in the future.

VI. References

Abashidze ED, Kipiani AA (1980) The use of sex pheromone traps of the codling moth with the aim of determining levels for its control and for the study of its population dynamics. Soobscheniya Akademii Nauk Gruzinskoi SSR **97**:713–716.

Alford DV, Carden PW, Dennis EB, Gould HJ, Vernon JDR (1979) Monitoring codling and tortrix moths in United Kingdom apple orchards using pheromone traps. Ann Appl Biol **91**:165–178.

Andersson CR (1979) Pheromone traps for spraying according to need. Plant Protection Conference 1979, Uppsala, Special Part 68–69.

Anglade P (1977) Variabilité des populations de la pyrale du maïs *Ostrinia nubilalis* (Hubn.) d'après les piègeages par phéromones. Ann Zool Ecol Anim **9**:590.

Arn H, Delley B, Baggiolini M, Charmillot P (1976) Communication disruption with sex attractant for control of the plum fruit moth *Grapholita funebrana*: A two-year study. Ent Exp Appl **19**:139–147.

Arsura E, Capizzi A, Piccardi P, Spinelli P (1979–1980) Some factors influencing the performance of pheromone traps for codling moth, oriental fruit moth and the two European grape vine moth species in Italy. Boll Zool Afr Bachic Ser. II **15**:15–28.

Audemard H (1979) Le piégeage du Carpocapse (*Laspeyresia pomonella* L.) avec la phéromone sexuelle de synthèse E-8, E-10 DDol dans la lutte raisonnée en verger de Pommiers en France. Ann Zool Ecol Anim **11**:565–586.

Audemard H (1982) Oriental fruit moth control by mating disruption technique: Trials of 1982. Unpublished internal IOBC report presented at IOBC/WPRS working group on pheromones, special meeting on mating disruption changins/myon, Switzerland, 28–29 September, 1982.

Audemard H, Charmillot PJ, Beauvais F (1979) Trois ans d'essais de lutte contre la carpocapse (*Laspeyresia pomonella* L.) par la méthode de confusion des males avec une phéromone sexuelle de synthèse. Ann Zool Ecol Animale **11**:641–658.

Audemard H, Fremond JC, Marboutie G, Gendrier JP, Reboulet JN (1976) Comparative study of the trapping of the oriental fruit moth *Grapholita molesta* of the peach tree with virgin females and with a synthetic pheromone. Rev Zool Agric Pathol Veg **75**:117–126.

Baker R (1979) Recent developments in pheromone chemistry: commercial possibilities. In: Proceedings 1979 British Crop Protection Conference Pests & Diseases, Brighton, pp 843–852.

Baker R, Herbert R, Howse PE, Jones OT, Franke W, Reith W (1980) Identification and synthesis of the major sex pheromone of the olive fly (*Dacus oleae*). JCS Chem Comm 52–53.

Baker R, Herbert RH, Lomer RA (1982) Chemical components of the rectal gland secretions of male *Dacus cuecurbitae,* the Melon fly. Experientia **38**: 232–233.

Bakke A (1982) Mass trapping of the spruce bark beetle *Ips typographus* in Norway as part of an integrated control programme. In: Insect Suppression with Controlled Release Pheromone Systems. Kydonieus AF, Beroza M (eds), CRC Press, Boca Raton, Florida, Vol 2, pp 17–25.

Baltensweiler W, Delucchi V (1979) The study of larch bud moth migration in the Engadine Valley by means of a parapheromone. Mitt Schweiz Ent Ges **52**:291–296.

Barrer PM (1976) The influence of delayed mating on the reproduction of *Ephestia cautella* (Walker) (Lepidoptera: Phycitidae). J Stored Prod Res **12**: 165–169.

Bedard WD, Wood DL (1981) Suppression of *Dendroctonus brevicomis* by using a mass-trapping tactic. In: Management of Insect Pests with Semiochemicals. Mitchell ER (ed), Plenum, New York/London, pp 103–114.

Beevor PS, Campion DG (1979) The field use of "inhibitory" components of lepidopterous sex pheromones and pheromone mimics. In: Chemical Ecology: Odour Communication in Animals. Ritter FJ (ed), Elsevier/North-Holland Biomedical Press, Amsterdam, pp 313–325.

Beevor PS, Campion DG, Moorhouse JE, Nesbitt BF (1973) Cross attractancy and cross-mating between the red bollworm *Diparopsis castanea* (Hmps.) and the Sudan bollworm *Diparopsis watersi* (Roths.) (Lep., Noctuidae). Bull Ent Res **62**:439–442.

Beevor PS, Dyck VA, Arida GS (1981) Formate pheromone mimics as mating disruptants of the striped rice borer moth *Chilo suppressalis* (Walker). In: Management of Insect Pests with Semiochemicals. Mitchell ER (ed), Plenum, New York/London, pp 305–311.

Berling R, Levinson AR, Levinson HZ, Naton E (1980) Pheromone and light traps and beating samples for determining the incidence of the apple-peel tortricid (*Adoxophyes reticulana* Hbn.). In: Proceedings International Symposium IOBC/WPRS on Integrated Control in Agriculture and Forestry. Ross K, Berger H (eds), Vienna, 8–12 October 1979, pp 397–398.

Bierl BA, Beroza M, Collier CW (1970) Potent sex attraction of the gypsy moth: Isolation, identification and synthesis. Science **170**:87–89.

Bierl BA, Beroza M, Staten RT, Sonnet PE, Adler VE (1974) The pink bollworm sex attractant. J Econ Ent **67**:211–216.

Birch MC (ed) (1974) Pheromones. Elsevier/North Holland, Amsterdam, p 495.

Birch MC, Miller JC, Paine TD (1982) Evaluation of two attempts to trap defined populations of *Scolytus multistriatus.* J Chem Ecol **8**:125–136.

Bishara I (1934) The cotton worm *Prodenia litura* F. in Egypt. Bull Soc Roy Entomol Egypte **18**:288–418.

Bjostad LB, Gaston LK, Noble LL, Moyer JH, Shorey HH (1980) Dodecyl acetate, a second pheromone component of the cabbage looper moth *Trichoplusia ni.* J Chem Ecol **6**:727–734.

Boness M (1976) Trials to control the fruit tree leaf roller *Archips podana* with pheromones. Z Angew Entomol **82**:104–107.

Boness M (1980) Die praktische Verwendung von Insektenpheromonen. In:

Chemie der Pflanzenschutz-und Schädlingsbekämpfungsmittel. Wegler R (ed), Springer-Verlag, Berlin, Vol VI, pp 165–184.

Boness M, Eiter K, Disselnkötter H (1977) Studies on sex attractants of Lepidoptera and their use in crop protection. Pflanzenschutz Nachr **30**:213–236.

Borden JH, Handley JR, McLean JA, Silverstein RM, Chong L, Slessor KN, Johnston BD, Schuler HR (1980) Enantiomer-based specificity in pheromone communication by two sympatric *Gnathotrichus* species (Coleoptera: Scolytidae). J Chem Ecol **6**:445–456.

Brand JM, Young JC, Silverstein RM (1979) Insect pheromones, a critical review of recent advances in their chemistry, biology and application. In: Progress in the Chemistry of Organic Natural Products. Herz W, Grisebach H, Kirby GW (eds), Springer-Verlag, New York, Vol 37, pp 1–190.

Brooks TW, Doane CC, Haworth JK (1979) Suppression of *Pectinophora gossypiella* with sex pheromones. In: Proceedings 1979 British Crop Protection Conference, Pests & Diseases. Brighton, pp 854–866.

Buchi R, Baldinger J, Blasser S, Brunetti R (1981) Versuche zur Bekämpfung des Maiszünslers, *Ostrinia nubilalis* Hbn. mit der Verwirrungstechnik. Mitt Schweiz Ent Ges **54**:87–98.

Burkholder WE (1981) Biomonitoring for stored product insects. In: Management of Insect Pests with Semiochemicals. Mitchell ER (ed), Plenum, New York/London, pp 29–40.

Butenandt AR, Beckmann R, Stamm D, Hecker E (1959) Über den Sexuallockstoff des Seidenspinners *Bombyx mori*. Reindarstellung und Konstitution. Z Naturforsch **14b**:283–284.

Cameron EA (1979) Disparlure and its role in gypsy moth manipulation. Mitt Schweiz Ent Ges **52**:333–342.

Campion DG, Hunter-Jones P, McVeigh LJ, Hall DR, Lester R, Nesbitt BF (1980) Modification of the attractiveness of the primary pheromone component of the Egyptian cotton leafworm *Spodoptera littoralis* (Boisd.) (Lepidoptera, Noctuidae) by secondary pheromone components and related chemicals. Bull Ent Res **70**:417–434.

Campion DG, Lester R, Nesbitt BF (1978) Controlled release of pheromones. Pest Sci **9**:434–440.

Campion DG, McVeigh LJ, Bettany BW, Hunter-Jones P (1981a) The use of sex pheromones for the control of Spiny bollworm *Earias insulana* and the American bollworm *Heliothis armigera* in cotton growing areas of Syria. FAO Unpublished Report, pp 10.

Campion DG, McVeigh LJ, Hunter-Jones P, Hall DR, Lester R, Nesbitt BF, Marrs GJ, Alder MR (1981b) Evaluation of microencapsulated formulations of pheromone components of the Egyptian cotton leafworm *Spodoptera littoralis* (Boisd.) in Crete. In: Management of Insect Pests with Semiochemicals. Mitchell ER (ed), Plenum, New York/London, pp 253–265.

Campion DG, Nesbitt BF (1981) Lepidopteran sex pheromones and pest management in developing countries. Tropical Pest Management **27**:53–61.

Campion DG, Nesbitt BF (1982a) Recent advances in the use of pheromones in developing countries with particular reference to mass-trapping for the control of the Egyptian cotton leafworm *Spodoptera littoralis* and mating disruption for the control of pink bollworm *Pectinophora gossypiella*. In:

Les Médiateurs Chimiques Agissant sur le Comportement des Insectes, Versailles 1981, (Les Colloques d'INRA 7). Institut National de la Recherche Agronomique, Paris, pp 335–342.

Campion DG, Nesbitt BF (1983) Sex pheromones for the control of stem borers. Insect Science and Its Application **4**:191–197.

Cardé RT (1979) Behavioural responses of moths to female-produced pheromones and the utilisation of attractant baited traps for population monitoring. In: Movement of Highly Mobile Insects: Concepts & Methodology in Research. Rabb RL, Kennedy GG (eds), North Carolina State University, Raleigh, Chap 22.

Cardé RT, Baker TC, Castrovillo PJ (1977) Disruption of sexual communication in *Laspeyresia pomonella* (codling moth), *Grapholita molesta* (oriental fruit moth) and *G. prunivora* (lesser apple worm) with hollow fibre attractant sources. Entomol Exp Appl **22**:280–288.

Cardé RT, Doane CC, Granett J, Hill AS, Kochansky J, Roelofs WL (1977) Attractancy of racemic disparlure and certain analogues to male gypsy moths and the effect of trap placement. Environ Ent **6**:765–767.

Caro JH, Freeman HP, Brower DL, Bierl-Leonhardt BA (1981) Comparative distribution and persistence of disparlure in woodland air after aerial application of three controlled–release formulations. J Chem Ecol **7**:867–880.

Charles JG, Wearing CH (1979) Codling moth control in gate-sales orchards. Orchardist of New Zealand **52**:171–174.

Charmillot PJ (1979) Efficacité du piège à carpocapse (*Laspeyresia pomonella* L.) appâté d'attractif sexuel synthétique. Ann Zool Econ Anim **11**:587–598.

Charmillot PJ (1980) Etude des possibilités d'application de la lutte par la technique de confusion contre le carpocapse *Laspeyresia pomonella* (L.) (Lep. Tortricidae). Thèse No. 6589 Polytechnique Fédérale, Zurich.

Charmillot PJ (1982a) Expérimentation de la technique de confusion contre Capua (*Adoxophyes orana* F.v.R.) en Suisse Romande. In: Les Médiateurs Chimiques Agissant sur le Comportement des Insectes, Versailles 1981, (Les Colloques d'INRA 7), Institut National de la Recherche Agronomique, Paris, pp 357–363.

Charmillot PJ (1982b) La lutte en verger contre *Laspeyresia pomonella* L. et *Adoxophyes orana* F.V.R. par la technique de confusion. Mitt Dtsch Ges Allg Angew Ent **2**:298–302.

Charmillot PJ, Baggiolini M (1975) Essai de lutte contre de carpocapse (*Laspeyresia pomonella* L.) par capture intensive des mâles a l'aide d'attractifs sexuels synthétiques. La Rech Agron Suisse **14**:71–77.

Cino U, Vita G (1980) Fruit fly control by chemical attractants and repellants. Boll Lab Ent Agr Filippo Silvestri Portici **37**:127–139.

Cogburn RR, Vick KW (1981) Distribution of Angoumois grain moth, almond moth, and Indian meal moth in rice fields and rice storages in Texas as indicated by pheromone traps. Environ Ent **10**:1003–1007.

Cranham JE (1978) Monitoring codling moth *Laspeyresia pomonella* with pheromone traps. Mitt Biol Bundesanst Land-Forstwirtsch Berlin-Dahlem **180**:31–33.

Critchley BR, Campion DG, McVeigh LJ, Hunter-Jones P, Hall DR, Cork A, Nesbitt BF, Marrs GJ, Jutsum AR, Hosny MM, Nasr El Sayed A (1983) Control of pink bollworm *Pectinophora gossypiella* (Saund.) in Egypt by communication-disruption using an aerially applied microencapsulated pheromone formulation. Bull Ent Res **73**:289–299.

Cross WH, Leggett JE, Hardee DD (1971) Improved traps for capturing boll weevils. Co-op Econ Insect Rep **21**:367–368.

Daterman GE (1982) Monitoring insects with pheromones: Trapping objectives and bait formulations. In: Insect Suppression with Controlled Release Pheromone Systems. Kydonieus AF, Beroza M (eds), CRC Press, Boca Raton, Florida, Chap 7.

David CT, Kennedy JS, Ludlow AR, Perry JN, Wall C (1982) A reappraisal of insect flight towards a distinct point source of wind-borne odour. J Chem Ecol **8**:1207–1215.

Dean P, Lingren PD (1982) Confusing and killing cotton pests. Agric Res **31**: 4–5.

Dejong DJ (1980) Monitoring techniques, forecasting systems and extension problems in relation to the summer fruit tortricid *Adoxophyes orana* (F.v.R.). EPPO Bull **10**:213–221.

Dissescu G (1978) Use of synthetic sex pheromones in oak forests infested by *Lymantria dispar* L. Zast Bilja **29**:105–109.

Djerassi C, Shih-Coleman C, Diekman J (1974) Insect control in the future: Operational and policy aspects. Science **186**:596–607.

Doane CC, Brooks TW (1981) Research and development of pheromones for insect control with emphasis on the pink bollworm. In: Management of Insect Pests with Semiochemicals. Mitchell ER (ed), Plenum, New York/London, pp 285–303.

Duran JE (1980) Practical problems in the use of traps for the Lepidoptera injurious to crops by means of synthetic sex pheromones. In: Proceedings International Symposium IOBC/WPRS on Integrated Control in Agriculture and Forestry, Vienna, October 1979. Russ K, Berger H (eds), pp 468–470.

Ferenczy J, Holzschuh C (1976) The sexual attractant disparlure as an aid in controlling *Lymantria monacha.* Allgemeine Forstzeitung **87**:109–112.

Flint HM, Merkle JR (1981) Early season movements of pink bollworm male moths between selected habitats. J Econ Ent **74**:366–371.

Flint HM, Smith RL, Bariola LA, Horn D, Forey D, Kuhn SJ (1976) Pink bollworm: Trap tests with gossyplure. J Econ Entomol **67**:738–740.

Funkhouser W (1979) Gossyplure H.F. Application. In: Proceedings of the 1979 Beltwide Cotton Production–Mechanization Conference. Brown JM (ed), National Cotton Council of America, Memphis, Tennessee, pp 76–77.

Furniss MM, Clausen RW, Martin GP, McGregor MD, Livingstone RL (1981) Effectiveness of Douglas-fir beetle antiaggregative pheromone applied by helicopter. General Tech Rpt Intermountain Forest and Range Expt Stn USDA Forest Service, No INT-011, p 6.

Furniss MM, Young JW, McGregor MD, Livingston RL, Hamel DR (1977) Effectiveness of controlled release formulations of MCH for preventing Douglas fir beetle infestation in felled trees. Can Entomol **109**:1063–1069.

Gaston LK, Kaae RS, Shorey HH, Sellers D (1977) Controlling the pink bollworm by disrupting sex pheromone communication between adult moths. Science **196**:904–905.

Gaston LK, Payne TL, Takahashi S, Shorey HH (1972) In: Olfaction and Taste 4. Wissenschaftliche Verlagsgesellschaft Stuttgart. Schneider D (ed), Correlation of Chemical Structure and Sex Pheromone Activity in *Trichoplusia ni* (Noctuidae), pp 167–173.

Gentry CR (1981) Peach tree borer (Lepidoptera: Sesiidae): Control by mass-trapping with synthetic pheromone. In: Tree Fruit and Nut Pest Management in the Southeastern United States. Johnson DT (Org.). Misc Pubs Ent Soc Amer **12**:15–19.

Gentry CR, Yonce CE, Bierl-Leonhardt BA (1982) Oriental fruit moth: Mating disruption trials with pheromone. In: Insect Suppression with Controlled Release Pheromone Systems. Kydonieus AF, Beroza M (eds), CRC Press, Boca Raton, Florida, Chap 8.

Glen DM, Brain P (1982) Pheromone-trap catch in relation to the phenology of codling moth (*Cydia pomonella*). Ann Appl Biol **101**:429–440.

Glick PA, Noble LW (1961) Airborne movement of the pink bollworm and other arthropods. Tech Bull U.S. Dep Agric No 1255, 20.

Gontarenko MA, Roshka GK, Kovolev BG, Oloi IN (1981) Pheromone traps for optimising control dates. Zashchita Rastenii **9**:26–32.

Graham HM, Martin DF, Ouye MT, Hardman RM (1966) Control of pink bollworms by male annihilation. J Econ Entomol **59**:590–593.

Graziano V, Viggiani G (1981) Four years observations on the flight and control of *Cydia molesta* (Busck.) and of *Anarsia lineatella* (Zell.) in peach orchards in Campania by means of synthetic pheromone traps. Annali Fac Sci Agr Universita Studi de Napoli Portici **15**:93–110.

Green N, Jacobson M, Keller JC (1969) Hexalure, an insect attractant discovered by empirical screening. Experientia **25**:682–683.

Gubbins KE, Campion DG (1982) Economic considerations relevant to the use of a pheromone mass trapping technique for the control of the Egyptian cotton leafworm *Spodoptera littoralis* in Egypt. Outlook Agric **11**:62–66.

Guerra AA, Garcia RD, Leal MP (1969) Suppression of populations of pink bollworm in field cages baited with sex attractant. J Econ Entomol **62**:741–742.

Hagley EA (1978) Sex pheromone and suppression of the codling moth (Lepidoptera, Olethreutidae). Can Ent **110**:781–783.

Haines CP, Read JS (1977) The effect of synthetic sex pheromones on fertilization in a warehouse population of *Ephestia cautella* (Walker) (Lepidoptera, Phycitidae). Rep Trop Prod Inst No L45, 10.

Hall DR (1981) Insect sex pheromones use in pest control. Zimbabwe Sci News **15**:230–233.

Hall DR, Beevor PS, Lester R, Nesbitt BF (1980) (*E,E*)-10,12-Hexadecadienal: A component of the female sex pheromone of the spiny bollworm, *Earias insulana* (Boisd.) (Lepidoptera, Noctuidae). Experientia **36**:152–153.

Hall DR, Nesbitt BF, Marrs GJ, Green A, St. J, Campion DG, Critchley BR (1982) Development of microencapsulated pheromone formulations. In: Insect Pheromone Technology: Chemistry and Applications. Leonhardt BA, Beroza M (eds), American Chemical Society Symposium Series No. 190, Washington, D.C., August 1981, pp 131–143.

Hartstack AW, Witz JA (1981) Estimating field populations of tobacco budworm moths from pheromone trap catches. Environ Ent **10**:908–914.

Haskell PT (1981) Can tropical crops be efficiently protected? Proceedings 1981 British Crop Protection Conference—Pests and Diseases, pp 741–748.

Haworth JK, Puck RP, Weatherston J, Doane CC, Ajesca S (1982) Research and development of a mating disruptant for control of the artichoke plume moth, *Platyptilia carduidactyla* (Riley) (Lepidoptera, Pterophoridae). In: Les Médiateurs Chimiques Agissant sur le Comportement des Insectes, Versailles 1981, (Les Colloques d'INRA 7), Institut National de la Recherche Agronomique, Paris, pp 343–356.

Henneberry TJ, Bariola LA, Flint HM, Lingren PD, Gillespie JM, Kydonieus AF (1981) Pink bollworm and tobacco budworm mating disruption studies on cotton. In: Management of Insect Pests with Semiochemicals. Mitchell ER (ed), Plenum, New York/London, pp 267–283.

Hirano C (1981) Evaluation of survey trapping locations for *Spodoptera litura* (Lepidoptera: Noctuidae) with sex pheromone traps. Jap J Appl Ent Zool **25**:272–275.

Hoppe T, Levinson HZ (1979) Befallserkennung und Populationsüberwachung vorratsschädlicher Motten (Phycitinae) in einer Schokoladenfabrik mit Hilfe pheromonbeköderter Klebefallen. Anz Schädlings Pflanz Umwelt **52**: 177–183.

Horak A, Hrdy I, Krampl F, Kalvoda L (1980) Field trials with pheromone traps for monitoring the pea moth *Cydia nigricana.* Ochrana Rostlin **16**:213–225.

Hosny MM, Iss-Hak RR, Nasr El Sayed A, El-Deeb YA, Critchley BR, Topper C. Campion DG (1979) Mass-trapping for the control of the Egyptian cotton leafworm *Spodoptera littoralis* (Boisd.) in Egypt. In: Proceedings 1979 British Crop Protection Conference, Pests & Diseases, Brighton, pp 395–400.

Hrdy I, Kuldova J, Krampl F, Marek J, Simko K (1979) Mapping of the oriental fruit moth (*Cydia molesta*) by means of pheromone traps. (in Czech) Sbornik Uvtiz-Ochrana Rostlin **15**:259–269.

Huber RT, Moore L, Hoffmann MP (1979) Feasibility study of area-wide pheromone trapping of male pink bollworm moths in a cotton insect pest management program. J Econ Entomol **72**:222–227.

Hummel HE, Gaston LK, Shorey HH, Kaae RS, Byrne KJ, Silverstein RM (1973) Clarification of the chemical status of the pink bollworm sex pheromone. Science **181**:873–875.

Iacob M, Iacob N, Dumitriu A (1980) Surveillance des populations de *Laspeyresia pomonella* L. par pièges à phéromones. Bull Acad Sci Agric For Bucarest **9**:63–74.

Ingram WR (1980) Studies of the pink bollworm, *Pectinophora gossypiella,* on sea island cotton in Barbados. Trop Pest Management **26**:118–137.

Isa AL (1981) Cotton pest problems in Egypt. In: Proceedings of Symposia IX International Congress of Plant Protection. Kommedahl T (ed), August 1979, Washington, D.C., Vol II, pp 545–547.

Iwaki S, Marumo S, Saito T, Yamada M, Katagiri K (1974) Synthesis and activity of optically active disparlure. J Am Chem Soc **96**:7842–7844.

Kaissling KE, Kasang G, Bestmann HJ, Stransky W, Vostrowsky O (1978) A new pheromone of the silkworm moth *Bombyx mori*—Sensory pathway and behavioural effect. Naturwissenschaften **65**:382–384.

Kanno H, Hattori M, Sato A, Tatsuki S, Ichiumi K, Kurihara M, Fukami J, Fujimoto Y, Tatsuno T (1980) Disruption of sex pheromone communication in the rice stem borer moth, *Chilo suppressalis* Walker (Lepidoptera:

Pyralidae), with sex pheromone components and their analogues. Appl Entomol Zool **15**:465–473.

Kennedy JS, Ludlow AR, Sanders CJ (1981) Guidance of flying male moths by wind-borne sex pheromone. Physiol Ent **6**:395–412.

Kennedy JW (1981) Practical application of pheromones in regulatory pest management programs. In: Management of Insect Pests with Semiochemicals. Mitchell ER (ed), Plenum, New York, pp 1–11.

Klassen W, Ridgway RL, Inscoe M (1982) Chemical attractants in integrated pest management programs. In: Insect Suppression with Controlled Release Pheromone Systems. Kydonieus AF, Beroza M (eds), CRC Press, Boca Raton, Florida, Vol 1, Chap 2, pp 13–130.

Klun JA, Collaborators (1975) Insect sex pheromones: Intraspecific pheromonal variability of *Ostrinia nubilalis* in North America and Europe. Environ Ent **4**:891–894.

Klun JA, Plimmer JA, Bierl-Leonhardt BA, Sparks AN, Chapman OL (1979) Trace chemicals: The essence of sexual communication systems in *Heliothis* species. Science **204**:1328–1330.

Knauf W, Bestmann HJ, Koschatzky KH, Suss J, Vostrowsky O (1979) Untersuchungen über die Lockwirkung synthetischer Sex-pheromone bei *Tortrix viridana* (Eichenwickler) und *Panolis flammea* (Kieferneule). Zeit Ange Ent **88**:307–312.

Knipling EF, McGuire JU Jr (1966) Population models to test theoretical effects of sex attractants used for insect control. US Dep Agric Inf Bull **308**:20.

Kobayashi M, Wada T, Inoue H (1981) A comparison of communication disruption technique and mass-trapping technique for controlling moths using sex pheromone of *Spodoptera litura* (F.) (Lepidoptera: Noctuidae). First Japan/USA Symposium on IPM, Tsukuba (Japan) September 1981, pp 32–40.

Konodrat'ev, Yu A, Lebedeva KV, Pyatnora Yu B (1978) Prospects for practical application of insect pheromones (in Russian). Zh. Vses. Khim O-va Im D I Mendeleeva **23**:179–188.

Kydonieus AF, Beroza M (eds) (1982) Insect Suppression with Controlled Release Pheromone Systems. CRC Press, Boca Raton, Florida, Vol 1, pp 274; Vol 2, pp 312.

Lanier GN (1981) Pheromone-baited traps and trap trees in the integrated management of bark beetles in urban areas. In: Management of Insect Pests with Semiochemicals. Mitchell ER (ed), Plenum, New York/London, pp 115–131.

Levinson HZ, Buchelos C Th (1981) Surveillance of storage moth species (Pyralidae, Gelechiidae) in a flour mill by adhesive traps with notes on the pheromone mediated flight behaviour of male moths. Zeit Ange Ent **92**:233–251.

Levinson HZ, Levinson AR (1979) Trapping of storage insects by sex and food attractants as a tool of integrated control. In: Chemical Ecology: Odour Communication in Animals. Ritter FJ (ed), Elsevier/North-Holland Biomedical Press, pp 327–341.

Lewis T, Macaulay EDM (1976) Design and elevation of sex attractant traps for pea moth *Cydia nigricana* (Steph.) and the effect of plume shape on catches. Ecol Ent **1**:175–187.

Lie R, Bakke A (1981) Practical results from the mass-trapping of *Ips typo-*

graphus in Scandinavia. In: Management of Insect Pests with Semiochemicals. Mitchell ER (ed), Plenum, New York/London, pp 175–181.

Livingstone RL, Daterman GE (1977) Surveying for Douglas-fir tussock moth with pheromone. Bull Ent Soc Am **23**:172–174.

Lloyd EP, McKibben GH, Knipling EF, Witz JA, Hartstack AW, Leggett JE, Lockwood DF (1981) Mass-trapping for detection, suppression and integration with other suppression measures against the boll-weevil. In: Management of Insect Pests with Semiochemicals. Mitchell ER (ed), Plenum, New York/London, pp 191–203.

Macaulay EDM, Lewis T (1977) Attractant traps for monitoring pea moth *Cydia nigricana* (S.). Ecol Ent **2**:279–284.

MacLellan CR (1976) Suppression of codling moth (Lepidoptera: Olethreutidae) by sex pheromone trapping of males. Can Ent **108**:1037–1040.

Madsen HF (1981) Monitoring codling moth populations in British Columbia apple orchards. In: Management of Insect Pests with Semiochemicals. Concepts & Practice. Mitchell EM (ed), Plenum, New York/London, pp 57–62.

Madsen HF, Carty BE (1979) Codling moth (Lepidoptera: Olethreutidae) suppression by male removal with sex pheromone traps in three British Columbia orchards. Can Ent **111**:627–630.

Madsen HF, Vakenti JM (1972) Codling moths: Female-baited and synthetic pheromone traps as population indicators. Environ Ent **1**:554–557.

Madsen HF, Vakenti JM (1973) Codling moth: Use of codlemone baited traps and visual detection of entries to determine need of sprays. Environ Ent **2**: 677–679.

Maini S, Bortolotti A, Pasqualini E (1982) Attempts to control *Laspeyresia pomonella* by mating disruption. Unpublished internal IOBC report presented at IOBC/WPRS working group on pheromones, special meeting on mating disruption changins/myon, Switzerland, 28–29 September, 1982.

Maksimovic M, Ljesov D, Prekajski P (1974) Comparative investigation on synthetic and natural sex lure of the gypsy moth and a trial of mass-trapping. Zast Bilja **25**:251–264.

Maksymov JK (1978) Surveillance of the nun moth *Lymantria monacha* L. (Lepidoptera, Lymantriidae) in the Swiss Alps by means of disparlure. Anzeiger Schädlingskunde Pflanzenschutz Umweltschutz **51**:70–75.

Mani E, Arn H, Wildbolz T, Hauri H (1978) A field test to control the plum fruit moth by communication disruption at high population density. Mitt Schweiz Ent Ges **51**:307–313.

Mani E, Wildbolz Th (1982) Control of codling moth by disruption: Trimmis 1979–81. Unpublished internal IOBC report presented at IOBC/WPRS working group on pheromones, special meeting on mating disruption changins/myon, Switzerland, 28–29 September, 1982.

Marks RJ (1976) Field evaluation of gossyplure, the synthetic sex pheromone of *Pectinophora gossypiella* (Saund.) (Lepidoptera, Gelechiidae) in Malawi. Bull Ent Res **66**:267–278.

Marks RJ (1977) Assessment of the use of sex pheromone traps to time chemical control of red bollworm *Diparopsis castanea* Hampson (Lepidoptera: Noctuidae) in Malawi. Bull Ent Res **67**:575–587.

Marks RJ, Hall DR, Lester R, Nesbitt BF, Lambert MRK (1981) Further studies on mating disruption of the red bollworm, *Diparopsis castanea* Hampson (Lepidoptera: Noctuidae) with microencapsulated mating inhibitor. Bull Ent Res **71**:403–418.

Marks RJ, Nesbitt BF, Hall DR, Lester R (1978) Mating disruption of the red bollworm of cotton *Diparopsis castanea* Hampson (Lepidoptera: Noctuidae) by ultra-low volume spraying with a microencapsulated inhibitor of mating. Bull Ent Res **68**:11–29.

Matvieskii AS (1980) Sex traps for counts of codling moth. Zashchita Rastenii **4**:44–45 (in Russian).

McLaughlin JR, Mitchell ER, Cross JH (1981) Field and laboratory evaluation of mating disruptants of *Heliothis zea* and *Spodoptera frugiperda* in Florida. In: Management of Insect Pests with Semiochemicals. Mitchell ER (ed), Plenum, New York/London, pp 243–251.

McLaughlin JR, Shorey HH, Gaston LK, Kaae RS, Stewart RD (1972) Sex pheromones of Lepidoptera XXXI. Disruption of sex pheromone communication in *Pectinophora gossypiella* with Hexalure. Environ Ent **1**:645–650.

McLean JA, Borden JH (1979) An operational pheromone-based suppression program for an ambrosia beetle *Gnathotrichus sulcatus,* in a commercial sawmill. J Econ Ent **72**:165–172.

McNally PS, Barnes MM (1980) Inherent characteristics of codling moth pheromone traps. Environ Ent **9**:538–541.

McVeigh EM, McVeigh LJ, Cavanagh GG (1983) A technique for tethering female moths of *Spodoptera littoralis* (Boisduval) (Lepidoptera: Noctuidae), to evaluate pheromone control methods. Bull Ent Res **73**:441–446.

Mineo G, Mirabello E, Del Busto T, Viggiani G (1980) Catture di adulti di *Prays citri* Mill. (Lep. Plutellidae) con trappole a feromoni e andamento delle infestazioni in limoneti della Sicilia Occidentale. Boll Lab Entomologia Agraria: Filippo Silvestri **37**:177–197.

Minks AK (1977) Trapping with behavior-modifying chemicals: Feasibility and limitations. In: Chemical Control of Insect Behavior. Theory & Application. Shorey HH, McKelvey JJ Jr (eds), Wiley, New York, Chap 23, pp 385–394.

Minks AK (1979) Present status of insect pheromones in agriculture and forestry. In: Proceedings of the International Symposium IOBC/WPRS on Integrated Control in Agriculture & Forestry. Russ K, Berger H (eds), Vienna, October 1979, pp 127–134.

Minks AK, Dejong DJ (1975) Determination of spraying dates for *Adoxophyes orana* by sex pheromone traps and temperature recordings. J Econ Ent **68**: 729–732.

Minks AK, Van Deventer P (1978) Phenological observations of elm bark beetles with attractant traps in the Netherlands during 1975 and 1976. Ned Bosg Tijdschr **50**:151–157.

Minks AK, Voerman S (1982) A three year study on mating disruption of the apple clearwing moth. Unpublished internal IOBC report presented at IOBC/WPRS working group on pheromones, special meeting on mating disruption changins/myon, Switzerland, 28–29 September, 1982.

Minks AK, Voerman S, Klun JE (1976) Disruption of pheromone communication with microencapsulated antipheromones against *Adoxophyes orana.* Ent Exp Appl **20**:163–169.

Mitchell EB, Hardee DD (1974) Infield traps: A new concept in survey and suppression of low populations of boll weevils. J Econ Entomol **67**:506–508.

Mitchell ER (1975) Disruption of pheromone communication among coexistent pest insects with multichemical formulations. Bioscience **25**:493–499.

Mitchell ER (ed) (1981) Management of Insect Pests with Semiochemicals. Concepts and Practice. Plenum, New York/London, pp 514.

Mitchell ER, McLaughlin JR (1982) Suppression of mating and oviposition by fall armyworm *Spodoptera frugiperda* and mating by corn earworm *Heliothis zea* in corn using the air permeation technique. J Econ Ent **75**:270–274.

Moffitt HR (1978) Control of the codling moth through mating disruption with the sex pheromone. Mitt Biol Bundesanst Land Forstwirtsch Berlin-Dahlem **180**:42–43.

Moorhouse JE, Yeadon R, Beevor PS, Nesbitt BF (1969) Method for use in studies of insect chemical communication. Nature (London) **223**:1174–1175.

Morner J (1981) Sex pheromones–A way to more effective control of the pea moth (in Swedish). Vaxtskyddsrapporter Jorduk **14**:22–36.

Murlis J, Bettany BW (1977) Night flight towards a sex pheromone source by male *Spodoptera littoralis* (Boisd.) (Lepidoptera, Noctuidae). Nature (London) **268**:433–435.

Murlis J, Bettany BW, Kelley J, Martin L (1982) The analysis of flight paths of the male Egyptian cotton leafworm moths *Spodoptera littoralis* (Boisd.) (Lepidoptera, Noctuidae) to a sex pheromone source in the field. Physiol Entomol **7**:435–441.

Myburgh AC, Madsen HF, Bosman IP, Rust DJ (1974) Codling moth (Lepidoptera: Olethreutidae): Studies on the placement of sex-attractant traps in South African orchards. Phytophylactica **6**:189–194.

Nakasuji F, Kiritani K (1978) Estimating the control threshold density of the tobacco cutworm *Spodoptera litura* (Lepidoptera: Noctuidae) on a corn crop, taro by means of pheromone traps. Prot Ecol **1**:23–32.

Negishi T, Ishiwatari T, Asano S (1977) Sex pheromone trapping for control of the oriental fruit moth *Grapholita molesta.* Jpn J Appl Entomol Zool **21**:210–215.

Negishi T, Ishiwatari T, Asano S, Fujikawa H (1980) Mass-trapping for the smaller tea tortrix control. Appl Ent Zool (Japan) **15**:113–114.

Nemoto H, Takahashi K, Kubota A (1980) Reduction of the population density of *Spodoptera litura* (Lepidoptera, Noctuidae) using a synthetic sex pheromone. 1. Experiment in a taro field. Jpn J Appl Entomol Zool **24**:211–216.

Nesbitt BF, Beevor PS, Cole RA, Lester R, Poppi RG (1973) Sex pheromones of two noctuid moths. Nature New Biol **244**:208–209.

Nesbitt BF, Beevor PS, Hall DR, Lester R, Dyck VA (1975) Identification of the female sex pheromones of the moth *Chilo suppressalis.* J Insect Physiol **21**:1883–1886.

Nordlund DA, Jones RL, Lewis WJ (eds) (1981) Semiochemicals, Their Role in Pest Control, Wiley, New York, pp 306.

Overhulser DL, Daterman GE, Sower LL, Sartwell C, Koerber TW (1980) Mating disruption with synthetic sex attractants controls damage by *Eucosoma sonomana* (Lepidoptera: Tortricidae, Olethreutidae) in *Pinus ponderosa* plantations. II. Aerially applied hollow fibre formulation. Can Entomol **112**:163–165.

Paradis RO, Trottier R, Maclellan CR (1979) Responses of the codling moth *Laspeyresia pomonella* to various trap designs tested in Eastern Canada. Ann Soc Entomol Que **24**:3–11.

Payne TL (1981) Disruption of Southern Pine Beetle infestations with attractants and inhibitors. In: Management of Insect Pests with Semiochemicals. Mitchell ER (ed), Plenum, New York/London, pp 365–383.

Peacock JW, Cuthbert RA, Lanier GN (1981) Deployment of traps in a barrier strategy to reduce populations of the European Elm Bark Beetle and the incidence of Dutch elm disease. In: Management of Insect Pests with Semiochemicals. Mitchell ER (ed), Plenum, New York/London, pp. 155–174.

Piccardi P (1980) Insect sex-communication and prospects for pheromones in pest management. Boll Zool **47**:397–408.

Pralavorio R, Jardak T, Arambourg Y, Renou M (1981) Utilisation du tetradécène *Z*7AL por la mise au point d'une methods de piègeage sexuel chez *Prays oleae* Bern. (Lep. Hyponomeutidae). Agronomie **1**:115–121.

Proverbs MD, Logan DM, Newton JR (1975) A study to suppress codling moth with sex pheromone traps. Can Ent **107**:1265–1269.

Quisumbing AR, Kydonieus AF (1982) Laminated structure dispensers. In: Insect Suppression with Controlled Release Pheromone Systems. Kydonieus AF, Beroza M (eds), CRC Press, Boca Raton, Florida, Vol 1, Chap 8, pp 213–236.

Rahn R (1980) L'utilisation d'une phéromone de synthèse contre la teigne du poireau. Phytoma No. 320, 22.

Ramos P, Campos M, Ramos JM (1981) A preliminary note on the relationship between the number of adult *Prays oleae* Bern. caught in pheromone traps and the resulting level of infestation. Experientia **37**:1282–1283.

Reichmuth C, Schmidt HU, Levinson AR, Levinson HZ (1978) The efficiency of adhesive pheromone traps for catching adults of *Ephestia elutella* (Hbn.) in slightly and heavily infested granaries. Zeit Ang Ent **86**:205–212.

Reynolds HT, Adkisson PL, Smith RF, Frisbie RE (1982) Cotton insect pest management. In: Introduction to Insect Pest Management. Metcalf RL, Luckmann WH (eds), Wiley, New York, pp 375–441.

Riedl H (1980) The importance of pheromone trap density and trap maintainance for the development of standardised monitoring procedures for the codling moth (Lepidoptera: Tortricidae). Can Ent **112**:655–663.

Riedl H, Croft BA (1974) A study of pheromone trap catches in relation to codling moth (Lepidoptera: Olethreutidae) damage. Can Ent **106**:525–537.

Riedl H, Croft BA, Howitt AJ (1976) Forecasting codling moth phenology based on pheromone trap catches and physiological time models. Can Ent **108**: 449–460.

Riedl H, Hoying SA, Barnett WA, Detar JE (1979) Relationship of within-tree placement of the pheromone trap to codling moth catches. Environ Ent **8**: 765–769.

Ritter FJ (ed) (1979) Chemical Ecology: Odour Communication in Animals. Elsevier/North-Holland Biomedical Press, Amsterdam, pp 427.

Roehrich R, Carles JP (1981) Mating disruption experiments in vineyards to control the grape vine moth *Lobesia botrana* (Schiff.). In: Les Médiateurs Chimiques Agissant sur le Comportement des Insectes. Versailles 1981 (Les Colloques d'INRA 7), pp 365–371.

Roelofs WL (1981) Attractive and aggregating pheromones. In: Semiochemicals: Their Role in Pest Control. Nordlund DA, Jones RL, Lewis WJ (eds), Wiley, New York, Chap 11, pp 215–235.

Roelofs WL, Cardé RT (1977) Responses of Lepidoptera to synthetic sex pheromone chemicals and their analogs. Ann Rev Ent **22**:377–399.

Roelofs WL, Comeau A, Hill A, Milicevic G (1971) Sex attractant of the codling moth: Characterization with electroantennogram technique. Science **174**: 297–299.

Rossiter B, Katsuki T, Sharpless KB (1981) Asymetric epoxidation provides shortest routes to four chiral alcohols which are key intermediates in synthesis of methymycin, erythromycin, leukotriene C-1 and disparlure. J Am Chem Soc **103**:464–465.

Rothschild GHL (1975) Control of oriental fruit moth (*Cydia molesta*) (Busck) (Lepidoptera, Tortricidae) with synthetic female pheromone. Bull Ent Res **65**:473–490.

Sanders CJ (1981a) Sex attractant traps: Their role in the management of spruce budworm. In: Management of Insect Pests with Semiochemicals. Mitchell ER (ed), Plenum Press, New York/London, pp 75–91.

Sanders CJ (1981b) Disruption of spruce budworm mating—State of the art. In: Management of Insect Pests with Semiochemicals. Mitchell ER (ed), Plenum, New York/London, pp 339–345.

Sanders CJ (1981c) Spruce budworm: Effects of different blends of sex pheromone components on disruption of male attraction. Experientia **37**:1176–1178.

Sanders CJ (1982) Disruption of male spruce bedworm orientation to calling females in a wind tunnel by synthetic pheromone. J Chem Ecol **8**:493–506.

Sanders CJ, Weatherston J (1976) Sex pheromone of the eastern spruce budworm: Optimum blend of *trans* and *cis*-11-tetradecenal. Can Ent **108**: 1285–1290.

Schruft G (1982) A study of control of the European grape berry moth *Eupoecilia ambiguella* by the mating disruption technique. Unpublished internal IOBC report presented at IOBC/WPRS working group on pheromones, special meeting on mating disruption changins/myon, Switzerland, 28–29 September, 1982.

Shani A (1982) Field studies and pheromone application in Israel. Third Israeli Meeting on Pheromone Research, Ben-Gurion University of the Negev. May 4, 1982. Abstracts of papers, pp 18–22.

Shani A, Klug JT (1980) Sex pheromone of Egyptian cotton leafworm (*Spodoptera littoralis*). Its chemical transformations under field conditions. J Chem Ecol **6**:875–881.

Shimada K (1980) Sex pheromone mass trapping for control of the smaller tea tortrix. Jap J Appl Ent Zool **24**:81–85.

Shorey HH, Gaston LK, Kaae RS (1976) Air permeation with gossyplure for control of the pink bollworm. In: Pest Management with Insect Sex Attractants. ACS Symposium Ser. 23, Beroza M (ed), American Chemical Society, Washington, D.C. pp 67–74.

Shorey HH, Kaae RS, Gaston LK, McLaughlin JR (1972) Sex pheromones of Lepidoptera XXX. Disruption of sex pheromone communication in *Trichoplusia ni* as a possible means of mating control. Environ Ent **1**:641–645.

Shorey HH, McKelvey JJ Jr (eds) (1977) Chemical Control of Insect Behavior: Theory & Application. Wiley, New York, p 414.

Showers WB, Reed GL, Robinson JF, DeRozari MB (1976) Flight and sexual activity of the European corn borer. Environ Entomol **5**:1099–1104.

Sifner F, Zdarek J (1982) Monitoring of stored food moths (Lepidoptera, Pyralidae) in Czechoslovakia by means of pheromone traps. Acta Ent Bohemoslov **79**:112–122.

Silveira Neto S, Precetti AACM, Braz AJBP, Santos PET (1981) Population fluctuation of *Grapholita molesta* (Busck.) (Lep., Olethreutidae) in peach and nectarine orchards by using sex pheromone. Anais Soc Ent (Brasil) **10**:43–49.

Silverstein RM (1981) Pheromones: Background and potential for use in insect pest control. Science **213**:1326–1332.

Silverstein RM, Rodin JO, Wood DL (1966) Sex attractants in frass produced by male *Ips confusus* in ponderosa pine. Science **154**:509–510.

Smetnik AI (1978) Use of pheromones against quarantine pests. Zashchita Rastenii (Mosc.) **9**:44–45 (in Russian).

Solomon ME (1978) Analysing the influence of temperature on pheromone trap catches of codling moth *Cydia pomonella.* Mitt Biol Bundesanst Land–Forstwirtsch Berlin-Dahlem **180**:34–35.

Sower LL (1982) Douglas-Fir Tussock moth disruption. In: Insect Suppression with Controlled Release Pheromone Systems. Kydonieus AF, Beroza M (eds), CRC Press, Boca Raton, Florida, Vol II, Chap 12, pp 165–173.

Sower LL, Daterman GE, Sartwell C, Cory HT (1979) Attractants for the western pine shootborer, *Eucosoma sonomana* and *Rhyacionia zozana* determined by field screening. Environ Entomol **8**:265–267.

Sower LL, Overhulser DL, Daterman GE, Sartwell C, Laws DE, Koerber TW (1982) Control of *Eucosoma sonomana* by mating disruption with synthetic sex attractant. J Econ Ent **75**:315–318.

Sower LL, Turner WK, Fish JC (1975) Population density dependent mating frequency among *Plodia interpunctella* (Lepidoptera: Phycitidae) in the presence of synthetic sex pheromone with behavioral observations. J Chem Ecol **1**:335–342.

Stenmark A (1978) Experiments with pheromones against tortricids in Swedish fruit orchards. (in Swedish). Vaxtskyddsrapporter Tradgard **2**:1–42.

Stenmark A (1979) Experiments with decamethrin against the codling moth (*Laspeyresia pomonella* L.) Vaxtskyddsnotiser **43**:61–63.

Sternlicht M (1982) Bionomics of *Prays citri* (Lepidoptera: Yponomeutidae) and their use in a model of control by male mass-trapping. Ecol Ent **7**:207–216.

Symmons P, Rosenberg LJ (1978) A model simulating mating behaviour of *Spodoptera littoralis*. J Appl Ecol **15**:423–437.

Tamaki Y (1974) Insect sex pheromones: Recent advances in Japan. Rev Plant Protection **7**:68–80.

Tamaki Y (1980) Insect sex pheromones and pest management: Recent advances in Japan. Japan Pesticide Information No. 37, 22–25.

Tanaka F, Yabuki S (1978) Forecasting oriental fruit moth, *Grapholita molesta* Busck. emergence time on the pheromone trap method by the estimate of temperature. Jap J Appl. Ent Zool **22**:162–168.

Taneja SL, Jayaswal AP (1981) Capture threshold of pink bollworm *Pectinophora gossypiella* moths on cotton *Gossypium hirsutum*. Trop Pest Management **27**:318–324.

Tanskii VI, Bulgak VD (1981) Effectiveness of using economic damage thresholds for the codling moth *Laspeyresia pomonella* (Lepidoptera, Tortricidae) and Tetranychid mites in the Crimea. (in Russian). Entomol Obozr **60**: 241–251.

Tatsuki S, Kanno H (1981) Disruption of sex pheromone communication in *Chilo suppressalis* with pheromones and analogues. In: Management of Insect Pests with Semiochemicals. Mitchell ER (ed), Plenum, New York/ London, pp 313–325.

Teich I, Neumark S, Jacobson M, Klug J, Shani A, Waters RM (1979) Mass-trapping of males of Egyptian cotton leafworm *Spodoptera littoralis* and large scale synthesis of Prodlure. In: Chemical Ecology, Odour Communication in Animals. Ritter FJ (ed), Elsevier/North-Holland Biomedical Press, Amsterdam, pp 343–350.

Touzeau J (1979) L'utilisation du piègeage sexuel pour les advertissements agricoles et la prévision des risques. Ann Zool Ecol Anim **11**:547–563.

Trammel K, Roelofs WL, Glass EH (1974) Sex pheromone trapping of males for control of red banded leafroller in apple orchards. J Econ Ent **67**:159–164.

Tumlinson JH, Heath RR, Teal PEA (1982) Analysis of chemical communication systems of Lepidoptera. In: Insect Pheromone Technology: Chemistry and Applications. Leonhardt BA, Beroza M (eds), ACS Symposium Series 190, American Chemical Society, Washington, D.C., pp 1–25.

Vakenti JM, Madsen HF (1976) Codling moth (Lepidoptera: Olethreutidae): Monitoring populations in apple orchards with sex pheromone traps. Can Ent **108**:433–438.

Vanwetswinkel G, Paternotte E (1982) Mating disruption experiments with high dosages of the sex pheromone of the summerfruit tortrix moth *Adoxophyes orana*. Unpublished internal IOBC report presented at IOBC/WPRS working group on pheromones, special meeting on mating disruption changins/myon, Switzerland, 28–29 September, 1982.

Vita G, Caffarelli V (1982) Essai de lutte par confusion contre *Lobesia botrana* Schiff. Unpublished internal IOBC report presented at IOBC/WPRS working group on pheromones, special meeting on mating disruption changins/myon, Switzerland, 28–29 September, 1982.

Vité JP, Gerken LB, Lanier GN (1976) Ulmensplintkäfer: Anlockversuche mit synthetischen Pheromonen im Oberrheintal. Z Pflanzenkrankh Pflanzenschutz **83**:166–171.

Vrkoch L, Kalvoda L, Skugravy V, Khokhmut R, Zhdyarek Ya (1981) The disorientation of males with the pheromones of the grey larch moth (*Zeiraphera diniana* Gn.) and the nun moth (*Lymantria monacha* L.) in the conditions of Czechoslovakia. Khemoretseptsiya Nasekomykh **6**:130–133.

Wall C, Perry JN (1978) Interactions between pheromone traps for the pea moth, *Cydia nigricana* (F.). Ent Exp Appl **24**:155–162.

Webb RE (1982) Mass-trapping of the gypsy moth. In: Insect Suppression with Controlled Release Pheromone Systems. Kydonieus AF, Beroza M (eds), CRC Press, Boca Raton, Florida, Chap 3, **2**:27–56.

Wilson HR, Trammel K (1980) Sex pheromone trapping for control of codling moth, Oriental fruit moth, lesser appleworm, and three tortricid leafrollers in a New York apple orchard. J Econ Ent **73**:291–295.

Wood DL (1979) Development of behavior modifying chemicals for use in forest pest management in the USA. In: Chemical Ecology, Odour Communication in Animals. Ritter FJ (ed). Elsevier/North-Holland Biomedical Press, Amsterdam, pp 261–279.

Yonce CE, Gentry CR (1982) Disruption of mating of peach tree borer. In: Insect Suppression with Controlled Release Pheromone Systems. Kydonieus AF, Beroza M (eds), CRC Press, Boca Raton, Florida, Chap 7, **2**:99–106.

Zhdarek Ya, Vrkoch Ya, Kalvoda L (1981) The use of pheromones in the study of seasonal and diurnal activity of the grey larch moth (*Zeiraphera diniana* Gn.). Khemoretseptsiya Nasekomykh **6**:105–108.

Zweig G, Cohen SL, Betz FS (1982) EPA registration requirements for biochemical pesticides with special emphasis on pheromones. In: Insect Suppression with Controlled Release Pheromone Systems. Kydonieus AF, Beroza M (eds), CRC Press, Boca Raton, Florida, Chap 5, **1**:159–168.

Index

Springer Series In Experimental Entomology
Editor: T.A. Miller

Insect Neurophysiological Techniques
By T.A. Miller
Neurohormonal Techniques in Insects
Edited by T.A. Miller
Sampling Methods in Soybean Entomology
Edited by M. Kogan and D. Herzog
Neuroanatomical Techniques
Edited by N.J. Strausfeld and T.A. Miller
Cuticle Techniques in Arthropods
Edited by T.A. Miller
Measurement of Ion Transport and Metabolic Rate in Insects
Edited by T.J. Bradley and T.A. Miller
Techniques in Pheromone Research
Edited by H.E. Hummel and T.A. Miller
Functional Neuroanatomy
Edited by N.J. Strausfeld
Neurochemical Techniques in Insect Research
Edited by H. Breer and T.A. Miller